Hamid R. Arabnia · Leonidas Deligiannidis ·
Farzan Shenavarmasouleh ·
Farid Ghareh Mohammadi · Soheyla Amirian
Editors

Emerging Trends in Health Informatics, Medical Devices, and Large Language Models

12th International Conference on Computational Science and Computational Intelligence, CSCI 2025
Las Vegas, NV, USA, December 3–5, 2025
Proceedings, Part V

Springer

Editors
Hamid R. Arabnia
University of Georgia
Athens, GA, USA

Leonidas Deligiannidis
Wentworth Institute of Technology
Boston, MA, USA

Farzan Shenavarmasouleh
Medialab Inc.
Lawrenceville, GA, USA

Farid Ghareh Mohammadi
Verify Radiology Images Consultants
Houston, TX, USA

Soheyla Amirian
Pace University
New York, NY, USA

ISSN 1865-0929 ISSN 1865-0937 (electronic)
Communications in Computer and Information Science
ISBN 978-3-032-28462-4 ISBN 978-3-032-28463-1 (eBook)
https://doi.org/10.1007/978-3-032-28463-1

This Springer imprint is published by the registered company Springer Nature Switzerland AG
The registered company address is: Gewerbestrasse 11, 6330 Cham, Switzerland

Communications in Computer and Information Science 2971

Rationale
The CCIS series is devoted to the publication of proceedings of computer science conferences. Its aim is to efficiently disseminate original research results in informatics in printed and electronic form. While the focus is on publication of peer-reviewed full papers presenting mature work, inclusion of reviewed short papers reporting on work in progress is welcome, too. Besides globally relevant meetings with internationally representative program committees guaranteeing a strict peer-reviewing and paper selection process, conferences run by societies or of high regional or national relevance are also considered for publication.

Topics
The topical scope of CCIS spans the entire spectrum of informatics ranging from foundational topics in the theory of computing to information and communications science and technology and a broad variety of interdisciplinary application fields.

Information for Volume Editors and Authors
Publication in CCIS is free of charge. No royalties are paid, however, we offer registered conference participants temporary free access to the online version of the conference proceedings on SpringerLink (http://link.springer.com) by means of an http referrer from the conference website and/or a number of complimentary printed copies, as specified in the official acceptance email of the event.

CCIS proceedings can be published in time for distribution at conferences or as post-proceedings, and delivered in the form of printed books and/or electronically as USBs and/or e-content licenses for accessing proceedings at SpringerLink. Furthermore, CCIS proceedings are included in the CCIS electronic book series hosted in the SpringerLink digital library at http://link.springer.com/bookseries/7899. Conferences publishing in CCIS are allowed to use our online conference service (Meteor) for managing the whole proceedings lifecycle (from submission and reviewing to preparing for publication) free of charge.

Publication process
The language of publication is exclusively English. Authors publishing in CCIS have to sign the Springer CCIS copyright transfer form, however, they are free to use their material published in CCIS for substantially changed, more elaborate subsequent publications elsewhere. For the preparation of the camera-ready papers/files, authors have to strictly adhere to the Springer CCIS Authors' Instructions and are strongly encouraged to use the CCIS LaTeX style files or templates.

Abstracting/Indexing
CCIS is abstracted/indexed in DBLP, Google Scholar, EI-Compendex, Mathematical Reviews, SCImago, Scopus. CCIS volumes are also submitted for the inclusion in ISI Proceedings.

How to start
To start the evaluation of your proposal for inclusion in the CCIS series, please send an e-mail to ccis@springer.com

Preface

It is our great pleasure to present this collection of research papers from the 12th International Conference on Computational Science and Computational Intelligence (CSCI 2025). This volume, titled "Emerging Trends in Health Informatics, Medical Devices, and Large Language Models", comprises selected papers accepted in the Research Tracks on Health Informatics and Medical Systems (CSCI-RTHI) and Large Language Models (CSCI-LLM). Due to the unprecedented response from authors, the Research Track on Large Language Models was elevated to a full International Conference (LLM). The CSCI and LLM conferences were held in Las Vegas, Nevada, USA, from December 3 to December 5, 2025.

The CSCI 2025 International Conference brought together papers from a diverse array of communities, including researchers from universities, corporations, and government agencies. Accepted papers are published by Springer Nature, and the proceedings showcase solutions to key challenges in various critical areas of Computational Science and Computational Intelligence.

Computational Science (CS) is the study of addressing problems that are impossible to solve (or difficult to solve) without computers. CS can be considered to be the bridge between computer science and other sciences. The field is interdisciplinary by nature and includes the use of advanced computing capabilities to understand and solve complex problems. In short, CS is the science of using computers to do science. Computational Intelligence (CI) is the study of computational methods that exhibit intelligence. These methods adapt to changing environments and changing goals. There is a significant overlap between the fields of CI and Artificial Intelligence (AI). However, there is also a difference: AI techniques often involve top-to-bottom methods (i.e., methods to the solutions are imposed from the top) whereas CI techniques often involve bottom-up methods (i.e., solutions emerge from unstructured beginnings). An important part of CI includes a set of Nature-inspired computational approaches to address complex problems for which traditional methods are infeasible. Computational Science and Computational Intelligence both share the same objective: finding solutions to difficult problems. However, the methods to reach solutions are different. The main objective of the CSCI Conference was to facilitate increased opportunities for cross-fertilization across CS and CI.

Considering the above broad outline, the CSCI 2025 International Conference was composed of the following focused Research Tracks:

Artificial Intelligence (CSCI-RTAI); Big Data and Data Science (CSCI-RTBD); Computational Science (CSCI-RTCS); Computational Intelligence (CSCI-RTCI); Computational Biology (CSCI-RTCB); Cyber Warfare, Cyber Defense, & Cyber Security (CSCI-RTCW); Signal & Image Processing, Computer Vision & Pattern Recognition (CSCI-RTPC); Smart Cities and Smart Mobility (CSCI-RTSC); Education - CS & CE

(CSCI-RTED); Health Informatics and Medical Systems (CSCI-RTHI); Mobile Computing, Wireless Networks, & Security (CSCI-RTMC); Software Engineering (CSCI-RTSE); Internet of Things & Internet of Everything (CSCI-RTOT); Social Network Analysis, Social Media, & Mining (CSCI-RTNA); Cloud Computing and Data Centers (CSCI-RTCC); Parallel & Distributed Computing (CSCI-RTPD); and Large Language Models (LLM). The scope of each track can be found at: https://www.american-cse.org/csci2025/topics and https://www.american-cse.org/LLM2025/topics.

An important objective of the CSCI 2025 International Conference and its associated Research Tracks is to foster opportunities for cross-fertilization between the fields of Computational Science and Computational Intelligence. The CSCI Conference is deeply committed to promoting diversity and eliminating discrimination, both in its role as a conference organizer and as a service provider. Our goal is to create an inclusive culture that respects and values differences, promotes dignity, equality, and diversity, and encourages individuals to reach their full potential. We are also dedicated, wherever possible, to organizing a conference that represents the global community. We sincerely hope that we have succeeded in achieving these important objectives.

The Steering Committee and the Program Committee would like to extend their gratitude to all the authors who submitted papers for consideration. CSCI 2025 received submissions from 48 countries, with approximately 48% of them coming from outside the United States. Each submitted paper underwent a rigorous peer-review process, with at least two experts (an average of 2.5 referees per paper) evaluating the submissions based on originality, significance, clarity, impact, and soundness. In cases where reviewers' recommendations were contradictory, a program committee member was tasked with making the final decision, often consulting additional referees for further guidance. The CSCI Conference followed the guidelines of COPE (Committee on Publication Ethics):

- Typical submissions underwent a single-blind peer review process, in which the authors remain unaware of the identities of the reviewers, while the reviewers are informed of the authors' identities.
- Papers authored by one or more members of the program committee, including co-chairs, were subjected to a double-blind peer review process, ensuring that neither the authors nor the reviewers were aware of each other's identities or affiliations.

The Research Track on Health Informatics and Medical Systems (CSCI-RTHI) received 168 submissions, of which 37 papers were accepted, resulting in a paper acceptance rate of 22%; we selected only 29 papers for this volume. The Research Track on Large Language Models (CSCI-LLM) received 294 submissions, of which 63 papers were accepted, resulting in a paper acceptance rate of 21%; we selected only 13 papers for this volume. Due to the unprecedented response from authors, CSCI-LLM was elevated to a full International Conference (LLM). A total of 42 papers are included in this volume.

We are deeply grateful to the many colleagues who contributed their time and effort to organizing the CSCI 2025 Conference. In particular, we extend our thanks to the members of the Program Committee, the Steering Committee, and the referees. We would also like to express our appreciation to the primary sponsor of the conference, the American Council on Science & Education. The list of members of the Program

Committee for each track can be found at: https://www.american-cse.org/csci2025/committees.

We extend our heartfelt gratitude to all the speakers and authors for their valuable contributions. We would also like to thank the following individuals and organizations for their support: the staff at the Luxor Hotel (an MGM property, conference/meeting department) and the staff of Springer, for their assistance in various aspects of the event.

We are pleased to present the proceedings of CSCI 2025 (selected papers of CSCI-RTHI Research Track and LLM Conference). These proceedings represent a collection of outstanding research contributions that reflect the diversity and depth of work in Health Informatics, Medical Systems and Devices, Large Language Models and applications.

Co-editors: Research Tracks on Health Informatics and Medical Systems (CSCI-RTHI) and Large Language Models (LLM) of CSCI 2025 International Conference:

December 2025

Hamid R. Arabnia
Leonidas Deligiannidis
Farzan Shenavarmasouleh
Farid Ghareh Mohammadi
Soheyla Amirian

Organization

Steering Committee – Co-chairs (CSCI 2025)

Publication Co-chairs and Co-editors

"Emerging Trends in Health Informatics, Medical Devices, and Large Language Models"

Research Track: Health Informatics and Medical Systems (CSCI-RTHI)

Research Track: Large Language Models (CSCI-LLM) - promoted to a Conference

Part of Proceedings of The 2025 International Conference on Computational Science and Computational Intelligence (CSCI)

Members of Steering Committee (CSCI 2025)

Mary Mehrnoosh Eshaghian-Wilner	University of California, Los Angeles, USA
David L. Foster	Kettering University, USA
Ching-Hsien (Robert) Hsu	ChungAsia University, Taiwan
Farid Ghareh Mohammadi	Verify Radiology Images Consultants, LLC, USA
James J. (Jong Hyuk) Park	SeoulTech, South Korea
Mohammad S. Obaidat	University of Jordan, Jordan
Gerald Schaefer	Loughborough University, UK
Farzan Shenavarmasouleh	Medialab Inc., USA
Fernando G. Tinetti	Universidad Nacional de La Plata, Argentina
Quoc-Nam Tran	Southeastern Louisiana University, USA
Shiuh-Jeng Wang	Central Police University, Taiwan
Layne T. Watson	Virginia Polytechnic Institute & State University, USA
Chao-Tung Yang	Tunghai University, Taiwan
Mary Yang	University of Arkansas for Medical Sciences, USA

Program Committee – Computational Science and Computational Intelligence (CSCI 2025)

Afrand Agah	West Chester University of Pennsylvania, USA
Bharat Bhushan Agarwal	IFTM University, India
Omaima Nazar Ahmad AL	Al-Zaytoonah University of Jordan, Jordan
Wasim A. Al – Hamdani	Kentucky State University, USA
Ismail Khalil Al Ani	Ittihad University, United Arab Emirates
Mehran Asadi	Lincoln University, USA
Travis Atkison	University of Alabama, USA
Azita Bahrami	IT Consult, USA
Mehdi Bahrami	Fujitsu Laboratories of America, Inc., USA
P. Balasubramanian	Nanyang Technological University, Singapore
Petra Saskia Bayerl	Erasmus University Rotterdam, The Netherlands
Jane M. Binner	University of Birmingham, UK
Juan-Vicente Capella-Hernandez	Universitat Politècnica de València, Spain
Juan Jose Martinez Castillo	Universidad Nacional Abierta, Venezuela
Rui Chang	Mount Sinai School of Medicine, USA
Dongsheng Che	East Stroudsburg University of Pennsylvania, USA
Jianhung Chen	Chung Hua University, Taiwan
Mu-Song Chen	Da-Yeh University, Taiwan
Xin (Thomas) Chen	University of Hawaii, USA

Steve C. Chiu	Idaho State University, USA
Mark Yul Chu	University of Texas Rio Grande Valley, USA
Jose Alfredo F. Costa	Universidade Federal do Rio Grande do Norte, Brazil
Arianna D'Ulizia	National Research Council of Italy, Italy
Zhangisina G. Davletzhanovna	Central Asian University, Kazakhstan
Wesley Deneke	Western Washington University, USA
Noel De Palma	Grenoble Alpes University, France
Mohsen Doroodchi	University of North Carolina at Charlotte, USA
Levent Ertaull	California State University, East Bay, USA
Mahmood Fazlali	Shahid Beheshti University, Iran
Boyuan Feng	University of California, Santa Barbara, USA
George A. Gravvanis	Democritus University of Thrace, Greece
Gheorghe Grigoras	"Gheorghe Asachi" Technical University of Iaşi, Romania
Ray Hashemi	Georgia Southern University, USA
Houcine Hassan	Universitat Politècnica de València, Spain
Abdolreza Hatamlou	Islamic Azad University, Iran
Bing He	Cisco System Inc., USA
Henry Hexmoor	Southern Illinois University at Carbondale, USA
Gahangir Hossain	Texas A&M University - Kingsville, USA
Guofeng Hou	AQR Capital Management, USA
Ren-Junn Hwang	Tamkang University, Taiwan
Naseem Ibrahim	Penn State Erie, The Behrend College, USA
Rabia Jafri	King Saud University, Saudi Arabia
Shahram Javadi	Azad University, Iran & University of Lisbon, Portugal
Young-Sik Jeong	Dongguk University, South Korea
Aleksandr Katkow	University of Computer Sciences & Skills, Poland
Byung-Gyu Kim	Sun Moon University, South Korea
Taihoon Kim	University of Tasmania, Australia
Sang-Wook Kim	Hanyang University, South Korea
Dattatraya Vishnu Kodavade	D.K.T.E Society's Textile & Engineering Institute, India
Elena B. Kozerenko	Russian Academy of Sciences, Moscow, Russia
Guoming Lai	Sun Yat-sen University, China
Hyo Jong Lee	Chonbuk National University, South Korea
Bo Liu	NEC Labs China, China
Eleanor Lockley	Sheffield Hallam University, UK
Jianbing Ma	Bournemouth University, UK
Julius Marpaung	Northeastern University, USA
Andrew Marsh	HoIP Telecom Ltd, UK

Juan Martinez	Universidad Gran Mariscal de Ayacucho, Venezuela
Praveen Meduri	California State University, Sacramento, USA
Ali Mostafaeipour	California State University, Fullerton, USA
Houssem Eddine Nouri	University of Tunis, Tunisia
Michael B. O'Hara	KB Computing, LLC, USA
Robert Ehimen Okonigene	Ambrose Alli University, Nigeria
Funminiyi Olajide	Nottingham Trent University, UK
Satish Penmatsa	University of Maryland - Eastern Shore, USA
Saman Parvaneh	Philips Research North America, USA
Ponalagusamy	National Institute of Technology, Tiruchirappalli, India
Laura L. Pullum	Oak Ridge National Laboratory, USA
Xuewei Qi	University of California, Riverside, USA
Junfeng Qu	Clayton State University, USA
Shahram Rahimi	University of Alabama, USA
Jaime Raigoza	California State University, Chico, USA
Arvind Ramanathan	Oak Ridge National Laboratory, USA
Iman M. Rezazadeh	University of California, Davis, USA
Om Prakash Rishi	University of Kota, India
Cristian Rodriguez Rivero	Universidad Nacional de Córdoba, Argentina
Seyed Roosta	Albany State University, USA
K. Martin Sagayam	Karunya Institute of Technology and Sciences, India
P. Sanjeevikumar	University of South-Eastern Norway, Norway
Benaoumeur Senouci	University of Southern Denmark, Denmark
Zhefu Shi	Microsoft Corporation, USA
Jawed Siddiqi	Sheffield Hallam University, UK
Akash Singh	IBM Corporation, USA
Anthony Skjellum	Auburn University, USA
Omer Muhammet Soysal	Southeastern Louisiana University, USA
Helman Stern	Ben-Gurion University of the Negev, Israel
Jonathan Z. Sun	University of Southern Mississippi, USA
Rahman Tashakkori	Appalachian State University, USA
Predrag Tosic	Microsoft, USA
Quoc-Nam Tran	Southeastern Louisiana University, USA
Jesus Vigo-Aguiar	University of Salamanca, Spain
Patrick Wang	East China Normal University, China & Northeastern University, USA
Weiqiang Wang	Opera Solutions LLC, USA
Yin Wang	Lawrence Technological University, USA
Alicia Nicki Washington	Duke University, USA

Contents

Health Informatics, Data Science, Prediction Methods, Applications and Tools

Poster Papers

Health Informatics, Medical Devices, Disease Diagnosis, and Applications

Sociodemographic Correlates of Tinnitus Occurrence Among Corona Health App Users During the COVID-19 Pandemic

Michael Winter[1,2(✉)], Winfried Schlee[3,4], Berthold Langguth[4], and Rüdiger Pryss[1,2]

[1] Institute of Clinical Epidemiology and Biometry, University of Würzburg, Würzburg, Germany
michael.winter@uni-wuerzburg.de
[2] Institute of Medical Data Science, University Hospital of Würzburg, Würzburg, Germany
[3] Eastern Switzerland University of Applied Sciences, St. Gallen, Switzerland
[4] Department of Psychiatry and Psychotherapy, University of Regensburg, Regensburg, Germany

Abstract. Tinnitus is a chronic condition whose perception is influenced by psychological, social, and health-related factors. The COVID-19 pandemic provided a unique situation in which many of these modulators were modified, while access to in-person care was limited. Digital health tools such as mobile applications offer new opportunities to investigate tinnitus in real-world populations. This study evaluated tinnitus occurrence among 2,694 adults from Germany who used the Corona Health app. Sociodemographic variables (age, gender, family status, educational and professional status), COVID-19 infection status, and mobile operating system (iOS vs Android) were considered. Associations were examined using a two-step regression approach: (1) logistic regression of any tinnitus occurrence versus none; (2) ordinal logistic regression of tinnitus occurrence. The results show that higher age, lower educational attainment and professional qualification, and a history of COVID-19 infection were associated with increased odds of reporting tinnitus. Among tinnitus participants, older age was the main predictor of higher tinnitus probability. The findings highlight the role of sociodemographic and health-related factors in tinnitus occurrence and demonstrate the feasibility of app-based platforms for large-scale, real-world tinnitus research.

Keywords: Tinnitus · COVID-19 · Digital Health · Corona Health

1 Introduction

Tinnitus, the perception of sound without an external source, affects a significant proportion of the adult population worldwide and can lead to substantial distress, sleep disturbances, and reduced quality of life [29]. The perception depends

Research Track: CSCI-RTHI.

H. R. Arabnia et al. (Eds.): CSCI 2025, CCIS 2971, pp. 3–15, 2027.
https://doi.org/10.1007/978-3-032-28463-1_1

on personality traits, lifestyle factors, and comorbid conditions [17]. Moreover, tinnitus perception fluctuates with external stressors and environmental changes that can disrupt established coping mechanisms. Given this sensitivity to stress and disruption, major societal events and health crises may have profound implications for individuals living with tinnitus.

The COVID-19 pandemic created unprecedented challenges for individuals with chronic health conditions, representing exactly the type of major disruption that could significantly impact tinnitus perception [16]. Heightened stress, social isolation, and limited access to in-person care have been suggested to increase tinnitus perception [25]. At the same time, the pandemic accelerated the adoption of digital health tools, offering new possibilities for symptom monitoring and self-management [13]. These digital platforms not only provided alternative care delivery methods but also created unique opportunities for large-scale data collection on symptom patterns and patient characteristics during this critical period. Such comprehensive data collection became essential for understanding the pandemic's true impact on tinnitus populations.

The Corona Health app was developed as part of the pandemic response to support affected individuals, representing one such digital platform that enabled data collection during this period [3]. Beyond its original purpose, it also provided an instrument to assess tinnitus occurrence (i.e., perception and probability) and collect information on users' sociodemographic and health-related characteristics. The app's dual functionality allowed researchers to capture real-time symptom data while simultaneously gathering detailed demographic and health profiles from a diverse user base. This rich dataset provided an opportunity to examine patterns and associations that had previously been difficult to study in traditional clinical settings, particularly during a global health crisis when conventional research approaches were severely limited.

While studies have investigated tinnitus during the pandemic, little is known about how the combination of sociodemographic factors, COVID-19 infection, and the used mobile operating systems relate to tinnitus occurrence. For this reason, this study aimed to evaluate tinnitus occurrence probability among 2,694 Corona Health app users during the COVID-19 pandemic, with a focus on age, gender, family status, educational and professional background, COVID-19 infection status, and operating system (iOS vs Android). By utilizing the comprehensive dataset collected through this digital platform, we sought to identify previously unexplored associations between user characteristics and tinnitus occurrence probability that could inform both clinical practice and future digital health interventions. Understanding these relationships is crucial for developing more targeted and personalized approaches to tinnitus management in the digital health era.

2 Related Work

Tinnitus is a prevalent condition that varies widely in its impact across individuals. Prior research has shown that its perception and probability is not determined solely by auditory factors [6] but is strongly influenced by psychological stress [12], emotional state [22], and sociodemographic variables such as age, gender, and educational background [2,21,31].

Given the influence of psychological and social factors on tinnitus occurrence, the COVID-19 pandemic provided a unique context in which many of these modulators were modified [5]. Heightened stress, social isolation, and reduced access to healthcare services during this period have been suggested to worsen tinnitus perception, and several studies have reported increased distress among patients with pre-existing tinnitus [4,5]. Moreover, evidence has raised the possibility that COVID-19 infection itself may contribute to the onset or aggravation of tinnitus [9].

The restrictions imposed during the pandemic not only intensified tinnitus-related stressors but also limited access to traditional in-person care. As a result, interest in digital health solutions increased markedly [23]. Mobile health applications and online platforms have emerged as accessible tools for the monitoring and self-management of tinnitus, offering educational resources, sound therapy, and symptom tracking [27]. Previous studies have demonstrated the feasibility of app-based approaches [20], yet evidence remains limited regarding their effectiveness in real-world populations, particularly during the pandemic context.

Despite growing interest in tinnitus research, only a few studies have comprehensively examined sociodemographic factors, COVID-19 effects, and technical aspects together in app-based populations [8]. This study addresses these gaps by evaluating tinnitus occurrence among Corona Health app users during the pandemic.

3 Materials and Methods

Corona Health: Data for this study were collected using the Corona Health app, a study- and sensor-based platform [30] developed to monitor physical and mental well-being during the COVID-19 pandemic [3]. The app served as a common platform for several questionnaire-based studies, each addressing different aspects of health and well-being. Upon onboarding, participants completed a baseline assessment, followed by scheduled follow-up assessments according to the respective study protocols. This structure enabled longitudinal data collection and helped to minimize recall bias. The app integrated ecological momentary assessment (EMA) with optional mobile sensing features [19,28]. The study was approved by the Ethics Committee of the University of Würzburg (Ref. No. 130/20-me). Corona Health was developed in compliance with the European Medical Device Regulation (MDR) and the General Data Protection Regulation [18], and was validated according to IEC 62304, IEC 82304, and GAMP 5 standards. The following Figs. 1, 2, 3 and 4 illustrate screenshots from the Corona Health App, showing the main menu, Mental Health for Adults study, questionnaire, and the end of the assessment.

For the present analysis, we considered data from the "Mental Health for Adults" study module. In this module, participants completed a baseline as well

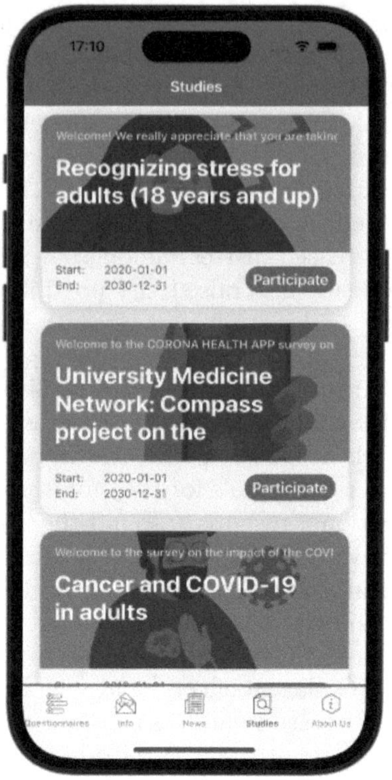

Fig. 1. Corona Health Main Menu

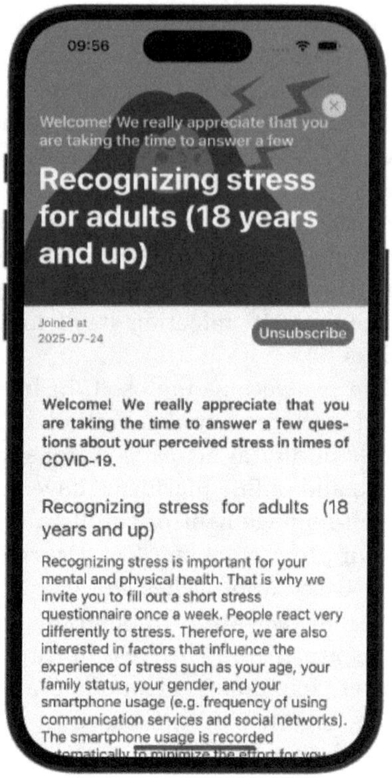

Fig. 2. Mental Health for Adults Study

as follow up assessments that included validated instruments such as Quality of Life (QOL) measures and the Patient Health Questionnaire (PHQ) [11]. For the purposes of this study, we focused exclusively on the baseline assessment and specifically on the item assessing tinnitus occurrence.

Measures: Tinnitus occurrence was assessed with a single baseline item asking participants how frequently they had been affected by tinnitus during the past week. Response options ranged from not at all, several days, more than half the days, to nearly every day.

Age (in years) and gender were recorded at baseline. Family status was categorized as partnership or single. Educational attainment was categorized into four levels: none, low (lower secondary school), middle (secondary school diploma or equivalent), and high (university entrance qualification such as high school diploma or advanced secondary school). Professional training was categorized into four levels: none, vocational (completed apprenticeship or comparable school-based training), higher level (advanced vocational training such as technician, or equivalent), and academic (degree from a university or university of

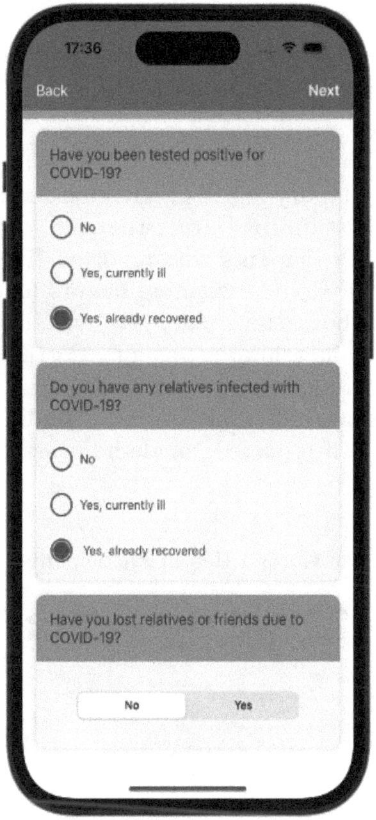

Fig. 3. Mental Health for Adults Study Questionnaire

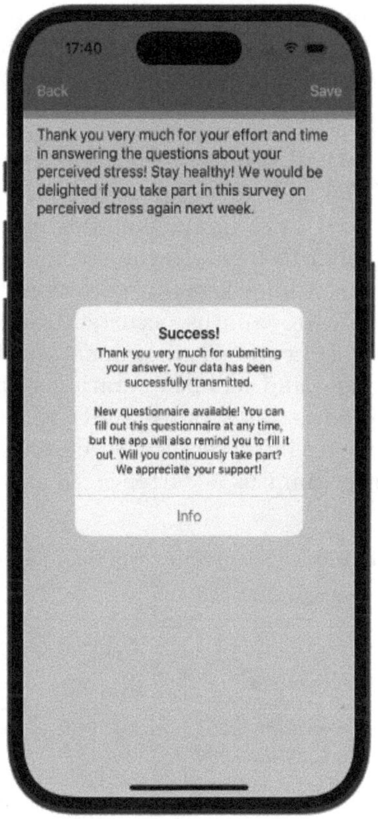

Fig. 4. Completion of Study Module

applied sciences). In addition, self-reported COVID-19 infection status (no/yes) and the mobile operating system (iOS vs Android) were documented.

Results of testing differences in these sociodemographics (i.e., age, gender family, educational, and professional status), positive COVID-19 infection, and mobile operating system (i.e., iOS vs Android) across tinnitus occurrence categories are presented in Table 1.

Participants: Participants were recruited through the Corona Health app and enrolled in the "Mental Health for Adults" study module. Eligible participants were adults aged 18 years or older who provided informed consent and completed the baseline assessment. Only questionnaires answered for oneself were considered (the app also offered the option to answer on behalf of another person). The study sample comprised exclusively participants residing in Germany. Only individuals who responded to the baseline tinnitus occurrence item were included in the analyses. The first contribution was recorded on 2020-07-21 and the last on 2024-10-30. Table 1 presents the sociodemographic characteristics of the considered Corona Health participants.

Statistical Analysis: Factors associated with tinnitus were examined using a two-step modeling approach. In the first step, a binary logistic regression model was used to estimate the odds of reporting any tinnitus during the past week compared to none. In the second step, among participants who reported tinnitus, an ordinal logistic regression model was applied to examine factors associated with tinnitus occurrence probability. In both steps, age, gender, family status, educational attainment, professional qualification, history of COVID-19 infection, and mobile operating system were included as predictors. Results are reported as odds ratios (OR) with 95% confidence intervals (CI). All findings should be interpreted with caution given the self-reported, single-item assessment of tinnitus occurrence.

Table 1. Sociodemographic Characteristics of the Corona Health Participants

		Tinnitus Occurrence				
Characteristic	All	None	Low	Medium	High	p
Participants (n/%)	2,694 (100)	1,719 (63.8)	558 (20.7)	137 (5.1)	280 (10.4)	.01
Age (m/SD)	41.2 (13.9)	40.69 (13.9)	39.75 (13.5)	42.72 (14.1)	46.45 (13.7)	.01
Gender (n/%)FemaleMaleDivers	1,455 (54.0) 1,217 (45.2) 22 (0.8)	907 (52.8) 802 (46.7) 10 (0.4)	335 (60.0) 217 (38.9) 6 (1.1)	74 (54.0) 62 (45.3) 1 (0.7)	139 (49.6) 136 (48.6) 5 (1.8)	.01
Family (n/%)PartnershipSingle	1,631 (60.5) 1,063 (39.5)	1,056 (61.4) 663 (38.6)	329 (59.0) 229 (41.0)	79 (59.6) 58 (42.3)	167 (59.6) 113 (40.4)	.63
Education (n/%)NoneLowMiddleHigh	30 (1.1) 173 (6.4) 636 (23.6) 1,855 (68.9)	18 (1.0) 88 (5.1) 364 (21.2) 1,249 (72.7)	6 (1.1) 43 (7.7) 144 (25.8) 365 (65.4)	1 (0.7) 10 (7.3) 49 (35.8) 77 (56.2)	5 (1.8) 32 (11.4) 79 (28.2) 164 (58.6)	.01
Profession (n/%)NoneVocationalHigher LevelAcademic	410 (15.2) 885 (32.9) 264 (9.8) 1135 (42.1)	257 (15.0) 512 (29.8) 149 (8.7) 801 (46.6)	97 (17.4) 203 (36.4) 60 (10.8) 198 (35.5)	25 (18.2) 52 (38.0) 21 (15.3) 39 (28.5)	31 (11.1) 118 (42.1) 34 (12.1) 97 (34.6)	.01
COVID19 (n/%)NoYes	2,498 (92.7) 196 (7.3)	1,614 (93.9) 105 (9.0)	508 (91.0) 50 (9.0)	124 (90.5) 13 (9.5)	252 (90.0) 28 (10.0)	.02
Mobile OS (n/%)iOSAndroid	889 (33.0) 1,805 (67.0)	589 (34.3) 1,130 (65.7)	169 (30.3) 389 (69.7)	35 (25.5) 102 (74.5)	96 (34.3) 184 (65.7)	.08

Note: Statistically significant differences (p-value ¡ .05) suggest potential differences in Sociodemographic Characteristics between Tinnitus Occurrence

4 Results

In the first step, comparing participants with any tinnitus occurrence during the past week to those without, higher age, lower educational attainment, lower professional qualification, and a history of COVID-19 infection emerged as significant predictors. Specifically, each additional year of age was associated with a 1.2% increase in the odds of reporting tinnitus (OR = 1.01, 95% CI: 1.01–1.02, p ¡ .001). Higher educational attainment (OR = 0.82, 95% CI: 0.72–0.93, p = .002) and higher professional qualification (OR = 0.87, 95% CI: 0.80–0.94, p = .001) were linked to reduced odds of reporting tinnitus. Although educational attainment and professional qualification were moderately correlated (r = .45), both remained independent predictors in the model. Other pairwise correlations among predictors were weak and did not suggest problematic multicollinearity. Participants with a history of COVID-19 infection had higher odds of reporting tinnitus compared to those without infection (OR = 1.50, 95% CI: 1.11–2.02, p = .008). Figures 5, 6, 7 and 8 visualize the effects of age, educational attainment, professional qualification, and COVID-19 infection on tinnitus occurrence probability.

In the second step, restricted to participants who reported tinnitus (n = 975), age was the only robust predictor of higher tinnitus occurrence probability. Each additional year of age increased the odds of belonging to a higher category (from "some days" toward "nearly every day") by 3% (OR = 1.03, 95% CI: 1.02–1.04, p ¡ .001).

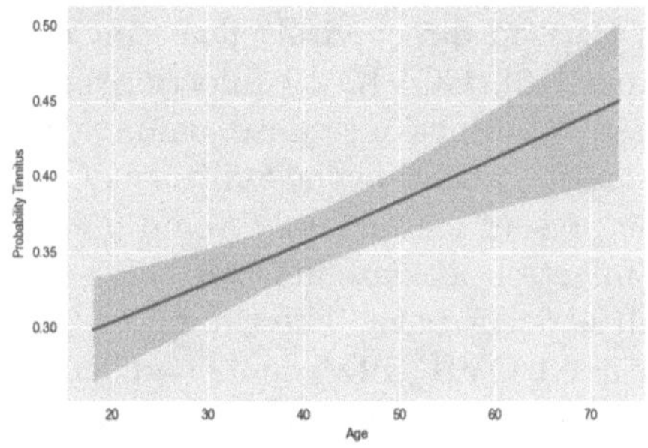

Fig. 5. Age

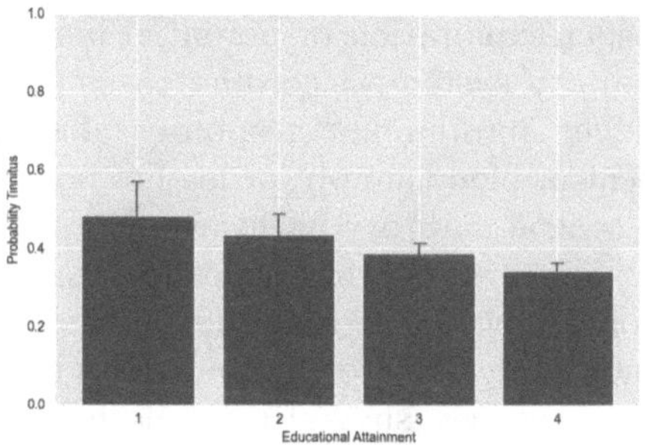

Fig. 6. Educational Attainment

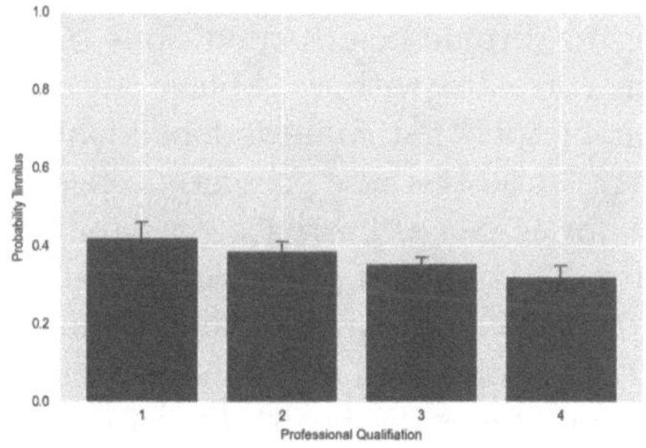

Fig. 7. Professional Qualification

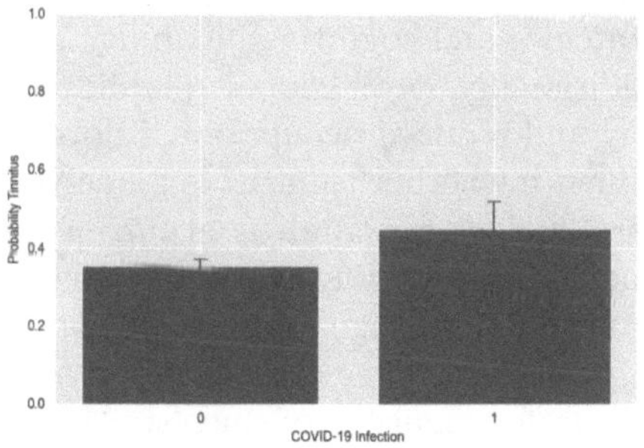

Fig. 8. COVID-19 Infection

5 Discussion

The association between higher age and greater likelihood of tinnitus occurrence, as well as higher tinnitus occurrence probability among those affected (OR = 1.01–1.03 per year), reveals a gradual but consistent increase with each year of life, as demonstrated in the literature [21]. The cumulative effects of lifelong noise exposure and reduced neural plasticity in older adults may constitute a catalyst for both tinnitus development and its persistence [24]. Clinically, this emphasizes the need for age-specific tinnitus management protocols that account for the complex nature of tinnitus in older populations, including considerations for cognitive decline, reduced technology adoption, and comorbid health conditions that may complicate treatment adherence.

The significant association between COVID-19 infection and increased odds of reporting any tinnitus occurrence (OR = 1.50, 95% CI: 1.11–2.02) warrants deeper exploration of potential mechanisms. Beyond direct viral effects on the auditory system, according to similar research [10], COVID-19 infection may trigger autoimmune responses affecting inner ear function, cause damage to cochlear blood supply, or lead to neuroinflammation affecting the auditory nerve or central auditory pathways. Notably, in the present sample only about 7% of participants reported a positive COVID-19 infection at baseline, yet this subgroup showed a meaningful elevation in tinnitus occurrence. The timing of this study (July 2020 to October 2024) spans multiple COVID-19 variants and vaccination phases, suggesting that the observed association may represent a robust relationship across different viral strains and population immunity levels.

The protective effects of higher education (OR = 0.82, 95% CI: 0.72–0.93) and professional qualification (OR = 0.87, 95% CI: 0.80–0.94) likely reflect a complex interplay of factors beyond simple socioeconomic status, according to similar research [7]. Higher-educated individuals may demonstrate superior self-advocacy skills, enabling them to access specialized care and evidence-based treatments more effectively. They may also possess better emotional regulation strategies and cognitive flexibility in adapting to chronic health conditions [15]. Additionally, professional qualifications often correlate with workplace autonomy and reduced occupational noise exposure, factors that could independently influence whether tinnitus is perceived at all [1]. Notably, these protective effects were evident for tinnitus occurrence but not for higher occurrence probability among those already affected, suggesting that education and professional status may play a stronger role in the prevention or onset of tinnitus rather than in its persistence once established. The gradient effect observed across educational and professional levels suggests that interventions targeting health literacy and coping skills development could be particularly beneficial for populations with lower socioeconomic resources.

The absent effects of gender and family status stand in contrast to some previous literature reporting higher tinnitus distress in women [2]. This finding may reflect the unique context of pandemic-related stress affecting all genders similarly, or it could indicate that app-based recruitment attracts more motivated participants regardless of gender, potentially reducing gender-based reporting differences. The non-significant family status effect is particularly interesting given the social isolation experienced during the pandemic, suggesting that the presence or absence of a partner may not have provided the expected buffering effect against tinnitus occurrence during this period of widespread social disruption.

The finding that mobile operating system (iOS vs Android) was not associated with tinnitus occurrence provides important validation for app-based research methodologies, particularly given that related research has documented systematic sociodemographic differences between iOS and Android users [20,32]. The absence of operating system effects in this study suggests that these underlying demographic differences did not translate into meaningful variations in tinnitus reporting patterns. However, this should be interpreted within the broader context of digital health equity. The study's requirement for smartphone ownership and digital literacy inherently selects for a population with certain socioeconomic characteristics, potentially creating a "digital divide" in research participation [14]. The fact that 68.9% of participants had high educational attainment suggests this selection effect was substantial, likely excluding populations who might show different tinnitus patterns or have limited access to digital health tools. This demographic skew may have masked potential operating system effects that could emerge in more socioeconomically diverse samples.

Another important consideration is the relatively high proportion of participants in our sample reporting tinnitus (about 36%). Epidemiological research typically estimates that around 15% of the general population experience chronic tinnitus, commonly defined as phantom sound lasting more than five minutes on most days [7]. In our data, this prevalence corresponds more closely to the medium and high groups. By contrast, participants in the low group may include individuals experiencing short or occasional tinnitus episodes, which are relatively common in the general population and not usually classified as chronic tinnitus. Thus, the single item measure employed in the Corona Health app may have broadened the definition of tinnitus, capturing both chronic and occasional experiences. This should be kept in mind when interpreting prevalence estimates and comparing them to epidemiological benchmarks.

Implications: These findings underscore the importance of adopting approaches for tinnitus assessment and management that integrate sociodemographic, health, and contextual factors. The differential patterns observed across tinnitus occurrence demonstrate that tinnitus cannot be adequately understood or treated through a purely biomedical lens, contributing to similar insights from related research [17]. This necessitates the development of stratified care pathways that account for individual risk profiles, with enhanced support systems for vulnerable populations and targeted interventions that apply existing strengths in more resilient groups [26].

The protective effects of education and professional qualification highlight opportunities for intervention through health literacy enhancement and skill-building programs. Importantly, these factors were associated with the occurrence of tinnitus but not its probability once present, suggesting that socioeconomic resources may be more relevant for prevention and early detection than for modifying the persistence of tinnitus. Healthcare systems should therefore consider implementing tinnitus management approaches that provide intensive, multidisciplinary support for high-risk populations (e.g., lower socioeconomic status) while offering self-directed digital therapeutics for individuals with higher baseline coping resources.

Furthermore, the association between COVID-19 infection and tinnitus occurrence has immediate clinical relevance. Post-COVID care protocols should routinely screen for auditory symptoms and consider tinnitus as a potential component of long-COVID syndrome. This finding also underscores the importance of maintaining audiological services during health emergencies and developing preparedness strategies for managing secondary health impacts of infectious disease outbreaks.

The successful implementation of this app-based research during a global pandemic validates digital health platforms as essential tools for both research and clinical care delivery. Beyond demonstrating feasibility, this study reveals the potential for real-time, population-level monitoring of health conditions during crises when traditional healthcare access is compromised. The scalability and ecological validity of mobile health assessments position them as valuable complements to traditional clinical evaluations, enabling continuous monitoring and just-in-time interventions that could transform chronic condition management. However, efforts must be made to address digital health equity to ensure these innovations benefit all population segments, not just those with existing technological access and literacy.

Future Work: Future analyses should utilize the longitudinal follow-up assessments available within the Corona Health platform to examine changes in tinnitus occurrence over time and identify temporal patterns in symptom progression or resolution. This would allow for the investigation of whether predictors of tinnitus onset differ from predictors of persistence. In addition, the timing of COVID-19 infection in relation to subsequent tinnitus development could be examined, clarifying whether tinnitus emerges immediately post-infection or develops gradually.

The platform also contains validated instruments on quality of life and mental health (QOL, PHQ), which were not analyzed here given the study's focus on the tinnitus occurrence item, but which could be incorporated in future research to explore relationships between tinnitus, psychological distress, and overall well-being.

Given the platform's EMA capabilities, additional work should also examine short-term fluctuations in tinnitus occurrence in relation to other mental health indicators. The baseline sociodemographic data could further be applied to develop predictive models for tinnitus occurrence and persistence using machine learning approaches on the longitudinal dataset. Moreover, cross-sectional analyses examining the interplay between tinnitus occurrence, professional status, and work-related quality of life measures could inform workplace interventions. Finally, the rich dataset should be explored for identifying distinct trajectories of tinnitus occurrence and their associated risk factors using growth mixture modeling approaches.

Limitations: Several limitations should be considered when interpreting these findings. First, tinnitus occurrence were assessed with a single self-report item rather than a validated multi-item instrument, which may limit precision and comparability with other studies (and did not allow integration with additional instruments such as QOL or PHQ). Second, the analysis relied on baseline data only, preventing conclusions about temporal dynamics or causality. Third, tinnitus was self-reported, which may introduce reporting bias. Fourth, the app-based design could lead to selection effects, as participation required digital literacy, smartphone ownership, and motivation to engage with the study. Fifth, educational attainment and professional qualification were moderately correlated, which should be considered when interpreting their independent effects. Importantly, the Corona Health app was primarily distributed in Germany, which limits the generalizability of the findings to other countries and cultural contexts.

6 Conclusion

In a sample of 2,694 adults recruited via the Corona Health app in Germany, tinnitus occurrence shaped by sociodemographic and health-related factors. Higher age and a history of COVID-19 infection were associated with increased odds of tinnitus occurrence, while higher educational attainment and professional qualification showed protective effects. Among participants with tinnitus, older age was the primary predictor of higher occurrence probability. These findings emphasize that tinnitus cannot be understood in isolation from broader social and health contexts and that distinct factors may shape its onset versus its persistence. They also demonstrate the potential of app-based platforms to capture large-scale, real-world data, even during periods of restricted healthcare access such as the COVID-19 pandemic. Future work should extend these insights through longitudinal analyses, validated multi-item measures, and broader international recruitment.

References

1. Allain, S., Naouri, D., Deroyon, T., Costemalle, V., Hazo, J.B.: Income and professional inequalities in chronic diseases: prevalence and incidence in France. Public Health **228**, 55–64 (2024)
2. Basso, L., et al.: Gender-specific risk factors and comorbidities of bothersome tinnitus. Front. Neurosci. **14**, 706 (2020)
3. Beierle, F., et al.: Corona health–a study-and sensor-based mobile app platform exploring aspects of the COVID-19 pandemic. Int. J. Environ. Res. Public Health **18**(14), 7395 (2021)

4. Beukes, E., Ulep, A.J., Eubank, T., Manchaiah, V.: The impact of COVID-19 and the pandemic on tinnitus: a systematic review. J. Clin. Med. **10**(13), 2763 (2021)
5. Beukes, E.W., et al.: Changes in tinnitus experiences during the COVID-19 pandemic. Front. Public Health **8**, 592878 (2020)
6. Biswas, R., et al.: Low evidence for tinnitus risk factors: a systematic review and meta-analysis. J. Assoc. Res. Otolaryngol. **24**(1), 81–94 (2023)
7. Biswas, R., Lugo, A., Akeroyd, M.A., Schlee, W., Gallus, S., Hall, D.: Tinnitus prevalence in Europe: a multi-country cross-sectional population study. Lancet Reg. Health Europe **12** (2022)
8. Chatterjee, N., Chattopadhyay, D., Chatterjee, I.: Management of tinnitus in COVID-19 outbreak-a comparative study between mindfulness based tinnitus stress reduction and tinnitus retraining therapy. Int. Tinnitus J. **25**(1), 29–33 (2021)
9. Daher, G.S., Nassiri, A.M., Vanichkachorn, G., Carlson, M.L., Neff, B.A., Driscoll, C.L.: New onset tinnitus in the absence of hearing changes following COVID-19 infection. Am. J. Otolaryngol. **43**(1), 103208 (2022)
10. De Luca, P., et al.: Auditory disturbances and SARS-CoV-2 infection: brain inflammation or cochlear affection? Systematic review and discussion of potential pathogenesis. Front. Neurol. **12**, 707207 (2021)
11. Eicher, S., Pryss, R., Baumeister, H., Hövener, C., Knoll, N., Cohrdes, C.: Quality of life during the COVID-19 pandemic-results of the corona health app study. J. Health Monit. **6**(Suppl 6), 2 (2021)
12. Elarbed, A., Fackrell, K., Baguley, D.M., Hoare, D.J.: Tinnitus and stress in adults: a scoping review. Int. J. Audiol. **60**(3), 171–182 (2021)
13. Fagherazzi, G., Goetzinger, C., Rashid, M.A., Aguayo, G.A., Huiart, L.: Digital health strategies to fight COVID-19 worldwide: challenges, recommendations, and a call for papers. J. Med. Internet Res. **22**(6), e19284 (2020)
14. Fox, G., Connolly, R.: Mobile health technology adoption across generations: Narrowing the digital divide. Inf. Syst. J. **28**(6), 995–1019 (2018)
15. Galenkamp, H., Van Oers, H.A., Kunst, A.E., Stronks, K.: Is quality of life impairment associated with chronic diseases dependent on educational level? Eur. J. Pub. Health **29**(4), 634–639 (2019)
16. Hacker, K.A., Briss, P.A., Richardson, L., Wright, J., Petersen, R.: COVID-19 and chronic disease: the impact now and in the future. Prev. Chronic Dis. **18**, E62 (2021)
17. Hobeika, L., et al.: Tinnitus risk factors and its evolution over time. Nat. Commun. **16**(1), 4244 (2025)
18. Holfelder, M., et al.: Medical device regulation efforts for mHealth apps during the COVID-19 pandemic—an experience report of corona check and corona health. J **4**(2), 206–222 (2021)
19. Kraft, R., Reichert, M., Pryss, R.: Mobile crowdsensing in ecological momentary assessment mHealth studies: a systematic review and analysis. Sensors **24**(2), 472 (2024)
20. Mehdi, M., et al.: Smartphone and mobile health apps for tinnitus: systematic identification, analysis, and assessment. JMIR Mhealth Uhealth **8**(8), e21767 (2020)
21. Oosterloo, B.C., Croll, P.H., Baatenburg de Jong, R.J., Ikram, M.K., Goedegebure, A.: Prevalence of tinnitus in an aging population and its relation to age and hearing loss. Otolaryngol. Head Neck Surg. **164**(4), 859–868 (2021)
22. Patil, J.D., Alrashid, M.A., Eltabbakh, A., Fredericks, S.: The association between stress, emotional states, and tinnitus: a mini-review. Front. Aging Neurosci. **15**, 1131979 (2023)

23. Ravichandran, A., Ferguson, M.A., Mulders, W.H., Choi, R.S., Bennett, R.J.: Understanding engagement with digital health interventions designed for adults with hearing loss and tinnitus: a mixed-method systematic review. Transl. Behav. Med. **15**(1), ibaf028 (2025)
24. Ruan, Q., Chen, B., Panza, F.: Which came first, age-related hearing loss with tinnitus or cognitive impairment? What are the potential pathways? J. Integr. Neurosci. **22**(5), 109 (2023)
25. Schlee, W., et al.: The effect of environmental stressors on tinnitus: a prospective longitudinal study on the impact of the COVID-19 pandemic. J. Clin. Med. **9**(9), 2756 (2020)
26. Schlee, W., et al.: Towards a unification of treatments and interventions for tinnitus patients: the EU research and innovation action UNITI. Prog. Brain Res. **260**, 441–451 (2021)
27. Searchfield, G.D., et al.: A state-of-art review of digital technologies for the next generation of tinnitus therapeutics. Front. Digit. Health **3**, 724370 (2021)
28. Shiffman, S., Stone, A.A., Hufford, M.R.: Ecological momentary assessment. Annu. Rev. Clin. Psychol. **4**(1), 1–32 (2008)
29. Swain, S.K.: Impact of tinnitus on quality of life: a review. Int. J. Adv. Med. **8**(7), 1006–10 (2021)
30. Vogel, C., Pryss, R., Schobel, J., Schlee, W., Beierle, F.: Developing apps for researching the COVID-19 pandemic with the trackyourhealth platform. In: 2021 IEEE/ACM 8th International Conference on Mobile Software Engineering and Systems (MobileSoft), pp. 65–68. IEEE (2021)
31. Waechter, S., Wilson, W.J., Magnusson, M., Brännström, K.J.: Extended high frequency hearing, but not tinnitus, is associated with every-day cognitive performance. Front. Psychol. **13**, 913944 (2022)
32. Winter, M., Schlee, W., Probst, T., Stach, M., Pryss, R.: Follow-up evaluation to explore disparities between android and IoS users utilizing the trackyourtinnitus mobile health platform. In: 2023 International Conference on Computational Science and Computational Intelligence (CSCI), pp. 1493–1497. IEEE (2023)

Synthetic Medical Image Generation with Conditional Generative Adversarial Networks for Enhanced Diagnostic Performance

Sayli Aphale[1(✉)], Aditya Kalapala[1], Ernest Tolliver[1], and Eugene John[2]

[1] Department of Electrical Engineering, The University of Texas at San Antonio, One UTSA Circle, San Antonio, TX 78249, USA
{sayli.aphale,aditya.akalapala,ernest.tolliver}@my.utsa.edu

[2] Department of Computer Engineering, The University of Texas at San Antonio, One UTSA Circle, San Antonio, TX 78249, USA
eugene.john@utsa.edu

Abstract. Medical imaging research is often hindered by limited datasets, class imbalance, high collection costs, and strict privacy and ethical constraints. To overcome these challenges, we propose a framework for synthetic medical image generation using a Conditional Deep Convolutional Generational Adversarial Network (CDCGAN) enhanced with the Wasserstein GAN and Gradient Penalty (WGAN-GP). The framework was applied across three modalities—breast cancer histopathology, dementia brain MRI, and COVID-19/pneumonia Chest X-rays to generate high-quality, class-conditional images that supplement real datasets. Performance was evaluated using modality-specific classifiers, including a Vision Transformer, a custom CNN, and a modified ResNet-34, achieving accuracies of 99.24%, 99.79%, and 95.86%, respectively. These results demonstrate the effectiveness of synthetic data in mitigating data scarcity, addressing class imbalance, and enhancing diagnostic accuracy in medical imaging applications.

Keywords: Synthetic Data Generation · Breast Cancer Histopathology Images · Dementia Brain MRI · COVID–19 · Pneumonia

Acronym: Health Informatics and Medical Systems (CSCI-RTHI).

1 Introduction

Medical data is abundant worldwide; however, a significant portion remains unavailable in usable digital formats due to privacy restrictions, high digitization costs, and regulatory barriers [1, 2]. Dataset collection is further constrained by the rarity of certain conditions and the expense of acquisition. Even widely available digital datasets are often inconsistent or insufficient for training reliable artificial intelligence (AI) models. Since AI research depends not only on advances in neural network architectures but also on large, diverse, and high-quality datasets, these limitations represent a critical barrier to progress.

H. R. Arabnia et al. (Eds.): CSCI 2025, CCIS 2971, pp. 16–32, 2027.
https://doi.org/10.1007/978-3-032-28463-1_2

Recent advances in deep learning and AI have demonstrated strong potential in medical image analysis. Frameworks such as Convolutional Neural Networks (CNNs) have been successfully applied to the classification of benign and malignant tumors in histopathology images [3, 4]. Similar methods have been extended to dementia detection using brain MRI [5] and the diagnosis of COVID-19 and pneumonia from Chest X-rays [6]. Despite these successes, real-world medical datasets often suffer from limited diversity and severe class imbalance, which restricts model generalization [7, 8]. Synthetic data generation offers a promising solution by producing realistic, class-specific samples to augment existing datasets. This approach improves training diversity, mitigates bias, and enhances the robustness of diagnostic models [9, 10].

In this study, we propose a Conditional Deep Convolutional Generative Adversarial Network (CDCGAN) [11, 12] to generate synthetic data across three domains: breast cancer histopathology, dementia brain MRI, and COVID-19/pneumonia Chest X-ray images. The fidelity of the generated data is evaluated using domain-specific classifiers. The remainder of this paper is organized as follows: Section 2 reviews related work, Sect. 3 describes the synthetic data generation and classification framework, Sect. 4 presents results and performance evaluation, Sect. 5 provides analysis, and Sect. 6 concludes the paper.

2 Related Work

Several studies have explored deep learning and generative approaches to overcome the challenges of limited medical imaging data. Desai et al. [13] used DCGANs to augment mammography datasets, validating synthetic images through ImageJ analysis and expert visual Turing tests. Khan et al. [14] proposed a CNN-based pneumonia detection model, where DCGAN-generated images enhanced performance, achieving accuracies of 95%, 91%, and 93% on synthetic, original, and combined datasets, respectively, surpassing those of VGG16 and ResNet50. Raza et al. [15] reviewed the DCGAN architecture and demonstrated its application as both a synthetic data generator and a binary classifier to distinguish real from fake images using a brain tumor MRI dataset, while also highlighting challenges of diversity and realism in GAN-generated images. Similarly, Maguolo and Nanni [16] highlighted fairness issues in COVID-19 CXR classification, demonstrating that networks often learn spurious patterns unrelated to disease presence. Castiglioni et al. [17] further examined CXR-based COVID-19 screening, warning against transfer learning biases and validating performance on a medium-sized dataset from an Italian hospital, emphasizing both the potential and limitations of AI-driven COVID-19 diagnosis.

3 Framework for Synthetic Data Generation and Classification

This study presents a framework for generating synthetic medical imaging data to address the challenge of data scarcity in three domains: breast cancer histopathology, dementia brain MRI, and COVID-19/pneumonia Chest X-rays. A Conditional Deep Convolutional GAN (CDCGAN), optimized with the Wasserstein GAN and Gradient Penalty (WGAN-GP), was developed to produce high-quality, class-conditional images. The fidelity and utility of the synthetic data were evaluated using three classification tasks: (1) a Vision Transformer (ViT) model to classify breast cancer histopathology images as benign or malignant, (2) a custom CNN to distinguish healthy from demented brain MRI scans, and (3) a ResNet34 model to categorize Chest X-rays into COVID-19, pneumonia, or normal cases.

3.1 Dataset Utilized for Synthetic Data Generation

This research utilized publicly available datasets to generate synthetic data. Breast Cancer Histopathology Images were sourced from the BreakHis Dataset [18]. Dementia Brain MRI images were obtained from the OASIS-2 Longitudinal Scan Dataset [19]. COVID-19 and normal images were also included. Pneumonia Chest X-ray Images were assembled from Kaggle [20].

BreakHis Dataset for Breast Cancer Histopathology Images. The BreakHis dataset comprises eight tumor subtypes, categorized into four benign and four malignant classes, grouped into two main classes: benign and malignant. It contains a total of 7,909 microscopic images in 3-channel RGB format, each with a resolution of 700 × 460 pixels, as shown in Fig. 1. The images were collected from 82 patients and captured at magnifications of 40X, 100X, 200X, and 400X. Among these, 2,480 samples are benign tumors, and 5,429 are malignant tumors. For this study, the data were divided into training (70%), testing (15%), and validation (15%) sets.

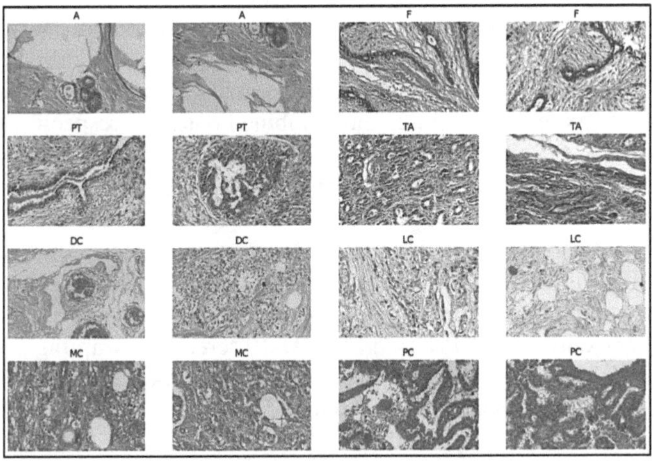

Fig. 1. Samples of BreakHis Dataset [18].

Pre-processing BreakHis Dataset. The histopathology images underwent a three-step preprocessing pipeline. First, Reinhard color normalization [21] was applied to achieve uniform color distribution across all samples. Next, an Enhanced Super-Resolution Generative Adversarial Network (ESRGAN) [22] was used to upscale low-resolution images into high-resolution versions, preserving fine structural details. Finally, Contrast Limited Adaptive Histogram Equalization (CLAHE) [23] was employed to enhance local contrast and improve visibility in regions of low contrast, resulting in clearer and more informative images for further analysis. Figure 2 gives the pre-processing pipeline for breast cancer histopathology images.

Reinhard Color Normalization. This method corrects color variations by selecting a suitable reference image and transferring its statistical characteristics to the rest of the dataset. Specifically, both the input and reference images are converted into the perceptual CIELab color space [24], where the mean and standard deviation of each channel are calculated. A randomly chosen source image served as the reference, and its color statistics were applied to transform all dataset images. This process ensured a consistent color distribution across the dataset, thereby reducing staining variability.

Enhanced Super-Resolution Generative Adversarial Network. In this study, a lightweight ESRGAN-inspired pipeline was applied to the BreakHis dataset, where images were resized into low-resolution inputs (128 × 128) and high-resolution ground truth (256 × 256). The generator network featured an up-sampling layer followed by convolutional and ReLU activation blocks, with a final Tanh activation. Training was conducted using the Mean Squared Error (MSE) loss, optimized with the Adam algorithm (learning rate = 0.0002) for 50 epochs. After training, the generator produced super-resolved images that maintained class consistency across benign and malignant categories. Evaluation using Peak Signal-to-Noise Ratio (PSNR) and Structural Similarity Index Measure (SSIM) yielded values of 55.09 dB and 0.9984, respectively, confirming high reconstruction fidelity and suitability for medical image analysis.

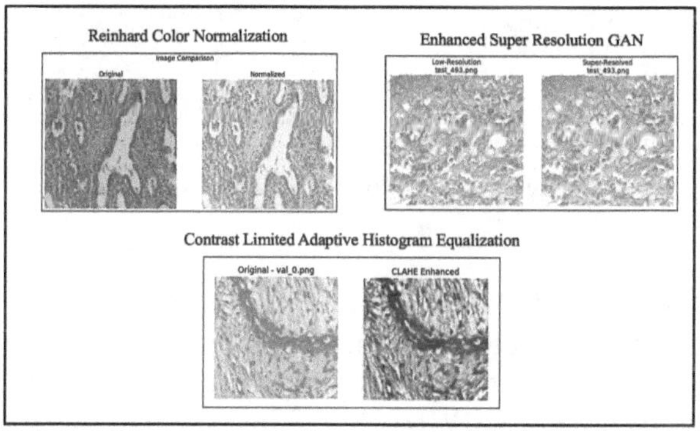

Fig. 2. Preprocessing of Breast Cancer Histopathology Images

Contrast Limited Adaptive Histogram Equalization. Contrast-limited adaptive Histogram Equalization was used to address the poorly defined boundaries, overlapping structures, and blurriness in the histopathology images. In this study, CLAHE was applied to each RGB image within the super-resolved histopathology dataset, covering both benign and malignant categories. CLAHE works by dividing an image into small contextual regions and performing histogram equalization within each, while limiting contrast enhancement to prevent excessive noise.

Longitudinal MRI Data for Dementia Analysis (OASIS-2). The OASIS-2 dataset contains 3-dimensional images of 150 subjects aged 60–96, with a total of 373 imaging sessions collected over multiple visits, at least a year apart. Each subject contributed three to four T1-weighted MRI scans per session, and all subjects were right-handed. The dataset was processed and converted into 2-dimensional, 256 × 256 images, categorized into two classes: healthy and demented. To preprocess the data, Selective Denoising was applied using masking with different thresholds for brain tissue and background. Then, CLAHE was used to enhance local contrast and improve image quality. The resulting images are shown in Fig. 3. The final dataset consisted of 6,835 MRI images, comprising 3,455 healthy and 3,380 demented samples.

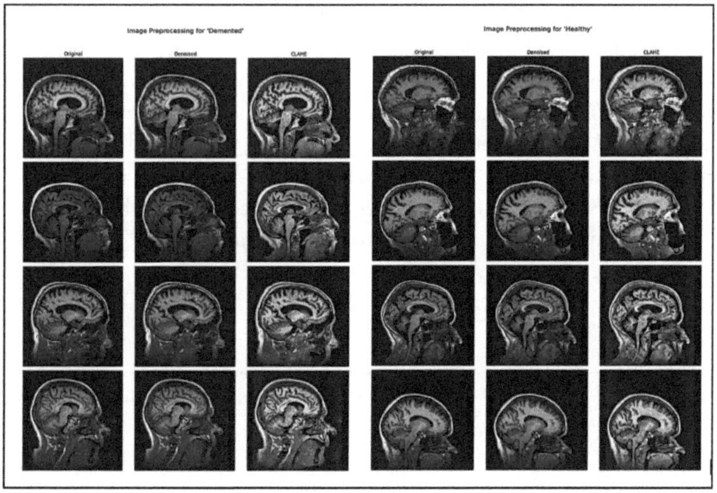

Fig. 3. Preprocessing of Demented and Healthy Brain MRI Images [19].

Chest X-ray Images of COVID-19, Pneumonia, and Normal. For COVID-19 and pneumonia classification, we utilized a publicly available Chest X-ray dataset from Kaggle, comprising 6,432 images split into training and testing sets. The images are categorized into three groups: COVID-19 (460 images), Pneumonia (3,418 images), and normal (1,266 images). The images were resized to 140 × 140 pixels and then center-cropped to 128 × 128 pixels to fit the model. Figure 4 presents the pre-processed COVID-19, Normal, and Pneumonia Chest X-ray Images that were provided as input to the DCGAN network.

The preprocessed images from the BreakHis Dataset, the OASIS-2 dataset, and the COVID-19 and Pneumonia dataset were used as input to the designed Conditional DCGAN.

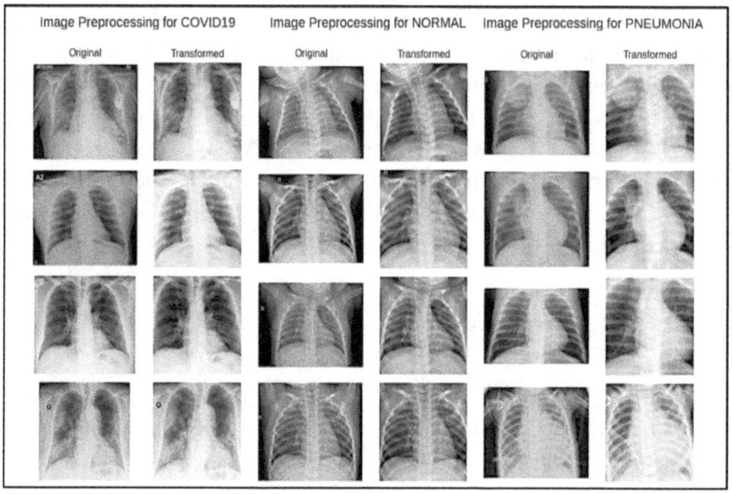

Fig. 4. Pre-processed COVID-19, Normal, and Pneumonia Chest X-ray Images [20].

3.2 Conditional Deep Convolutional Generative Adversarial Network

The synthetic image generation process was divided into three stages. First, a Conditional DCGAN architecture was developed, specifically designed for colored histopathology images and subsequently adapted for grayscale dementia MRI images and Chest X-rays, enabling it to handle multiple modalities. Second, the model was trained using WGAN-GP [25] optimization, carried out separately for each dataset to ensure stability and high-quality output. Finally, the trained model was used to generate synthetic datasets across all three modalities, which were then used to enhance training, improve generalization, and support robust diagnostic performance.

Conditional DCGAN Architecture. The Conditional DCGAN developed in this study enhances the standard DCGAN by integrating the Wasserstein GAN with Gradient Penalty (WGAN-GP), enabling more stable training and higher-quality image generation. The model was designed to produce synthetic images across three modalities: breast cancer histopathology, dementia brain MRI scans, and Chest X-rays labeled as COVID-19, normal, or pneumonia. An RGB version was initially implemented for histopathology and later adapted into a grayscale version for dementia and Chest X-ray data. Table 1 provides the detailed architecture of the designed Conditional GAN network.

To enable class-specific image generation, label conditioning was applied in both the generator and discriminator. In the generator, class labels were transformed into dense vectors using an embedding layer and concatenated with the input noise vector, allowing the model to learn class-aware latent features and generate images aligned with the desired category. The generator architecture included multiple ConvTranspose2d layers with batch normalization and ReLU activations that progressively upsample the latent vector to produce images. An early dropout layer was added to enhance diversity and reduce overfitting, especially important for medical datasets with limited size. On the discriminator (critic) side, class labels were embedded into full-resolution spatial maps and concatenated with the input image as an additional channel, allowing the discriminator to evaluate image realism within the context of the specified class. The discriminator was designed without a final sigmoid activation, as WGAN-GP employed the Wasserstein loss with a gradient penalty to enforce Lipschitz continuity. Unlike traditional GANs that rely on weight clipping, WGAN-GP used a gradient penalty, which provided smoother gradients, more stable convergence, and higher-quality image generation. This made the architecture well-suited for complex medical imaging tasks across diverse modalities.

Table 1. Conditional DCGAN Architecture.

Aspect	Your Conditional DCGAN with WGAN - GP
Generator Input Conditioning	Random noise/image + class label embedding
Discriminator Input	Image + spatially expanded label map
Generator Architecture	ConvTranspose2D layers + label embedding and concatenation
Regularization	Dropout (0.3) added in Generator
Loss Function/GAN Variant	Wasserstein Loss with Gradient Penalty
Label Conditioning	Label Embedding + Concatenation
Training Objective	Maximize critic difference (real + fake scores)
Output Activation	None (for WGAN-GP)
Image Modality	Specialized for high-res histopathology images
Output Resolution	128 × 128 – better suited for histopathology images
Number of classes	2 (Benign and Malignant)

Training Loop of DCGAN with WGAN-GP. The preprocessed data were provided as input to the Conditional DCGAN with WGAN-GP and trained for multiple epochs with distinct batch sizes for every dataset. Training initialized two networks: the generator, which produced synthetic images from noise vectors and class labels, and the critic, which distinguished real from synthetic images without using a sigmoid activation. Both were optimized with Adam ($lr = 1e - 4$, $\beta = 0.0, 0.9$). During each epoch, the critic was updated multiple times relative to the generator. In each iteration, the generator created class-conditioned images, while the critic evaluated real and generated samples. The Wasserstein loss was computed as the difference between the real and fake outputs, with a gradient penalty added to ensure Lipschitz continuity. This total loss was then used to update the critic. The generator was then optimized to maximize the critic's scores, encouraging realistic outputs. At defined intervals, the generator entered evaluation mode to produce labeled synthetic images for monitoring, then resumed training. For histopathology, the network operated in RGB mode, whereas for dementia MRI and Chest X-rays, both generator and critic were adapted to grayscale.

Synthetic Data Generation. The overall training process aimed to improve stability, diversity, and conditional control over generated images. It effectively captured the complex distribution of medical images, enriching the datasets with high-quality synthetic samples suitable for further classification tasks.

Generated Histopathology Images of Breast Cancer. In histopathology, a total of 1,736 benign and 3,800 malignant real images were used as input for the Conditional DCGAN for class-specific training. The network was trained over multiple runs of 200, 250, 500, 750, and 1000 epochs, resulting in a total of 15,187 synthetic histopathology images, comprising 7,601 benign and 7,586 malignant samples. Figure 5 shows the Synthetic Breast Cancer Histopathology Images Generated Using Conditional DCGAN Trained with the BreakHis Dataset.

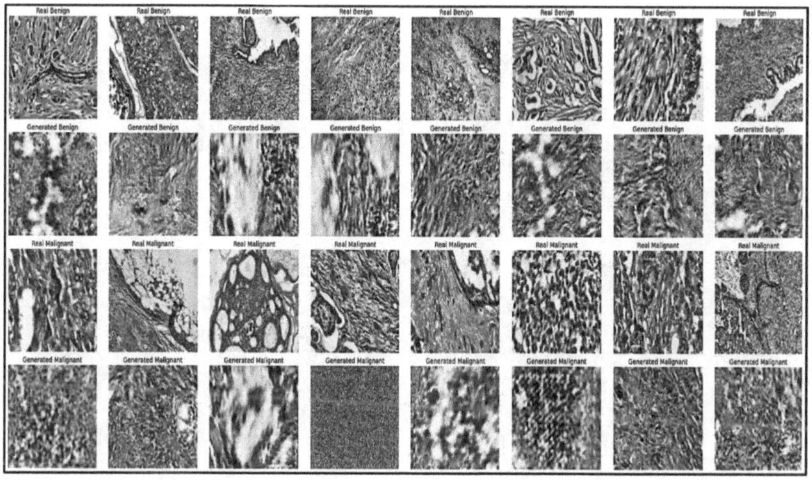

Fig. 5. Real and Generated Breast Cancer Histopathology Images

Generated Brain MRI Scans of Demented and Healthy Subjects. For dementia, the training set included 3,455 healthy and 3,380 demented MRI images. The Conditional DCGAN was adapted to grayscale mode for this modality and trained for 1000 epochs using the same learning rate of 1e−4. This method ensured the creation of high-quality synthetic dementia images that preserved structural fidelity while enhancing dataset diversity for classification tasks. Figure 6 gives real vs. generated healthy and dementia data. One thousand images for each class were generated, combined with real images, and used as input for the classifier.

Generated Chest X-ray images for COVID-19, Normal, and Pneumonia. The Conditional DCGAN architecture proposed in this paper was modified to enhance the stability and image quality of the Chest X-ray Images. The standard batch normalization was replaced with conditional batch normalization, equalized convolutions were added for consistent learning rates, controlled noise was introduced into feature maps, and modular upsample blocks were used to reduce checkerboard artifacts. On the discriminator side, a projection-based design was implemented to integrate class information directly into the adversarial process, combining unconditional and conditional outputs. A ResBlock-Down module with spectral normalization, a self-attention mechanism for capturing long-range dependencies, and global sum pooling instead of fully connected layers was applied to linear and embedding layers for improved training stability. Figure 7 displays real and generated Chest X-ray images for COVID-19, Normal, and Pneumonia after the modification as mentioned earlier. All Chest X-ray images were resized and center-cropped to 128 × 128 pixels before use, resulting in a dataset of 6,432 samples, which was provided as input to the DCGAN network. During GAN training, 15 synthetic images were generated every 5 epochs for monitoring and visualization, resulting in a total of 3,000 images across the full training process.

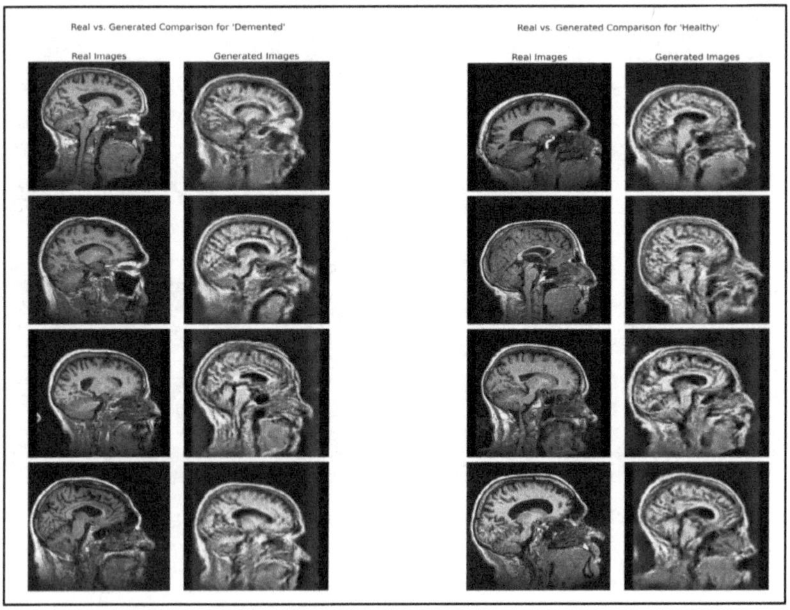

Fig. 6. Real and Generated Dementia Data

Overall, the proposed generative approach effectively produced high-quality, class-conditional synthetic images across histopathology, dementia brain MRI, and COVID-19/pneumonia Chest X-rays.

3.3 Classifier Network Frameworks

To assess data, the synthetic images were combined with real data and classified using three different models: a Vision Transformer [26, 27] for Breast Cancer Histopathology Images, a Convolutional Neural Network for Dementia MRI classification, and a modified ResNet-34 for COVID-19, Normal, and Pneumonia Chest X-ray images.

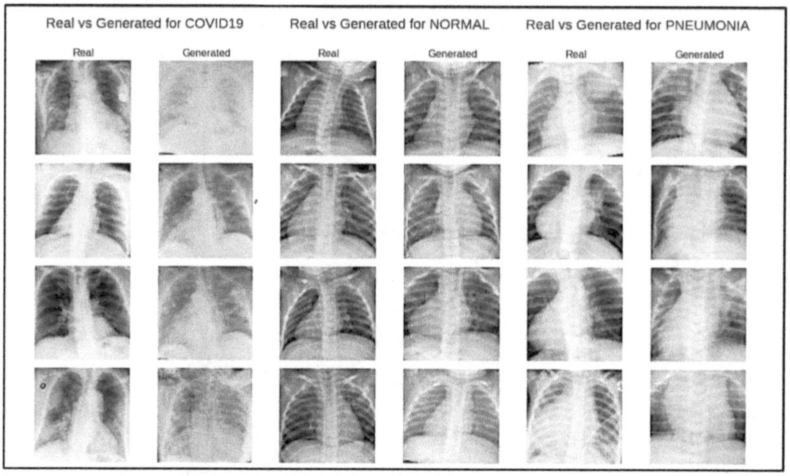

Fig. 7. Real and Generated Chest X-ray Data

Vision Transformer for Breast Cancer Detection. A Vision Transformer (ViT-Base/16, 384 × 384) pretrained with augreg_in21k_ft_in1k weights from the Timm library was used for binary classification of histopathology images. Each image was divided into non-overlapping 16 × 16 patches, resulting in 576 patch tokens plus a [CLS] token, which were embedded into a 768-dimensional space and processed through 12 Transformer encoder layers. Each layer included multi-head self-attention (12 heads), feed-forward networks with GELU activation, residual connections, layer normalization, and stochastic depth (drop_path = 0.1). The final hidden state of the [CLS] token was passed through a linear classifier to predict benign versus malignant cases. The training dataset combined real and synthetic histopathology images, while the validation and test sets used only real data to prevent synthetic leakage. Images were resized to 384 × 384, with augmentations including random flips, rotations of $\leq 10^{\circ}$, and normalization using ImageNet mean and standard deviation. Validation and test images were only resized and normalized. To address class imbalance, we employed a WeightedRandomSampler based on inverse class frequencies, with a batch size of 4. Generalization was improved through label smoothing ($\varepsilon = 0.2$), MixUp, and CutMix ($\alpha = 0.4$), which boosted robustness against staining and tissue variability. Training was conducted with AdamW (lr = 1e−5, weight decay = 1e−4) and Cosine Annealing Warm Restarts ($T_0 = 5$, Tmult = 2) for up to 50 epochs. Early stopping (patience = 10) based on validation accuracy was used, and the checkpoint with the highest accuracy was saved for final evaluation.

Convolutional Neural Network for Dementia Classification. A Convolutional Neural Network [28] is used for binary classification of brain MRI images to distinguish between healthy and demented cases. The input consisted of preprocessed grayscale MRI scans (256 × 256) from the OASIS-2 dataset. Each image was enhanced through selective denoising and CLAHE to improve structural visibility. Table 2 provides the layer-wise details of the custom CNN model. The CNN architecture included convolutional layers with ReLU activation and batch normalization, followed by max-pooling to extract spatial features. These features were flattened and fed into fully connected layers with dropout to reduce overfitting, and the final output layer classified images into two categories (healthy vs. dementia).

Table 2. CNN Architecture for Dementia Classification.

Model Layers	Specifications of Layers
Conv2d	filter size = 32, kernel size = 4, strides = 2, padding = 1, activation = "leakyRelu"
Conv2d	filter size = 64, kernel size = 4, strides = 2, padding = 1, activation = "leakyRelu"
BatchNorm2d	64
Conv2d	filter size = 128, kernel size = 4, strides = 2, padding = 1, activation = "leakyRelu"
BatchNorm2d	128
Conv2d	filter size = 256, kernel size = 4, strides = 2, padding = 1, activation = "leakyRelu"
BatchNorm2d	256
Flatten	Flatten the layers
Dense	units = 1, activation = "sigmoid"

ResNet-34 for COVID-19, Normal, and Pneumonia Detection. The ResNet-34 Keras pre-defined network is utilized to classify the COVID-19, Normal, and Pneumonia Chest X-ray images. These transformations ensured consistent input dimensions and intensity distributions across both real and synthetic images. The Classifier was trained on 128 × 128, 6,432 samples, plus an additional 1,000 samples per class.

The following section presents the experimental results along with the performance evaluation metrics, offering a detailed assessment of the proposed framework.

4 Results and Performance Analysis

After training, the modeled were evaluated on independent validation and test sets across all three imaging domains: histopathology, dementia, and COVID-19. Performance was assessed using Accuracy, Precision, Recall, F1-score, classification reports, and confusion matrices, enabling a comprehensive evaluation of both class-level and overall predictive effectiveness.

4.1 Vision Transformer Classifier for Breast Cancer Histopathology Images

Both the raw BreakHis dataset and the generated synthetic data were used as inputs to the ViT classifier. With combined training data, the model achieved an overall accuracy of 99.24% and a test loss of 0.5057 on real test data, compared to 98.13% accuracy when trained only on real data. The classification report is summarized in Table 3, where the support column indicates the number of test samples per class. The model achieved 99.24% accuracy for both macro and weighted averages across the two classes. Notably, the malignant class reached a recall of 100%, confirming the model's reliability in correctly identifying all malignant cases. The confusion matrix (Fig. 8) further illustrates this, showing that the ViT-based classifier correctly classified 813 out of 815 malignant samples. Importantly, the model eliminated false negatives, misclassifying only seven benign samples as malignant, which strengthens its clinical reliability. Incorporating synthetic data also mitigated the initial 50% class imbalance between benign and malignant samples, reducing it to approximately 8–9%. This balance significantly improved the classifier's ability to identify the minority benign class and enhanced overall diagnostic accuracy.

Table 3. Classification Report: Histopathology Images

	Precision	Recall	F1-Score	Support
Benign	0.99	0.98	0.99	372
Malignant	0.99	1.00	0.99	815
Accuracy			0.99	1187
Macro avg	0.99	0.99	0.99	1187
Weighted avg	0.99	0.99	0.99	1187

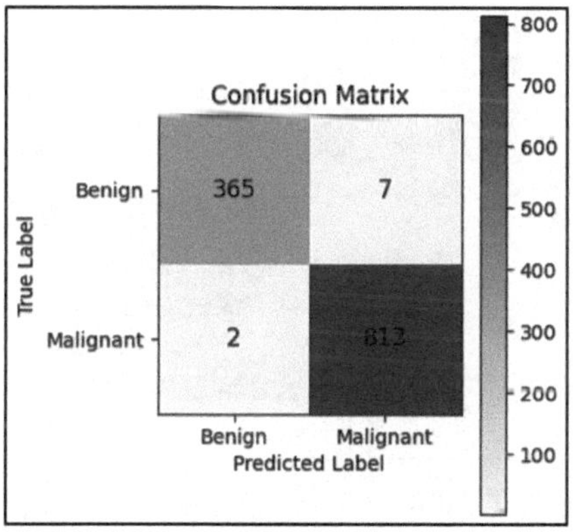

Fig. 8. Confusion Matrix: Histopathology Images

4.2 Convolutional Neural Network Model for Dementia Detection

The custom CNN model was trained on a dataset comprising 5,021 real training images and 2,000 synthetic images (1,000 per class), for a total of 7,021 training samples. Training was performed using the Adam optimizer with a learning rate of 1e−4, a batch size of 64, over 50 epochs. Validation was conducted after each epoch, and the checkpoint with the highest validation accuracy was retained for final testing. On the independent dementia brain MRI test set, the CNN achieved an overall accuracy of 99.69% when trained only on real data, and a slightly higher accuracy of 99.79% when trained on the combined dataset. Although the gain appears modest, the confusion matrix (Fig. 9) highlights the impact of synthetic data: all healthy cases were correctly identified, and only two demented cases were misclassified. This indicates that incorporating synthetic images improved the model's learning capacity, resulting in greater specificity and overall diagnostic accuracy for dementia detection. The classification report (Table 4) further confirmed these findings, with both macro and weighted averages at 1.00, underscoring the robustness and reliability of the CNN in distinguishing healthy subjects from those with dementia.

Table 4. Classification Report: Dementia Detection

	Precision	Recall	F1-Score	Support
Healthy	1.00	1.00	1.00	472
Demented	1.00	1.00	1.00	486
Accuracy			1.00	958
Macro avg	1.00	1.00	1.00	958
Weighted avg	1.00	1.00	1.00	958

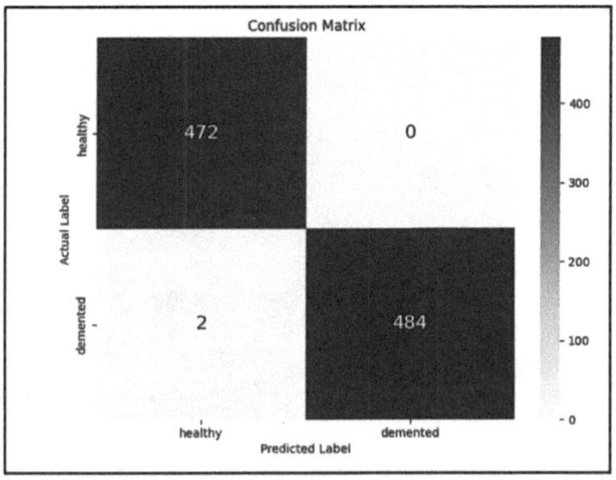

Fig. 9. Confusion Matrix: Dementia Detection

4.3 ResNet-34 for COVID-19, Normal, and Pneumonia Chest X-ray Classification

The classifier was trained on a combined dataset of real and synthetic images, with 1,000 synthetic samples generated per class, resulting in a total of 3,000 synthetic images. Training was conducted for 25 epochs and evaluated on a real test set. The optimization process employed the Cross-Entropy loss function and the Adam optimizer with a learning rate of 1.2×10^{-4}. During training, both loss and accuracy for the training and validation sets were monitored at each epoch, and the model corresponding to the lowest validation loss was retained for further analysis.

Table 5. Classification Report: Detection of COVID-19, Normal, and Pneumonia

	Precision	Recall	F1-Score	Support
COVID-19	1.00	0.94	0.97	116
Normal	0.95	0.90	0.92	317
Pneumonia	0.96	0.98	0.97	847
Accuracy			0.96	1280
Macro avg	0.96	0.94	0.95	1280
Weighted avg	0.96	0.96	0.96	1280

The model achieved an overall accuracy of 95.16% with a loss of 0.176 when trained exclusively on real data, and an accuracy of 95.94% with a reduced loss of 0.138 when trained on the combined real and synthetic dataset (Table 5). While these results indicate only a modest improvement in overall accuracy, the confusion matrix (Fig. 10) reveals a marked reduction in misclassifications for the combined data, particularly in lowering false positives for the minority class (COVID-19). Pneumonia cases were classified with the highest consistency across both training conditions. Overall, while the inclusion of synthetic data produced only a modest gain in overall accuracy, it had a meaningful impact in reducing false positives for the minority class (COVID-19), underscoring its practical value in improving diagnostic reliability.

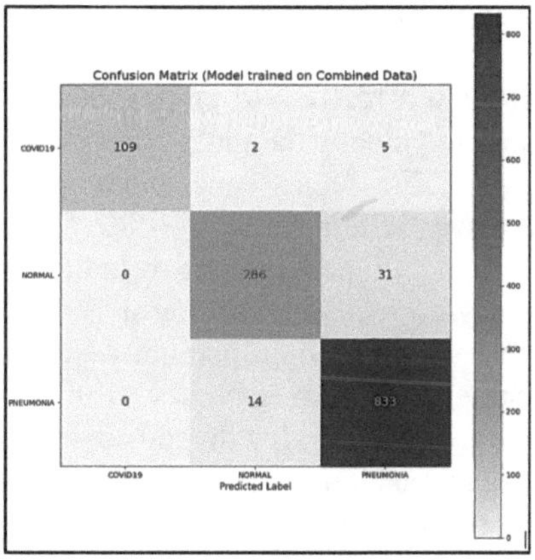

Fig. 10. Confusion Matrix: COVID-19, Normal, and Pneumonia detection

5 Discussion

To evaluate the effectiveness of the proposed framework, we conducted a comparative analysis with existing state-of-the-art studies across multiple medical imaging domains. Narin et al. [29] investigated COVID-19 detection using a GAN-augmented

CNN applied to Chest X-ray images, reporting an accuracy of 93.94%. In contrast, our framework, which incorporates high-fidelity synthetic data generation, achieved a classification accuracy of 95.86% on Chest X-rays, thereby demonstrating a measurable improvement in diagnostic performance. For breast cancer histopathology, the proposed Vision Transformer (ViT) model achieved an accuracy of 99.24%. Comparable studies, including Deebani et al.'s [30] multi-scale transfer learning approach (99.2%) and Hayat et al.'s [31] EfficientNetV2–ViT hybrid (99.83%), reported similarly strong results; however, these methods provided limited attention to interpretability, an aspect central to clinical adoption. In the context of dementia diagnosis, Gajjar et al. [32] evaluated generative models such as GANs, VAEs, and Diffusion Models for MRI synthesis, achieving classification accuracies up to 92.42% with DenseNet architectures. Taken together, these comparisons highlight that the proposed framework not only achieves state-of-the-art accuracy across diverse diagnostic tasks but also advances methodological robustness by integrating synthetic data generation strategies specifically designed to mitigate data scarcity and class imbalance. Importantly, by coupling performance with interpretability and data efficiency, the framework provides a viable pathway toward clinically translatable AI systems that can support reliable, equitable, and scalable medical decision-making.

6 Conclusion

This study demonstrates the effectiveness of integrating high-quality synthetic medical images with real datasets to enhance diagnostic performance across multiple imaging domains. By leveraging a Conditional Deep Convolutional GAN (CDCGAN) optimized with WGAN-GP, we generated synthetic samples for breast cancer histopathology, dementia brain MRI, and COVID-19/pneumonia Chest X-rays. A dedicated preprocessing pipeline further improved the fidelity and consistency of inputs, ensuring that synthetic data complemented real-world clinical datasets effectively. Across all three domains, the inclusion of synthetic data improved classification performance and strengthened model robustness. For breast cancer detection, the Vision Transformer (ViT) achieved 99.24% accuracy, with a recall of 100% for malignant cases, eliminating false negatives and reinforcing clinical reliability. In dementia detection, the custom CNN reached 99.79% accuracy, correctly identifying all healthy cases and misclassifying only two demented cases, thereby improving specificity. For COVID-19 and pneumonia classification, ResNet-34 achieved 95.95% accuracy, with synthetic data reducing false positives for the minority class and improving diagnostic reliability despite only modest gains in overall accuracy. In summary, this work highlights the practical value of synthetic data generation in advancing medical imaging AI. By coupling robust preprocessing with synthetic augmentation, the proposed framework not only achieves state-of-the-art accuracy but also improves specificity, interpretability, and clinical reliability. These findings underscore the potential of synthetic data–driven methods to support scalable, equitable, and clinically translatable AI systems for next-generation healthcare.

References

1. Conduah, A.K., Ofoe, S., Siaw-Marfo, D.: Data privacy in healthcare: global challenges and solutions. Digital Health **11**, 20552076251343959 (2025)
2. Paul, M., Maglaras, L., Ferrag, M.A., Almomani, I.: Digitization of healthcare sector: a study on privacy and security concerns. ICT Express **9**(4), 571–588 (2023)
3. Jiang, B., Bao, L., He, S., Chen, X., Jin, Z., Ye, Y.: Deep learning applications in breast cancer histopathological imaging: diagnosis, treatment, and prognosis. Breast Cancer Res. **26**(1), 137 (2024)
4. Laçi, H., Sevrani, K., Iqbal, S.: Deep learning approaches for classification tasks in medical X-ray, MRI, and ultrasound images: a scoping review. BMC Med. Imaging **25**(1), 156 (2025)
5. Rejusha, T.R., Vipin Kumar, K.S.: Artificial MRI image generation using deep convolutional GAN and its comparison with other augmentation methods. In: 2021 International Conference on Communication, Control and Information Sciences (ICCISc), vol. 1, pp. 1–6. IEEE (2021)
6. Singh, A.R.: Feature controlled synthetic medical image generation using conditional generative adversarial network. In: 2025 International Conference on Inventive Computation Technologies (ICICT), pp. 136–142. IEEE (2025)
7. Walsh, R., Tardy, M.: A comparison of techniques for class imbalance in deep learning classification of breast cancer. Diagnostics **13**(1), 67 (2022)
8. Carriero, A., Groenhoff, L., Vologina, E., Basile, P., Albera, M.: Deep learning in breast cancer imaging: state of the art and recent advancements in early 2024. Diagnostics **14**(8), 848 (2024)
9. Rai, H.M., Dashkevych, S., Yoo, J.: Next-generation diagnostics: the impact of synthetic data generation on the detection of breast cancer from ultrasound imaging. Mathematics **12**(18), 2808 (2024)
10. Gopinath, K., et al.: Synthetic data in generalizable, learning-based neuroimaging. Imaging Neurosci. **2**, 1–22 (2024)
11. Radford, A., Metz, L., Chintala, S.: Unsupervised representation learning with deep convolutional generative adversarial networks. arXiv preprint arXiv:1511.06434 (2015)
12. https://docs.pytorch.org/tutorials/beginner/dcgan_faces_tutorial.html
13. Desai, S.D., Giraddi, S., Verma, N., Gupta, P., Ramya, S.: Breast cancer detection using gan for limited labeled dataset. In: 2020 12th International Conference on Computational Intelligence and Communication Networks (CICN), pp. 34–39. IEEE (2020)
14. Atbaş, B., Calşşkan, E., Erdem, O.: AI-supported pneumonia diagnosis: performance of deep learning models and synthetic data generation. In: 2024 Signal Processing: Algorithms, Architectures, Arrangements, and Applications (SPA), pp. 115–120. IEEE (2024)
15. Divya, S., Padma Suresh, L., John, A.: Medical MR image synthesis using dcgan. In: 2022 First International Conference on Electrical, Electronics, Information and communication technologies (ICEEICT), pp. 01–04. IEEE (2022)
16. Maguolo, G., Nanni, L.: A critic evaluation of methods for COVID-19 automatic detection from X-ray images. Inform. Fus. **76**, 1–7 (2021)
17. Tartaglione, E., Barbano, C.A., Berzovini, C., Calandri, M., Grangetto, M.: Unveiling covid-19 from chest x-ray with deep learning: a hurdles race with small data. Int. J. Environ. Res. Public Health **17**(18), 6933 (2020)
18. Spanhol, F., Oliveira, L.S., Petitjean, C., Heutte, L.: A dataset for breast cancer histopathological image classification. IEEE Trans. Biomed. Eng. **63**(7), 1455–1462 (2016)
19. Marcus, D.S., Wang, T.H., Parker, J., Csernansky, J.G., Morris, J.C., Buckner, R.L.: Open Access Series of Imaging Studies (OASIS): cross-sectional MRI data in young, middle aged, nondemented, and demented older adults. J. Cogn. Neurosci. **19**(9), 1498–1507 (2007)
20. https://www.kaggle.com/datasets/prashant268/chest-xray-covid19-pneumonia

21. Reinhard, E., Adhikhmin, M., Gooch, B., Shirley, P.: Color transfer between images. IEEE Comput. Graph. Appl. **21**(5), 34–41 (2002)
22. Wang, X., et al.: Esrgan: enhanced super-resolution generative adversarial networks. In Proceedings of the European Conference on Computer Vision (ECCV) Workshops (2018)
23. Zuiderveld, K.: Contrast limited adaptive histogram equalization. In: Heckbert, P.S. (ed.) Graphics Gems IV, pp. 474–485. Academic Press, Boston (1994)
24. Misue, K., Kitajima, H.: Design tool of color schemes on the CIELAB space. In: 2016 20th International Conference on Information Visualisation (IV), pp. 33–38. IEEE (2016)
25. Gulrajani, I., Ahmed, F., Arjovsky, M., Dumoulin, V., Courville, A.C.: Improved training of wasserstein gans. Adv. Neural Inform. Process. Syst. **30** (2017)
26. Dosovitskiy, A., et al.: An image is worth 16x16 words: transformers for image recognition at scale. arXiv preprint arXiv:2010.11929 (2020)
27. Vaswani, A., et al.: Attention is all you need. Adv. Neural Inform. Process. Syst. **30** (2017)
28. LeCun, Y., Bengio, Y., Hinton, G.: Deep learning. Nature **521**(7553), 436–444 (2015)
29. Sakib, S., Tazrin, T., Fouda, M.M., Md Fadlullah, Z., Guizani, M.: DL-CRC: deep learning-based chest radiograph classification for COVID-19 detection: a novel approach. IEEE Access **8**, 171575–171589 (2020)
30. Deebani, W., Aziz, L., Aziz, A., Basri, W.S., Alawad, W.M., Althubiti, S.A.: Synergistic transfer learning and adversarial networks for breast cancer diagnosis: benign vs. invasive classification. Sci. Reports **15**(1),7461 (2025)
31. Hayat, M., Ahmad, N., Nasir, A., Tariq, Z.A.: Hybrid deep learning efficientnetv2 and vision transformer (effnetv2-vit) model for breast cancer histopathological image classification. IEEE Access (2024)
32. Gajjar, P., et al.: An empirical analysis of diffusion, autoencoders, and adversarial deep learning models for predicting dementia using high-fidelity MRI. IEEE Access **12**, 131231–131243 (2024)

Beyond Accuracy: Deep Learning for Clinically Usable and Accessible Capsule Endoscopy Diagnostics

Aram Kurdi and Adrian N. Evans(✉)

Department of Electronic and Electrical Engineering, University of Bath, Bath BA2 7AY, UK
{ak3225,a.n.evans}@bath.ac.uk

Abstract. Capsule endoscopy is a non-invasive diagnostic tool that improves patient comfort and can be the only option for individuals unable to undergo conventional gastrointestinal endoscopy. A pill-sized camera is swallowed and captures tens of thousands of images, including of the small bowel that is unreachable by other types of endoscopy. However, image review is performed manually and is very resource-intensive. While deep learning has achieved high accuracy in its image classification, shortcut learning and poor interpretability have limited its clinical acceptance. This study develops and evaluates a novel overlay-aware cleaning framework to standardize images and reduce non-anatomical elements. An application-aware, systematic, orientation-preserving data augmentation strategy was implemented to improve generalizability while maintaining realistic imaging characteristics. Multiple convolutional neural networks were benchmarked under matched training conditions and DenseNet169 was selected for its accuracy, stability and computational efficiency. Interpretability, assessed using Grad-CAM++, showed attention concentrated on relevant anatomy rather than artifacts. Addressing interpretability and bias strengthens clinical credibility and supports deployment in high-volume capsule endoscopy workflows.

Keywords: Capsule Endoscopy · Deep Learning · Interpretability · Grad-CAM

1 Introduction

Capsule endoscopy (CE) is a non-invasive diagnostic tool that not only improves patient comfort compared to traditional gastrointestinal (GI) endoscopy, but can be the only available option for many patients [1]. A swallowed pill-camera captures tens of thousands of images as it traverses the GI tract, including small-bowel regions unreachable by standard endoscopy [2–4]. CE is now an established tool for suspected bleeding, inflammation and neoplasia and improves patient experience, with 68% of patients preferring it over conventional procedures and reporting fewer adverse effects [5, 6]. However, its clinical adoption is limited by time-intensive manual review (often weeks to report) [7] and reading costs contributing ~35% of the total procedure [8].

Deep learning (DL), and particularly convolutional neural networks (CNNs), offer a promising solution to automated CE image classification, potentially reducing clinician workload and enabling faster, more scalable diagnosis [9]. CNNs' strength lies in

H. R. Arabnia et al. (Eds.): CSCI 2025, CCIS 2971, pp. 33–47, 2027.
https://doi.org/10.1007/978-3-032-28463-1_3

extracting hierarchical features efficiently through local connections and weight sharing [10]. Across GI imaging tasks, reported DL accuracies often exceed 95% [11–14], and a recent prospective trial showed AI-assisted CE reading improved diagnostic yield (73.7% vs. 62.4%) while reducing the mean review time from 33.7 to 3.8 min [14].

While AI adoption is accelerating across sectors, its integration into medicine remains slow, with no universal standard (e.g., ISO) for assessing clinical reliability. In practice, there are a number of barriers limiting the clinical adoption of DL-based algorithms to automatically classify CE sequences on a frame-by-frame basis, hence, reducing the manual review time, and this has limited its widespread adoption [13]. The most significant barrier is that the black box nature of DL algorithms fails to provide the clinical trust needed for fully automated diagnostic classification. This lack of interpretability undermines the level of clinical trust required for automated diagnostic use, leaving accuracy alone as insufficient to bridge the gap between research performance and clinical acceptance. In addition, ethical and legal challenges persist, including concerns over data privacy, algorithmic bias, limited transparency and liability for diagnostic errors [15].

Motivated by this clinical need, this work targets interpretability as a prerequisite for trustworthy automation. Rather than treating capsule endoscopy classification as a purely numerical optimization task, the focus is on ensuring that model decisions are grounded in anatomically meaningful features. This work presents a DL pipeline for classifying anonymized CE images into four clinically relevant categories: Normal, Polyps, Ulcerative Colitis (UC) and Esophagitis. A novel preprocessing framework removes overlays and borders to standardize images and improve anatomical focus and an application-aware, orientation-preserving augmentation strategy supports generalizability. Model explanations are provided with Grad-CAM++, demonstrating attention concentrated on meaningful GI structures; even misclassifications reflect plausible diagnostic ambiguity under image-only conditions. These findings emphasize the need for and possibility of AI systems that combine high accuracy with transparent, clinically-aligned reasoning to be viable in real-world diagnostic workflows.

The remainder of this paper is organized as follows. Section 2 introduces the CE dataset and highlights the challenges posed by artifacts and overlays. Section 3 reports baseline performance using an EfficientNetB2 model, establishing a benchmark capability. The final methodology is described in Sect. 4, including preprocessing, model selection and augmentation. Section 5 presents experimental results, with emphasis on interpretability via Grad-CAM++. Section 6 concludes with key findings and implications for trustworthy AI in CE and directions for future work are outlined in Sect. 7.

2 Dataset Description

The dataset used in this project is the publicly available, anonymized "WCE Curated Colon Disease Dataset Deep Learning" sourced from real CE procedures [16]. It contains 11,954 images evenly distributed across four classes: normal, polyps, UC and esophagitis. These classes cover diverse and clinically relevant conditions arising in different regions of the GI tract, corresponding to pathologies typically assessed in both upper GI endoscopy and colonoscopy.

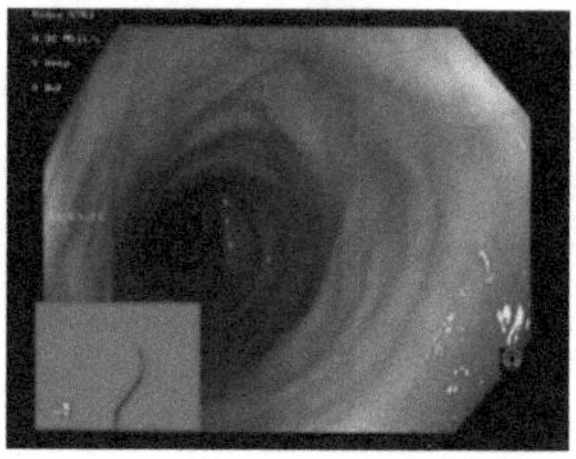

Fig. 1. Normal class image with multiple overlays from WCE Curated Colon Disease Dataset Deep Learning [16].

A manual review revealed four main image formats: clean images without artifacts, images with text on black sidebars, images with various text overlays and images containing multiple graphical overlays. One of the major issues is a green box overlay, see Fig. 1, that consistently occurs in the bottom left corner of all images within the normal class, the majority of images in the UC and polyps classes and none of the esophagitis class. These artifacts pose a risk of bias, potentially causing the model to learn non-anatomical features. Additionally, the image resolutions vary significantly between classes, necessitating standardized resizing during preprocessing to ensure consistency.

3 Baseline Performance of DL-Based Architectures for Capsule Endoscopy Image Classification

To establish a performance benchmark, a baseline model built on the EffecientNetB2 (EB2) CNN architecture was used. The EB2 architecture is not only characterized by high accuracy and computational efficiency, but its capabilities to manage the intricate features present in medical imaging make it a very suitable choice for this application [17]. The baseline implementation was adopted from an open-source Kaggle notebook entitled "95% test acc - EfficientNetB2 - capsule endoscopy" by user khattab1412" [18], chosen because it was specifically developed for the dataset used in this project and therefore provides a reliable reference point.

3.1 Baseline Model Data Preprocessing

Basic preprocessing was applied to the CE dataset. First, all images were normalized using ImageNet mean ([0.485, 0.456, 0.406]) and standard deviation ([0.229, 0.224, 0.225]) values, which is standard practice in transfer learning with pretrained models. This normalization centers pixel values and reduces lighting and color bias, helping the model focus on relevant anatomical features. However, normalization is sensitive to image content, for example artifacts such as overlays can impact the resulting pixel distributions and potentially bias the model beyond the human's ability to identify, see Figs. 2 and 3. All images were resized to 224 × 224 pixels to match EB2's input requirements. To improve robustness, random data augmentation was applied to training images, including transformations such as horizontal flipping and zoom, though some augmentations may introduce unrealistic variations.

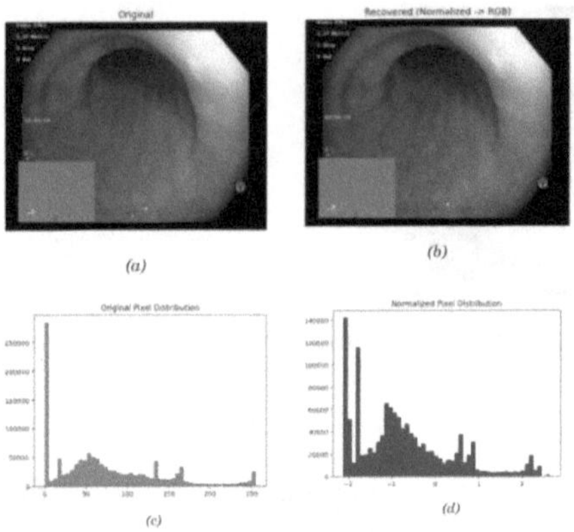

Fig. 2. (a) Original Image with Green Box Present, (b) Image After Standardization, (c) Pixel Distribution of Original Image with a peak present near the middle of the distribution, (d) Pixel Distribution of Normalized Image with a prominent and exaggerated peak (Color figure online)

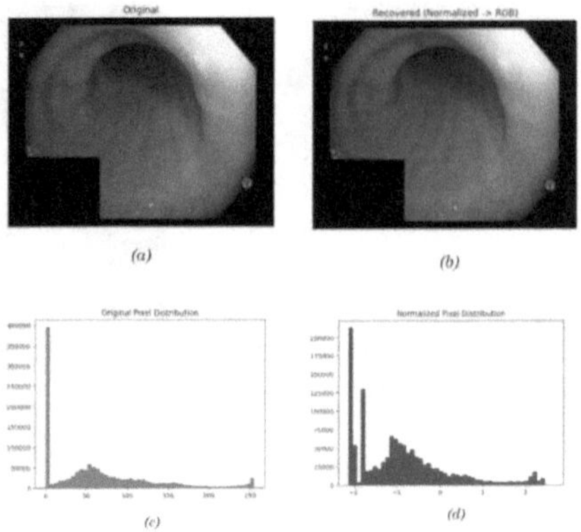

Fig. 3. (a) Green Box-Free and Text-Free Image, (b) Image After Standardization, (c) Pixel Distribution of Green Box-Free and Text-Free Image without Presence of Second Peak Like Fig. 2c, (d) Pixel Distribution of Normalized Image without Presence of Second Peak like Fig. 2d (Color figure online)

3.2 Baseline Model Development and Performance

The baseline model used transfer learning by loading the EB2 architecture pretrained on ImageNet with an input shape of (224, 224, 3). The original classification head was removed to accommodate the four-class CE task. The EB2 backbone layers were frozen to preserve pretrained weights and avoid overfitting on the relatively small dataset. A custom classification head was added, consisting of Gaussian noise and dropout layers for regularization, global average pooling for dimensionality reduction, dense layers for feature learning, batch normalization for stability and a softmax output layer for multi-class classification. The model was compiled with the Adam optimizer and categorical cross-entropy loss and trained for up to 30 epochs with validation monitoring. Three callbacks optimized performance: ReduceLROnPlateau (adjusted learning rate on loss plateaus), EarlyStopping (patience of 6 epochs) and ModelCheckpoint (retained the best weights).

The accuracy achieved on the training, validation and test sets was 96.14%, 94.77% and 96.42% respectively, indicating good generalization with only mild overfitting. Learning curves showed stable accuracy but occasional validation loss fluctuations, suggesting limited variability in the validation set. Training and validation recall and AUC quickly reached high, stable values with minimal divergence, confirming effective learning.

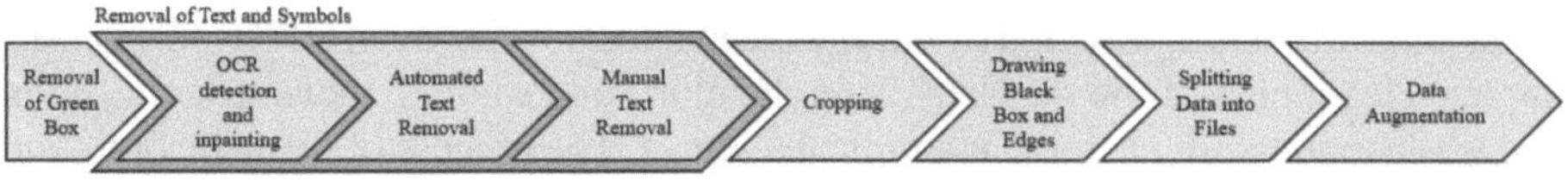

Fig. 4. Data Preprocessing Pipeline used to produce the publicly available dataset [20].

4 Final Model Development

4.1 Data Preprocessing

The data preprocessing pipeline is arguably the most critical component of this study and is shown in Fig. 4. AI models can nowadays be constructed relatively easily using prebuilt libraries and widely available online resources [19]. However, it is the initial phase, that is ensuring the data fed into the model is free from bias in any form, that truly strengthens the model and brings the deployment of such systems in real-world clinical settings one step closer. The final dataset used has been made publicly available [20] and consists of 700 Normal, 693 Esophagitis, 660 Polyps and 667 UC pre-processed images.

To neutralize the biasing occurrence of the green box, a custom OpenCV-based function, mask_green_box_rect, was used to detect and remove it via HSV color-space conversion, through employing green-range masking ([35, 50, 50]–[90, 255, 255]) and morphological filtering with a 7×7 kernel to refine the mask. The largest detected contour identifies the box and was overlaid with a solid black rectangle, ensuring its visual absence while preserving surrounding anatomical detail, see Fig. 5.

Textual overlays such as timestamps and symbols, predominantly located in the leftmost 40% of images, posed a risk of bias by introducing non-anatomical cues. To remove these artifacts, a custom remove_text_left() function was developed using Easy-OCR with a restricted search region, reducing false positives and computational cost. Multiple OCR confidence thresholds (50%-0.01%) were tested on 800 images, representing roughly one-quarter of the dataset, and the lowest threshold with minimal false positives was selected to maximize coverage. Detected text regions were then removed using Navier–Stokes inpainting, which smoothly propagated surrounding textures into masked areas, preserving anatomical detail and avoiding artificial boundaries.

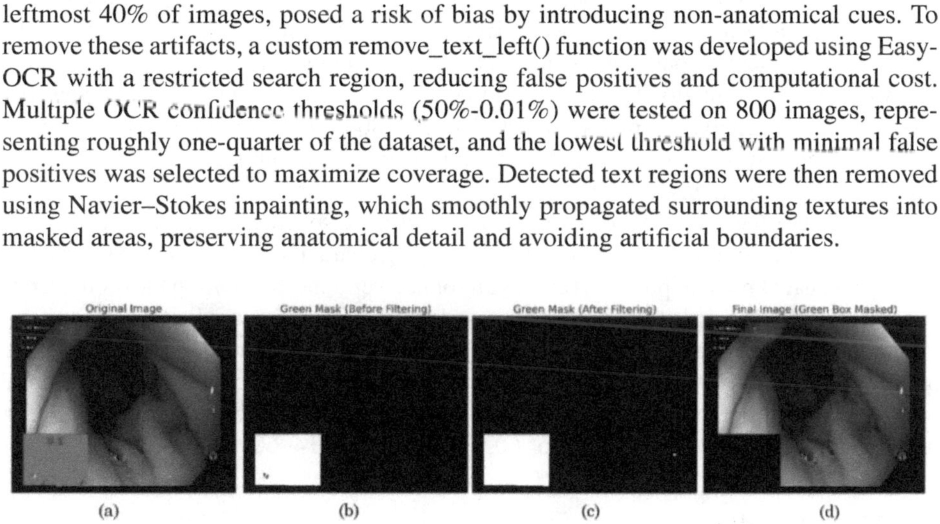

Fig. 5. Process For Removal of Green Box: (a) Original Image, (b) Green Range Masking, (c) Morphological Filtering, (d) Green Box Elimination. (Color figure online)

Residual artifacts undetected by OCR were manually removed in GIMP 3.0 to ensure complete cleanliness of the dataset. Using the clone and smudge tools, text was eliminated while maintaining natural anatomical textures. The process of manual corrections was distributed randomly across classes to avoid introducing class-specific biases due to human correction patterns. This combination of automated OCR-based removal and targeted manual refinement ensured that all images were free from text and symbols without compromising the integrity of clinical features.

Final image standardization comprised of three steps, to ensure consistent visual structure and eliminate residual non-anatomical cues. First, images were tightly cropped to anatomical content using a customized crop_tightly_around_red() function, which detects red-toned tissue in HSV space and excludes low-saturation or dark regions. External contours were then used to define a slightly shrunken bounding box, standardizing spatial alignment and avoiding border artifacts. Next, all images were resized to 224×224 pixels via OpenCV's INTER_AREA interpolation, a method that preserves fine details when downsampling while introducing minimal aliasing; although aspect ratio distortion can occur, preliminary testing showed no detrimental effect on feature extraction. Finally, to neutralize the class-specific bias introduced by the recurring green box, an 80×80 pixels black square was overlaid in the bottom-left corner of all images, with manual review ensuring that no medical regions were obscured and selective rotation applied when necessary. Triangular corner overlays were also added to replicate original capsule endoscopy borders and conceal icons such as magnets, with dimensions refined through trial-and-error on 800 samples.

Data augmentation, used on this dataset, is a technique used to artificially expand a training set, increasing DL models' generalizability, without the need for data collection [21]. Transformations such as zooming or flipping were excluded, as they would produce anatomically implausible images given the fixed position of the light source and characteristic reflection patterns found in CE. Instead, augmentation leveraged CE's natural variation in orientation by applying a systematic rotation strategy. Each image was rotated around each axis by major angles (0°, 90°, 180°, 270°) combined with $\pm 7°$ perturbations, generating fine-grained orientation diversity without significant anatomical distortion or cropping loss.

To preserve the integrity of preprocessing artifacts, specifically the black box and triangular corners introduced to replicate CE imaging characteristics, these overlays were kept in consistent relative positions across all rotated versions, preventing the model from exploiting their placement for classification. Augmentation was applied exclusively to the training set, following best practice to avoid biasing evaluation [22]. Processing was parallelized using Joblib library's Parallel and delayed functions enabling simultaneous rotation of images across all available CPU cores. The total number of images used for training hence increased by a factor of 60, from 1901 to 114060.

4.2 Model Selection

Architecture Selection Strategy. Five established CNN architectures (EfficientNetB0, ResNet50, DenseNet121, InceptionV3, VGG16) were evaluated to identify the best performer for CE classification. All models employed transfer learning with ImageNet pre-trained weights and were trained under identical conditions. A standardised classification head, comprising a Global Average Pooling layer, Dropout (rate 0.3) and softmax output, was appended. Models were optimised using the Adam optimizer (learning rate 0.0001) with categorical cross-entropy loss. Images were normalized to [0,1] and loaded dynamically in batches to optimize memory use.

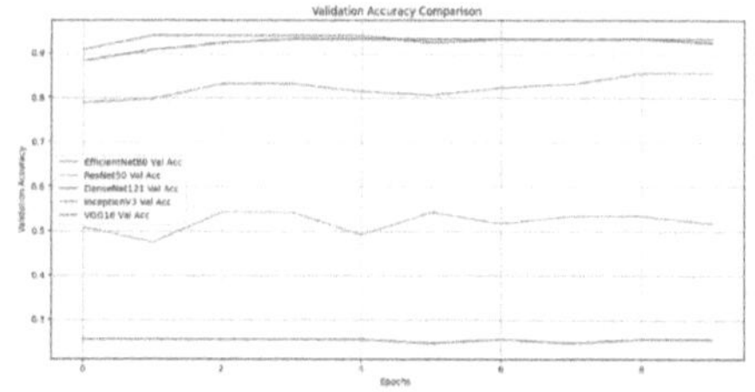

Fig. 6. Validation Accuracy Comparison Across Architectures.

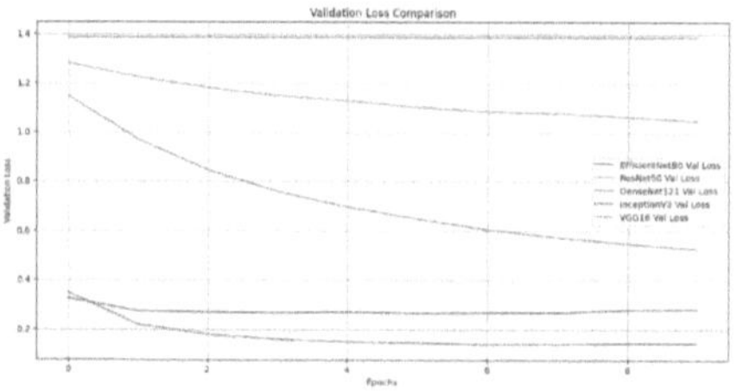

Fig. 7. Validation Loss Comparison Across Architectures

Comparative Evaluation of Architectures. Each model was trained for 10 epochs, with validation accuracy and loss monitored per epoch. Validation data remained unseen during training to provide unbiased generalization estimates. DenseNet121 achieved the highest validation accuracy (~94%) and the lowest validation loss (~0.1), reflecting confident and accurate classification. Its stable learning curve indicated effective transfer learning and efficient feature reuse, see Figs. 6 and 7. InceptionV3 had comparable accuracy but higher loss (~0.3), indicating less reliable predictions. VGG16 showed moderate performance, consistent with its shallower design, while ResNet50 and EfficientNetB0 underperformed, exhibiting unstable learning and poor accuracy. These results underscored the importance of empirical evaluation across architectures. DenseNet121 was selected for further intra-architecture exploration.

Intra-Architecture Version Comparison. Three DenseNet variants (DenseNet121, DenseNet169, DenseNet201) were compared to identify the optimal configuration. All exceeded 92% validation accuracy, with DenseNet169 outperforming the others by achieving ~96.7% accuracy and the smoothest, lowest validation loss (~0.10), see Figs. 8 and 9. DenseNet201 closely followed, with slightly lower accuracy and marginally higher loss. DenseNet121, while still strong (~95.8% accuracy), showed a greater loss variability, suggesting slower convergence. Additional metrics (precision, recall, F1-score, AUC) all exceeded 0.975 for all variants. Although DenseNet201 marginally outperformed DenseNet169 across metrics (<0.9% difference), DenseNet169's superior training stability and convergence led to its selection as the final architecture, balancing accuracy, robustness and computational efficiency.

The final model was built using transfer learning, leveraging DenseNet169 pretrained on the ImageNet dataset. The use of pretrained weights allows the model to retain general low-level features (such as edge detectors and shape recognizers) learned from natural images and adapt them to the specialized task of medical image classification. This method is particularly effective when training data is limited, as is often the case in healthcare applications [23]. The DenseNet169 base was loaded without its fully connected classification layers (include_top = False) and a new head was added consisting of a Global Average Pooling layer to reduce spatial dimensions by averaging across feature maps, a Dropout layer (rate = 0.3), used to combat overfitting by randomly deactivating 30% of the neurons during each training iteration and a final Dense layer with softmax activation, outputting probabilities across the four classes.

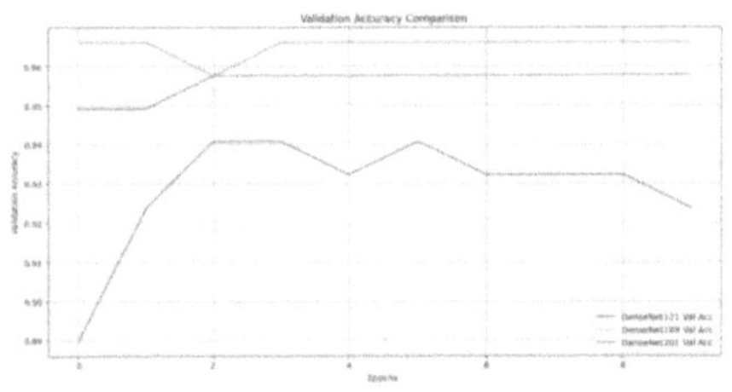

Fig. 8. Validation Accuracy for DenseNet121, DenseNet210 and DenseNet169.

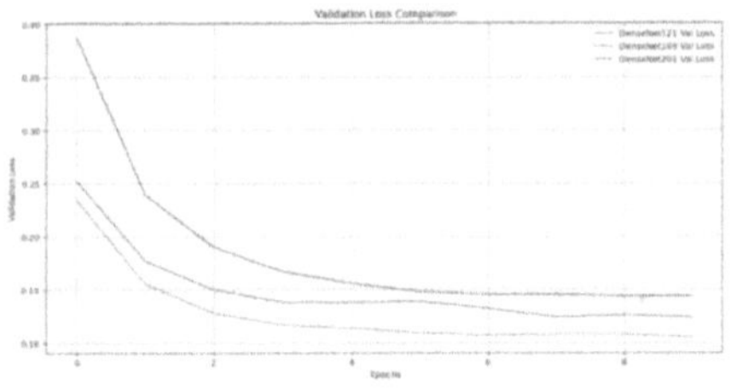

Fig. 9. Validation Loss for DenseNet121, DenseNet210 and DenseNet169.

The model was compiled using the Adam optimizer with a learning rate of 0.0001 and categorical cross-entropy loss, appropriate for multi-class tasks. Training ran for up to 70 epochs with early stopping (patience of 4 epochs) to prevent overfitting, halting when validation loss plateaued. Additionally, model checkpointing preserved weights from the best validation loss epoch. Input images were pre-processed using DenseNet-specific normalization (preprocess_input), standardizing pixel values to align with ImageNet training distributions.

5 Classification Results and Discussion

5.1 Final Model Performance and Optimization

The DenseNet169 model (Sect. 4.2.2) was trained on the augmented dataset for 70 epochs, with early stopping restoring the best weights at epoch 7. Transfer learning from ImageNet enabled rapid convergence, with accuracy rising from 25.0% to 76.7% in the first epoch. Final training achieved 97.47% accuracy and 98.53% validation accuracy (loss = 0.04489). On the test set (n = 412), the model reached 98.54% accuracy, precision, recall and F1-score (all equal to 0.9854) and an AUC of 0.9995. Class-level performance was near-perfect for Normal and Esophagitis (100% correct), with minimal errors in Polyps (2 False Positives) and Ulcerative Colitis (1 False Positive and 2 False Negatives), see Fig. 10.

The confidence histogram provides an additional layer of insight into the model's decision-making process, see Fig. 11. The distribution of confidence scores shows a clear separation between correct and incorrect predictions, with the majority of correct predictions occurring at very high confidence levels (close to 1.0). Of the six misclassified test images, four had low confidence scores, while two were predicted with high confidence. These high-confidence misclassifications, both belonging to the UC class, were manually reviewed to assess their cause and clinical plausibility. Input was sought from a medical intern with clinical experience, who provided contextual insights into the challenges of interpreting such cases. The first misclassification (85.15% confidence), see Fig. 12, was inspected by the medical intern who noted: "This could pass as normal to the untrained eye; the ulcers are quite mild and peripheral. Without clinical context or high-resolution focus, they can be missed." The second (91.23% confidence), misclassified as Esophagitis, see Fig. 13, displayed mucosal erythema common to both conditions. The intern noted, "If you don't know the anatomical location… it's almost impossible to be sure… both diagnoses are visually justifiable here", reflecting the challenge of distinguishing visually similar inflammatory features and the importance of location and contextual awareness.

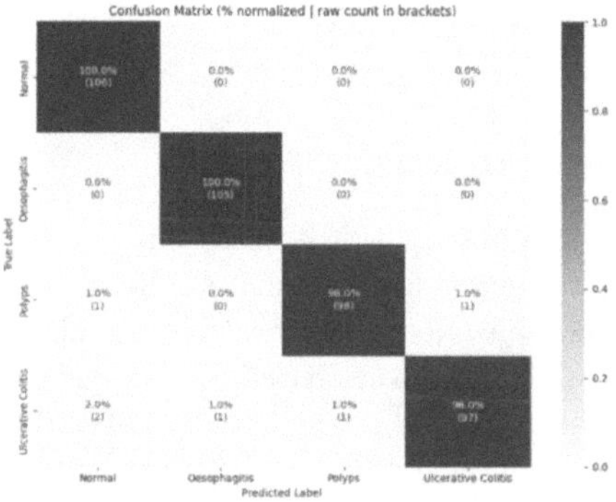

Fig. 10. Confusion matrix showing classification results of the final DenseNet169 model across four classes. Values indicate both the percentage of correct or incorrect predictions and the raw number of test images per class ($n = 412$). The model achieved perfect classification for Normal and Esophagitis, with minor misclassifications in Polyps and Ulcerative Colitis.

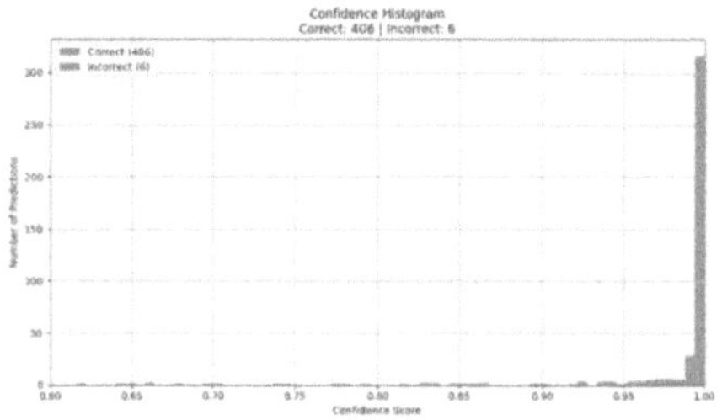

Fig. 11. Confidence Score Histogram for Final Model Predictions. This histogram illustrates the distribution of confidence scores for all predictions made by the final model on the test set ($n = 412$), distinguishing between correct (green) and incorrect (red) outcomes. (Color figure online)

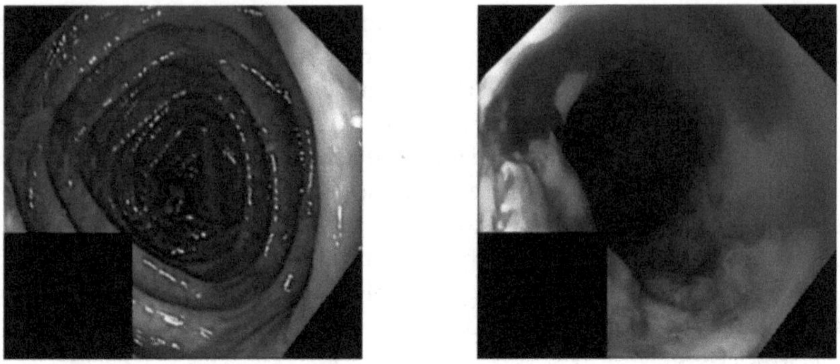

(a) (b)

Fig. 12. Images incorrectly classified: (a) is UC classified as Normal and (b) is UC classified as Esophagitis.

5.2 Visual Interpretability and Clinical Trustworthiness

In medical imaging applications, especially in diagnostic tasks such as this one, high numerical accuracy alone is insufficient. For DL models to be trusted in clinical environments, their decision-making process must be transparent, interpretable and aligned with anatomical reasoning. To assess whether predictions were based on clinically relevant features, Gradient-weighted Class Activation Mapping++ (Grad-CAM++) was employed. Grad-CAM++ improves on Grad-CAM by using higher-order gradients for more precise, class-specific localization, even when multiple features contribute to a decision [24–26]. As illustrated in Figs. 13, 14 and 15, these representative examples show how good preprocessing enhances decision transparency and reduces reliance on irrelevant artifacts. For each test image, the predicted class was used as the target. Heatmaps were upsampled and overlaid with a fixed color scale. This sampling reveals the hierarchical progression of evidence from primitive cues to lesion-centric features. The class-wise comparative analysis focuses on the heatmaps produced at the final convolutional layer.

Class-wise Comparative Analysis. *Polyps* (Fig. 13) In the baseline model, Grad-CAM++ heatmaps often highlight artifacts such as text overlays, timestamps and light reflections, rather than the polyp itself. The final model shows consistent attention around the polyps' anatomical region, particularly at its edges and surrounding mucosa, likely using contrast and shadow patterns caused by elevation. While some risk remains from shadow-based cues, this represents a clear improvement in localization and interpretability.

Ulcerative Colitis (Fig. 14) UC images in the final model display activation around ulcerated or discolored mucosa, including yellowish areas indicative of exudate. The baseline model frequently focuses on reflections or user interface elements, occasionally missing ulcerations entirely. The final model distributes attention more contextually across affected mucosa, suggesting reduced artifact reliance and stronger clinical relevance.

Esophagitis This class is visually subtle, with diffuse mucosal erythema and textural changes. The baseline model sometimes attends to masked image corners or timestamps; biases linked to acquisition format. The final model consistently centers on inflamed esophageal mucosa, largely ignoring non-informative overlays (Fig. 15). This shift indicates improved robustness and generalization, despite Grad-CAM++'s limited capture of deeper-layer color cues.

Normal Unlike humans, who identify normality by the absence of visible abnormalities, DL models distinguish the Normal class by learning its unique visual patterns from labelled examples across all categories [27]. The final model correctly identifies smooth, uniformly colored mucosa, ignoring artifacts that often mislead the baseline model (Fig. 15). While easier to classify, the risk remains if artifacts or lighting conditions mimic abnormal cues; underscoring the value of clean training data.

Across all classes, the final model consistently focuses on anatomically relevant regions rather than artifacts, reflecting a clear improvement in interpretability and trustworthiness after training data cleaning. Heatmaps are smoother and more distributed due to DenseNet169's deeper convolutional layer compared to EfficientNet-

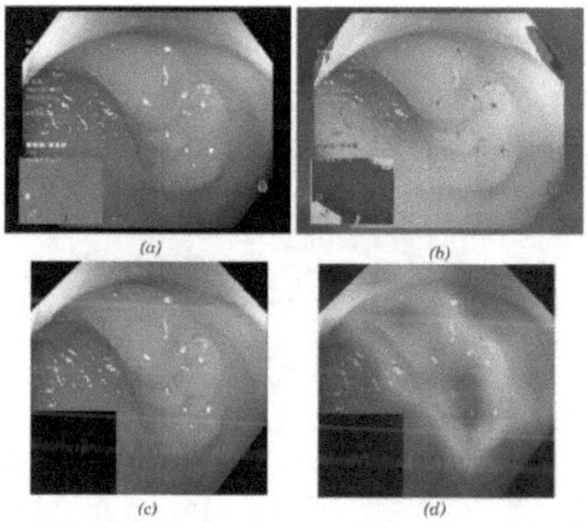

Fig. 13. (a) Polyps Example Original Image, (b) Grad-CAM++ Heatmap of Baseline Model, (c) Pre-processed Image, (d) GradCAM++ Heatmap of Final Model.

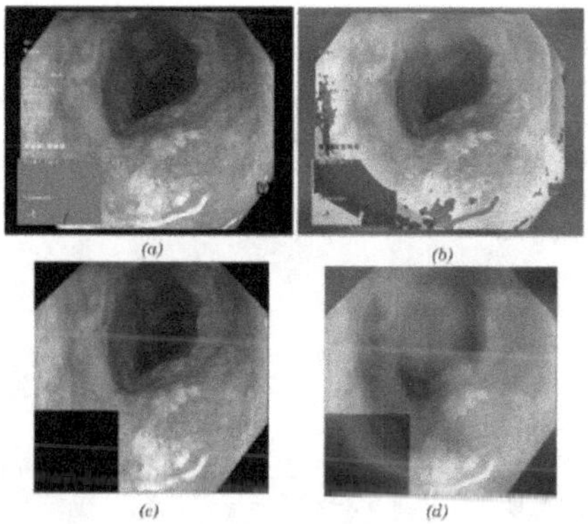

Fig. 14. (a) UC Example Original Image, (b) Grad-CAM++ Heatmap of Baseline Model, (c) Pre-processed Image, (d) GradCAM++ Heatmap of Final Model.

B2 (e.g., Figs. 13 and 14), with high-confidence predictions producing milder activation intensities. Minor attention along padded borders is attributable to Grad-CAM++ interpolation rather than model bias. These results confirm that preprocessing effectively shifted model attention toward genuine clinical features, reducing reliance on dataset-specific noise.

Layer-wise Comparison. Layer-wise Grad-CAM++ analysis of DenseNet-169 (conv2/conv3/conv4/conv5 endpoints) was further conducted to show how the model gradually shifts from simple visual cues to clinically meaningful features when classifying capsule endoscopy images. At the earliest level, conv2, the maps mainly including the frame corners as DenseNet169's initial layers focus on high-contrast, high-frequency signals. These are low-level image features generally common across all classes. By conv3, the focus becomes broader: in Normal cases, the fold structure and lumen boundary start to appear; in Polyps, the curved shape of the protrusion is outlined; in Esophagitis, irregular surface streaks and disrupted texture are visible; while in UC the model begins to register more general surface irregularity, though not yet specific signs, consistently highlighting bright reflections, bubble edges and small texture details.

At conv4 (conv4_block32_concat), the maps become broadly saturated, with diffuse red activation across most of the mucosal surface and decreased focus on the non-anatomical corners. This suggests that the model is aggregating mid-level features into a more global representation, without strongly discriminating between anatomical structures. The apparent uniformity reflects a stage where many feature maps are active but not yet distilled into the class-specific attention seen at the final conv5 layer. By conv5, non-anatomical cues are suppressed and attention strictly concentrates on disease-relevant regions. This progressive reduction of high-frequency, non-anatomical artifacts reflects the dense connectivity of DenseNet-169: early layers make fine-grained features available to deeper layers, while later layers integrate broader contextual representations and selectively utilise only those features that contribute to class discrimination.

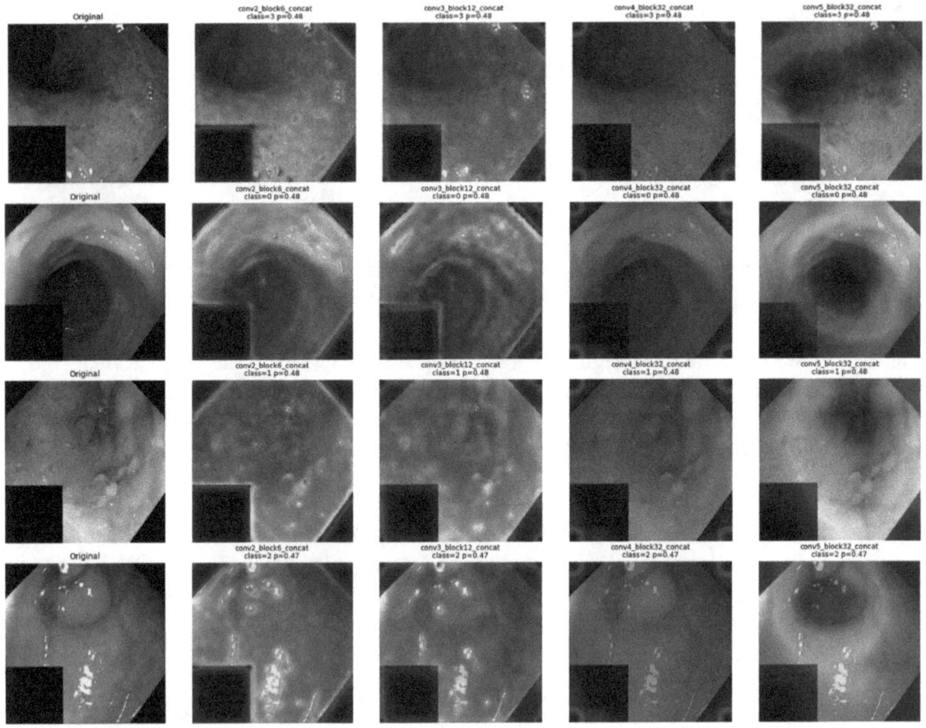

Fig. 15. Grad-CAM++ visualizations across DenseNet-169 layers. Columns show the original image (left) and activations from the earlier to final layers. Rows correspond to (top to bottom): Ulcerative Colitis, Normal, Esophagitis and Polyps.

6 Conclusions

This paper demonstrates that DL can be used not only to achieve high accuracy in capsule endoscopy diagnostics, but also to advance the field toward clinical usability and broader accessibility. By moving beyond accuracy alone, this work highlights how robust data preprocessing, artifact removal and interpretability tools like Grad-CAM++ can be used to build trustworthy AI models that align with the needs of both patients and clinicians.

A key contribution of this study is the development of an overlay cleaning pipeline using OCR and inpainting, allowing for the use of previously unusable data. This not only improved classification performance but also reduced reliance on non-clinical features, addressing a major barrier to real-world deployment. The manually cleaned dataset has been made publicly available for use in further research [20]. The resulting model, based on DenseNet169, achieved an F1- score of 0.9854 and demonstrated strong alignment between model attention and clinically relevant anatomy across all classes. Taken together, these results show that careful control of input data, combined with explicit interpretability analysis by healthcare professionals, can close part of the gap between research-grade performance and clinically acceptable behavior. Ultimately, this work positions AI-enhanced capsule endoscopy as both a practical and forward-looking solution for scalable gastrointestinal care, reducing the need for invasive and uncomfortable diagnostic procedures.

7 Future Work

The findings of this work suggest several potential avenues for advancing CE analysis and deployment. Model optimization could benefit from advanced strategies such as hyperparameter tuning, architecture ensembling, multi-scale learning and inclusion of additional data from diverse capsule devices. While DenseNet169 offers strong performance and efficiency, its deployment on resource-constrained medical systems may necessitate further optimisation or the exploration of lighter model architectures to ensure real-time applicability and reduced computational overhead.

One promising direction is the extension from 2D frame-level classification to immersive 3D gastrointestinal reconstructions using techniques such as 3D Gaussian Splatting [28, 29]. This may improve lesion localization, anatomical orientation and review efficiency by allowing physicians to navigate reconstructed GI environments similar to the more widely adopted traditional endoscopy which clinicians are better trained on reviewing. The overlay-removal approach demonstrated could also enable the reuse of archived capsule datasets for predictive diagnostics [30]. Such systems could support dynamic dataset updating and risk-based patient alerts without the need to acquire entirely new datasets.

Finally, employing such models also aligns with emerging full-length Mouth-to-Anus capsule workflows, where AI may act as a supplementary observer, flagging overlooked abnormalities across the GI tract and improving diagnostic coverage without extending review time [31, 32].

Acknowledgements. The authors extends sincere gratitude to Dr. Jonathan Graham-Harper-Cater for his invaluable guidance throughout this project. Special thanks are due to Dr. Zaid Nsour for his clinical insights and contributions, which greatly strengthened the diagnostic relevance of this work. Appreciation is also expressed to future medical professionals, Layan Jnaidy and Ghaid Abu Sharar, for their support and commitment.

References

1. Romero-Vázquez, J.: Capsule endoscopy in patients refusing conventional endoscopy. World J. Gastroenterol. **20**(23), 7424–7430 (2014)
2. University Hospital Southampton: Capsule endoscopy. UHS Patient Information Leaflet (2023). https://www.uhs.nhs.uk/Media/UHS-website-2019/Patientinformation/Endoscopy/Capsule-endoscopy-1311-PIL.pdf
3. Cleveland Clinic: Capsule endoscopy: Ulcerative colitis symptoms and Crohn's disease test. https://my.clevelandclinic.org/health/diagnostics/16829-capsule-endoscopy
4. Mayo Clinic: Capsule endoscopy. https://www.mayoclinic.org/tests-procedures/capsule-endoscopy/about/pac-20393366
5. Hale, M., McAlindon, M.E.: Capsule endoscopy as a panenteric diagnostic tool. Br. J. Surg. **101**(3), 148–149 (2014)
6. Rasmussen, M.S., Hansen, L.Ø., Deding, U., Ellebæk, M.B., Kjeldsen, J., Bjørsum-Meyer, T.: Applicability of colon capsule endoscopy for monitoring ulcerative colitis: a systematic review. Scand. J. Gastroenterol. **60**, 1–7 (2025)
7. GOV.UK: Bowel cancer screening: having a colonoscopy. https://www.gov.uk/government/publications/bowel-cancer-screening-colonoscopy/bowel-cancer-screening-having-a-colonoscopy-fit. Accessed 10 Jan 2025
8. Rondonotti, E., et al.: Cost estimation of small bowel capsule endoscopy based on real-world data: inpatient or outpatient procedure? Dig. Liver Dis. **42**(11), 798–802 (2010)
9. Messmann, H., et al.: Expected value of artificial intelligence in gastrointestinal endoscopy: ESGE position statement. Endoscopy **54**(12), 1211–1231 (2022)
10. Sun, J.: Introduction to Deep Learning for Healthcare, 1st edn. Springer, Cham (2022)
11. Esteva, A., et al.: Dermatologist-level classification of skin cancer with deep neural networks. Nature **542**(7639), 115–118 (2017)
12. Zachariah, R., Rombaoa, C., Samarasena, J., Suraweera, D., Wong, K., Karnes, W.: The potential of deep learning for gastrointestinal endoscopy: a disruptive new technology. Artific. Intell. Med. 223–245 (2021)
13. Soffer, S., et al.: Deep learning for wireless capsule endoscopy: a systematic review and meta-analysis. Gastrointest. Endosc. **92**, 831–839 (2020)
14. Park, S.J., et al.: Artificial intelligence–assisted reading of small bowel capsule endoscopy in a prospective study. Lancet Digital Health **6**(6), e388–e397 (2024)
15. Preim, B.: Visual Computing for Medicine. Morgan Kaufmann, Burlington (2014)
16. Montalbo, F.J.: WCE curated colon disease dataset for deep learning. https://www.kaggle.com/datasets/francismon/curated-colon-dataset-for-deep-learning. Accessed 30 May 2024
17. Zulfiqar, F., Bajwa, U.I., Mehmood, Y.: Multi-class classification of brain tumor types from MR images using EfficientNets. Biomed. Signal Process. Control **82**, 104529 (2023)
18. Khattab, M.: EfficientNetB2 capsule endoscopy classification (95% test accuracy). https://www.kaggle.com/code/khattab1412/95-test-acc-efficientnetb2-capsule-endoscopy. Accessed 10 Jan 2025
19. Erickson, B.J., Korfiatis, P., Akkus, Z., Kline, T., Philbrick, K.: Toolkits and libraries for deep learning. J. Digit. Imaging **30**(4), 400–405 (2017)
20. Kurdi, A.: WCE curated colon disease dataset (text removed). https://www.kaggle.com/datasets/aramkkk/manually-text-removed-test-validation-wcecurated

21. Islam, T., Hafiz, M.S., Jim, J.R., Kabir, M.M., Mridha, M.F.: A systematic review of deep learning data augmentation in medical imaging. Healthcare Anal. **5**, 100340 (2024)
22. Shorten, C., Khoshgoftaar, T.M.: A survey on image data augmentation for deep learning. J. Big Data **6**(1), 60 (2019)
23. Kim, H.E., Cosa-Linan, A., Santhanam, N., Jannesari, M., Maros, M.E., Ganslandt, T.: Transfer learning for medical image classification: a literature review. BMC Med. Imaging **22**, 69 (2022)
24. Selvaraju, R.R., Cogswell, M., Das, A., Vedantam, R., Parikh, D., Batra, D.: Grad-CAM: visual explanations from deep networks via gradient-based localization. Int. J. Comput. Vision **128**(2), 336–359 (2020)
25. Chattopadhyay, A., Sarkar, A., Howlader, P., Balasubramanian, V.: Grad-CAM++: generalized gradient-based visual explanations for deep convolutional networks. In: Proceedings of the IEEE Winter Conference on Applications of Computer Vision (WACV), pp. 839–847 (2018)
26. Lerma, M.A., Lucas, M.: Grad-CAM++ is equivalent to Grad-CAM with positive gradients. arXiv preprint arXiv:2103.14591 (2021)
27. Tang, Y., Tang, Y., Han, M., Xiao, J., Summers, R.M.: Deep adversarial one-class learning for normal and abnormal chest radiograph classification. In: Medical Imaging 2019: Computer-Aided Diagnosis, pp. 1–8. SPIE (2019)
28. Wu, T. et al.: Recent advances in 3D Gaussian splatting. arXiv preprint arXiv:2403.11134 (2024)
29. Li, T., Dai, Y., Zhang, J., Zhang, W.: Design of 3D reconstruction system for monocular endoscopy based on real-time radiance fields. In: Proceedings of the Chinese Control Conference, pp. 7528–7533 (2024)
30. Vannier, M., Staab, E., Clarke, L.: Medical image archives—present and future. In: Medical Imaging: Image Processing, pp. 565–570. Springer, Berlin (2002)
31. Phillips, F., Beg, S.: Video capsule endoscopy: pushing the boundaries with software technology. Transl. Gastroenterol. Hepatol. **6**, 17 (2021)
32. George, A.A., et al.: Artificial intelligence in capsule endoscopy: development status and future expectations. Mini-invasive Surg. **8**, 12 (2024)

Cross-Modal Knowledge Distillation from Wearable Physiology to Wi-Fi CSI for Sensor-Less Agitation Detection

Awan-Ur Rahman[1(✉)], Tanbin Islam Rohan[1], Sumit Chakravarty[1], Mary Dioise Ramos[2], and Nazmus Sakib[1]

[1] Kennesaw State University, Georgia, USA
awanrahman55@gmail.com, {schakra2,nsakib1}@kennesaw.edu
[2] School of Nursing, LSU Health New Orleans, New Orleans, USA
mramo1@lsuhsc.edu

Abstract. The rising prevalence of Alzheimer's disease and related dementias (AD/ADRD) is overwhelming caregivers and straining existing care systems, making advanced remote monitoring essential for patient safety. As the elderly population grows rapidly, the need for innovative and scalable solutions becomes urgent. Agitation stands out as one of the most urgent and complex symptoms to manage in patients with AD/ADRD, requiring constant oversight and timely intervention. Traditional monitoring methods can be intrusive or impractical at scale. To address this gap, we introduce a privacy-preserving and nonintrusive monitoring framework that relies solely on ambient Wi-Fi signals. By transferring key physiological markers from wearables to the Wi-Fi CSI space, our framework enables sensitive and early detection of agitation and related behaviors without compromising privacy. Experimental results demonstrate that our framework can differentiate between eight different types of activities, along with the agitation state of the patients, with robust and meaningful accuracy. With scopes for further advancements, this framework has the potential to transform AD/ADRD care and monitoring, and improve outcomes for people living with AD/ADRD.

Keywords: Agitation monitoring · Alzheimer's disease · Knowledge distillation · Deep learning

1 Introduction and Motivation

Alzheimer's disease and related dementias (AD/ADRD), a neurological disorder characterized by progressive cognitive decline, often results in a progressive deterioration of everyday functioning and memory impairment [13,24]. While early-onset Alzheimer's can strike a person as young as 40, Alzheimer's disease mainly affects those over 65. The elderly population in the US is increasing gradually and is expected to sum up to 20.2% of the total population by 2050,

H. R. Arabnia et al. (Eds.): CSCI 2025, CCIS 2971, pp. 48–63, 2027.
https://doi.org/10.1007/978-3-032-28463-1_4

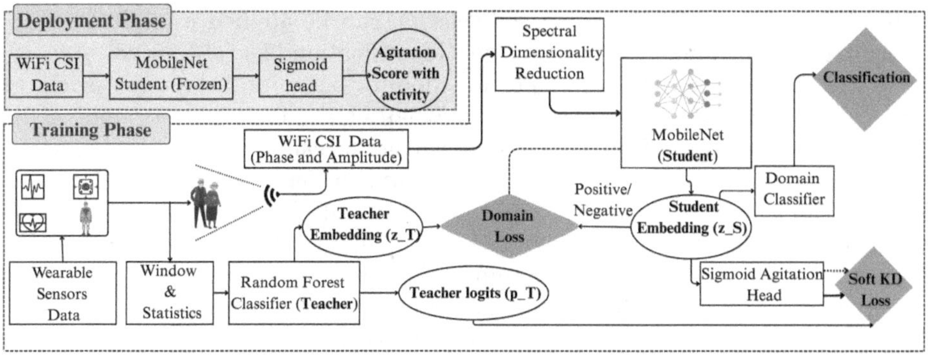

Fig. 1. Architecture of the proposed knowledge-distillation framework.

while in 2013, it was around 13% of the total population [10]. In 2024, it was projected that 6.93 million of older Americans (over 65 years old) had lived with AD/ADRD, and the figure is predicted to go up to 13.8 million by 2060 [20]. By 2060, an estimated 30% of the world's population is expected to be elderly [2]. As the disease progresses, people with AD/ADRD experience physical and psychological impairments that make it harder for them to carry out daily tasks. Consequently, cognitive decline causes people with AD/ADRD to experience erratic emotions, trepidation, or depression, while the loss of personal autonomy can cause agitation and a deep sense of helplessness. About 40–60% of AD/ADRD patients have agitation, with the percentage rising to approximately 75% in moderate-to-severe stages. [3,25]. Agitated individuals may self-harm, scream, pound walls, or even attempt suicide, which significantly increases the strain on caregivers. These traits demonstrate that agitation is a serious clinical problem, and consistent agitation episode monitoring is essential for both patient safety and caregiver health. Existing remote monitoring solutions are often erroneous, too pricey for widespread usage, uncomfortable for users, or a breach of privacy. To address this pressing issue, we have developed a novel cross-modal knowledge distillation technique that maps wearable physiological cues onto contact-free Wireless Fidelity (Wi-Fi) signals. Figure 1 shows the framework, which will be explained in more detail in the methodology section. To develop such a framework, we have investigated the following research questions:

RQ1: How can we embed the physiological patterns captured by wearables into the Wi-Fi Channel State Information (Wi-Fi CSI) signals, so the algorithm can monitor agitation without requiring a wearable device?

RQ2: How can we reduce subcarriers and still retain important activity details to optimize computation cost?

Before investigating these questions, analyzing existing technologies and their application in agitation recognition is required to identify research gaps and establish grounds for future innovation. Researchers have demonstrated in recent years that deep learning and machine learning algorithms, in conjunction with wearable technology and vision-based sensors, can be applied in digital health, particularly in the area of human activity identification [2,11,14,15,23]. According to previous studies, these gadgets can monitor behavioral changes associated with agitation in dementia in real time. Khan et al., for example, built a full IoT pipeline around the Empatica E4 wristband, streaming electrodermal activity, three-axis motion, blood-volume pulse, and skin temperature at 64 Hz to an Android gateway to agitation detection [9]. They collected more than 600 days of data from 20 residents, split the signals into one-minute windows, and found that a small two-block Transformer (32-unit embeddings) performed best, matching earlier random-forest baselines while needing no handcrafted features [6]. However, any discomfort that drives patients to take off their wearables significantly diminishes their effectiveness, as these devices only function while worn. Pragadeeswaran et al. further developed the concept with a two-tier mesh: a fixed sensor recorded temperature, humidity, light, and noise, while a wrist node sent motion and heart rate data; a modest two-layer neural network then turned 60-second statistical and spectral characteristics into real-time agitation scores for caregiver alerts. After cleaning each 60-second segment, a two-layer feedforward ANN (with batch normalization and dropout) ingests mean, variance, spectral entropy, jerk energy, and low-frequency (LF)/high-frequency (HF) heart rate variability (HRV) to issue real-time agitation probabilities that can trigger caregiver alerts or automated environmental soothing [17]. However, the strategy at some point compromises people's comfort and convenience because it still relies on wearables that they must continuously wear and maintain. Shehroz S. Khan et al. have developed a spatio-temporal convolution autoencoder to detect agitation episodes from a video stream [7]. Despite the effectiveness of their approach, majority of individuals are still concerned about using standard cameras because they capture their faces and regular activities. Together, these projects demonstrate that neural network pipelines driven by wearables, video cameras, or hybrid sensor meshes can detect impending agitation with a reasonable degree of accuracy. However, they also highlight the need for models that generalize across devices, environments, and participants.

Notwithstanding the usefulness of optical and wearable sensors in detecting agitation episodes, these technologies have several drawbacks. One of the biggest concerns for people who use these devices is the ongoing discomfort of wearing them all the time. Their persistent presence on the body can be uncomfortable, which may lead to anxiety or even skin irritation. Conventional continuous camera-based systems cause users to worry about their privacy. These factors could drive consumers to stop using these gadgets, which could undermine their objective. To overcome these deficiencies, this research addressed these significant gaps in the current methodologies by presenting a comprehensive data processing and AI-driven detection framework for monitoring agitation which can answer our research questions. We have developed the cross-modal knowledge distillation framework, which combines cutting-edge technologies to improve the detection of agitation. Together with eight different activity categories that agitated patients display, this framework seeks to efficiently monitor agitation episodes while lowering computational costs. The architecture by design protects user privacy and eliminates the discomfort of prolonged usage of wearable gadgets because it uses Wi-Fi routers as the primary sensor modality.

2 Methodology

2.1 Data Acquisition

This research used two datasets: Wireless Fidelity Channel State Information (Wi-Fi CSI) and physiological data. The datasets employed are Wi-Gitation and Multimodal Wearable Sensor Datasets for Agitation Detection in Dementia Care Environments [8,21]. Each dataset provides complementary information for detecting agitation: Wi-Gitation to environmental sensing through Wi-Fi signals, and the second dataset to physiological aspects and motion (captured by wearables) in real people with dementia. Table 1 explains the summary of these two datasets. The Wi-Gitation data was acquired from 23 healthy participants (both male and female) who were asked to execute seven predetermined activities of agitation (disturbed walking, flipping objects, kicking, rubbing table with hands, repetitive standing and sitting, tapping on the table and wringing hands) and two calm/normal state activities (normal walking and sitting normally) in the controlled environment of a home-like setting in two different places. The height and weight of the participants vary from 157–190 cms and 46–120 kgs, respectively. Moreover, the participants' mean Body Mass Index (BMI (kg/m^2)) is 22.25 ± 3.90, and they are all aged between 18 and 64. Each subject was captured under different distances to several Wi-Fi receivers (from 0.5 m to 8 m), which were expected to generate different signal strengths, reflections, and distortions. In contrast, the Multimodal Wearable Sensor Dataset was collected over 24 months (2017–2019) from a cohort of 20 individuals with dementia residing in a dedicated dementia care facility. Participants used the Empatica

Table 1. Agitation-monitoring datasets

Dataset	Purpose/Setting	Participants	Behaviour Labels & Modalities
WiGi-tation	Contact-free agitation recognition with WiFi-CSI using physical activity	18 older adults simulate the activities	*Labels:* 9 classes—normal walking, sitting normally, sit-to-stand, tapping, rubbing the table, wringing hands, disturbed walk, flipping objects, and kicking. *Modalities:* CSI (Intel 5300)
Physiological Data	Agitation detection using physiological signals	600 recording days of 20 patients in a nursing home	*Labels:* Binary—*agitation* (182 episodes) vs *normal*. *Modalities:* Empatica E4 (EDA, BVP, TEMP, ACC)

Table 2. Wearable dataset features

Feature	**Purpose**
Mean	The mean records the average level of the signal across the whole window; when an individual is agitated, this average remains higher than their typical calm level.
Standard deviation (SD)	High variability are frequently associated with unpredictable movements or abrupt changes in the body. A high SD across heart-rate, EDA, or ACC signals indicates repeated activity spikes and is an early sign of agitation.
Minimum	Minimum values can indicate pauses or relaxations, such as the greatest decrease in skin conductance or acceleration. If these lowest points remain higher than usual, then it is an indication of agitation.
Maximum	Records peaks, such as an abrupt kick or spike in EDA, which is crucial for identifying agitation bursts.
Skewness	Explains whether the signal has more sporadic extreme occurrences on one side, such as spikes, than declines. A positive skew indicates repeated sudden spikes signaling agitation.
Kurtosis	High kurtosis means the signal has many sudden spikes instead of smooth changes, which helps us spot agitation.
Energy mean	Focuses on high-magnitude activity at quieter times, such as intense motions or powerful physiological reactions.

E4 wristband, including four types of physiological signals: triaxial accelerometer (ACCX, ACCY, ACCZ), blood volume pulse (BVP), electrodermal activity (EDA), and skin temperature (TEMP). A total of 600 days were recorded, and episodes of agitation were annotated using a two-step process based on nurse observation and video confirmation. The dataset contains 231 normal days, 140 days with whole labeled agitation, and 64 days with partial labels, with over 230000 one-minute windows.

2.2 Cross-Modal Knowledge Distillation Framework

This framework introduces a cross-modal/cross-domain knowledge distillation approach to agitated state recognition based on Wi-Fi CSI by transferring semantic knowledge from wearable physiological data. Figure 1 describes the methodology of the proposed framework with multi-task learning: temperature-scaled Kullback-Leibler divergence (KL) is applied to the teacher's softened logits in place of any explicit contrastive term, and a soft knowledge distillation loss is applied to match teacher logits. This loss creates a bridge between agitation categories and provides the student with extensive, class-aware feedback. In the detailed architecture (Fig. 1), teacher embeddings z_{T} serve as *anchors*, and a temperature-scaled Kullback–Leibler divergence pulls student embeddings z_{S} toward the same anchor whenever their labels match, while allowing them to drift apart otherwise. Temperature weighting thereby preserves the semantic meaning of agitation across the highly dissimilar CSI and wearable feature spaces.

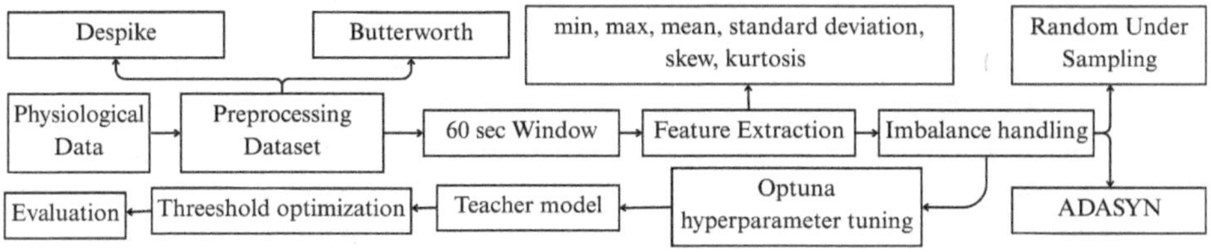

Fig. 2. RandomForest (teacher) model architecture.

For the training stage, physiological streams are filtered by applying a Butterworth filter and a spike-removal step to reduce motion artifacts and high-frequency noise. After that, the denoised signals are split into consistent 60 s frames so that every sample has the same temporal context. Basic statistics such as min, max, mean, standard deviation, skewness, and kurtosis are computed for each window to produce a compact feature vector. Table 2 explains these features and their purposes. The pipeline then performs class-imbalance correction by combining Adaptive Synthetic Sampling(ADASYN) with random under-sampling of the majority class because calm/normal segments vastly outnumber agitated ones. ADASYN adds new minority cases rather than eliminating the majority ones [4]. It creates synthetic samples along the line segments connecting these "hard" points to their neighbors after identifying problematic minority instances, those located close to the class boundary, using a k-nearest neighbor approach. By smoothing the class boundary and increasing minority density close to decision frontiers, this adaptive focus lowers the likelihood of overfitting to noise. To maximize cross-validated performance, an Optuna Bayesian search optimizes the hyper-parameters of a tree-based learner on this balanced feature set. Figure 2 depicts the methodology of the teacher model. Random Forest was chosen because it employs ensemble learning with many decision trees, which lowers variance and overfitting, making it ideal for highly imbalanced classification applications. Furthermore, it facilitates balanced sub-sampling and class weighting, which allow the model to concentrate more on the minority class during training [5,16,19].

After the teacher model is trained, its softened predictions are leveraged via knowledge distillation to supervise and train the student model. By reducing a temperature-scaled Kullback-Leibler divergence between "teacher" and a Wi-Fi CSI "student," we graft wearable physiology onto Wi-Fi CSI data using (Eq. 1).

$$\mathcal{L}_{\text{KD}} = \tau^2 \sum_{c} p_T(c \mid w) \log \frac{p_T(c \mid w)}{p_S(c \mid H)}, \qquad \tau = 2, \tag{1}$$

where $p_T(c \mid w)$ denotes the teacher's probability for class c in wearable window w, and $p_S(c \mid H)$ is the student's prediction for the aligned CSI segment H. The temperature τ "softens" both distributions to keep gradients informative even for rare classes, while the prefactor τ^2 balances the KL term against the concurrent hard cross-entropy loss.

For each 60 s window, the teacher maps EDA, BVP, skin temperature, and tri-axial accelerometry to a soft probability vector over eight classes. The teacher's two-class probabilities, along with the wearable readings, were attached to the corresponding CSI tensor with its true label. Normal walking and normal sitting are two everyday behaviors that we purposefully combined into one Calm label during development. This choice is the result of both pragmatic and ethical considerations. The project adheres to the ethical principle of privacy first: caregivers don't want a moment-by-moment breakdown of everyday motions, but rather reassurance that everything is normal once a patient is involved in any normal, non-agitated activity. By keeping the alert stream brief, we think a single calm state will help the caregivers concentrate on situations that truly need an intervention. On the other hand, each agitated behavior (kicking, flipping objects, wringing hands, tapping, repeating sitting-to-standing, rubbing the table, and disturbed walking) is distinct. Caregivers need to act according to these activities to mitigate the damage. That's why we have focused on these granular-level activities when individuals are in agitated states, so that caregivers can respond promptly. The synchronised CSI slice (80 sub-carriers at roughly 80 packets s^{-1}) is reshaped into a two-channel tensor of amplitude and phase and passed through the MobileNet and Temporal Convolutional Networks(TCN) backbone separately. Before using these models, we have split them into 70% training, 15% test and 15% validation dataset. We have selected the MobileNet and TCN for their lightweight nature, which makes them more robust in low-resource and real-time applications. The Wi-Gitation CSI data consists of the amplitude and phase of 270 subcarriers, which we reduced to 42 features to have features of the same dimensionality as wearable data and efficient processing. We have compressed the original 270 sub-carriers down to 80 to lower the computational cost by ranking them with six different criteria: (i) phase variance, (ii) amplitude variance, (iii) saliency maps, (iv) principal-component loading strengths, (v) ANOVA F-statistics, and (vi) Integrated Gradients. Saliency and IG scores were derived from a CNN–LSTM network(CSINet), the full configuration of which is detailed in Table 3. A CNN-LSTM network was used because it effectively captures temporal relationships and spatial aspects in CSI signals while maintaining vital internal representations, which is essential for reducing spectral dimensionality without losing significant information [1, 12, 22, 26].

A key real-world problem that our framework tackles is the cross-domain of the data: The data (wearable dataset and Wi-Fi dataset) come from a completely different set of participants, and it was recorded in two unique physical setups; thus, there are no overlapping subjects between them. Soft agitation probabilities produced by the physiology-based teacher serve as an objective reference that captures genuine stress responses, untethered to any specific participant or recording site. A temperature-scaled Kullback–Leibler divergence compels the Wi-Fi CSI student to mimic these scores, steering its radio-derived features toward the same biologically meaningful patterns and away from environment or device-specific quirks. By anchoring learning to continuous physiological outputs rather than heuristic thresholds, the combined teacher–student scheme removes

Table 3. CSINet layer summary

Layer	Input	Output	Params
CSINet (CSINet)	[1,2,3,270]	[1,3]	–
Sequential (conv)	[1,2,3,270]	[1,128,3,270]	–
Conv2d (0)	[1,2,3,270]	[1,64,3,270]	1,216
BatchNorm2d (1)	[1,64,3,270]	[1,64,3,270]	128
LeakyReLU (2)	[1,64,3,270]	[1,64,3,270]	–
Dropout (3)	[1,64,3,270]	[1,64,3,270]	–
Conv2d (4)	[1,64,3,270]	[1,128,3,270]	73,856
BatchNorm2d (5)	[1,128,3,270]	[1,128,3,270]	256
LeakyReLU (6)	[1,128,3,270]	[1,128,3,270]	–
Dropout (7)	[1,128,3,270]	[1,128,3,270]	–
AdaptiveAvgPool2d	[1,128,3,270]	[1,128,3,1]	–
LSTM (lstm)	[1,3,128]	[1,3,256]	659,456
Sequential (classifier)	[1,768]	[1,3]	–
Linear (0)	[1,768]	[1,128]	98,432
ReLU (1)	[1,128]	[1,128]	–
Dropout (2)	[1,128]	[1,128]	–
Linear (3)	[1,128]	[1,3]	387
Sequential (domain)	[1,768]	[1,2]	–
Linear (0)	[1,768]	[1,100]	76,900
ReLU (1)	[1,100]	[1,100]	–
Linear (2)	[1,100]	[1,2]	202
Linear (regressor)	[1,768]	[1,42]	32,298
Total params			943,131
Trainable params			943,131
Non-trainable params			0
Total mult-adds (MB)			63.00
Input size (MB)			0.01
Forward/backward pass size (MB)			2.50
Params size (MB)			3.77
Estimated total size (MB)			6.28
FLOPs per sample			63.58M

subjective bias and yields representations that remain stable across both sensor domains. The result is a scientifically grounded and domain-invariant agitation detector that transfers reliable knowledge from wearables to CSI, eliminating the need for wearables. In addition to improving generalizability in unseen surroundings, this cross-modal integration makes low-cost, privacy-preserving monitoring possible in actual contexts. The architecture ultimately opens the door to scaled agitation detection without the need for intrusive sensors. It helps to ensure that even mild or early-stage agitation events can be accurately detected with just WiFi CSI signals by maintaining the physiological fidelity of stress responses.

3 Results and Discussion

3.1 Reducing Spectral Dimensionality

The six panes in Fig. 3 summarise complementary relevance metrics computed for every one of the 270 Wi-Fi sub-carriers: (i) phase variance, (ii) amplitude variance, (iii) model saliency, (iv) PCA loading magnitudes, (v) ANOVA F-scores, and (vi) Integrated Gradients (IG). To compress the CSI tensor while preserving the most discriminative information, we adopt a multi-criterion ranking scheme that fuses these traces and selects the top 80 sub-carriers, roughly one sixth of the raw dimensionality, used by MobileNet(student model). First, each metric trace is min–max normalised to the range $[0, 1]$ so that no single scale dominates. Second, the traces are summed with equal weights to produce a composite saliency curve S_k ($k = 1 \ldots 270$); empirically, we found that equal weighting balances purely statistical cues (variance, F-score), unsupervised structure (PCA loading), and supervised importance (saliency, IG). The resulting S_k shows sharp peaks at the sub-carriers highlighted in every individual panel; most prominently around indices 5–15 and 25–35 where phase variance, amplitude variance, and saliency all spike, and a secondary plateau between 140 and 200 where PCA and IG indicate moderate but consistent relevance. Third, non-maximum suppression with a five-sub-carrier radius is applied to avoid picking clusters that

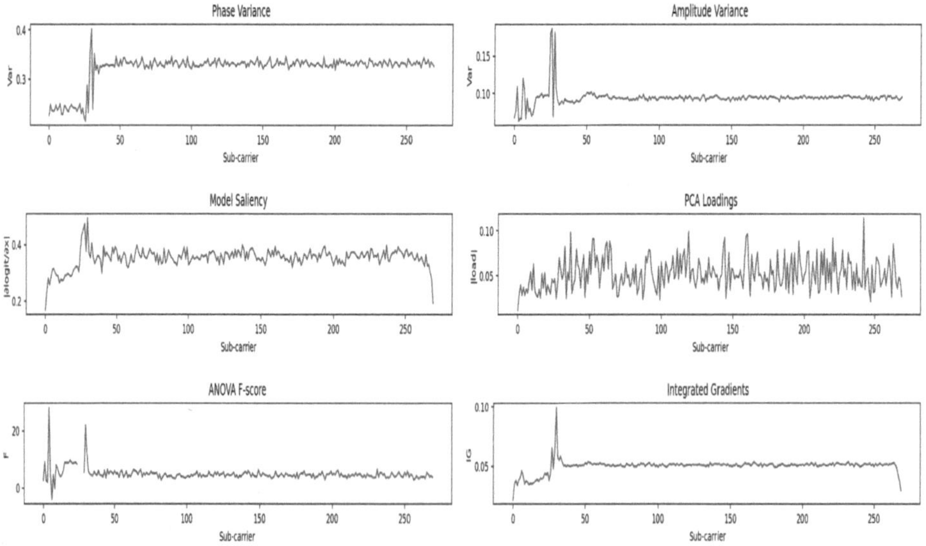

Fig. 3. Evaluation metrics for spectral dimensionality reduction.

convey redundant information; we retain only the highest-scoring carrier within each local neighbourhood. Finally, the carriers are sorted by S_k and the top 80 unique indices are retained as the reduced CSI channel list. This list preserves: (a) low-index carriers that exhibit the largest physical channel dynamics (high phase/amplitude variance), (b) carriers on which the trained model relies most heavily for agitation discrimination (high saliency and IG), and (c) carriers that capture orthogonal variance directions in the raw tensor (high PCA loadings and ANOVA scores). The result is a 74% reduction in input width and a sparsified tensor of shape $2 \times 3 \times 80$ instead of $2 \times 3 \times 270$.

3.2 Lightweight Knowledge-Distilled Teacher Network

For the teacher model, we first extracted the wearable physiological streams – electrodermal activity, skin temperature, blood-volume pulse, and tri-axial accelerometry – recorded at 64 Hz using Empatica E4 wristbands. Following signal synchronization, motion artifact removal, normalization to participant baselines, and derivation of window-level statistics (mean, variance, skew, kurtosis, spectral entropy, HRV metrics), we tested four classical machine learning models to predict soft agitation probabilities: a k-nearest neighbors (KNN) model with dynamic time-warping distance; a Gaussian-kernel support vector machine (SVM); a 300-tree Random Forest (RF); and a class-weighted Balanced Random Forest (BRF). Cross-validating in a patient stratified split, RF came ahead by a good margin, producing the largest Area Under the Receiver Operating Characteristic (AUROC) curve (0.895 ± 0.02). Consequently, we used its probabilistic predictions to guide knowledge-distillation into the WiFi-CSI student model. Tables 4 and 5 depicts the difference between different methodologies.

Table 4. Comparison of classifier performance

Model	ACC	F1	BAL	AUC
Proposed Methodology	**0.969**	**0.544**	**0.771**	**0.941**
LightGBM (5-fold)	0.952	0.230	0.740	0.895

Table 5. Comparison of AUC performance with different feature sets

Model	Features	AUC
Random Forest	67	0.870
Random Forest	37	0.857
Proposed methodology	**42**	**0.895**

The evaluation yielded a 96.9% overall accuracy, 52% precision, and 56% recall for the minority class, a balanced accuracy of 77%, and an AUROC of 0.94.

These figures explain how this classifier outperforms every other algorithm we examined, including Balanced Random Forest, K-Nearest Neighbours, and SVM. Headline accuracy is naturally inflated because the dataset is heavily skewed toward calm segments. The real accomplishment is that our model identifies 131 agitation windows, about 1.8% of the total, while preserving near-perfect detection of calm periods. Figure 4 explains the confusion matrix for the proposed model. The confusion matrix shows the model can distinguish between calm and agitated classes very efficiently.

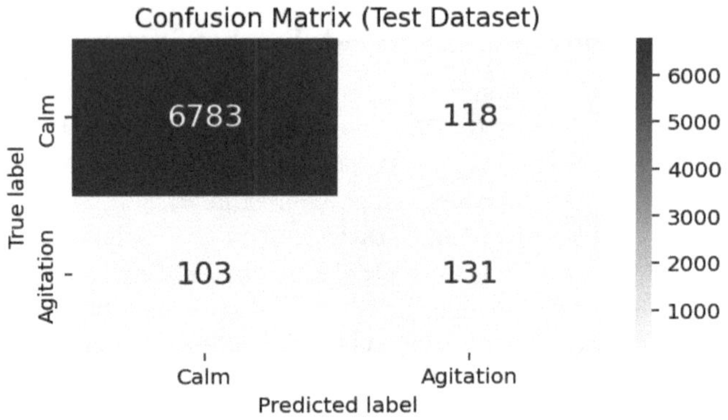

Fig. 4. Confusion matrix of Random Forest Classifier with ADYSN.

3.3 Lightweight Knowledge-Distilled Student Network

The student network trained on WiFi Channel-State-Information (CSI) achieves an overall accuracy of 52.9% and a macro-F1 of 0.543 across eight activity classes. Figure 5 depicts the confusion matrix for the multi-class classification. While those headline numbers may appear modest next to binary agitation detectors, these accuracy metrics are obtained in a highly imbalanced, fine-grained taxonomy that includes a dominant "calm" class and several low-frequency agitation behaviors (tapping on the table and wringing hands). Class imbalance can be addressed by using data augmentation to produce more samples for the minority class [18]. We deliberately did not apply any data augmentation in the dataset due to user privacy and clinical confidentiality. Furthermore, we want to maintain the true correlation between physiological data and real Wi-Fi CSI data so that the student model learns actual cross-modal relationships instead of learning patterns from synthetic data.

The classifier accurately captures 467 windows from approximately nine hundred of our "Calm" windows, which make up just around 15% of the test set. The rest of the windows are misclassified as non-violent agitation (rubbing the table and repetitive sitting-standing) states. This accuracy enables caregivers to

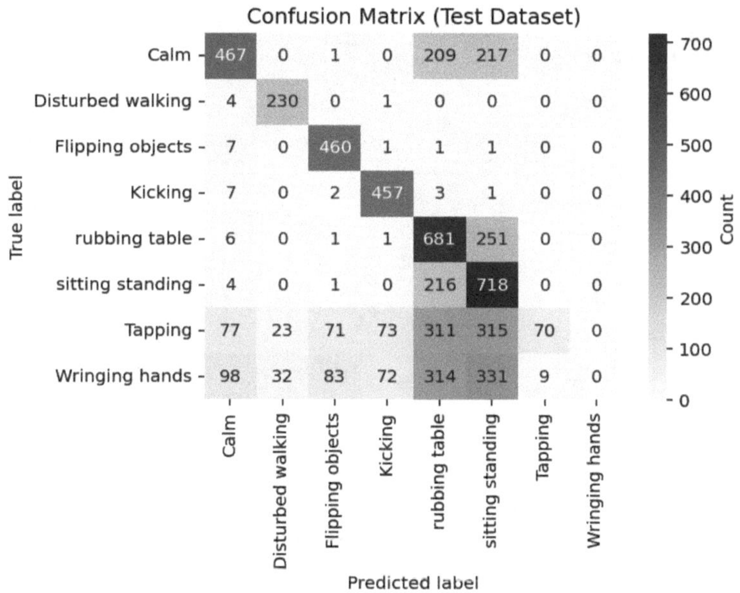

Fig. 5. Confusion matrix of student model with knowledge distillation.

put their trust in green-flag rooms and divert attention to other areas. Importantly, violent classes like "Kicking" and "Flip objects," which had respective recall rates of 97.02% and 97.87%. This implies that almost all violent incidents are immediately and clearly alerted, which is a safety necessity that cannot be compromised. Moderate agitation classes, such as "Rubbing table," "Disturbed walking," and "Sit-to-stand," have recall rates of 72–98%, making them effective yellow-flag warnings that occur well before escalation. The "Tapping" and "Wringing hands" parts of the model are its weakest points. The tapping and wringing are the fine-grained activities that include subtle and repetitive hand gestures, which are difficult to differentiate from one another. Together, these behaviors make up hardly 2% of the body and medically result in micro-motion patterns that resemble benign fidgeting. In contrast to large-scale behaviors like walking or kicking, they generate weak, inconsistent signals. Due to these relatively inconspicuous activities, the model produces misclassification for these activities. Due to these misclassifications of these activities and also for the higher imbalanced dataset, the model shows a low macro-F1 score. The practical cost of such a label is minimal when weighed against the risk of failing to detect violent behaviors, even though precision may drop.

The loss curve corroborates that the network has genuinely learned discriminative characteristics: training loss falls steeply from 1.8 to 0.6 within the first 100 epochs and then converges smoothly, indicating stable optimization rather than memorization. Figure 6 demonstrates training losses over 500 epochs for our proposed methodology. No late-epoch uptick is visible, suggesting little over-fitting despite the imbalance. Even though it was slightly overfitting, we found no significant changes in accuracy, precision, or recall.

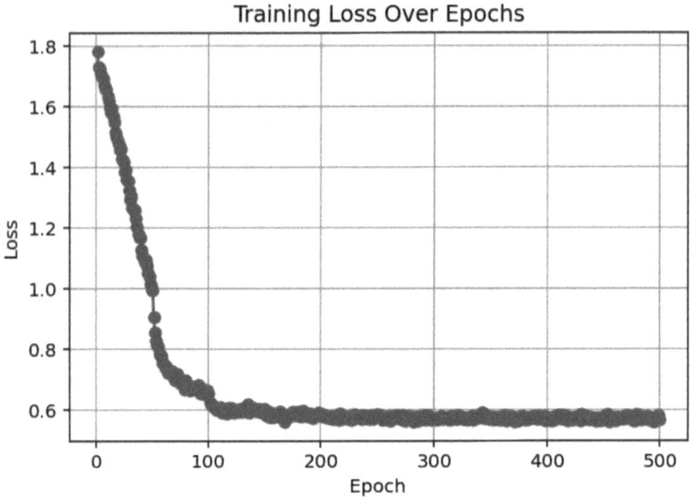

Fig. 6. Curve of training losses for student model.

Finally, our proposed student Wi-Fi CSI network's output confirms that wearable physiological patterns have been successfully encoded into the spatial structure of Wi-Fi data. Different subjects in different domains were the primary problem with this encoding. And by employing knowledge distillation, which has been thoroughly covered in the methodology part, we have effectively solved this problem. Through distillation, latent cues once captured only by wearables now manifest in CSI features, allowing the model to distinguish calm from agitated states with no physical contact. Beyond this binary flag, it also recognizes fine-grained behaviors, such as walking, tapping, kicking, and more, so caregivers know exactly what a patient is doing and can intervene appropriately, whether that means a simple comfort check or an urgent safety response. Compared with only a calm-versus-agitated alert warning, our Wi-Fi CSI student model offers a care provider richer information by correctly characterizing eight distinctive behaviors, including calm (normal walking, normal sitting), repeated sit-to-stand, tapping the table or kicking, and more. The caregivers can continue to engage in other things that shift attention when the patient is in a calm state. Early fidgeting is shown through the subtle non-aggressive sit and stand, which allows for a quick comfort test. In contrast, physically aggressive behaviors like kicking elicit a quick, safety-focused response. Thus, the final model delivers room-scale, privacy-preserving activity and agitation monitoring without wearables.

4 Conclusion and Future Works

Our research has investigated the two research questions we asked before beginning our study. Firstly, by reducing the number of subcarriers by almost 70%, the computational cost is significantly reduced. This is much more beneficial when we want to deploy the student model in a lightweight device such as a Raspberry Pi. Secondly, the outputs of the student model show that we can efficiently embed the physical characteristics into the Wi-Fi signal data. Using this framework, we can reduce the use of wearable devices, which will increase patient comfort and user privacy. Moving forward, we plan to investigate personalized modeling to comprehend how distinct physiological characteristics (such as skin conductance and heart rate variability) uniquely impact CSI signal distortions, even though our current framework uses knowledge distillation to transfer general physiological knowledge into CSI signals. This will facilitate subject-specific calibration, enhance the ability to attribute agitation to the appropriate person in situations involving multiple people, and progress the fine-grained mapping between the dynamics of external signals and interior physiological states. We also intend to implement this framework in facility-based care settings for the contactless remote patient monitoring systems.

References

1. Garza, D.T., Yilmaz, M.: Artificial intelligence and challenges in a research-teaching model. In: International Conference on Computational Science and Computational Intelligence, pp. 131–143. Springer, Cham (2024). https://doi.org/10.1007/978-3-031-94943-2_10
2. Gonçalves, D.A., Barbosa, R.S.: Human activity recognition using neural networks. In: International Conference on Computational Science and Computational Intelligence, pp. 334–347. Springer, Cham (2024). https://doi.org/10.1007/978-3-031-94937-1_27
3. Halpern, R., Seare, J., Tong, J., Hartry, A., Olaoye, A., Aigbogun, M.S.: Using electronic health records to estimate the prevalence of agitation in alzheimer disease/dementia. Int. J. Geriatr. Psychiatry **34**(3), 420–431 (2019)
4. He, H., Bai, Y., Garcia, E.A., Li, S.: ADASYN: adaptive synthetic sampling approach for imbalanced learning. In: 2008 IEEE International Joint Conference on Neural Networks. IEEE World Congress on Computational Intelligence, pp. 1322–1328. IEEE (2008)
5. Imtinan Uddin, J., Toha, T.H., Qin, H.: Enhancing binary classification accuracy using hybrid ensemble methods. In: International Conference on Computational Science and Computational Intelligence, pp. 215–231. Springer, Cham (2024). https://doi.org/10.1007/978-3-031-94937-1_18
6. Khan, S.S., Lau, G., Ye, B., Newman, K., Mihailidis, A., Iaboni, A.: Multimodal sequential deep learning for agitation detection in people living with dementia. In: 2024 IEEE Conference on Pervasive and Intelligent Computing (PICom), pp. 131–136. IEEE (2024)
7. Khan, S.S., et al.: Unsupervised deep learning to detect agitation from videos in people with dementia. IEEE Access **10**, 10349–10358 (2022)

8. Khan, S.S., et al.: A novel multi-modal sensor dataset and benchmark to detect agitation in people living with dementia in a residential care setting. ACM Trans. Comput. Healthcare **6**(3), 1–19 (2025)
9. Khan, S.S., et al.: Daad: a framework for detecting agitation and aggression in people living with dementia using a novel multi-modal sensor network. In: 2017 IEEE International Conference on Data Mining Workshops (ICDMW), pp. 703–710. IEEE (2017)
10. Kong, X., Meng, Z., Nojiri, N., Iwahori, Y., Meng, L., Tomiyama, H.: A HOG-SVM based fall detection IoT system for elderly persons using deep sensor. Procedia Comput. Sci. **147**, 276–282 (2019)
11. Mosin, A., Min, C.H., Salamy, H.: Using a wrist-worn accelerometer to detect off-task activities while studying or working. In: International Conference on Computational Science and Computational Intelligence, pp. 131–143. Springer, Cham (2024). https://doi.org/10.1007/978-3-031-94953-1_12
12. Mujtaba, D.F., Mahapatra, N.R.: Automatic speech recognition APIs: models, features, and challenges. In: International Conference on Computational Science and Computational Intelligence, pp. 33–48. Springer, Cham (2024). https://doi.org/10.1007/978-3-031-94940-1_3
13. Nandi, A., et al.: Cost of care for alzheimer's disease and related dementias in the united states: 2016 to 2060. NPJ Aging **10**(1), 13 (2024)
14. Ni, L., Cha, T., Cha, S.H.: Preferential ensemble method with exponentially weighted scores and its application to facial expression classification. In: International Conference on Computational Science and Computational Intelligence, pp. 324–331. Springer, Cham (2024). https://doi.org/10.1007/978-3-031-90341-0_24
15. Olugboja, A.: Data science in healthcare systems: a comprehensive review and future directions. In: International Conference on Computational Science and Computational Intelligence, pp. 178–191. Springer, Cham (2024). https://doi.org/10.1007/978-3-031-94953-1_15
16. Olugboja, A.: Optimizing predictive performance: a comparative study of machine learning algorithms for classification tasks. In: International Conference on Computational Science and Computational Intelligence, pp. 143–152. Springer, Cham (2024). https://doi.org/10.1007/978-3-031-90341-0_11
17. Pragadeeswaran, S., Mohankumar, N., Puviarasi, R., Nandhini, N., Sivaraman, A., Rajmohan, M.: Detecting and predicting agitation episodes in dementia patients using IoT and artificial neural networks. In: 2024 International Conference on Emerging Technologies and Innovation for Sustainability (EmergIN), pp. 644–649. IEEE (2024)
18. Pulliam, K., Griffin, T., Stringfellow, C.: Generating synthetic datasets to improve deep learning model training. In: International Conference on Computational Science and Computational Intelligence, pp. 297–307. Springer, Cham (2024). https://doi.org/10.1007/978-3-031-94940-1_24
19. Rafiq, H., Manandhar, P., Rodriguez-Ubinas, E.: Sequential disaggregation of residential energy consumption using random forest regression. In: International Conference on Computational Science and Computational Intelligence, pp. 300–312. Springer, Cham (2024). https://doi.org/10.1007/978-3-031-90341-0_22
20. Rajan, K.B., Weuve, J., Barnes, L.L., McAninch, E.A., Wilson, R.S., Evans, D.A.: Population estimate of people with clinical alzheimer's disease and mild cognitive impairment in the united states (2020–2060). Alzheimer's & dementia **17**(12), 1966–1975 (2021)

21. Sharma, N., Klein Brinke, J., Braakman Jansen, L.A., Havinga, P.J., Le, D.V.: Wi-gitation: Replica Wi-fi CSI dataset for physical agitation activity recognition. Data **9**(1), 9 (2023)
22. Soysal, Ö.M., Kelvin, I.E., Oztemel, M.E.: Investigation of the effect of signal enhancement in use of handcrafted and ML-based feature map representation for EEG biometrics. In: International Conference on Computational Science and Computational Intelligence, pp. 194–207. Springer, Cham (2024). https://doi.org/10.1007/978-3-031-94940-1_16
23. Sudharsan, B., Waxman, P.: Humanizing ai-enhancing user engagement in health applications with personality-driven ai design. In: International Conference on Computational Science and Computational Intelligence, pp. 130–143. Springer (2024)
24. Tufail, H., Ahad, A., Puspitasari, I., Shayea, I., Coelho, P.J., Pires, I.M.: Deep learning in smart healthcare: a GAN-based approach for imbalanced alzheimer's disease classification. Procedia Comput. Sci. **241**, 146–153 (2024)
25. Wong, B., Wu, P., Ismail, Z., Watt, J., Goodarzi, Z.: Detecting agitation and aggression in persons living with dementia: a systematic review of diagnostic accuracy. BMC Geriatr. **24**(1), 559 (2024)
26. Xiao, Z., He, Y., Hao, P.: Research on estimation algorithm of track irregularity of heavy-haul railway based on vehicle dynamic response. In: International Conference on Computational Science and Computational Intelligence, pp. 218–231. Springer, Cham (2024). https://doi.org/10.1007/978-3-031-94940-1_18

Determinants of Pandemic App Acceptance: Evidence from the Corona Health App

Michael Winter[1,2(✉)], Marc Holfelder[3], Lena Mulansky[1,2], Johannes Schobel[4], Carsten Vogel[1,2], Lorina Buhr[5], Silke Schicktanz[5], Julian Haug[1,2], Fabian Haug[1,2], Helmut Greger[6], Peter Heuschmann[1,2], and Rüdiger Pryss[1,2]

[1] Institute of Clinical Epidemiology and Biometry, University of Würzburg, Würzburg, Germany
michael.winter@uni-wuerzburg.de
[2] Institute of Medical Data Science, University Hospital of Würzburg, Würzburg, Germany
[3] LA2 GmbH, Erlangen, Germany
[4] Institute DigiHealth, Neu-Ulm University of Applied Sciences, Neu-Ulm, Germany
[5] Department of Medical Ethics and History of Medicine, University Medical Center Göttingen, Göttingen, Germany
[6] Service Center Medical Informatics, University Hospital Würzburg, Würzburg, Germany

Abstract. Pandemic apps were crucial during the COVID-19 pandemic, but their success depended on how willingly the public accepted and used them continuously over time. We examined the acceptance of pandemic apps in Germany using data from 508 participants in the COMPASS study module of the Corona Health App. Our analysis took into account what motivated people to use the apps, how much they trusted such apps, their attitudes toward data privacy and data management, and how well they were informed about the regulations. Our results show that people primarily used pandemic apps to protect themselves and others. Those who did not use them did so because of doubts about the apps' effectiveness and privacy issues. Trust was highest in government-funded and government institutions in app publishing, but significantly lower in private companies. Participants generally had no issues with basic permissions such as Bluetooth and GPS, but were hesitant about more invasive data collection. While most believed that pandemic apps were useful, they were divided on metadata collection. We also found that people knew little about the regulations (i.e., the Medical Devices Regulation), but those who were aware of them and believed that the apps complied with the regulations tended to have more trust, even though they believed that compliance was expensive. The findings show that trust, transparency, and good governance are crucial for public acceptance of pandemic apps.

Keywords: Pandemic Apps · COVID-19 · Digital Health · Acceptance Study

---

Research Track: CSCI-RTHI.

H. R. Arabnia et al. (Eds.): CSCI 2025, CCIS 2971, pp. 64–79, 2027.
https://doi.org/10.1007/978-3-032-28463-1_5

1 Introduction

The COVID-19 pandemic increased the widespread adoption of digital health interventions within public health frameworks, with mobile apps emerging as central tools in the response effort [24]. These pandemic apps facilitated contact tracing, symptom monitoring, and vaccination verification, providing scalable mechanisms for disease surveillance and population health management. The rapid deployment of these technologies highlighted the significant potential of digital health solutions to address critical needs during public health crises.

However, a notable disconnect persisted between technological capabilities and real-world impact [22]. The effectiveness of pandemic apps depended heavily on achieving broad adoption and sustained engagement, yet patterns varied across different countries and application types. Factors including public trust, privacy concerns, and perceived usefulness limited user acceptance [17]. Additionally, The release of separate, incompatible apps showed poor coordination and weak governance, which reduced their overall effectiveness.

Overcoming these limitations necessitates a deeper understanding of the determinants that influence public acceptance of pandemic apps [28]. While existing research has emphasized the critical role of trust, transparency, and robust data governance frameworks [4], further evidence spanning various application categories remains scarce. Acceptance research provides essential groundwork for developing interventions that achieve both technical functionality and societal alignment, thereby enhancing readiness for future health crises.

This analysis examines exploratively factors influencing pandemic app acceptance through data obtained via the Corona Health app's COMPASS study module [5]. By analyzing adoption patterns, usage motivations, trust and privacy preferences, and regulatory attitudes, this research offers descriptive insights into user perspectives based on responses from 508 German participants. The results provide theoretical contributions and practical guidance to enhance digital health interventions to achieve population-wide engagement and effectiveness.

2 Background on COMPASS

The COVID-19 pandemic highlighted the potential and the limitations of mobile health apps in public health management. While smartphone-based tools proved critical for communication, monitoring, and research, their effectiveness was constrained by challenges related to user trust, data protection, interoperability, and regulatory compliance. To address these gaps, the COMPASS initiative (Coordination on Mobile Pandemic Apps Best Practice and Solution Sharing) was established within Germany's Network University Medicine (NUM)[1].

COMPASS aimed to develop a structured, sustainable, and interoperable framework for pandemic-related apps[2]. The initiative brought together nine German university hospitals alongside scientific and industry partners, combining expertise in medical informatics, public health, and regulatory science. Its

[1] https://www.netzwerk-universitaetsmedizin.de/en; last accessed at: 09/26/2025.

[2] https://num-compass.science/en/; last accessed at: 09/26/2025.

development principles were in open-source collaboration, ensuring transparency, adaptability, and wide accessibility.

The objective of COMPASS focused on creating standardized app components, best-practice guidelines, and modular technical platforms to support rapid and compliant deployment of pandemic apps. In addition to legal and ethical guidance, COMPASS prioritized research compatibility, interoperability, and adherence to international data standards, including the FAIR (Findable, Accessible, Interoperable, Reusable) principles [18].

The practical effectiveness of this comprehensive approach was demonstrated through real-world validation via several study applications (e.g., [18,29]), which collectively generated nearly one million data points from over 25,000 participants. By integrating governance, regulatory, ethical, and technological considerations, COMPASS established a blueprint for mobile health solutions that extend beyond pandemic use cases. It offers a model for developing scalable, secure, and research-ready applications, thereby contributing to long-term preparedness and resilience in digital health infrastructures.

Note that a national telephone-based survey on acceptance and attitudes towards pandemic apps for research was also conducted [6].

3 Related Work

Extensive research has explored COVID-19 health apps, particularly focusing on contact tracing, exposure notification, and symptom monitoring [12,15]. However, most studies examine only subsets of available apps and overlook the comprehensive ecosystem of mHealth technologies implemented during the pandemic [8,16]. When broader application collections are analyzed, studies typically neglect critical factors such as interoperability, governance frameworks, and sustainable healthcare workflow integration. Various initiatives comparable to COMPASS have emerged [1], comprising technical protocols for proximity tracing and digital health platforms for information sharing. Although these initiatives offer valuable insights, they remain fragmented and lack the comprehensive interdisciplinary approach that COMPASS provides. The literature consistently emphasizes persistent adoption challenges, including privacy concerns, regulatory ambiguity, and insufficient public trust [3,23]. These obstacles demonstrate the need for structured, scalable frameworks like COMPASS that integrate technical infrastructure with patient-reported outcome measures to enable both immediate pandemic response and long-term digital health readiness that is socially accepted and ethically justified.

4 Materials and Methods

Corona Health: This analysis utilized data gathered through the Corona Health app[3], a research-oriented platform incorporating sensor technology [26]

[3] https://www.corona-health.ike-b.de/; last accessed at: 09/26/2025.

designed to track physical and psychological health as well as information regarding the acceptance of pandemic apps throughout the COVID-19 crisis [5]. This application functioned as a unified framework for multiple questionnaire-driven investigations, with each study focusing on distinct dimensions of health, quality of life, and acceptability. Ethical approval for the research was obtained from the University of Würzburg's Ethics Committee (Reference Number: 130/20-me). The Corona Health app was constructed in adherence to both European Medical Device Regulation (MDR) requirements and General Data Protection Regulation guidelines [11], with validation performed according to IEC 62304, IEC 82304, and GAMP 5 certification standards.

For the current analysis, we only utilized data from the "Study on University Medicine Network: COMPASS" study module [5]. The module addressed pandemic app acceptance through a comprehensive questionnaire comprising five distinct parts: (1) Pandemic apps - General, covering basic usage patterns, trust factors, and data sharing preferences; (2) Pandemic apps - Usage, examining technical aspects, permissions, and user behavior; (3) Pandemic apps - Development, exploring perspectives on app creation, economic considerations, and regulatory awareness; (4) MDR - General, assessing trust in regulatory compliance; and (5) MDR - Development, investigating technical knowledge of regulatory requirements. The complete questionnaire is presented in the Appendix, organized by their respective parts and sequential numbering within each section. Detailed answers from the participants and related code book can be found on Zenodo[4].

Measures: The questionnaire comprised a total of 60 questions distributed across the five parts (i.e., 15 in part 1, 20 in part 2, 13 in part 3, 5 in part 4, and 7 in part 5. It should be noted that parts of the questionnaire used in the mobile app were created as part of another study, which can be found here [6]. For the purposes of the current analysis, we focus only on a subset of questions identified as one of most relevant for understanding adoption, usage behavior, and perceptions of regulation. Specifically, we concentrate on:

Measures Part 1: Our analysis centered on key questions designed to measure app adoption and trust. We first established a baseline adoption rate by asking participants if they currently use a pandemic app. To understand the reasons behind their choices, we then asked why they either used or avoided these apps. This revealed key motivators, such as protection and solidarity, alongside major barriers (e.g., privacy concerns, lack of trust). To further explore the role of trust, we asked which types of organizations (e.g., governments, health authorities, and technology companies) participants considered reliable publishers.

Measures Part 2: We emphasized privacy, trust, and usefulness. We assessed data privacy acceptance by asking participants what permissions (e.g., GPS, Bluetooth) they would be willing to grant for scientific studies. A separate question examined their willingness to grant additional permissions to official pan-

[4] https://zenodo.org/records/17208748; last accessed at: 09/26/2025.

demic apps, which directly reflected their trust in authorities versus their privacy concerns. We also evaluated the perceived usefulness of these apps in combating a pandemic. Finally, participants were asked to weigh the justification of metadata collection (e.g., GPS, phone usage data), highlighting the perceived trade-off between public health benefits and personal privacy.

Measures Part 3: This part considers expectations and governance aspects of app development. Participants answered about the future role of pandemic apps, indicating whether they see them as essential, limited, or temporary. Opinions on Google and Apple's blocking of non-official pandemic apps capture views on platform control versus public safety. Another item assessed whether the allocation of pandemic apps by official institutions is fair and sensible, reflecting perceptions of transparency and trust. Finally, participants estimated the demand for additional pandemic apps, shedding light on whether a single official app is seen as sufficient or if further innovation is expected.

Measures Part 4: This part examines the role of regulation in shaping awareness and trust. Participants were first asked whether they are familiar with the MDR, providing a baseline of regulatory knowledge. They were then asked if they place greater trust in apps developed in accordance with MDR and whether they trust such apps more than other apps, directly addressing the impact of certification on user confidence.

Measures Part 5: This part focuses on the practical implications of MDR compliance. Participants were asked to estimate the additional effort required for MDR-compliant app development and the expected costs of certification, highlighting both workload and economic considerations.

Participants: This analysis included 508 German participants who were recruited via the Corona Health app and enrolled in the COMPASS module. To be included, participants had to be at least 18 years old and provide informed consent by digital approval in the app. We considered data for this analysis collected from December 23, 2020, to January 27, 2024. Demographics for this sample are provided in Table 1.

Statistical Analysis: Results were summarized descriptively, with categorical variables shown as absolute frequencies and mean values reported where relevant. Since not all items were mandatory, response counts for particular questions may differ from the total sample of 508 participants. Missing data were not imputed but are reflected in the reported distributions. This ensures that the findings presented remain faithful to the collected data while acknowledging natural variation in participant responses.

5 Results and Discussion

The first set of findings (see Table 2) presents adoption levels, obstacles, and trust in pandemic app providers. A large majority of participants (463) reported

Table 1. Sample Demographics of COMPASS Participants

Characteristic	Category (n = 508)	n	(%)
Gender			
	Female	302	(59.5%)
	Male	197	(38.8%)
	Diverse	5	(1.0%)
	No answer	4	(0.8%)
Age (years)			
	Mean (SD)	42.2	(14.6)
Education			
	No diploma	6	(1.2%)
	Lower secondary	38	(7.5%)
	Intermediate	113	(22.2%)
	Upper secondary	141	(27.8%)
	Bachelor's degree	39	(7.7%)
	Master's/Diplom/etc.	125	(24.6%)
	Doctorate	26	(5.1%)
	In school	8	(1.6%)
	No answer	12	(2.4%)
COVID-19 infection in personal circle			
	Yes	262	(51.6%)
	No	239	(47.0%)
	No answer	7	(1.4%)

actively using such apps. Their main motivations were the containment of infectious diseases (392), protection of others (296), personal protection (230), and contributing to research efforts (245), which are in line with related literature [9]. Instances of non-use were relatively uncommon (42), and when reported, they were mainly linked to doubts about app effectiveness (29), concerns about third-party data use (15), or worries about surveillance (11)

Patterns of trust revealed a strong tendency toward government agencies and publicly funded institutions when acting as app providers or as custodians of collected app data, with substantially lower trust in private-sector entities. In detail, 451 participants considering state-funded research institutions and 385 considering federal or state government trustworthy, while private-sector actors received much lower levels of trust (e.g., 93 for private health insurance companies, 78 for international software companies) [10]. This highlights how user confidence is closely tied to institutional affiliation, with public organizations being regarded as more dependable stewards of sensitive health data compared to commercial companies.

Table 2. Part 1 - Pandemic Apps - General

Are you currently using a pandemic app on your smartphone?	
Yes/No/NA	463/42/3
Why are you using a pandemic app on your smartphone?	
Containment of infectious diseases	392
Research into infectious disease	245
Protect myself from infection	230
Protect others from infection	296
My personal circle has asked me to use the app	28
Other reasons	33
Why are you **not** using a pandemic app on your smartphone?	
Doubtful app usefulness for disease containment	29
App and data are too tedious to handle	8
Concerned about third-party data use without consent	15
Lack background on data collection and use	6
Feels monitored by the app	11
Fear of app reducing smartphone battery life	10
Past pandemic app use caused technical problems	7
Other reasons	34
Which pandemic app publishers do you consider trustworthy?	
State-funded research institutions	451
Privately funded research institutions	134
Federal or state government	385
Regional health authority	307
State-funded clinics	220
Privately funded clinics	84
Statutory health insurance companies	250
Private health insurance companies	93
German software companies	186
International software companies	78

Overall, the results illustrate that adoption is largely driven by collective and individual protection goals but limited by concerns over effectiveness and privacy. At the same time, the pronounced trust in governmental and publicly funded institutions suggests that acceptance efforts should build on the credibility of the public sector to address persisting fears of surveillance and usability issues.

The second set of results (see Table 3) highlights user attitudes toward permissions, perceived usefulness, and metadata collection. Most participants expressed willingness to grant permissions such as Bluetooth (387) and GPS (308), which are linked to core pandemic app features. In contrast and in accordance with literature [27], permissions perceived as more intrusive, such as microphone (26) or camera (36), received minimal approval. This differentiation suggests a pragmatic acceptance of data access when tied to infection monitoring, but resistance when the connection to public health goals is less transparent.

When asked about granting additional permissions to official pandemic apps, 302 participants indicated readiness to comply, while 62 rejected and 144 were undecided. As also shown in [20], the usefulness of apps in combating pandemics was affirmed, with 477 participants supporting their value and only 31 rejecting it. However, responses regarding metadata collection revealed division: 312 considered it justified, whereas 196 opposed such practices, reflecting persistent tensions between public health benefits and individual privacy protection.

The findings from Part 2 demonstrate that participants are not opposed to data sharing but their acceptance is based on perceived necessity, transparency, and proportionality. The combination of endorsement of app usefulness with ambivalence toward metadata collection underscores the importance of designing data practices that are functionally essential and publicly justifiable.

Table 3. Part 2 - Pandemic Apps - Usage

What permissions are you willing to grant to an app for scientific studies?			
Bluetooth	387	GPS	308
Microphone	26	Camera	36
Storage	132	Call logs	15
Calendar	45	Contacts	68
Body Sensors	237	Device Data	200
Others	61		
Are you ready to grant more permissions to official pandemic apps?			
Yes/No/Not Sure		302/62/144	
Do you think apps are useful for combating a pandemic?			
Yes/No		477/31	
Do you consider the collection of metadata from pandemic apps justified?			
Yes/No		312/196	

The third set of results (see Table 4) examines perspectives on app development, governance, and demand. Answers revealed no preference for a single technical implementation, with support distributed across web-based (72), native (45), hybrid (46), and cross-platform approaches (52). This dispersion indicates an absence of user expectations regarding technical design, suggesting that accessibility and functionality may outweigh the specifics of implementation [21].

With respect to platform control, participant views on Google and Apple's restriction of non-official pandemic apps were divided: 71 regarded these policies as meaningful, 49 as ethically acceptable, 33 as unnecessary, and 23 as distorting competition. These differing perspectives capture broader tensions between safeguarding the public, corporate gatekeeping, and fostering innovation [2].

People also disagreed about whether pandemic apps were distributed fairly by official institutions: 81 considered the allocation fair and sensible, 21 did not, and 55 were unsure. Interest in additional pandemic apps was moderately high, with the average estimated demand at 72.5 (on a 0–100 scale), suggesting that existing offerings were not considered fully adequate. These results are in line with similar findings obtained from related work [19].

Taken together, the Part 3 results highlight ambivalence about governance structures and market control while signaling continued openness to further app development. Acceptance therefore appears contingent not only on technical quality but also on perceptions of fairness, transparency, and the balance between public authority and private influence.

Table 4. Part 3 - Pandemic Apps - Development

Where do you see the development of a pandemic app?			
Web App	72	Native	45
Hybrid	46	Cross-Platform	52
Others	31		
What is your position on Google and Apple's blocking of pandemic-related apps for developers without contacts to official institutions?			
Meaningful	71	Distorting competition	23
Unnecessary	33	Ethically Acceptable	49
Do you find the official allocation of pandemic apps fair and sensible?			
Yes/No/Not Sure		81/21/55	
How high do you estimate the demand for additional pandemic apps?			
0 = Low; 100 = High		72.5	

Table 5. Part 4 & 5 - MDR - General & Development

Are you familiar with the MDR?	
Yes/No	180/328
Do you have greater trust in apps developed in accordance with MDR?	
Yes/No/Does Not Matter	161/35/85
Do you trust apps developed according to the MDR more than "other apps"	
Yes/No/Does Not Matter	174/58/80
How much additional effort do you estimate for MDR-compliant app development?	
0 = Low; 100 = High	63.61
How much do you estimate the additional costs of MDR certification to be?	
0 = Low; 100 = High	63.78

The final set of results in Part 4 and 5 (see Table 5) addresses awareness and perceptions of the MDR, and the anticipated implications for app development. Most participants reported unfamiliarity with MDR (328), while 180 indicated knowledge of it, reflecting a limited baseline understanding of regulatory frameworks among the population [11]. A considerable number expressed greater trust in apps developed under MDR compliance: 161 participants reported greater trust, 35 did not, and 85 stated it did not matter. Similarly, 174 participants trusted MDR-compliant apps more than other apps, while 58 did not and 80 said it made no difference [10]. This suggests that without detailed regulatory knowledge, users interpret certification as a meaningful quality marker.

Moreover, participants estimated the additional effort of MDR-compliant app development at an average of 63.61 (on a 0–100 scale) and the additional costs of certification at an average of 63.78. These expectations indicate recognition of regulatory processes as resource-intensive, yet they also highlight an acknowledgment of the safeguards such compliance provides.

In summary, the Part 4 and 5 findings show a notable trust for MDR-certified apps, despite low general awareness of the regulation itself. This trust advantage, however, is counterbalanced by perceptions of increased workload and costs in development. Together, the results illustrate a dual challenge: utilizing MDR compliance as a user-trust mechanism while mitigating concerns about its economic and technical burden on developers.

Key Takeaway: The results highlight that pandemic app adoption is driven by a balance of protective motivations, privacy concerns, and institutional trust, confirming prior observation [9]. The strong orientation toward public-sector providers is consistent with earlier studies indicating that governments and health authorities are regarded as more credible institutions of sensitive data than commercial actors [25]. This preference reflects longstanding trust structures in digital health, where the legitimacy of public institutions helps alleviate fears of data misuse. As in previous work, privacy emerges as a key barrier, especially around metadata collection, where participants carefully weighed potential benefits against risks of surveillance [7]. Similar findings were observed from a German national survey study [6], where participants also reported higher institutional trust and identified privacy concerns as main barriers to app adoption. Both studies indicate that acceptance depends largely on transparent governance and trustworthy institutional backing.

Participants' readiness to allow device permissions (e.g., Bluetooth, GPS) coupled with their rejection of more invasive access aligns with earlier results showing that acceptance depends on whether data types are perceived as essential for infection control [27]. This reflects a broader privacy scope, in which users assess data sensitivity against expected health gains. Similarly, the ambivalence observed toward platform governance mirrors earlier debates on corporate gatekeeping [25]. While some considered such restrictions to enhance safety and reliability, others regarded them as barriers to competition and innovation, reinforcing longstanding critiques of platform power during the pandemic.

Finally, the positive association of MDR compliance with trust despite limited public awareness resonates with literature on regulatory certification as an indicator for reliability [14]. Even in the absence of detailed knowledge, regulatory signals appear to function as trust-enhancing heuristics. This finding parallels broader evidence from health technology adoption, where certification and compliance frameworks reassure users about safety and quality [13]. Taken together, the results confirm established acceptance factors (e.g., trust, privacy) regarding governance and regulatory perceptions. In particular, this analysis confirms prior literature by demonstrating that institutional affiliation and certification serve as an anchors of trust, whereas unresolved tensions in governance and data practices continue to limit full acceptance of pandemic apps.

Implications: The results offer several implications pandemic and health-related apps. First, the high reliance on public-sector institutions indicates that government agencies and publicly funded research bodies should take a leading role in communication and app provision. Relying on their higher baseline of trust can help counteract skepticism and promote adoption of such apps.

Second, the differentiated acceptance of data permissions underscores the importance of transparency in data governance. Developers should communicate why specific data types are required and ensure that permissions are limited to essential functions. Providing users with control over data sharing and ensuring visible safeguards against misuse may help reduce privacy-related resistance.

Third, the ambivalence toward platform governance suggests a need for more transparent collaboration between technology companies, health authorities, and regulators. While restrictive policies by platform providers may enhance safety, they should be accompanied by mechanisms that maintain fairness and openness for developers, avoiding perceptions of monopolistic control.

Finally, the trust associated with MDR compliance highlights the value of regulatory certification not only as a technical safeguard but also as a communication tool. Policymakers and developers should emphasize compliance in public messaging, while simultaneously addressing concerns about the associated costs and development burden. Clear communication of regulatory processes may enhance credibility while reducing misperceptions about feasibility.

Limitations: Several limitations must be taken into account when interpreting the findings. First, the analysis was restricted to a subset of questions from the broader questionnaire, meaning that potentially relevant aspects of pandemic app acceptance were not included. Second, as many questions were not mandatory, numbers varied across items, which may introduce bias if non-participation differ from participation. Third, the analysis relied exclusively on self-reported data, which can be subject to recall inaccuracies and social desirability bias. Fourth, the app-based design implies selection effects, as only individuals with smartphone access, sufficient digital literacy, and willingness to engage with the Corona Health app could participate. Further limitation constitutes that the term "pandemic apps" covers a heterogenous set of different types of apps for different purposes, thus potential differences in acceptance weren't addressed by the survey. Finally, the sample was drawn entirely from Germany, limiting the generalizability of the results to other countries where trust, privacy concerns, and adoption dynamics may differ.

6 Conclusion and Future Work

This analysis examined a set of determinants of pandemic app acceptance using data from 508 German participants of the Corona Health app's COMPASS study module. Results highlight that adoption is primarily motivated by protection goals but constrained by privacy concerns and skepticism about effectiveness. Trust was strongly tied to public-sector institutions, while private actors received limited confidence. Users were generally willing to grant functional permissions but resisted intrusive access, and although regulatory knowledge was low, MDR compliance served as a trust-enhancing signal. These findings underscore trust, transparency, and governance as central conditions for broad acceptance.

Future research should build on these insights by examining longitudinal dynamics of app acceptance, as trust and privacy concerns may shift over time or across different phases of a health crisis. Comparative studies across countries and cultural contexts are also needed to test the generalizability of results. Moreover, further work is required to evaluate how regulatory frameworks can balance trust enhancement with manageable burdens for developers.

Appendix

See

Part	#	Question
		Pandemic apps - General
Part 1	1	Do you use a smartphone?
Part 1	2	Are you currently using a pandemic app on your smartphone?
Part 1	3	Why are you currently using a pandemic app on your smartphone?
Part 1	4	Why are you NOT currently using a pandemic app on your smartphone?
Part 1	5	Do you plan to continue using a pandemic app on your smartphone in the future?
Part 1	6	Do you plan to use a pandemic app on your smartphone in the future?
Part 1	7	Which pandemic app publishers do you consider trustworthy?
Part 1	8	Where do you think the data collected by a pandemic app should be stored?
Part 1	9	Would you provide data collected using a pandemic app to government research institutes if it would then only be used in coded form?
Part 1	10	Which of the data collected and coded using a pandemic app would you make available for medical research?

Part	#	Question
Part 1	11	Would it be very important, somewhat important, not so important, or not important at all to you to receive a detailed list of which research institutes are evaluating the data from your pandemic app?
Part 1	12	Why wouldn't you make the data collected and coded using a pandemic app available?
Part 1	13	How would you make data collected using a pandemic app available to government research institutes for further use in coded form?
Part 1	14	How do you rate the following statement: "The use of pandemic apps is a social obligation."
Part 1	15	What sources do you use to inform yourself about the novel coronavirus?
		Pandemic apps - Usage
Part 2	1	On which devices do you typically use apps?
Part 2	2	On which devices do you think apps should be available?
Part 2	3	Where do you use apps?
Part 2	4	How important is it to you that an app is fully functional without an internet connection?
Part 2	5	How well do you navigate the common app stores (Google Play Store and Apple App Store)? Please consider processes such as searching for and finding apps, installation procedures, and the information and functions you're looking for.
Part 2	6	How do you keep your apps up to date?
Part 2	7	Do you always keep your operating system up to date (i.e., do you regularly install updates)?
Part 2	8	What permissions are you willing to grant to an app for scientific studies?
Part 2	9	How do you grant these permissions to existing apps?
Part 2	10	Are you ready to grant more permissions to official pandemic apps?
Part 2	11	For which permissions is it important for you to know exactly when or how often the app accesses the corresponding sensors or data?
Part 2	12	Do you think the collection of metadata (GPS, volume, mobile phone usage data, etc.) is useful for pandemic apps?
Part 2	13	Do you consider the collection of metadata from pandemic apps justified?
Part 2	14	Is it important to you that apps that access sensitive data (e.g., pandemic apps) are made available by official authorities or the government rather than through app stores?
Part 2	15	How often do you use the bookmark function in your browser?
Part 2	16	Do you have shortcuts to specific websites on your home screen?
Part 2	17	If so, which websites are they? (Please list them separated by commas)
Part 2	18	In your opinion, how much funding is justified for the development of an official pandemic app? (Reference value: Corona Warn App approx. €15,000,000)
Part 2	19	Do you think apps are useful for combating a pandemic?

Part	#	Question
Part 2	20	Is it important to you that an app to combat a pandemic is available in your native language?
		Pandemic apps - Development
Part 3	1	Where do you see the future of app development in general?
Part 3	2	Where do you see the development of a pandemic app?
Part 3	3	What is your position on Google and Apple's blocking of pandemic-related apps for developers without contacts to official institutions?
Part 3	4	How high do you estimate the demand for additional pandemic apps?
Part 3	5	Do you think apps are useful for combating a pandemic?
Part 3	6	Do you consider the (current) allocation of (pandemic) apps by official institutions to be fair and sensible?
Part 3	7	How should future contracts for such apps be awarded?
Part 3	8	How much do you estimate the general effort required to develop and maintain a useful pandemic app?
Part 3	9	How do you estimate the public response to a pandemic app?
Part 3	10	How do you assess the external impact (PR, public associations with the manufacturer) on your "company" that could arise from the development of such a pandemic app?
Part 3	11	How high do you estimate the risk of economic damage in the event of a possible failure of the app?
Part 3	12	How high do you estimate the risk of damage to your company's image in the event of a possible app failure?
Part 3	13	Are you familiar with the Medical Devices Regulation (MPV)?
		MDR - General
Part 4	1	Do you have greater trust in apps developed with MDR?
Part 4	2	Do you trust apps developed according to the MDR more than "other apps"?
Part 4	3	What do you consider more relevant when developing an app to combat a pandemic?
Part 4	4	Do you consider the Federal Government's Corona Warning App to be a medical device?
Part 4	5	Do you think that the Corona Warn App was developed in accordance with MDR and certified accordingly?
		MDR - Development
Part 5	1	Are you familiar with the MDR?
Part 5	2	Do you know the requirements that MDR-compliant apps must meet?
Part 5	3	Have you already developed apps that are MDR compliant?
Part 5	4	How much additional effort do you estimate for MDR-compliant app development?
Part 5	5	How much documentation (in pages) do you estimate has been produced for the "Corona Health" app?
Part 5	6	How much do you estimate the additional costs of MDR certification to be?
Part 5	7	In what ways does MDR compliance provide a higher level of security for you as a pandemic app developer?

References

1. Agarwal, S., et al.: Guidelines for reporting of health interventions using mobile phones: mobile health (mHealth) evidence reporting and assessment (mera) checklist. BMJ **352** (2016)
2. Alanzi, T.: A review of mobile applications available in the app and google play stores used during the COVID-19 outbreak. J. Multi. Healthcare 45–57 (2021)
3. Amiri, P., Nadri, H., Bahaadinbeigy, K.: Facilitators and barriers of mHealth interventions during the COVID-19 pandemic: systematic review. BMC Health Serv. Res. **23**(1), 1176 (2023)
4. Austin, L.M.: COVID-19 and the data governance gap. Annu. Rev. Law Soc. Sci. **19**(1), 235–253 (2023)
5. Beierle, F., et al.: Corona health–a study-and sensor-based mobile app platform exploring aspects of the COVID-19 pandemic. Int. J. Environ. Res. Public Health **18**(14), 7395 (2021)
6. Buhr, L., Schicktanz, S., Nordmeyer, E.: Attitudes toward mobile apps for pandemic research among smartphone users in Germany: national survey. JMIR Mhealth Uhealth **10**(1), e31857 (2022)
7. Fahey, R.A., Hino, A.: COVID-19, digital privacy, and the social limits on data-focused public health responses. Int. J. Inf. Manage. **55**, 102181 (2020)
8. Garavand, A., Ameri, F., Salehi, F., Talebi, A.H., Karbasi, Z., Sabahi, A.: A systematic review of health management mobile applications in COVID-19 pandemic: features, advantages, and disadvantages. Biomed. Res. Int. **2024**(1), 8814869 (2024)
9. Golinelli, D., Boetto, E., Carullo, G., Nuzzolese, A.G., Landini, M.P., Fantini, M.P.: Adoption of digital technologies in health care during the COVID-19 pandemic: systematic review of early scientific literature. JMIR **22**(11), e22280 (2020)
10. Gunawan, J., Choffnes, D., Hartzog, W., Wilson, C.: The COVID-19 pandemic and the technology trust gap. Seton Hall L. Rev. **51**, 1505 (2020)
11. Holfelder, M., et al.: Medical device regulation efforts for mHealth apps during the COVID-19 pandemic—an experience report of corona check and corona health. J **4**(2), 206–222 (2021)
12. Holl, F., Schobel, J., Swoboda, W.J.: Mobile apps for COVID-19: a systematic review of reviews. Healthcare **12**, 139 (2024)
13. Iqbal, J.D., Biller-Andorno, N.: The regulatory gap in digital health and alternative pathways to bridge it. Health Policy Technol. **11**(3), 100663 (2022)
14. Karopoulos, G., et al.: A survey on digital certificates approaches for the COVID-19 pandemic. IEEE Access **9**, 138003–138025 (2021)
15. Kondylakis, H., et al.: COVID-19 mobile apps: a systematic review of the literature. J. Med. Internet Res. **22**(12), e23170 (2020)
16. Mansouri, F., Darvishpour, A.: Mobile health applications in the COVID-19 pandemic: a scoping review of the reviews. Med. J. Isl. Rep. Iran **37**, 8 (2023)
17. Manteghinejad, A., Javanmard, S.H.: Challenges and opportunities of digital health in a post-COVID19 world. J. Res. Med. Sci. **26**(1), 11 (2021)
18. Oehm, J.B.: Integration of patient-reported outcome data collected via web applications and mobile apps into a nation-wide COVID-19 research platform using fast healthcare interoperability resources: development study. JMIR **26** (2024)
19. Oldeweme, A., Märtins, J., Westmattelmann, D., Schewe, G.: The role of transparency, trust, and social influence on uncertainty reduction in times of pandemics: empirical study on the adoption of COVID-19 tracing apps. JMIR **23**(2) (2021)

20. Pandit, J.A., Radin, J.M., Quer, G., Topol, E.J.: Smartphone apps in the COVID-19 pandemic. Nat. Biotechnol. **40**(7), 1013–1022 (2022)
21. Petracca, F., Ciani, O., Cucciniello, M., Tarricone, R.: Harnessing digital health technologies during and after the COVID-19 pandemic: context matters. J. Med. Internet Res. **22**(12), e21815 (2020)
22. Niakan Kalhori, S.R., et al.: Digital health solutions to control the COVID-19 pandemic in countries with high disease prevalence: literature review. JMIR **23**(3), e19473 (2021)
23. Saeidnia, H.R., Kozak, M., Ausloos, M., Lund, B.D., Ghorbi, A., Mohammadzadeh, Z.: Evaluation of COVID-19 m-health apps: an analysis of the methods of app usability testing during a global pandemic. Inf. Med. Unl. **41**, 101310 (2023)
24. Sharma, S., et al.: Digital health innovation: exploring adoption of COVID-19 digital contact tracing apps. IEEE Trans. Eng. Manage. **71**, 12272–12288 (2020)
25. Storeng, K.T., de Bengy Puyvallée, A.: The smartphone pandemic: how big tech and public health authorities partner in the digital response to COVID-19. Glob. Public Health **16**(8–9), 1482–1498 (2021)
26. Vogel, C.: Developing apps for researching the COVID-19 pandemic with the track-yourhealth platform. In: MobileSoft, pp. 65–68. IEEE (2021)
27. Vokinger, K.N., Nittas, V., Witt, C.M., Fabrikant, S.I., Von Wyl, V.: Digital health and the COVID-19 epidemic: an assessment framework for apps from an epidemiological and legal perspective. Swiss Med. Wkly. **150**, w20282 (2020)
28. Wang, Q., et al.: Integrating digital technologies and public health to fight COVID-19 pandemic: key technologies, applications, challenges and outlook of digital healthcare. Int. J. Env. Res. Pub. Health **18**(11), 6053 (2021)
29. Winter, M., et al.: Exploring the usability of the German COVID-19 contact tracing app in a combined eye tracking and retrospective think aloud study. In: EMBC, pp. 2215–2221. IEEE (2021)

An Explainable Hybrid AI Framework for Enhanced Tuberculosis and Symptom Detection

Neel Patel[1], Alexander Wong[2], and Ashkan Ebadi[2,3](✉)

[1] Mechanical and Mechatronics Engineering, University of Waterloo, Waterloo, ON N2L 3G1, Canada

[2] Systems Design Engineering, University of Waterloo, Waterloo, ON N2L 3G1, Canada
ashkan.ebadi@nrc-cnrc.gc.ca

[3] Digital Technologies, National Research Council Canada, Toronto, ON M5T 3J1, Canada

Abstract. Tuberculosis remains a critical global health issue, particularly in resource-limited and remote areas. Early detection is vital for treatment, yet the lack of skilled radiologists underscores the need for artificial intelligence (AI)-driven screening tools. Developing reliable AI models is challenging due to the necessity for large, high-quality datasets, which are costly to obtain. To tackle this, we propose a teacher-student framework which enhances both disease and symptom detection on chest X-rays by integrating two supervised heads and a self-supervised head. Our model achieves an accuracy of 98.85% for distinguishing between COVID-19, tuberculosis, and normal cases, and a macro-F1 score of 90.09% for multilabel symptom detection, significantly outperforming baselines. The explainability assessments also show the model bases its predictions on relevant anatomical features, demonstrating promise for deployment in clinical screening and triage settings.

Keywords: Tuberculosis · Symptoms · Lung Disease · Self-supervised Learning · Explainable Neural Network · Rapid Screening

1 Introduction

Tuberculosis (TB) continues to be a major global health challenge, despite being both preventable and curable [1]. Transmitted via aerosols of transmission of mycobacterium tuberculosis, TB remains particularly burdensome in low– and middle–income regions, ranking second only to COVID-19 among infectious causes of death in 2022 [2, 3]. The risk of developing active TB is highest within the first two years post-infection–approximately 5% progress during this period, with an additional ~5% later in life–highlighting the critical need for timely diagnosis and treatment [4, 5]. Untreated pulmonary TB has a high fatality rate (pre-chemotherapy estimates around 70% for smear-positive cases), whereas current 4–6 month treatment regimens achieve a global treatment success rate of ~88% for drug-susceptible TB in 2022 [3]. Despite reaching record notifications of 7.5 million in 2022 and 8.2 million in 2023, global efforts still fall short of the End TB Strategy milestones [2, 6].

H. R. Arabnia et al. (Eds.): CSCI 2025, CCIS 2971, pp. 80–95, 2027.
https://doi.org/10.1007/978-3-032-28463-1_6

Chest radiography (CXR) is the workhorse for TB screening [7] due to its rapidity and widespread availability, yet the quality of interpretation varies with reader expertise and workload, particularly in high-burden settings [8]. Reflecting this reality, recent World Health Organization (WHO) guidance supports the use of computer-aided detection (CAD) to interpret digital CXR for TB screening and triage in people aged 15 and older [9, 10].

Identifying and screening populations at risk is crucial for effectively controlling the spread of the TB disease [11]. To bridge the data and trust gaps that limit robust CAD implementation, we have developed an explainable, semi-supervised framework for CXR that performs both disease triage and radiographic symptom detection within a unified model. Our approach is based on a "distillation for self-supervision" paradigm (DISTL) [12], employing a ViT-Small teacher-student backbone optimized with a DINO [13] self-supervised head, alongside two supervised heads for learning disease and symptoms. We leverage CheXpert-pretrained [14] CXR weights for domain-specific initialization and utilize multi-crop training to expose the network to both global lung context and fine local patterns. The model's decision-making process is further validated by comparing Gradient-weighted Class Activation Mapping (Grad-CAM) [15] saliency maps to dataset bounding boxes for symptom findings, ensuring alignment with plausible thoracic anatomy.

Our experiments on a comprehensive dataset, including TB, COVID-19, normal controls, and seven symptom labels, demonstrate that the proposed model outperforms strong CNN baselines across both tasks. It achieves near-ceiling performance in disease classification and substantial gains in symptom detection, particularly for small or subtle findings, with Grad-CAM overlays focusing within annotated lung and pleural regions. These results indicate that integrating self-supervised distillation with joint disease–symptom supervision produces features that are both discriminative and clinically relevant, supporting deployment in real-world TB screening workflows where sensitivity, specificity, and explainability are crucial.

2 Data

A composite chest-radiograph corpus was assembled from four open-access sources to cover disease, healthy control, and finding-level tasks. Tuberculosis-positive images (n = 2,141) were collected from the Montgomery (n = 58) [16], Shenzhen (n = 336) [16], Belarus TB-Portals (n = 1,047) [17] and Rahman et al. (n = 700) [18] cohorts. Normal controls (n = 3,500) were taken from the Rahman collection [18]. Additionally, 4,000 positive COVID-19 images were collected from the COVIDx-CXR4 repository [19, 20]. Finally, 15,000 images bearing at least one of seven radiological findings—infiltration, effusion, atelectasis, nodule, mass, pneumothorax, and consolidation—were sampled from the NIH ChestXray14 archive [21] (see Table 1).

Table 1. Data distribution, diseases and symptoms.

No	Source	Normal	TB	COVID	Symptoms
1	Montgomery [16]	0	58	0	—
2	Shenzhen [16]	0	336	0	—
3	Belarus [17]	0	1,047	0	—
4	Rahman et al. [18]	3,500	700	0	—
5	COVIDx-CXR [19]	0	0	4,000	—
6	NIH ChestXray [21]	—	—	—	15,000
Total		3,500	2,141	4,000	15,000
After quality control		3,163	1,904	2,358	12,024

Table 2 presents the distribution of symptoms within the dataset, along with their total counts and the percentage of images in which they appear. The total percentage exceeds 100% since the dataset allows for multi-label classification, meaning an image can exhibit more than one symptom simultaneously.

Table 2. Symptoms distribution.

Symptom	Total Count	% of Images
Infiltration	4,379	36.4%
Effusion	3,351	27.9%
Atelectasis	3,311	27.5%
Nodule	2,105	17.5%
Mass	1,881	15.6%
Pneumothorax	1,740	14.5%
Consolidation	1,465	12.2%
Total	18,232	151.6%*

* Sum exceeds 100% due to the multi-label nature.

Table 3 provides information on the distribution of images based on the number of symptoms they exhibit, with a total of 12,024 images. As seen, the major

ity of images (62.6%) contain a single symptom, while fewer images contain multiple symptoms, ranging from two symptoms (26.3%) to seven symptoms (0.01%).

Table 3. Multi-symptoms distribution.

	Count	Percentage
Single symptom	7,530	62.6%
Two symptoms	3,157	26.3%
Three symptoms	1,030	8.6%
Four symptoms	251	2.1%
Five symptoms	43	0.4%
Six symptoms	12	0.1%
Seven symptoms	1	0.01%
Total	12,024	100.0%

3 Methods

Figure 1 presents a high-level conceptual flow of the data analysis process, beginning with data collection and integration, followed by segmentation, and concluding with the model training phase. The components depicted in the figure are described in detail in this section, highlighting the methodologies and techniques employed at each stage.

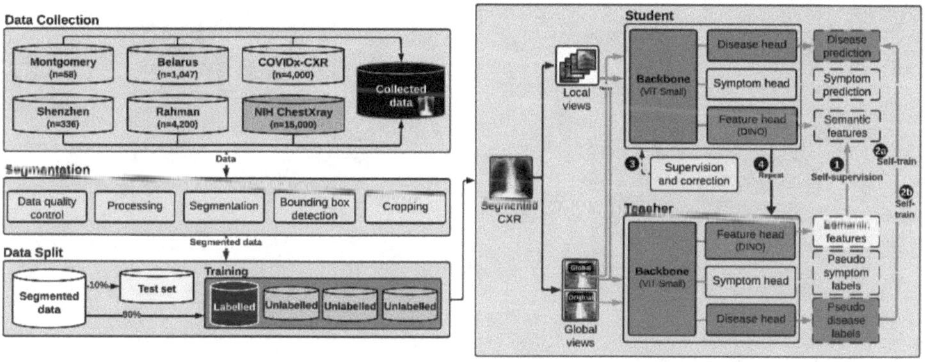

Fig. 1. High-level conceptual flow of the analysis.

3.1 Lung Segmentation

We employed U-Net [22], a symmetric encoder–decoder convolutional network with skip connections that preserves fine edge details while maintaining global context, making it the standard choice for biomedical image segmentation. At the start of the process, we loaded a pretrained checkpoint [23] and kept it in evaluation mode throughout the pipeline. Each radiograph is converted to an 8-bit greyscale format and resized to 225 $\times$ 225 pixels before being processed by the network; the output per-pixel logits are then transformed into a binary mask using arg-max and nearest-neighbour upsampling.

We conducted rigorous quality control to ensure the integrity of our segmentation masks. Masks are retained only if (i) lung pixels occupy between 8% and 90%of the image, and (ii) there are at least two external contours present. Accepted masks define the tightest bounding box around the two largest contours. The corresponding region was cropped from the original CXR to focus subsequent analysis on parenchymal tissue while minimizing background artifacts. Any image failing these automatic checks or showing residual issues during subsequent manual review (such as embedded lead markers) was discarded. This procedure resulted in the elimination of 337 normal, 237 tuberculosis, 1,642 COVID-19, and 2,976 symptom-labelled images, leaving a curated dataset of 1,904 tuberculosis, 3,163 normal, 2,358 COVID-19, and 12,024 symptom CXRs, totalling 19,449 images. Figure 2 presents sample images from the dataset, showcasing both original and cropped versions, demonstrating how the segmentation and cropping process isolates relevant regions while reducing background noise.

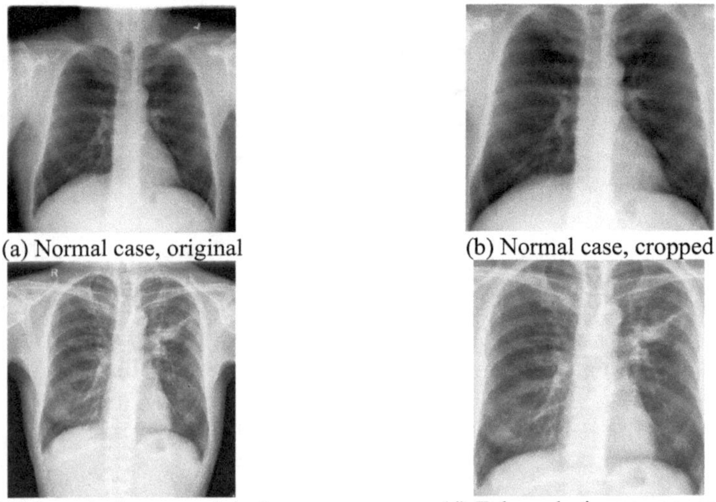

(a) Normal case, original (b) Normal case, cropped

(c) Tuberculosis case, original (d) Tuberculosis case, cropped

Fig. 2. Sample images in the dataset.

3.2 Data Split

The curated dataset was split on a patient-wise basis: 10% was reserved as an independent unseen test set, and the remaining 90% constituted the training pool. Within this training pool, 30% was treated as labelled data, and the rest was evenly divided into three unlabeled folds for self-supervised distillation.

3.3 Network Architecture

We adopt the DISTL framework [12], configuring a teacher-student network architecture using a ViT-Small backbone for both teacher and student models. Images are transformed into 384-dimensional tokens via a strided convolution with a patch size of 8, and processed through 12 transformer blocks, featuring 6 heads, LayerNorm ($\varepsilon = 10^{-6}$), and a stochastic depth of 0.1 applied solely to the student network. Each backbone operates within a multi-crop pipeline, where the student receives two global crops (the original and a 256-pixel crop with a scale range of (0.75, 1.0)) along with multiple local crops (defaulting to 8 at 128 pixels with a scale range of (0.2, 0.6)), while the teacher processes only the two global crops.

Three distinct heads are attached to the backbone: (1) a DINO [13] projection head, consisting of a 3-layer multilayer perception (MLP) with GELU activation, featuring a *hidden_dim* of 2,048, *bottleneck_dim* of 256, and a weight-normalized output layer with *out_dim* of 65,536, which is used for the self-supervised learning objective; (2) a disease classification head, a 3-layer MLP with ReLU activation as: 384 → 256 → 256 → 3, designed to produce logits for a three-class disease classifier; and (3) a parallel symptom classification head, another 3-layer MLP with ReLU activation configured as 384 → 256 → 256 → 7, which outputs logits for a multi-label set of radiological symptoms, i.e., infiltration, effusion, atelectasis, nodule, mass, pneumothorax, and consolidation. Both networks are initialized with ViT-Small weights pre-trained on the CheXpert dataset [14].

3.4 Training Strategy

Our training strategy integrates self-supervision and self-training techniques. The DINO framework is used to align student and teacher features across different views, while knowledge distillation is employed to transfer the teacher's predictions to the student. This is achieved using temperature-scaled KL divergence (with a temperature $\tau = 2.0$) for the disease classification head, and focal binary cross-entropy (BCE) loss (with a gamma value of $\gamma = 2.0$) applied to sigmoid-activated teacher logits for the symptom classification head. During the initial training epochs ($<$ *ssl_epoch*), these losses are combined with a mixing coefficient λ (default set to 0.5), gradually transitioning to pure distillation without the DINO loss. The teacher model's parameters are updated using an exponential moving average (EMA) of the student's parameters, following a cosine momentum schedule ranging from 0.9995 to 1.0. Additionally, every 500 iterations, a labeled "correction" phase is introduced, applying cross-entropy loss to the disease head and focal BCE to the symptom head. This phase incorporates per-label *pos_weight* based on prevalence, clipped at 20.0, and employs a weighted sampler that increases the sampling frequency (by 3 times) of images containing at least one positive symptom. This comprehensive approach allows the model to effectively learn both disease states and radiological symptoms within a unified teacher--student pipeline.

3.5 Network Training Process

The network training process is structured to extract features and perform multi-disease classification and radiological symptom detection. We leverage the DINO head for feature extraction, the CLS head for multi-disease classification, and a symptom head, tailored to detect seven specific radiological findings.

Phase 1: Supervised Pretraining. In the initial phase, we focus on developing robust baseline representations using solely labelled data. A three-tiered data augmentation strategy is employed to enhance robustness under varied imaging conditions. The first global crop undergoes minimal transformation, primarily normalization using ImageNet [24] statistics. The second global crop receives moderate augmentation, including random resized cropping (256 × 256 pixels, scale 0.75–1.0), horizontal flipping ($p = 0.5$), rotation (±15°), auto-contrast, equalization, and Gaussian blur (each with $p = 0.3$). Local crops (8 crops of 128 × 128 pixels, scale 0.2–0.6) undergo stronger augmentation with higher probabilities ($p = 0.5$) to capture fine-grained pathological details crucial for symptom detection. The student minimizes a weighted combination of losses: cross-entropy with inverse-frequency class weights for disease classification and focal binary cross-entropy (Focal BCE, $\gamma = 2.0$) for symptom detection. The combined loss is formulated as:

$$\mathcal{L} = 0.25\mathcal{L}_{disease} + 0.75\mathcal{L}_{symptom} \tag{1}$$

prioritizing symptom learning. Symptom detection incorporates per-label positive weights derived from class prevalence, clipped at 50.0, with a 1.5 × additional boost for rare symptoms, such as pneumothorax and consolidation, capped at 75.0. Optimization is performed using AdamW with cosine annealing, starting from a learning rate of 5×10^{-5} to 10^{-6}, and a fixed weight decay of 0.01.

Phase 2: Semi-Supervised DISTL Training. In this phase, both teacher and student models are initialized from Phase 1. The DISTL framework is fully utilized, leveraging both labelled and unlabeled data through three successive training runs (folds), each incorporating progressively larger portions of unlabeled data. We use two primary loss mechanisms:

1. **DINOLoss:** Aligns feature representations between teacher and student across views. The teacher processes only global views, while the student processes both global and local views. The loss is computed using cross-entropy between softmax outputs after centring and sharpening with temperature scheduling.
2. **Knowledge distillation:** Transfers the knowledge from the teacher to the student via two paths: KL divergence with temperature scaling ($\tau = 2.0$) for disease classification, and Focal BCE for symptom detection, comparing student logits against sigmoid-activated teacher predictions. The distillation loss is given by:

$$\mathcal{L}_{distill} = \frac{1}{2}(\mathcal{L}_{KL} + \mathcal{L}_{symptom}) \tag{2}$$

During early epochs (*ssl_epoch*), the total loss is ($\lambda = 0.5$):

$$\mathcal{L} = \lambda \mathcal{L}_{DINO} + (1 - \lambda)\mathcal{L}_{distill} \quad (3)$$

after which the training transitions to pure distillation ($\mathcal{L} = \mathcal{L}_{distill}$). Student parameters are updated by backpropagation, while teacher parameters are updated using EMA with a cosine momentum schedule (0.9995 to 1.0). Across folds, the student is warm-started from its latest checkpoint, and the teacher from its EMA weights, stabilizing training as the proportion of unlabeled data increases.

Phase 3: Correction Phase. Every 500 iterations, a supervised correction phase is implemented to counteract pseudo-label noise and maintain model accuracy. During this phase, the student is trained with ground-truth labels using cross-entropy for disease classification and Focal BCE for symptom detection, with positive weights clipped at 20.0. To address class imbalance, a weighted sampler increases the selection probability of images containing at least one positive symptom by threefold.

3.6 Performance Evaluation

For performance evaluation, disease classification is assessed using precision, recall, F1-score, and accuracy metrics across three categories: COVID-19, Normal, and Tuberculosis. Symptom detection is evaluated in a multi-label context, employing per-symptom and average precision, recall, and F1-scores. We benchmark our model against four convolutional neural network (CNN) baselines–vanilla CNN, VGG16, ResNet-18, and ResNet-50–all of which are fine-tuned following the same multi-class classification protocol.

4 Results

In this section, we present results from the held-out test split for two key tasks: (1) three-class disease classification, and (2) multilabel detection of seven radiological symptoms. Metrics are calculated per class/label on the test set, unless otherwise specified, and we also provide macro-averages–unweighted means across classes/labels–to facilitate comparability.

4.1 Three-Class Disease Classification

Table 4 showcases a comparison between our teacher-student ViT model and four CNN baselines, all trained and evaluated on the same dataset, with the highest values highlighted in bold. Among the baselines, VGG16 emerges as the strongest, achieving an overall accuracy of 96.84% and a macro-F1 score of 96.86%. In contrast, our model significantly outperforms these benchmarks with an accuracy of 98.85% and a macro-F1 score of 98.89%, marking improvements of +2.01 and +2.03 percentage points over the best baseline, respectively. When examining class-wise performance, our model delivers F1 scores of 98.94% for normal cases, 99.44% for tuberculosis, and 98.30% for COVID-19, surpassing the best baseline in each class by +2.08, +2.26, and + 1.75 percentage points, respectively. Additionally, macro precision and recall metrics are enhanced, with our model achieving 98.94% and 98.85%, compared to the best baseline's 96.99% and 96.75%. In terms of recall, our model demonstrates marked improvements across all three classes: Normal sees an increase of +1.05 (98.94% vs. 97.89%), TB rises by +0.57 (98.88% vs. 98.31%), and COVID-19 improves by + 1.28 (98.72% vs. 97.44%). We attribute these improvements to: (1) a lung-focused preprocessing approach that reduces non-parenchymal artifacts, and (2) the integration of multi-crop self-supervision and teacher-student distillation, which enhances the stability of representation learning from limited labels.

Table 4. Model performance comparison on 3-class disease classification.

Model	Class	Precision	Recall	F1	Accuracy
Vanilla CNN	Normal	73.41%	93.31%	82.17%	80.60%
	TB	91.84%	75.84%	83.08%	
	COVID	85.64%	68.80%	76.30%	
VGG16	Normal	95.86%	97.89%	96.86%	96.84%
	TB	97.73%	96.63%	97.18%	
	COVID	97.39%	95.73%	96.55%	
ResNet18	Normal	95.02%	94.01%	94.51%	92.10%
	TB	96.75%	83.71%	89.76%	
	COVID	86.21%	96.15%	90.91%	
ResNet50	Normal	98.86%	91.55%	95.06%	95.26%
	TB	90.67%	98.31%	94.34%	
	COVID	95.00%	97.44%	96.20%	
Our model	Normal	**98.94%**	**98.94%**	**96.94%**	**98.85%**
	TB	**100.00%**	**98.88%**	**99.44%**	
	COVID	**97.88%**	**98.72%**	**98.30%**	

4.2 Symptom-Wise Multilabel Detection

Table 5 presents a detailed summary of per-symptom precision, recall, and F1 scores for the seven radiological findings, with the highest values highlighted in bold. Our model demonstrates significant advancements in multi-label detection, achieving a macro-F1 score of 90.09%, along with substantial improvements in macro precision (86.90%) and macro recall (93.60%), registering absolute gains of +31.76, +30.50, and +28.13 percentage points over the best-performing baseline, respectively. Notably, improvements are consistent across all symptoms, with the most significant F1 score gains seen in detecting smaller or more subtle anomalies: nodule (+40.58; 87.05% vs. 46.47%), effusion (+37.17; 89.02% vs. 51.85%), mass (+35.37; 85.08% vs. 49.71%), consolidation (+28.98; 84.49% vs. 55.51%), infiltration (+27.03; 93.36% vs. 66.33%), and atelectasis (+25.49; 94.75% vs. 69.26%). Pneumothorax also improves strongly (+14.75; 96.91% vs. 82.16%). Recall gains are particularly marked for nodule (+30.77), effusion (+23.81), mass (+25.56), and infiltration (+14.89). These consistent improvements highlight the efficacy of the proposed approach, underscoring the effectiveness of incorporating multi-crop local views for detailed pathology detection, class-balanced sampling with positive re-weighting for symptom classification, and the DISTL-style self-training approach to leverage both labelled and unlabeled data.

Table 5. Model performance comparison on multilabel symptom detection.

Model	Symptom	Precision	Recall	F1
Vanilla CNN	Pneumothorax	67.07%	57.89%	62.15%
	Consolidation	42.20%	81.11%	55.51%
	Nodule	30.00%	46.15%	36.36%
	Mass	14.36%	58.89%	23.09%
	Atelectasis	60.58%	80.84%	69.26%
	Effusion	26.67%	28.57%	27.59%
	Infiltration	44.13%	71.76%	54.65%
VGG16	Pneumothorax	84.44%	80.00%	82.16%
	Consolidation	58.82%	44.44%	50.63%
	Nodule	36.23%	54.95%	43.67%
	Mass	51.81%	47.78%	49.71%
	Atelectasis	64.07%	74.68%	68.97%
	Effusion	46.67%	58.33%	51.85%
	Infiltration	52.75%	73.28%	61.34%
ResNet18	Pneumothorax	46.15%	50.53%	48.24%
	Consolidation	45.83%	61.11%	52.38%
	Nodule	23.03%	41.76%	29.69%
	Mass	53.25%	45.56%	49.10%

(*continued*)

Table 5. (*continued*)

Model	Symptom	Precision	Recall	F1
	Atelectasis	52.24%	83.12%	64.16%
	Effusion	27.80%	67.86%	39.45%
	Infiltration	59.57%	74.81%	66.33%
ResNet50	Pneumothorax	51.92%	56.84%	54.27%
	Consolidation	40.82%	44.44%	42.55%
	Nodule	37.33%	61.54%	46.47%
	Mass	27.84%	60.00%	38.03%
	Atelectasis	54.31%	87.99%	67.16%
	Effusion	29.44%	63.10%	40.15%
	Infiltration	43.59%	84.35%	57.48%
Our model	Pneumothorax	**94.95%**	**98.95%**	**96.91%**
	Consolidation	**81.44%**	**87.78%**	**84.49%**
	Nodule	**82.35%**	**92.31%**	**87.05%**
	Mass	**84.62%**	**85.56%**	**85.08%**
	Atelectasis	**90.29%**	**99.68%**	**94.75%**
	Effusion	**86.52%**	**91.67%**	**89.02%**
	Infiltration	**88.14%**	**99.24%**	**93.36%**

4.3 Symptom Classification Explainability

To evaluate whether the model accurately identifies actual patterns in CXR images, we generated Grad-CAM [15] maps from the final transformer block for each positive symptom logit and overlaid them on the corresponding CXR images (Fig. 3). In each image, the green rectangle represents the dataset-provided ground-truth bounding box for that symptom. We qualitatively assess the alignment of model saliency with these annotations, examining whether the highlighted areas correspond to the expected locations and appearance of each finding on chest radiographs.

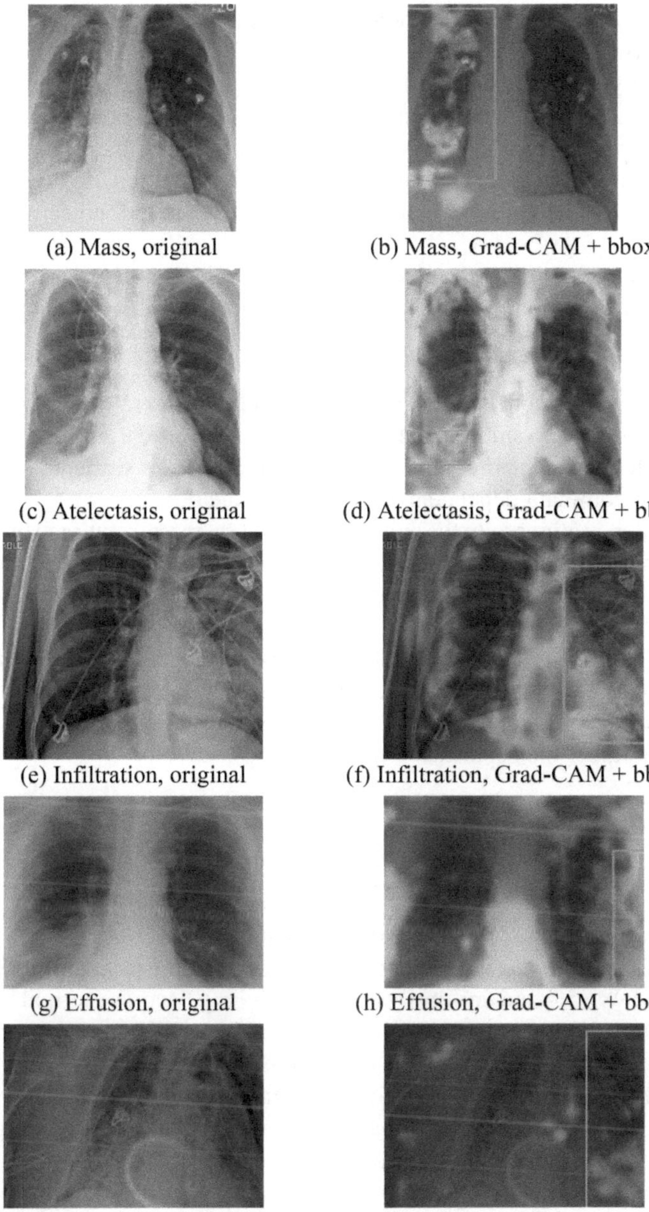

(a) Mass, original (b) Mass, Grad-CAM + bbox

(c) Atelectasis, original (d) Atelectasis, Grad-CAM + bbox

(e) Infiltration, original (f) Infiltration, Grad-CAM + bbox

(g) Effusion, original (h) Effusion, Grad-CAM + bbox

(i) Pneumothorax, original (j) Pneumothorax, Grad-CAM + bbox

Fig. 3. Symptom-level explainability examples. Each row represents a different symptom. The left image depicts the original chest X-ray, while the right image shows the Grad-CAM overlay highlighting areas of model focus. The green rectangle indicates the dataset-provided ground-truth bounding box for each symptom.

Figure 3b. The heatmap reveals a compact, high-intensity focus within the annotated lung region, tapering off sharply at the edges. This pattern is consistent with a discrete, rounded pulmonary opacity, characteristic of a mass, rather than diffuse parenchymal change. Saliency is primarily contained within the ground-truth bounding box.

Figure 3d. The strongest saliency is observed at the lung base on the annotated side near the diaphragm, with activation overlapping the bounding box. The elongated, band-like basal pattern is typical of plate (discoid) atelectasis and indicative of regional volume loss.

Figure 3f. The activation appears as a mottled, patchy distribution throughout the annotated lateral lung region, lacking a single round hotspot. This distribution could be characteristic of air-space disease, where alveolar filling leads to patchy opacities instead of a solitary nodule.

Figure 3h. Saliency is concentrated along the dependent lateral pleural region within the annotation, displaying a smooth, meniscus-like slope toward the costophrenic angle, which aligns with the expected contour of pleural fluid. Minor scattered signals above do not disrupt the dominant dependent pattern within the box.

Figure 3j. Saliency aligns with the apico-lateral pleural margin outlined by the narrow vertical annotation, while the adjacent, more lucent lung on the same side shows reduced activation. This corresponds to the expected location of the visceral pleural line on a pneumothorax film.

As shown in these five random examples, the model's attention concentrates on clinically meaningful regions that largely coincide with the dataset's boxes. This qualitative concordance indicates that the symptom classification head leverages radiographically relevant evidence rather than spurious artifacts.

5 Discussion and Conclusion

This study introduces a DISTL-style teacher-student framework built on a ViT-Small backbone with two supervised heads for disease and symptom classification and a self-supervised DINO head. On the three-class disease classification task, the model achieves a high accuracy of 98.85% and uniformly high class-wise F1 scores (normal 98.94%, TB 99.44%, COVID-19 98.30%; see Table 4). For the multi-label symptom detection task, it excels all seven findings with a macro-F1 score of 90.09% (Table 5), outperforming four widely used CNN baselines. These advancements are attributed to three key design elements: (1) the joint optimization of disease and symptom objectives, which regularizes the backbone toward radiographic primitives; (2) the DISTL schedule that mixes self-supervision with distillation before transitioning to pure distillation, improving sample efficiency under limited labels; and (3) multi-crop training that exposes the network to both global lung context and fine-grained local views.

The symptom head proves particularly effective for detecting subtle or small targets, significantly boosting F1 scores for nodule (+40.6), effusion (+37.2), mass (+35.4), consolidation (+29.0), infiltration (+27.0), and atelectasis (+25.5), while also improving pneumothorax (+14.8) compared to the best CNN baseline. These improvements suggest that features learned for symptom recognition enhance disease classification, providing superior discrimination for TB/COVID than using disease labels alone. This supports workflows where disease triage is complemented by machine-generated radiographic evidence (e.g., "infiltration present" or "effusion present"), aiding in prioritizing readings and documenting findings.

Qualitative explainability analyses indicate that predictions are rooted in anatomically relevant areas. Grad-CAM maps, when compared to dataset bounding boxes (Fig. 3), concentrate over focal intraparenchymal regions for "mass", basal band-like opacities for "atelectasis", patchy parenchymal changes for "infiltration", dependent lateral pleura for "effusion", and the apico-lateral pleural margin for "pneumothorax". Although some images exhibit minor off-box saliency (e.g., small superior hotspots), the predominant activation typically lies within or near the annotated region, supporting symptom-specific reasoning instead of non-diagnostic artifacts.

To conclude, the proposed single teacher-student ViT with joint disease and symptom supervision achieves state-of-the-art performance in TB/COVID/normal classification while simultaneously detecting diverse radiographic findings. Beyond aggregate metrics, the alignment between attention maps and annotated regions indicates that the model utilizes radiographically meaningful evidence. These properties render the approach a promising candidate for real-world screening and triage pipelines, where both reliable sensitivity for tuberculosis and actionable symptom readouts are crucial.

6 Limitations and Future Work

This study has several limitations that warrant attention. First, the evaluation is retrospective and draws from multiple sources, which, despite patient-wise splits, may lead to distribution shifts due to label noise and inter-dataset heterogeneity in terms of projection, positioning, equipment, and acquisition protocols. Second, the symptom supervision relies on image-level labels and dataset-provided boxes that may be coarse, potentially resulting in partial localization without guaranteeing pixel-accurate delineation. Third, our analysis focuses on frontal CXRs, with no exploration of lateral views or computed tomography (CT) scans. Lastly, while the EMA teacher and pseudo-labelling help stabilize training, the model's performance might be sensitive to certain thresholds and the relative weighting between disease and symptom losses.

To address these limitations, future work will focus on several key areas: (1) conducting prospective, multi-site validation with site-specific calibration to assess the model's generalizability and robustness across different clinical settings; (2) developing explicit localization heads trained with boxes or weak masks to complement Grad-CAM visualizations; (3) implementing an adaptive curriculum for unlabeled data that utilizes confidence or uncertainty-aware sampling to improve model training efficiency; and (4) integrating clinical metadata, such as age, symptoms, and prior imaging, to improve model calibration and robustness.

Acknowledgments. This study was funded by the New Beginnings program of the National Research Council of Canada.

Disclosure of Interests. The authors have no competing interests to declare that are relevant to the content of this article.

References

1. Patel, N., Wong, A., Ebadi, A.: Empowering tuberculosis screening with explainable self-supervised deep neural networks. In: 2024 International Conference on Machine Learning and Applications (ICMLA), pp. 794–797 (2024). https://doi.org/10.1109/ICMLA61862.2024.00113
2. World Health Organization, "Global Tuberculosis Report (2023). https://www.who.int/teams/global-programme-on-tuberculosis-and-lung-health/tb-reports/global-tuberculosis-report-2023. Accessed 6 Jan 2026
3. World Health Organization. Global Tuberculosis Report 2024 (2024). https://www.who.int/teams/global-programme-on-tuberculosis-and-lung-health/tb-reports/global-tuberculosis-report-2024. Accessed 6 Jan 2026
4. United Nations. The Sustainable Development Goals Report 2022. https://unstats.un.org/sdgs/report/2022/. Accessed 6 Jan 2026
5. CDC. Clinical Overview of Tuberculosis, Tuberculosis (TB). https://www.cdc.gov/tb/hcp/clinical-overview/index.html. Accessed 6 Jan 2026
6. World Health Organization. The end TB strategy. https://www.who.int/publications/i/item/WHO-HTM-TB-2015.19. Accessed 6 Jan 2026
7. Herrera Diaz, M., Haworth-Brockman, M., Keynan, Y.: Review of evidence for using chest X-rays for active tuberculosis screening in long-term care in Canada. Front. Public Health **8** (2020). https://doi.org/10.3389/fpubh.2020.00016
8. Rajpurkar, P., et al.: CheXaid: deep learning assistance for physician diagnosis of tuberculosis using chest x-rays in patients with HIV. NPJ Digit. Med. **3**, 115 (2020). https://doi.org/10.1038/s41746-020-00322-2
9. World Health Organization. Use of computer-aided detection software for tuberculosis screening. World Health Organization (2025). https://www.who.int/publications/b/79103. Accessed 6 Jan 2026
10. World Health Organization. WHO consolidated guidelines on tuberculosis: module 2: screening: systematic screening for tuberculosis disease. https://www.who.int/publications/i/item/9789240022676. Accessed 6 Jan 2026
11. Wong, A., Lee, J.R.H., Rahmat-Khah, H., Sabri, A., Alaref, A., Liu, H.: TB-Net: a tailored, self-attention deep convolutional neural network design for detection of tuberculosis cases from chest X-ray images. Front. Artific. Intell. **5** (2022). https://www.frontiersin.org/articles/10.3389/frai.2022.827299. Accessed 9 Jan 2026
12. Park, S., et al.: Self-evolving vision transformer for chest X-ray diagnosis through knowledge distillation. Nat. Commun. **13**(1), 3848 (2022). https://doi.org/10.1038/s41467-022-31514-x
13. Caron, M., et al.: Emerging properties in self-supervised vision transformers. In: 2021 IEEE/CVF International Conference on Computer Vision (ICCV), pp. 9630–9640 (2021). https://doi.org/10.1109/ICCV48922.2021.00951
14. Irvin, J., et al.: CheXpert: a large chest radiograph dataset with uncertainty labels and expert comparison. In: Proceedings of the Thirty-Third AAAI Conference on Artificial Intelligence and Thirty-First Innovative Applications of Artificial Intelligence Conference and Ninth AAAI Symposium on Educational Advances in Artificial Intelligence, in AAAI'19/IAAI'19/EAAI'19, vol. 33, pp. 590–597. Honolulu, Hawaii, USA: AAAI Press (2019). https://doi.org/10.1609/aaai.v33i01.3301590
15. Selvaraju, R.R., Cogswell, M., Das, A., Vedantam, R., Parikh, D., Batra, D.: Grad-CAM: visual explanations from deep networks via gradient-based localization. In: 2017 IEEE International Conference on Computer Vision (ICCV), pp. 618–626 (2017). https://doi.org/10.1109/ICCV.2017.74

16. Jaeger, S., Candemir, S., Antani, S., Wáng, Y.-X.J., Lu, P.-X., Thoma, G.: Two public chest X-ray datasets for computer-aided screening of pulmonary diseases. Quant. Imaging Med. Surg. **4**(6), 475–477 (2014). https://doi.org/10.3978/j.issn.2223-4292.2014.11.20
17. TB Portals Program. Drug resistant tuberculosis X-rays. https://www.kaggle.com/datasets/raddar/drug-resistant-tuberculosis-xrays. Accessed 6 Jan 2026
18. Rahman, T., et al.: Reliable tuberculosis detection using chest X-ray with deep learning, segmentation and visualization. IEEE Access **8**, 191586–191601 (2020). https://doi.org/10.1109/ACCESS.2020.3031384
19. Wu, Y., Gunraj, H., Tai, C.A., Wong, A.: COVIDx CXR-4: an expanded multi-institutional open-source benchmark dataset for chest X-ray Image-based computer-aided COVID-19 diagnostics, arXiv: arXiv:2311.17677 (2023). https://doi.org/10.48550/arXiv.2311.17677
20. Wang, L., et al.: Actualmed COVID-19 chest X-ray dataset initiative. GitHub. https://github.com/agchung/Actualmed-COVID-chestxray-dataset. Accessed 6 Jan 2026
21. Wang, X., Peng, Y., Lu, L., Lu, Z., Bagheri, M., Summers, R.M.: ChestX-Ray8: hospital-scale chest x-ray database and benchmarks on weakly-supervised classification and localization of common thorax diseases, Presented at the 2017 IEEE Conference on Computer Vision and Pattern Recognition (CVPR), pp. 3462–3471. IEEE Computer Society (2017). https://doi.org/10.1109/CVPR.2017.369
22. Ronneberger, O., Fischer, P., Brox, T.: U-Net: convolutional networks for biomedical image segmentation. Presented at the Medical Image Computing and Computer-Assisted Intervention (MICCAI) (2015). https://lmb.informatik.uni-freiburg.de/Publications/2015/RFB15a/. Accessed 16 Feb 2021
23. Ovcharenko, I.: Lung segmentation for chest X-Ray images (2019). Jupyter Notebook. https://github.com/IlliaOvcharenko/lung-segmentation. Accessed 6 Jan 2026
24. Deng, J., Dong, W., Socher, R., Li, L.-J., Li, K., Fei-Fei, L.: ImageNet: a large-scale hierarchical image database. In: 2009 IEEE Conference on Computer Vision and Pattern Recognition, pp. 248–255 (2009). https://doi.org/10.1109/CVPR.2009.5206848

Evaluating Tree-Oriented Machine Learning Models in Cardiovascular Disease Diagnosis

Sandhya Dharshini Sasikumar[1](✉) and Siva Sivatha Sindhu[2]

[1] West Windsor-Plainsboro High School South, Princeton Junction, NJ 08550, USA
sandhya.dh0520@gmail.com

[2] Nomura Securities International, New York, NY 10019, USA

Abstract. Cardiovascular diseases (CVDs) are the leading cause of death globally and the World Health Organization (WHO) reports that these diseases claim approximately 17.9 million lives each year. This highlights the urgent need for effective CVD prevention and treatment strategies. The purpose of this study is to determine if tree based machine learning algorithm can be used for predicting cardiovascular disease. The study also identifies the key attributes associated with CVD by analyzing the average impurity decrease. Furthermore, our research seeks to hypothetically determine which tree-based machine learning algorithm provides the most accurate predictions for CVD diagnosis. Based on our preliminary analysis utilizing a heart disease dataset from Kaggle, which incorporates various key clinical factors, our results indicate that the random forest model demonstrates superior predictive performance compared to other tree-based machine learning models. Additionally, it's evident from the experiment that ST slope, exercise angina and chest pain types can be used reliably in predicting CVD diagnosis. The model's performance was evaluated using 10-fold cross-validation to ensure its generalizability to unseen CVD data patterns. The promising results prove that random forest based machine learning approach trained with key attributes like ST slope, exercise angina and chest pain type can be used in diagnosis of CVD, thereby preventing premature deaths.

Keywords: Cardio Vascular Disease · Artificial Intelligence · Health Care

1 Introduction

The heart is one of the most important organs in our body. Given its significance, it's fitting that heart diseases are the leading cause of death globally, accounting for 30% of deaths worldwide [1]. Despite its importance, heart disease remains a major health concern. Currently, one way to diagnose heart disease includes looking for common symptoms such as chest pain, shortness of breath, swelling, fatigue, and nausea. Other diagnostic tests used by healthcare workers include physical examinations, lab tests, electrocardiograms (ECG), echocardiograms, stress tests, electron beam computed tomography, coronary angiography, and cardiac catheterization. Approximately 55% of heart disease patients pass away within the first three years, and the treatment costs for heart disease account for around 4% of annual healthcare costs.

H. R. Arabnia et al. (Eds.): CSCI 2025, CCIS 2971, pp. 96–104, 2027.
https://doi.org/10.1007/978-3-032-28463-1_7

AI has been rapidly advancing globally, and Machine Learning (ML) and AI have become topics of interest in many fields and in the healthcare industry, including cardiology [2]. ML algorithms can analyze vast amounts of cardiovascular disease (CVD) medical data to identify patterns and predict the onset of diseases. This paper examines various tree-based machine learning approaches, including Decision Stump, Hoeffding Tree, J48, LMT, REPTree, Random Tree, and Random Forests [3]. Tree-based models are proposed due to their several advantages in predicting cardiovascular diseases (CVD). These models are easy to interpret and visualize, making it simpler for healthcare professionals to understand the decision-making process and trust the model's predictions (Interpretability). They can capture complex, non-linear relationships between features, which is often the case with medical data (Handling Non-linear Relationships). Tree-based models can provide insights into the importance of different features, helping to identify which health factors are most influential in predicting CVD (Feature Importance). They are generally robust to outliers and missing values, making them suitable for real-world medical datasets that may have such issues (Robustness to Outliers). Additionally, these models can handle large datasets efficiently, which is beneficial when working with extensive medical records and health data (Scalability). Therefore, in this work, CVD diagnosis is analyzed using tree-based machine learning approaches. Based on extensive experimental evaluation and 10-fold cross-validation, we propose that random forest is the best approach among all other tree-based models. Random forest provide promising results as they combine multiple trees to improve predictive performance and reduce over-fitting, leading to more accurate and reliable predictions compared to other tree-based models. Also, we suggest ST slope, exercise angina and chest pain types are responsible for highest accuracy in predicting CVD.

2 Proposed Methodology and Related Work

The goal of this proposed work (i) Identifying if tree based machine learning algorithm can be used in diagnosis of CVD. (ii) Identifying key factor/attributes that predict CVD (iii) Identifying best tree based machine learning model among Decision Stump, Hoeffding Tree, J48, LMT, REPTree, Random Tree, and Random Forests which predicts CVD with higher accuracy. We used Weka models written in Java and conducted our experiment. The heart dataset from Kaggle with 1190 instances having both normal and disease pattern is fed into the tree based machine learning models. In order to identify crucial factors we used attribute importance based on average impurity decrease.

Our performance comparison on different tree based machine learning model show that Random forest is efficient in diagnosing heart disease. Our random forest model follows an ensemble approach combining multiple decision trees in detecting the heart disease. Each tree in the random forest is built using random subset from the whole dataset and with a random subset of features in order to introduce diversity among the trees, so that the model is robust and avoid over fitting. Our random forest uses a technique called bagging in order to achieve this. In the training, each tree is constructed by recursively partitioning the data based on the feature and optimal feature is selected based on Information gain. This process is repeated till there are minimum numbers of samples in each leaf node. We used this trained model to make CVD predictions using the vote from each tree for predicting a patient having CVD or not. A research work published in National Library of Medicine which conducted a similar study of CVD detection using random forest but they used different medical data with fewer instances (498) to train and test the model [4]. Support vector machine and k-nearest neighbor algorithms of machine learning are also used to predict and classify the patient with heart disease with satisfactory results [5]. Our experimental study includes data from 1190 patients building a larger patient population which leads to more varied and precise data, which in turn improved the overall performance and better detection accuracy.

In order to perform a more fair comparison, we also compared with other tree models like random tree, REP etc., using the same data set, out of all models random forest out performed. The results are provided in Table 4. There are some limitations in our study. These limitations include data size (1190) and feature size (12). Our future study includes large patient's data set with more number of features in identifying heart disease and also using enhanced AI [6]. Thus our model can significantly enhance the CVD diagnostic process by automating classification using CVD patient data. Its high accuracy ensures effectiveness, enabling early detection and reducing diagnosis time. By eliminating the need for extensive physicians input, the model is particularly impactful in regions with delayed diagnostic services. Prompt and accurate diagnosis facilitated by the model could contribute to improved CVD survival rates globally.

3 Materials and Methods

We used the most comprehensive dataset from Kaggle, a data science platform [7]. The dataset used in this study was formed by combining different heart disease datasets already available independently but not combined before. The five datasets combined are: Cleveland (303 instances), Hungarian (294 instances), Switzerland (123 instances), Long Beach VA (200 instances), Stalog Heart Data Set (270 instances) with total of 1190 instances. Also, the dataset created is combined over 11 common features which make it the largest heart disease dataset available for research purposes. We loaded the combined preprocessed heart disease dataset (heart_statlog_clevland_hungay.csv) in.csv format with target class as discrete value into WEKA tree models which is written in Java [8]. WEKA provides set of hyper parameters that allow us to fine tune our model. As selecting the appropriate hyper parameter is vital in optimizing the performance of the model. The first crucial parameter we considered is the no. of trees in the forest. Generally more trees can improve performance but with increased computational time. So we started with number of trees with 50 and increased till 150 by tens. We noticed that our model performed well for 100 trees.

Next, parameter we considered for tuning is the depth of the tree, as deeper trees can learn complex patterns but sometimes may result in over fitting. So we tried different experiments with value between 5 and 20 and obtained good results between 12 and 15 as the depth. The number of feature that is considered for each split is decided using the formula $\log_2(N) + 1$, if < 1, where N is the total number of features and split quality is measured using information gain. We also analyzed the feature importance which refers to a measure of how much each individual attribute in the dataset contributes to the prediction made by our model, essentially telling which features have the most significant impact on the final outcome; Among all attributes in multiple experiments conducted "ST slope" with 12.53 had higher attribute importance value indicating a greater influence on the prediction and next attribute which has higher attribute importance ranking is "exercise angina" with 9.81. The size of the tree is 275 with max depth of 15 for the results shown in Tables 1, 2, 3, 4, 5 and 6. We used 10-fold cross-validation in order to get an evaluation result and estimate of the error. In 10-fold cross validation the dataset is divided into 10 equal parts (called "folds"), then the machine learning tree model is trained on 9 of those parts while testing it on the remaining one, repeating this process 10 times with each fold being used as the test set once, and finally averaging the results to get a more robust evaluation of our model's performance. Ideally the tree model's generalization ability is assessed by exposing it to different subsets of the data for testing. Portion of evolved tree during the experiment conducted is shown below.

```
max heart rate < 134.5
|  oldpeak < 0.65
|  |  ST slope < 1.5
|  |  |  cholesterol < 79.5
|  |  |  |  max heart rate < 113 : 1 (4/0)
|  |  |  |  max heart rate >= 113
|  |  |  |  |  resting bp s < 156
|  |  |  |  |  |  oldpeak < 0.35 : 0 (7/0)
|  |  |  |  |  |  oldpeak >= 0.35 : 1 (1/0)
|  |  |  |  |  resting bp s >= 156 : 1 (1/0)
|  |  |  cholesterol >= 79.5
|  |  |  |  chest pain type < 1.5 : 1 (3/0)
|  |  |  |  chest pain type >= 1.5
|  |  |  |  |  exercise angina < 0.5 : 0 (60/0)
|  |  |  |  |  exercise angina >= 0.5
|  |  |  |  |  |  cholesterol < 219 : 1 (4/0)
|  |  |  |  |  |  cholesterol >= 219 : 0 (9/0)
|  |  ST slope >= 1.5
|  |  |  oldpeak < 0.05
|  |  |  |  age < 41.5
|  |  |  |  |  chest pain type < 3 : 0 (5/0)
|  |  |  |  |  chest pain type >= 3 : 1 (2/0)
|  |  |  |  age >= 41.5
|  |  |  |  |  chest pain type < 3.5
|  |  |  |  |  |  resting ecg < 0.5
|  |  |  |  |  |  |  max heart rate < 95.5 : 1 (4/0)
|  |  |  |  |  |  |  max heart rate >= 95.5
|  |  |  |  |  |  |  |  max heart rate < 97.5 : 0 (1/0)
|  |  |  |  |  |  |  |  max heart rate >= 97.5 : 1 (2/0)
|  |  |  |  |  |  resting ecg >= 0.5 : 1 (12/0)
|  |  |  |  |  chest pain type >= 3.5 : 1 (55/0)
```

4 Results and Discussion

The experimented tree based models prediction diagnosis varied from 78% to 94% proving hypothetically that tree based models can be used in predicting CVD. Also, among all the tree based models Random Forest performed well with an accuracy of 94% provided promising results than other model proving it's a better model to be used for CVD diagnosis (Fig. 1).

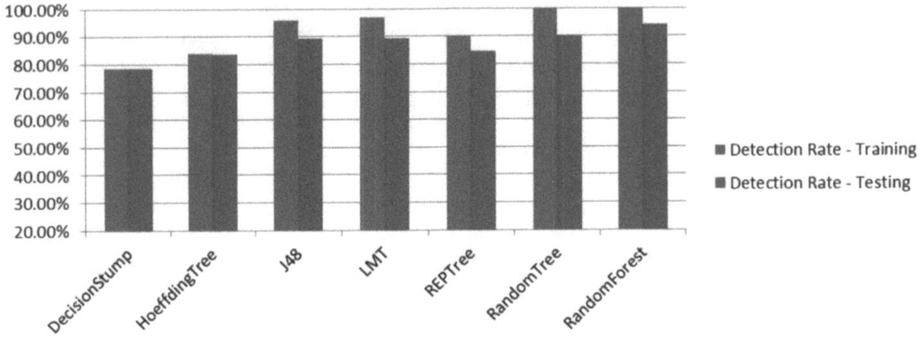

Fig. 1. Detection Rate comparison of random forest with other machine learning tree based models showing higher detection rate of 94% in testing and 100% in training.

Although random forest and random tree both performed well in training with the accuracy of 100%, random tree is only able to predict with the accuracy of 89% during the testing phase. Additionally, we are able to prove that ST slope (rate of change in the ST segment of an electrocardiogram), exercise angina (chest pain that occurs when the heart muscle is strained during exercise) and chest pain types (risk level of chest pain) are the life-threatening factors responsible for prediction of CVD disease (Fig. 2). Among all features ST slope ranked top with the value of 12.53, exercise angina with 9.81 and chest pain type with 4.64. Therefore, medical professionals, like doctors, should take these factors into account when evaluating patients. The results obtained are shown in Fig. 2

Various performance metrics are used to evaluate the performance of the different tree based models. Accuracy (correctly classified disease) is good indices of performance, as they measure what percentage the model is able to detect CVD and how many misdiagnoses made in the course of action. Other metrics include Kappa statistics which is used in assessing the degree to which two or more raters, examining the same data, agree when it comes to assigning the data to categories. Mean Absolute Error (MAE) is the average absolute differences between the predicted value and the corresponding observation or actual value. Root Mean Squared Error (RMSE) is quadratic scoring rule which measures the average magnitude of the error by measuring the difference between forecast value and corresponding observed values are each squared and then averaged over the sample.

Finally, the square root of the average is taken. Relative Absolute Error (RAE) is similar to the relative squared error in the sense that it is also relative to a simple predictor,

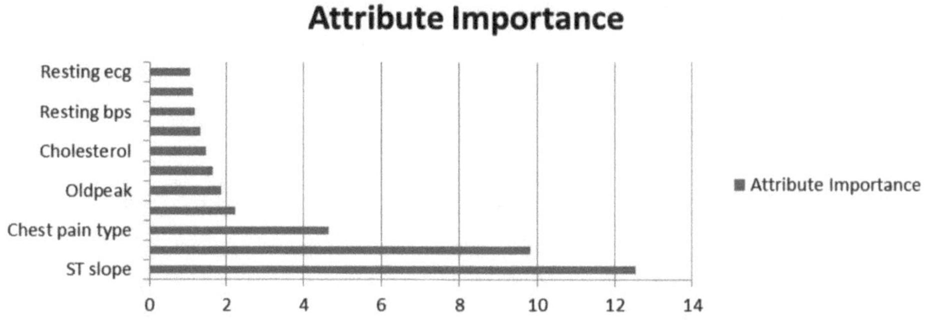

Fig. 2. Factors responsible for higher accuracy based on attribute importance

Table 1. Training dataset evaluation of tree models in detection performance

Machine Learning	Correctly classified	Incorrectly classified	Kappa statistic
DecisionStump	78.49%	21.51%	0.5669
HoeffdingTree	84.03%	15.97%	0.682
J48	96.05%	3.95%	0.9206
LMT	96.97%	3.03%	0.9392
REPTree	90.08%	9.92%	0.8012
RandomTree	100%	0%	1
RandomForest	100%	0%	1

which is just the average of the actual values. Relative Squared Error (RSE) calculates the total squared error and normalizes it by dividing the total squared error of the simple predictor. The results of these metrics for various models are provided in Tables 1, 2, 3, 4, 5 and 6.

The comparative evaluation of tree-based classifiers reveals distinct performance profiles across training and testing datasets. RandomForest and RandomTree consistently achieved perfect classification accuracy on the training set (Table 1), with 100% correctly classified instances and a Kappa statistic of 1.0, indicating ideal agreement. However, such flawless performance may suggest overfitting, especially when contrasted with their slightly reduced accuracy during 10-fold cross-validation and testing phases (Tables 4 and 5). Among the interpretable models, LMT and J48 demonstrated robust performance, with over 96% accuracy on the training set and high Kappa statistics (0.9392 and 0.9206, respectively). Their generalization capability was further supported by strong cross-validation results, where J48 maintained an 89.33% accuracy and a Kappa of 0.7858, outperforming LMT in consistency.

Error metrics (Tables 2 and 5) further underscore RandomForest's efficiency, yielding the lowest mean absolute error (0.0529 training, 0.1487 testing) and root mean squared error (0.0863 training, 0.2357 testing). These values reflect minimal deviation between

Table 2. Performance evaluation with training dataset of random forest with six other tree classifiers based on True Positive (TP) Rate, False Positive Rate (FP) Rate, Precision, Recall, F-Measure, Matthews Correlation Coefficient (MCC), Receiver Operating Characteristic (ROC) Area and Precision Recall Curve (PRC) Area

Machine Learning	Mean absolute error	Root mean squared error	Relative absolute error	Root relative squared error
DecisionStump	0.3377	0.4109	67.75%	82.31%
HoeffdingTree	0.1709	0.3581	34.29%	71.73%
J48	0.0708	0.1882	14.21%	37.70%
LMT	0.0774	0.1776	15.54%	35.58%
RandomForest	0.0529	0.0863	10.62%	17.29%
RandomTree	0	0	0.00%	0.00%
REPTree	0.1639	0.2863	32.89%	57.35%

Table 3. Error rate evaluation with training dataset of random forest with six other classifiers

Machine Learning	TP Rate	FP Rate	Precision	Recall	F-Measure	MCC	ROC Area	PRC Area
DecisionStump	0.785	0.22	0.785	0.785	0.784	0.568	0.783	0.723
HoeffdingTree	0.84	0.153	0.847	0.84	0.84	0.688	0.918	0.916
J48	0.961	0.042	0.961	0.961	0.96	0.921	0.98	0.975
LMT	0.97	0.032	0.97	0.97	0.97	0.94	0.991	0.991
RandomForest	1	0	1	1	1	1	1	1
RandomTree	1	0	1	1	1	1	1	1
REPTree	0.901	0.099	0.901	0.901	0.901	0.801	0.942	0.929

predicted and actual outcomes, reinforcing its predictive reliability.In terms of precision-recall dynamics (Tables 3 and 6), RandomForest again led with perfect scores across all metrics during training and near-perfect scores during testing (e.g., TP Rate = 0.94, Precision = 0.941, ROC Area = 0.971). J48 and LMT followed closely, maintaining high F-measures and ROC areas above 0.91, indicating strong discriminative power.

Interestingly, DecisionStump consistently underperformed across all evaluations, with lower accuracy, higher error rates, and reduced ROC and PRC areas. This highlights the limitations of simplistic models in capturing complex patterns within the dataset.Overall, RandomForest emerges as the most effective classifier, balancing accuracy, error minimization, and generalization. However, models like J48 and LMT offer competitive performance with added interpretability, making them suitable alternatives in contexts where model transparency is essential.

Table 4. 10-fold cross validation evaluation of tree models in detection performance

Machine Learning	Correctly classified	Incorrectly classified	Kappa statistic
DecisionStump	78.49%	21.51%	0.5669
HoeffdingTree	83.70%	16.30%	0.6735
J48	89.33%	10.67%	0.7858
LMT	89.33%	10.67%	0.5878
REPTree	84.71%	15.29%	0.6936
RandomTree	89.83%	10.17%	0.7965
RandomForest	94.03%	5.97%	0.8801

Table 5. Error rate evaluation with testing dataset of random forest with six other classifiers

Machine Learning	Mean absolute error	Root mean squared error	Relative absolute error	Root relative squared error
DecisionStump	0.3379	0.4112	67.79%	82.37%
HoeffdingTree	0.1806	0.3704	36.24%	74.19%
J48	0.1365	0.3072	27.38%	61.54%
LMT	0.1369	0.3054	27.48%	61.17%
RandomForest	0.1487	0.2357	29.85%	47.21%
RandomTree	0.1017	0.3189	20.40%	63.88%
REPTree	0.2028	0.3477	40.70%	69.65%

Table 6. Performance evaluation with test dataset of random forest with six other tree classifiers based on True Positive (TP) Rate, False Positive Rate (FP) Rate, Precision, Recall, F-Measure, Matthews Correlation Coefficient (MCC), Receiver Operating Characteristic (ROC) Area and Precision Recall Curve (PRC) Area

Machine Learning	TP Rate	FP Rate	Precision	Recall	F-Measure	MCC	ROC Area	PRC Area
DecisionStump	0.785	0.22	0.785	0.785	0.784	0.568	0.764	0.721
HoeffdingTree	0.837	0.162	0.838	0.837	0.837	0.674	0.903	0.899
J48	0.893	0.108	0.893	0.893	0.893	0.786	0.918	0.895
LMT	0.893	0.108	0.893	0.893	0.893	0.786	0.931	0.92
RandomForest	0.94	0.062	0.941	0.94	0.94	0.881	0.971	0.967
RandomTree	0.898	0.1	0.899	0.898	0.898	0.797	0.899	0.859
REPTree	0.847	0.152	0.848	0.847	0.847	0.694	0.9	0.876

5 Conclusions

This work proposed a tree based machine learning approach more suitable for CVD analysis. The system also identifies the key metrics that helps medical professionals in diagnosis of CVD in patients. This also has a greater impact on reducing the computational complexity of the classifier. The experiments were conducted using a more comprehensive and reliable dataset from Kaggle, which integrates diverse heart disease data, resulting in a significantly improved diagnostic accuracy.

One limitation of this study is its exclusive reliance on a single Kaggle dataset. While this dataset provides a valuable benchmark, its relatively modest size and limited demographic diversity restrict the extent to which our findings can be generalized to broader, global populations. Future research would benefit from validation using larger and more representative datasets drawn from varied sources and regions, thereby increasing the robustness and wider applicability of the findings.

Acknowledgments. I would like to express my deepest gratitude to my beloved grandmother, whose journey with heart disease has been a source of both inspiration and learning throughout my research. Her courage in the face of her condition, combined with her willingness to share her experiences and symptoms, has profoundly deepened my understanding of the disease. She has not only helped me grasp the complexities of heart disease but has also inspired me to pursue this project with a genuine desire to make a meaningful contribution to the field. I would like to dedicate this work to her and to all those who continue to fight against heart disease.

References

1. Cardio Vascular Diseases. World Health Organization. https://www.who.int/health-topics/cardiovascular-diseases. Accessed Nov 2024
2. Sindhu, S.S.S., et al.: Decision tree based light weight intrusion detection using a wrapper approach. Springer Expert Systems with Applications, vol. 39, pp. 129–141 (2012). https://doi.org/10.1016/j.eswa.2011.06.013
3. Breiman, L.: Random forests. ACM Mach. Learn. **45**, 5–32 (2001). https://doi.org/10.1023/A:1010933404324
4. Su, X., et al.: Prediction for cardiovascular diseases based on laboratory data: An analysis of random forest model. J. Clin. Lab. Anal. **34** (2020). https://doi.org/10.1002/jcla.23421
5. Ghazal, T.M., et al.: Heart disease prediction using machine learning. In: IEEE International Conference on Business Analytics for Technology and Security (2023)
6. Yan, K., et al.: Machine learning for AI-enhanced healthcare and medical services: new development and promising solution. IEEE/ACM Trans. Comput. Biol. Bioinf. **18**, 850–851 (2021). https://doi.org/10.1109/TCBB.2021.3050935
7. Cardio Vascular Diseases Data Set. Kaggle. https://www.kaggle.com/datasets/sid321axn/heart-statlog-cleveland-hungary-final?resource=download. Accessed Nov 2024
8. Waikato environment for knowledge analysis. WEKA version 3.8.6. https://ml.cms.waikato.ac.nz/weka/ Accessed Nov 2024

Early Prediction of Aggressive Prostate Cancer Using Machine Learning and Electronic Health Records

Wanting Cui(✉) and Joseph Finkelstein

Arizona Telemedicine Program, University of Arizona, Tucson, AZ 85724, USA
wantingcui@arizona.edu

Abstract. The prognosis of prostate cancer depends heavily on stage, making early identification of aggressive disease especially important. While most cases are detected at a localized stage with over 99% five-year survival, about 9% present with metastatic disease, where survival falls to 37%. Early risk stratification is therefore critical for guiding management.

In this study, we investigated whether machine learning models applied to electronic health records (EHRs) could predict cancer aggressiveness up to six months before diagnosis. We analyzed prostate cancer patients diagnosed between 2010 and 2019, dividing them into aggressive and non-aggressive groups. Aggressiveness was defined by receipt of chemotherapy, hormone therapy, prostatectomy, or radiation therapy within nine months of diagnosis. Patients' prior diagnoses, procedures, laboratory results (including PSA), demographic characteristics, and Charlson comorbidity scores were aggregated as predictors. The final cohort included 764 non-aggressive and 437 aggressive prostate cancer patients.

Model performance was moderate, with XGBoost and Balanced Random Forest (BRF) achieving similar accuracies (~0.56–0.58) and balanced precision–recall metrics. Important predictors included comorbidity scores, PSA levels, obesity, hypercholesterolemia, and diseases of the male genital organs. These findings suggest that machine learning applied to EHRs may support early prediction of aggressive prostate cancer and inform timely clinical interventions.

Keywords: Machine Learning · Electronic Health Records · Aggressive Prostate Cancer

1 Introduction

Prostate cancer is one of the most diagnosed cancers among men in the United States. According to CDC, over 255,000 new cases was reported in 2022 and an American cancer society estimated 313,000 new cases will be expected in 2025 [1, 2]. Prostate cancer predominantly affects older men. The average age at diagnosis is around 67 years. About 60% of cases occur in men aged 65 or older, while the disease is rare in individuals under 40 [2].

Despite its high incidence, prostate cancer generally has a favorable prognosis. The 5-year relative survival rate across all stages is about 97%. Approximately 70% of prostate

H. R. Arabnia et al. (Eds.): CSCI 2025, CCIS 2971, pp. 105–114, 2027.
https://doi.org/10.1007/978-3-032-28463-1_8

cancer cases are detected at a localized stage. Localized and regional cases tend to have good outcomes, where more than 99% of patients survive at least five years. In contrast, about 9% of cases are diagnosed at a distant (metastatic) stage, and these cases the 5-year relative survival rate drops to 37% [1, 3, 4].

Risk stratification is essential to identifying aggressive prostate cancer and guiding management. Patients are classified into low-, intermediate-, or high-risk groups based on three key clinical measures: prostate-specific antigen (PSA) level, Gleason grade/Grade Group, and clinical tumor stage. Low-risk disease typically involves a PSA less than 10 ng/mL, a Gleason Grade Group 1, and a tumor confined to one part of the prostate. Intermediate risk includes features such as PSA levels between 10–20 ng/mL, Gleason Grade Groups 2–3, or larger tumors still confined to the prostate. High-risk disease is defined by PSA greater than 20 ng/mL, Gleason Grade Groups 4–5, or locally advanced tumors extending beyond the prostate. Additional tools such as multiparametric MRI and prostate cancer risk calculators can refine these categories [5, 6].

Treatment strategies for prostate cancer are closely associated with both risk category and stage at diagnosis. Men with low-risk disease are often managed with active surveillance, since their cancers are unlikely to progress quickly. In contrast, intermediate- and high-risk disease is treated more aggressively, due to the higher probability of progression or metastasis. Standard options include radical prostatectomy or radiation therapy, frequently combined with androgen deprivation therapy (ADT) in higher-risk cases [7, 8].

There are several studies that used machine learning (ML) for cancer research and care [9–11]. ML techniques, ranging from simple regression models to advanced neural networks, can process large amounts of data to find patterns that are not easily seen with traditional methods. These patterns may help identify patients at risk earlier, guide treatment choices, and improve the use of healthcare resources. ML can also combine different types of information, such as electronic health records, genetic data, and medical notes, which strengthens both the accuracy of diagnosis and the personalization of treatment in cancers.

In our previous studies, we developed a pipeline for early prediction of prostate cancer diagnoses using healthcare utilization pattern [12, 13]. In this study, we aimed to explore using healthcare utilization pattern to predict the aggressiveness of prostate cancer patients 6 month prior to their actual diagnoses.

2 Method

This study used data from the All of Us Research Program, a national project that collects health information from a diverse group of participants [14]. This study used Version 7 of the controlled tier.

Prostate cancer patients were identified through a cohort-building tool by selecting "Prostate Cancer" under conditions. The first diagnosis date was obtained by linking the condition occurrence and person tables. The study included patients diagnosed between 2010 and 2019, excluding those with missing age or gender data or with a prior cancer diagnosis before their first prostate cancer record.

The final cohort was divided into two groups: aggressive and non-aggressive prostate cancer. Aggressive cases were defined as those who received chemotherapy, hormone therapy, prostatectomy, or radiation therapy within nine months of diagnosis. Those without any of these treatments during that period were classified as non-aggressive.

Patients were classified as receiving treatment within nine months of their first prostate cancer diagnosis using multiple coding systems. RxNorm codes identified chemotherapy (e.g., cisplatin, docetaxel) and hormone therapy (e.g., leuprolide, enzalutamide). CPT codes were used to capture prostatectomy procedures and radiation therapy.

After defining the cohort, we extracted demographics and calculated the Charlson Comorbidity Index (CCI). We also examined prior conditions relevant to prostate cancer, including urinary system disorders, sexually transmitted infections, and diseases of male genital organs. Additional comorbidities of interest included hypertension, hypercholesterolemia, obesity, testicular hypofunction, and elevated white blood cell count. Finally, we identified prostate-related symptoms such as hematuria and urinary retention, and quantified total urinary symptoms using ICD-10 codes.

Procedure data were analyzed using CPT4 codes, which were grouped into 165 categories with the CCS-Services and Procedures tool [15]. Surgical procedures were flagged as narrow, broad, or neither. To improve classification, further groupings were added using American Academy of Professional Coders guidance (AAPC), capturing evaluation and management services as well as new patient visit lengths [16]. Laboratory data were organized into 80 subgroups and 5 parent categories using a LOINC dictionary. Test frequencies and abnormal results were recorded, and prostate-specific antigen (PSA) values were summarized by most recent level, highest level, and average change over time.

In a previous study, we developed a machine learning model to predict whether a patient would receive a prostate cancer diagnosis within the following six months. Building on that approach, the present study aims to predict whether a future prostate cancer diagnosis will be classified as aggressive or non-aggressive, also within a six-month window. The target variable was binary, indicating whether a patient was diagnosed with aggressive prostate cancer.

Three machine learning algorithms were used: XGBoost, balanced random forest, and a neural network. We examined two healthcare utilization windows by collecting all medical activities that occurred 7 to 18 months (1-year window) and 7 to 24 months (1.5-year window) prior to the diagnosis date. To be included in the analysis, patients were required to have at least three medical activities recorded within the specified window.

The dataset was randomly divided into a training set (70%) and a testing set (30%). During model training, hyperparameter tuning was conducted using 5-fold cross-validation. Model performance was evaluated using accuracy, precision, recall, and F1 score. The best-performing model for each algorithm was then applied to the testing set, and performance was again assessed using the same metrics. All analyses were conducted in Python 3.8 within an Anaconda Jupyter Notebook environment.

3 Result

There were 764 non-aggressive prostate cancer patients and 437 aggressive prostate cancer patients in the dataset. For aggressive cancer patients, 210 patients underwent prostatectomy, 186 received radiation, 159 received hormone therapy, and 3 received chemotherapy (Fig. 1). Prostatectomy was the most common treatment, given alone to 186 patients. Combinations of prostatectomy with radiation (88 patients) or hormone therapy (53 patients) were also frequent. Only a small number received chemotherapy, either alone or in combination.

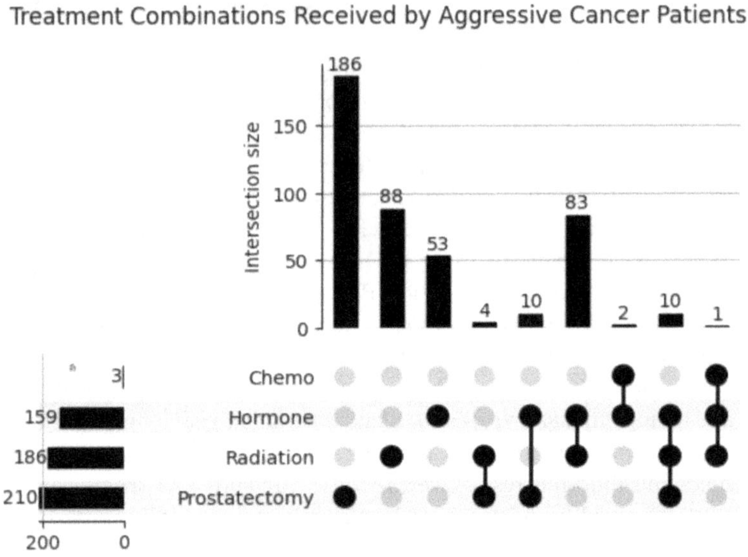

Fig. 1. Treatment combination received by aggressive cancer patients.

The age at diagnosis for non-aggressive patients was 67.13 ± 8.08 years, while the age for aggressive cancer patients was 66.21 ± 7.91 years. In terms of race, approximately 80% of non-aggressive patients were White and 17% were Black or African American. Similarly, among aggressive cancer patients, 78% were White and 19% were Black or African American. There were no significant differences between the two cohorts in age at diagnosis or race.

The average comorbidity score for non-aggressive prostate cancer patients was 1.29 ± 1.86, compared to 1.56 ± 2.22 for aggressive cancer patients. The comorbidity score of aggressive cancer patients was significantly higher than that of non-aggressive cancer patients (Table 1).

Table 2 compared the prevalence of selected pre-existing conditions and PSA test values between non-aggressive and aggressive prostate cancer patients. In terms of prior conditions, a higher proportion of aggressive prostate cancer patients had diseases of the male genital organs (61.1%) compared to non-aggressive patients (48.2%). In addition, obesity was more prevalent among aggressive patients (27.0%) than among non-aggressive patients (18.2%). Hypercholesterolemia was also more common in the aggressive group, affecting 32.7% of patients compared to 24.4% in the non-aggressive group. There were statistically significant differences across all three conditions.

Regarding PSA test results, aggressive prostate cancer patients had significantly higher values. The maximum PSA value was 7.38 ± 11.64 in the aggressive group compared to 5.02 ± 6.27 in the non-aggressive group (p = 0.01). Similarly, the most recent PSA value was higher in aggressive patients at 6.96 ± 11.29, compared to 4.73 ± 6.14 in non-aggressive patients (p = 0.01).

Table 1. Patients' demographic information.

	Non-Aggressive (n = 764)		Aggressive (n = 437)		P value
	mean	std	mean	std	
Age at Dx	67.13	8.08	66.21	7.91	0.06
Comorbidity	1.29	1.86	1.56	2.22	0.04
	count	**percent**	**count**	**percent**	
Race					
Black or African American	129	16.88%	82	18.76%	0.41
Others	23	3.01%	13	2.97%	0.97
White	612	80.10%	342	78.26%	0.45

Table 2. Summary statistics of prior conditions and PSA levels.

	Non-Aggressive		Aggressive		P value
Condition	count	percent	count	percent	
Urinary System Disorder	127	16.62%	82	18.76%	0.35
STI	16	2.09%	9	2.06%	0.97
Male Genital Organ Disease	368	48.17%	267	61.10%	0.00
Elevated White Blood Cell	21	2.75%	16	3.66%	0.38
Hypertension	442	57.85%	276	63.16%	0.07
Obesity	139	18.19%	118	27.00%	0.00
Hypercholesterolemia	186	24.35%	143	32.72%	0.00
	mean	**std**	**mean**	**std**	
PSA Max Value	5.02	6.27	7.38	11.64	0.01
PSA Most Recent Value	4.73	6.14	6.96	11.29	0.01

Table 3 presented the performance of three machine learning models- XGBoost (XGB), Balanced Random Forest (BRF), and Neural Network—on predicting aggressive prostate cancer using two different time intervals for healthcare utilization: 7–18 months and 7–24 months prior to diagnosis.

For the 7–18 month window, XGBoost achieved the highest accuracy (0.56), while BRF had slightly better precision and recall (0.56 each). The Neural Network performed the worst across all metrics, with an accuracy of 0.46. For the 7–24 month window, model performance improved slightly. BRF and XGBoost had similar accuracy (0.57 and 0.58, respectively) and balanced precision and recall. However, Neural Network performance declined further, especially in accuracy (0.40) and F1 score (0.33), despite a relatively high precision (0.58). Overall, XGBoost and BRF outperformed the Neural Network in both time intervals.

Table 3. Model results.

Time interval: 7 months – 18 months				
	Accuracy	Precision	Recall	F1
XGB	0.56	0.55	0.55	0.55
BRF	0.54	0.56	0.56	0.54
Neural Network	0.46	0.53	0.52	0.46
Time interval: 7 months – 24 months				
XGB	0.58	0.56	0.56	0.56
BRF	0.57	0.57	0.57	0.56
Neural Network	0.40	0.58	0.52	0.33

4 Discussion

In this study, we investigated the feasibility of using machine learning (ML) models to predict whether a future prostate cancer diagnosis would be classified as aggressive or non-aggressive, using healthcare utilization data.

The analysis showed that several clinical factors were significantly associated with aggressive prostate cancer. Aggressive cancer patients had higher comorbidity scores, more frequent histories of obesity, hypercholesterolemia, and diseases of the male genital organs. In addition, PSA test results were significantly elevated in the aggressive group. These findings were consistent with the clinical understanding that higher PSA values often correlated with more advanced or aggressive forms of prostate cancer. In contrast, demographic factors such as age and race did not differ significantly between aggressive and non-aggressive groups.

XGBoost, Balanced Random Forest (BRF), and a Neural Network were evaluated across two different healthcare utilization windows: 7–18 months and 7–24 months before diagnosis. The predictive performance of these models was moderate, with XGBoost and BRF showing similar accuracies (~0.56–0.58) and balanced precision-recall metrics. In contrast, the neural network model had extremely poor performance. Thus, tree-based ensemble methods appeared to be better suited for this binary classification task than neural networks, as the relatively small sample size limited the effectiveness of the neural network approach.

Although the predictive accuracy of the models was modest, the study showed that it was possible to distinguish aggressive from non-aggressive prostate cancer patients before formal diagnosis using routinely available medical data. Such a system has the potential to be used to support early triage, guide personalized screening frequency, or identify high-risk individuals for further clinical evaluation.

There were several limitations. First, we only obtained structured electronic health data, and clinical notes that might contain valuable information on symptoms, family history, and genetic markers were not available. In addition, we defined aggressive cancer solely based on medical activities after prostate cancer diagnoses. We were unable to obtain treatment intent or stage at diagnosis, which would have allowed for a more precise definition of "aggressiveness."

Our study comports with previous studies demonstrating the value of real-world data for predictive modeling [17, 18]. In future studies, we will include other important data sources such as procedures, medications, and laboratory results [19, 20]. Natural language processing will be used to extract unstructured data elements including symptoms, health behaviors, and social determinants of health [21, 22]. Since tree-based methods performed particularly well in this study, we will continue to expand on these approaches and compare them with other advanced machine learning methods, such as gradient boosting machines (e.g., XGBoost, LightGBM, CatBoost) and ensemble deep learning models [23, 24]. Application of Large Language Models can significantly improve accuracy of data extraction from unstructured clinical notes [25, 26]. In order to develop scalable predictive approaches, particular attention will be given to bias reduction and explainability [27, 28]. Once sufficient scalability and accuracy is achieved, the algorithms for predicting aggressive prostate cancer can be embedded in electronic health records as a part of clinical decision support that can be triggered by each new medical encounter [29, 30]. Information about the patients with heightened risk of aggressive prostate cancer will be conveyed to providers to consider prioritization of definitive clinical work-up for cancer diagnosis [31, 32]. This would allow timely commencement of pharmacological and lifestyle interventions depending on the results of definitive work-up [33, 34]. Lifestyle interventions may include physical telerehabil itation [35, 36], chatbots for patient engagement [37, 38], and interactive applications providing health education and social support [39, 40].

5 Conclusion

Our study showed that routinely collected healthcare data can be used to build models that differentiate aggressive from non-aggressive prostate cancer before diagnosis. While model performance was modest, tree-based ensemble methods consistently provided more reliable predictions than neural networks. Features such as comorbidity scores, PSA levels, and histories of obesity, hypercholesterolemia, and male genital diseases emerged as particularly informative for prediction. Future research should focus on using clinical notes and richer longitudinal data to improve predictive accuracy.

References

1. Centers for Disease Control and Prevention: U.S. Cancer Statistics Prostate Cancer Stat Bite. U.S. Department of Health and Human Services (2025). https://www.cdc.gov/united-states-cancer-statistics/publications/prostate-cancer-stat-bite.html
2. American Cancer Society: Prostate cancer key statistics. https://www.cancer.org/cancer/types/prostate-cancer/about/key-statistics.html
3. American Cancer Society: Prostate cancer survival rates. https://www.cancer.org/cancer/types/prostate-cancer/detection-diagnosis-staging/survival-rates.html
4. Kratzer, T.B., Mazzitelli, N., Star, J., Dahut, W.L., Jemal, A., Siegel, R.L.: Prostate cancer statistics, 2025. CA Cancer J. Clin. 1–13 (2025)
5. PDQ® Adult Treatment Editorial Board: PDQ Prostate Cancer Treatment. National Cancer Institute, Bethesda, MD. https://www.cancer.gov/types/prostate/patient/prostate-treatment-pdq. Accessed 16 Sept 2025
6. Parker, C., Castro, E., Fizazi, K., Heidenreich, A., Ost, P., Procopio, G., et al.: Prostate cancer: ESMO clinical practice guidelines for diagnosis, treatment and follow-up. Ann. Oncol. **31**(9), 1119–1134 (2020)
7. Litwin, M.S., Tan, H.J.: The diagnosis and treatment of prostate cancer: a review. JAMA **317**(24), 2532–2542 (2017)
8. Mohler, J.L., Antonarakis, E.S., Armstrong, A.J., D'Amico, A.V., Davis, B.J., Dorff, T., et al.: Prostate cancer, version 2.2019, NCCN clinical practice guidelines in oncology. J. Natl. Compr. Canc. Netw. **17**(5), 479–505 (2019)
9. Nagy, M., et al.: Machine learning in oncology: what should clinicians know? JCO Clin. Cancer Inform. **4**, 799–810 (2020)
10. Placido, D., Yuan, B., Hjaltelin, J.X., et al.: A deep learning algorithm to predict risk of pancreatic cancer from disease trajectories. Nat. Med. **29**(5), 1113–1122 (2023)
11. Riviere, P., Tokeshi, C., Hou, J., et al.: Claims-based approach to predict cause-specific survival in men with prostate cancer. JCO Clin. Cancer Inform. **3**, 1–7 (2019)
12. Huo, X., Finkelstein, J.: Prostate cancer prediction using classification algorithms. J. Clin. Oncol. **40**(16_suppl), e13590 (2022)
13. Cui, W., Halwani, A., Li, C., Finkelstein, J.: Predicting prostate cancer diagnosis using machine learning analysis of healthcare utilization patterns. Stud. Health Technol. Inform. **323**, 6–10 (2025)
14. All of Us Research Program. https://allofus.nih.gov/
15. Healthcare Cost and Utilization Project (HCUP): CCS-Services and Procedures. Agency for Healthcare Research and Quality, Rockville, MD (2021). www.hcup-us.ahrq.gov/toolssoftware/ccs_svcsproc/ccssvcproc.jsp
16. Codify by AAPC. https://www.aapc.com/codes/cpt-codes-range/

17. Paranjpe, I., Russak, A.J., De Freitas, J.K., et al.: Retrospective cohort study of clinical characteristics of 2199 hospitalised patients with COVID-19 in New York City. BMJ Open **10**(11), e040736 (2020)
18. Cui, W., Kawamoto, K., Morgan, K., Finkelstein, J.: Reducing diagnostic uncertainty in emergency departments: the role of large language models in age-specific diagnostics. In: 2024 IEEE 12th International Conference on Healthcare Informatics (ICHI), pp. 525–527. Orlando, FL, USA (2024)
19. Thyvalikakath, T.P., Duncan, W.D., Siddiqui, Z., et al.: Appl. Clin. Inform. **11**(2), 305–314 (2020). https://doi.org/10.1055/s-0040-1709506. PMID: 32349142; PMCID: PMC7190390
20. Locke, B.W., Richards, W.W., Gouripeddi, R., et al.: The consistency of hypercapnic respiratory failure case definitions in electronic health record data. Chest. S0012-3692(25)05126-8 (2025). https://doi.org/10.1016/j.chest.2025.08.002. PMID: 40885535; PMCID: PMC12739763
21. Shah-Mohammadi, F., Finkelstein, J.: Combining NLP and machine learning for differential diagnosis of COPD exacerbation using emergency room data. Stud. Health Technol. Inform. **305**, 525–528 (2023). https://doi.org/10.3233/SHTI230549. PMID: 37387083
22. Shah-Mohammadi, F., Cui, W., Finkelstein, J.: Comparison of ACM and CLAMP for entity extraction in clinical notes. In: The Annual International Conference of the IEEE Engineering in Medicine and Biology Society 2021, pp. 1989–1992 (2021). https://doi.org/10.1109/EMBC46164.2021.9630611. PMID: 34891677
23. Finkelstein, J., Smiley, A., Echeverria, C., Mooney, K.: AI-driven prediction of symptom trajectories in cancer care: a deep learning approach for chemotherapy management. Bioengineering (Basel) **11**(11), 1172 (2024). https://doi.org/10.3390/bioengineering11111172. PMID:39593830;PMCID:PMC11592055
24. Huo, X., Kohli, M., Finkelstein, J.: Machine learning approaches for predicting mortality in metastatic castration-resistant prostate cancer. Stud. Health Technol. Inform. **328**, 26–30 (2025). https://doi.org/10.3233/SHTI250666. PMID: 40588874
25. Bai, Y., Cui, W., Finkelstein, J.: Utilizing open-source large language models to extract genitourinary symptoms from clinical notes. Stud Health Technol Inform. **328**, 16–20 (2025). https://doi.org/10.3233/SHTI250664. PMID: 40588872
26. Reategui-Rivera, C.M., Finkelstein, J.: Evaluation of the performance of a large language model to extract signs and symptoms from clinical notes. Stud Health Technol Inform. **323**, 71–75 (2025). https://doi.org/10.3233/SHTI250051. PMID: 40200448
27. Smiley, A., Villarreal-Zegarra, D., Reategui-Rivera, C.M., Escobar-Agreda, S., Finkelstein, J.: Methodological and reporting quality of machine learning studies on cancer diagnosis, treatment, and prognosis. Front. Oncol. **15**, 1555247 (2025). https://doi.org/10.3389/fonc.2025.1555247.PMID:40297817;PMCID:PMC12034563
28. Smiley, A., Reategui-Rivera, C.M., Villarreal-Zegarra, D., Escobar-Agreda, S., Finkelstein, J.: Exploring artificial intelligence biases in predictive models for cancer diagnosis. Cancers (Basel) **17**(3), 407 (2025). https://doi.org/10.3390/cancers17030407.PMID:39941776; PMCID:PMC11816222
29. Kawamoto, K., Finkelstein, J., Del Fiol, G.: Implementing machine learning in the electronic health record: checklist of essential considerations. Mayo Clin. Proc. **98**(3), 366–369 (2023). https://doi.org/10.1016/j.mayocp.2023.01.013. PMID: 36868743
30. Shah-Mohammadi, F., Finkelstein, J.: Accuracy evaluation of GPT-assisted differential diagnosis in emergency department. Diagnostics (Basel) **14**(16), 1779 (2024). https://doi.org/10.3390/diagnostics14161779.PMID:39202267;PMCID:PMC11354035
31. Finkelstein, J., Gabriel, A., Schmer, S., Truong, T.T., Dunn, A.: Identifying facilitators and barriers to implementation of AI-assisted clinical decision support in an electronic health record system. J. Med. Syst. **48**(1), 89 (2024). https://doi.org/10.1007/s10916-024-02104-9. PMID:39292314;PMCID:PMC11410896

32. Finkelstein, J., Parvanova, I., Xing, Z., Truong, T.T., Dunn, A.: Qualitative assessment of implementation of a discharge prediction tool using RE-AIM framework. Stud. Health Technol. Inform. **302**, 596–600 (2023). https://doi.org/10.3233/SHTI230212. PMID: 37203755
33. Bedra, M., McNabney, M., Stiassny, D., Nicholas, J., Finkelstein, J.: Defining patient-centered characteristics of a telerehabilitation system for patients with COPD. Stud. Health Technol. Inform. **190**, 24–26 (2013). PMID: 23823363
34. Finkelstein, J., Wood, J.: Implementing home telemanagement of congestive heart failure using Xbox gaming platform. The Annual International Conference of the IEEE Engineering in Medicine and Biology Society 2011, pp. 3158–3163 (2011). https://doi.org/10.1109/IEMBS.2011.6090861. PMID: 22255010
35. Castro, H.K., Cross, R.K., Finkelstein, J.: Using a Home Automated Telemanagement (HAT) system: experiences and perceptions of patients with inflammatory bowel disease. AMIA Annual Symposium Proceedings 2006, p. 872 (2006). PMID: 17238492; PMCID: PMC1839272
36. Jeong, I.C., Karpatkin, H., Finkelstein, J.: Physical telerehabilitation improves quality of life in patients with multiple sclerosis. Stud. Health Technol. Inform. **284**, 384–388 (2021). https://doi.org/10.3233/SHTI210752. PMID: 34920553
37. Reategui-Rivera, C.M., Smiley, A., Finkelstein, J.: LLM-based chatbot to reduce mental illness stigma in healthcare providers. In: 2025 IEEE 15th Annual Computing and Communication Workshop and Conference (CCWC), pp. 00001–00007. Las Vegas, NV, USA (2025). https://doi.org/10.1109/CCWC62904.2025.10903778
38. Villarreal-Zegarra, D., Reategui-Rivera, C.M., Paredes-Gonzales, Y., Centeno-Terrazas, G., Finkelstein, J.: Assessment of GPT-based conversational agents aimed to reduce healthcare provider stigma. Stud. Health Technol. Inform. **328**, 4–8 (2025). https://doi.org/10.3233/SHTI250661. PMID: 40588869
39. Finkelstein, J., Lapshin, O., Cha, E.: Feasibility of promoting smoking cessation among methadone users using multimedia computer-assisted education. J. Med. Internet Res. **10**(5), e33 (2008). https://doi.org/10.2196/jmir.1089.PMID:18984556;PMCID:PMC2630834
40. Barron, J., Bedra, M., Wood, J., Finkelstein, J.: Exploring three perspectives on feasibility of a patient portal for older adults. Stud. Health Technol. Inform. **202**, 181–184 (2014). PMID: 25000046

Early Identification of Dementia via Machine Learning Analysis of Health Care Utilization Patterns

Wanting Cui(✉) and Joseph Finkelstein

Arizona Telemedicine Program, University of Arizona, Tucson, AZ 85724, USA
wantingcui@arizona.edu

Abstract. Early detection of Alzheimer's disease and related dementias (ADRD) remains a major unmet need in clinical care, as delays in diagnosis often limit opportunities for timely intervention and care planning. This study aimed to develop and evaluate machine learning models to predict ADRD one year in advance using electronic health record (EHR) data from the All of Us Research Platform. Patients aged 60 years or older whose first ADRD diagnosis occurred between 2012 and 2019 were included. Data were structured using a three-year observation period followed by a one-year gap before the prediction window, allowing the models to predict ADRD one year ahead based on prior health activity. Predictors included diagnostic information, healthcare utilization measures, demographic variables, visit types, and the Charlson Comorbidity Index. Three models, including logistic regression, random forest, and a multi-layer perceptron (MLP), were trained and compared. The random forest model performed best, achieving an AUC of 0.90 and accuracy of 0.82, outperforming both logistic regression and MLP. The results showed that ADRD patients had significantly higher overall healthcare utilization, including greater numbers of diagnoses and visits. Key predictors included depressive disorders, hypertension, lipid disorders, chest pain, and sleep wake disorders.

Keywords: Alzheimer's Disease and Related Dementias (ADRD) · Machine Learning · Early Prediction

1 Introduction

Early detection of dementia and cognitive decline remains a critical unmet need in clinical care [1]. Delays in diagnosis often lead to missed opportunities for intervention and planning [2]. By leveraging machine learning (ML) to analyze longitudinal health care utilization patterns in electronic health records (EHRs), we can identify subtle signals of cognitive impairment well before clinical presentation [3]. Health care utilization, such as frequency of primary care visits, diagnostic tests, medication changes, and hospitalization patterns, reflects emerging cognitive and functional changes [4]. Traditional screening methods rely on cognitive assessments conducted infrequently in primary care, limiting the window for early action [5]. ML models trained on rich EHR data can

H. R. Arabnia et al. (Eds.): CSCI 2025, CCIS 2971, pp. 115–123, 2027.
https://doi.org/10.1007/978-3-032-28463-1_9

potentially detect complex, nonlinear patterns and risk trajectories, enabling proactive management, targeted referrals, and personalized care pathways [6]. However, limited information exists on feasibility of this approach. Further studies are necessary to identify in EHRs distinct electronic phenotypes indicative of incipient cognitive decline especially in older adults.

Previous studies have proposed using chronic conditions identified from EHR data to predict early ADRD onset [7–10]. However, most of these studies focused solely on chronic conditions and did not include other types of clinical information. In this study, we aimed to include a broader range of diagnostic categories as well as visit information to diversify the predictive features. The goal of this study was to develop and evaluate machine learning models to predict Alzheimer's disease and related dementias (ADRD) one year in advance using electronic health record data. We also aimed to identify key factors and healthcare utilization patterns associated with ADRD risk.

2 Method

The All of Us Research Platform is a large-scale initiative by the National Institutes of Health (NIH) that collects health data from participants across the United States [11]. For this study, we used the Controlled Tier V8 dataset. ADRD conditions were defined using a list of ICD-9 codes provided by a prior study [9]. We converted these codes into their ICD-10 equivalents and used both ICD-9 and ICD-10 codes to identify ADRD cases. We identified patients whose first ADRD diagnosis occurred between 2012 and 2019. To ensure diagnostic validity, eligible patients were required to have at least two ADRD diagnoses within one year of each other. In addition, patients were restricted to those aged 60 years or older and were required to have at least three valid diagnoses during the observation window in order to be included in the study cohort.

For the control group, we selected patients who never received an ADRD diagnosis. To match the case group design, each control was randomly assigned a pseudo-diagnosis date between 2012 and 2019. Controls were subject to the same inclusion requirements as cases: they had to be 60 years or older and have at least three valid diagnoses during the observation window.

Following cohort construction, we applied 1:1 propensity score matching to balance cases and controls on age, gender, race, and diagnosis year. In the end, there were 805 ADRD patients and 805 matched controls.

ADRD is a slowly developing condition. Thus, we structured the data using a three-year observation period followed by a one-year gap before the prediction window. This design allowed us to predict ADRD diagnoses one year in advance based on the prior three years of health activity. For example, if a patient was first diagnosed with ADRD on March 1, 2017, the model used health activities recorded between March 1, 2013, and March 1, 2016, to generate predictions.

The predictive variables were derived from diagnostic information and healthcare utilization. All ICD codes were converted to the ICD-10 format and then mapped into clinically meaningful categories using the Clinical Classifications Software Refined (CCSR) dictionary, which groups ICD-10 codes into over 400 diagnostic categories [12]. To reduce sparsity, we kept only those categories represented by at least 100 patients. This filtering resulted in 108 diagnostic categories included as predictors. For each patient, we counted the number of occurrences of each diagnostic category, normalized these counts by the total number of conditions per patient, and calculated the total number of condition occurrences. We also calculated visit-based features, including the total number of visits, outpatient visits, inpatient admissions, office visits, emergency room visits, and telehealth visits. In addition, we included an age-adjusted Charlson comorbidity score as a predictor.

Finally, three demographic variables—age, gender, and race—were included. Altogether, the final feature set consisted of diagnostic categories, visits, comorbidity score, and demographics (118 features). We trained and compared three machine learning models for the prediction task: Logistic Regression, Random Forest, and a Multi-Layer Perceptron (MLP). Each model was run 10 times to ensure stability of results. Model performance was evaluated by calculating the average accuracy, F1 score, and area under the receiver operating characteristic curve (AUC). In addition, we identified the top predictive features from the best performing model. All models were implemented using scikit-learn in Python.

3 Results

The study included 805 ADRD patients and 805 non-ADRD patients after matching propensity scores. The average age was 72.02 ± 7.67 years old for ADRD patients and 72.39 ± 7.87 years for non-ADRD patients. Gender distribution was also similar, with males comprising 52.80% of ADRD cohort and 52.92% of non-ADRD cohort, while females accounted for 47.12% of ADRD cohort and 47.08% of non-ADRD cohort. In terms of race, the majority of patients were White (63.98% of ADRD cohort and 67.45% of non-ADRD cohort), followed by Black or African American patients (13.91% of ADRD cohort and 13.17% of non-ADRD cohort).

For predictive modeling, we compared the performance of logistic regression, random forest, and multilayer perceptron (MLP) models (Table 1). Logistic regression was the baseline model. It achieved modest discrimination, with an accuracy of 0.60, F1 score of 0.54, and AUC of 0.62. The MLP model performed slightly better, with an accuracy of 0.63, F1 score of 0.58, and AUC of 0.69. In contrast, the random forest showed the highest overall performance, with an accuracy of 0.82, F1 score of 0.82, and AUC of 0.90.

Table 1. Model Performance.

	Accuracy	F1	AUC
Logistic Regression	0.60 ± 0.01	0.54 ± 0.02	0.62 ± 0.02
Random Forest	0.82 ± 0.01	0.82 ± 0.02	0.90 ± 0.01
Multilayer Perceptron	0.63 ± 0.04	0.58 ± 0.10	0.69 ± 0.05

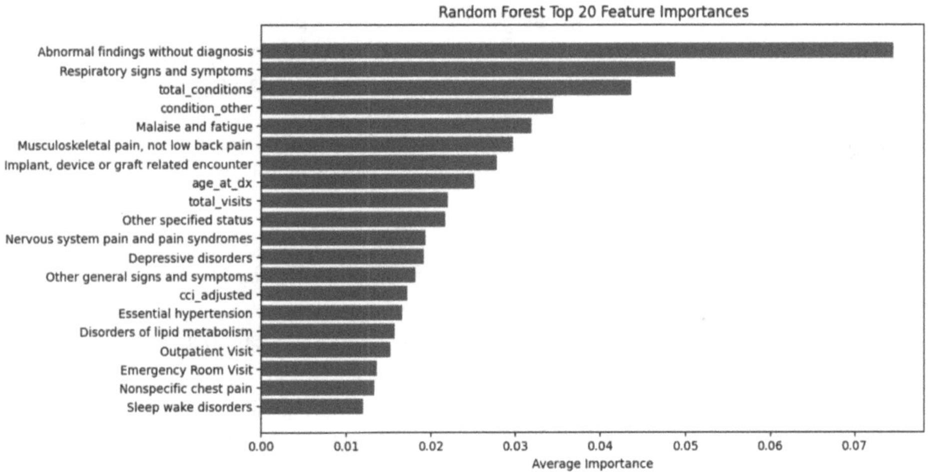

Fig. 1. Top 20 important features of random forest models.

Since random forest was the best performing model, we further investigated the top 20 features generated from the random forest model. The top predictors were abnormal findings without diagnosis and respiratory signs and symptoms, followed by total conditions, condition other, and malaise and fatigue (Fig. 1). Table 2 showed frequencies of visits for the top 20 features. The average ADRD patients and non-ADRD patients showed significant differences across all features (all $p < 0.001$). On average, ADRD patients had a higher use of healthcare utilization compared to non-ADRD patients. For example, ADRD patients had more total conditions and more total visits. Specific conditions such as depressive disorders, essential hypertension, disorders of lipid metabolism, non-specific chest pain and sleep wake disorders were more prevalent among ADRD patients. Similarly, ADRD patients had higher numbers of outpatient and emergency room visits. In contrast, several features were more common among non-ADRD patients, including abnormal findings without diagnosis, respiratory signs and symptoms, malaise and fatigue, musculoskeletal pain not involving the low back, and implant/device-related encounters.

Table 2. Summary statistics of the top 20 features.

	ADRD patients		Non-ADRD patients		
Features	mean	std	mean	std	p-value
Abnormal findings without diagnosis	0.58	2.02	2.97	17.14	<0.001
Respiratory signs and symptoms	0.37	1.22	1.96	4.54	<0.001
Total Conditions	171.70	252.40	112.11	217.61	<0.001
Condition Other	36.67	76.10	20.34	50.09	<0.001

(continued)

Table 2. (*continued*)

	ADRD patients		Non-ADRD patients		
Features	mean	std	mean	std	p-value
Malaise and fatigue	0.04	0.29	0.99	3.30	<0.001
Musculoskeletal pain, not low back pain	2.33	6.39	4.22	16.00	<0.001
Implant, device or graft related encounter	0.00	0.00	0.51	2.40	<0.001
Other specified status	0.00	0.09	0.30	1.33	<0.001
Nervous system pain and pain syndromes	0.13	1.14	0.66	2.84	<0.001
Depressive disorders	5.10	17.84	1.09	4.88	<0.001
Other general signs and symptoms	0.36	1.40	1.20	4.87	<0.001
Essential hypertension	8.85	12.40	5.90	13.40	<0.001
Disorders of lipid metabolism	8.74	13.41	5.69	10.10	<0.001
Nonspecific chest pain	1.36	3.45	0.52	1.75	<0.001
Sleep wake disorders	3.17	9.34	1.50	4.44	<0.001
Outpatient Visit	37.10	48.26	22.94	30.34	<0.001
Emergency Room Visit	1.66	4.01	0.67	2.07	<0.001
Total Visits	55.89	61.32	36.41	37.91	<0.001

4 Discussion

In this study, we developed and compared machine learning models to predict Alzheimer's disease and related dementias (ADRD) diagnosis one year ahead using electronic health record (EHR) data. Among the three models tested, the random forest achieved the best performance, with an AUC of 0.90, which was significantly better than the performance of logistic regression and multilayer perceptron. The dataset contained more than 6,500 unique ICD-10 codes, which we grouped into about 400 broader categories. To reduce sparsity, we selected the less sparse groups as predictors, but many variables still showed high correlation. Random forest performed well in this setting, showing that it can handle high-dimensional EHR data with many correlated features. In contrast, logistic regression was more limited in this context and achieved only modest discrimination. While the MLP model performed slightly better than logistic regression, it was still less effective than random forest. Neural networks often require much larger sample sizes to reach their full potential, so the relatively small sample size in our study may have limited the performance of the MLP model.

In addition, the random forest model provided insights into key features associated with ADRD risk. Our analysis of the top 20 predictors showed that ADRD patients had higher healthcare utilizations compared to non-ADRD patients. Higher numbers of total conditions, outpatient visits, and emergency room visits were consistently observed among ADRD patients which suggested that healthcare service patterns may be important markers of dementia risk. In addition, ADRD patients had higher frequencies of depressive disorders, essential hypertension, disorders of lipid metabolism, nonspecific chest pain, and sleep-wake disorders. These findings were consistent with prior studies showing that cardiovascular, metabolic, and psychiatric conditions may contribute to ADRD onset and progression.

There were several limitations to this study. First, many of the variables in the dataset were highly correlated, which may have affected model performance and stability. Although we grouped more than 6,500 ICD-10 codes into broader categories and selected less sparse groups, collinearity remained an issue. In future work, we plan to explore additional strategies to reduce multicollinearity, such as dimensionality reduction or regularization techniques, while still maintaining interpretability of individual predictors [13]. Second, our analysis focused only on diagnoses and visit types, which did not provide a full picture of healthcare utilization.

Our study comports with previous studies demonstrating the value of real-world data for predictive modeling [14, 15]. In future studies, we will include other important data sources such as procedures, medications, and laboratory results [16, 17]. Natural language processing will be used to extract unstructured data elements including symptoms, health behaviors, and social determinants of health [18, 19]. Since tree-based methods performed particularly well in this study, we will continue to expand on these approaches and compare them with other advanced machine learning methods, such as gradient boosting machines (e.g., XGBoost, LightGBM, CatBoost) and ensemble deep learning models [20, 21]. Application of Large Language Models can significantly improve accuracy of data extraction from unstructured clinical notes [22, 23]. In order to develop scalable predictive approaches, particular attention will be given to bias reduction and explainability [24, 25]. Once sufficient scalability and accuracy is achieved, the algorithms for identifying early ADRD diagnoses can be embedded in electronic health records as a part of clinical decision support that can be triggered by each new medical encounter [26, 27]. Information about the patients with heightened risk of ADRD diagnosis will be conveyed to providers to consider prioritization of definitive clinical work-up for ADRD [28, 29]. This would allow timely commencement of pharmacological and lifestyle interventions to slow down or prevent further cognitive decline [30, 31]. Lifestyle interventions may include geriatric telerehabilitation [32, 33], chatbots with cognitive exercises [34, 35], and interactive applications providing health education and social support [36, 37].

5 Conclusion

In conclusion, random forest performed significantly better than logistic regression and MLP in predicting ADRD one year in advance, achieving an AUC of 0.90. The model showed that ADRD patients had significantly higher overall healthcare usage, such as a greater total number of diagnoses and visits. In addition, depressive disorders, hypertension, lipid disorders, chest pain, and sleep-wake disorders were important predictors. In future studies, we will expand the feature set and implement other tree-based machine learning approaches to further improve prediction and understanding of factors associated with ADRD.

References

1. Liu, Y., Jun, H., Becker, A., Wallick, C., Mattke, S.: Detection rates of mild cognitive impairment in primary care for the United States Medicare population. J. Prev. Alzheimers Dis. **11**(1), 7–12 (2024)
2. He, Z., Dieciuc, M., Carr, D., et al.: New opportunities for the early detection and treatment of cognitive decline: adherence challenges and the promise of smart and person-centered technologies. BMC Digit. Health **1**, 7 (2023)
3. Javeed, A., Dallora, A.L., Berglund, J.S., et al.: Machine learning for dementia prediction: a systematic review and future research directions. J. Med. Syst. **47**(1), 17 (2023)
4. Fu, M., Sankararaman, S., Pasaniuc, B., et al.: Identifying common disease trajectories of Alzheimer's disease with electronic health records. EBioMedicine **118**, 105831 (2025)
5. Hampel, H., Li, G., Mielke, M.M., et al.: The impact of real-world evidence in implementing and optimizing Alzheimer's disease care. Med. **6**(5), 100695 (2025)
6. Cui, W., Halwani, A., Li, C., Finkelstein, J.: Predicting prostate cancer diagnosis using machine learning analysis of healthcare utilization patterns. Stud. Health Technol. Inform. **323**, 6–10 (2025)
7. Li, Q., Yang, X., Xu, J., et al.: Early prediction of Alzheimer's disease and related dementias using real-world electronic health records. Alzheimer's Dement. **19**(8), 3506–3518 (2023)
8. Hu, X., et al.: Self-explainable graph neural network for Alzheimer disease and related dementias risk prediction: algorithm development and validation study. JMIR Aging **7**(1), e54748 (2024)
9. Schliep, K.C., et al.: Predicting the onset of Alzheimer's disease and related dementia using electronic health records: findings from the Cache County Study on Memory in Aging (1995–2008). BMC Med. Inform. Decis. Mak. **24**(1), 316 (2024)
10. Davis-Ajami, M.L., et al.: Mortality and discharge location of intensive care patients with Alzheimer disease and related dementia. Am. J. Crit. Care **32**(4), 249–255 (2023)
11. All of Us Research Program. https://allofus.nih.gov/. Accessed 23 Oct 2025
12. Clinical Classifications Software (CCS) for ICD-10-PCS (beta version). Healthcare Cost and Utilization Project (HCUP). July 2025. Agency for Healthcare Research and Quality, Rockville, MD. hcup-us.ahrq.gov/toolssoftware/ccs10/ccs10.jsp. Accessed 23 Aug 2025
13. Lin, Z., Ruszczynski, A.: An integrated transportation distance between kernels and approximate dynamic risk evaluation in markov systems. arXiv:2311.06645 [math.OC] (2023)
14. Paranjpe, I., Russak, A.J., De Freitas, J.K., et al.: Retrospective cohort study of clinical characteristics of 2199 hospitalised patients with COVID-19 in New York City. BMJ Open **10**(11), e040736 (2020)

15. Cui, W., Kawamoto, K., Morgan, K., Finkelstein, J.: Reducing diagnostic uncertainty in emergency departments: the role of large language models in age-specific diagnostics. In: 2024 IEEE 12th International Conference on Healthcare Informatics (ICHI), pp. 525–527. Orlando, FL, USA (2024)
16. Thyvalikakath, T.P., Duncan, W.D., Siddiqui, Z., et al.: Appl. Clin. Inform. **11**(2), 305–314 (2020). https://doi.org/10.1055/s-0040-1709506. PMID: 32349142; PMCID: PMC7190390
17. Locke, B.W., Richards, W.W., Gouripeddi, R., et al.: the consistency of hypercapnic respiratory failure case definitions in electronic health record data. Chest S0012-3692(25)05126-8 (2025). https://doi.org/10.1016/j.chest.2025.08.002. PMID: 40885535; PMCID: PMC12739763
18. Shah-Mohammadi, F., Finkelstein, J.: Combining NLP and machine learning for differential diagnosis of COPD exacerbation using emergency room data. Stud. Health Technol. Inform. **29**(305), 525–528 (2023). https://doi.org/10.3233/SHTI230549. PMID: 37387083
19. Shah-Mohammadi, F., Cui, W., Finkelstein, J.: Comparison of ACM and CLAMP for entity extraction in clinical notes. In: Annual International Conference of the IEEE Engineering in Medicine and Biology Society. 2021, pp. 1989–1992 (2021). https://doi.org/10.1109/EMBC46164.2021.9630611. PMID: 34891677
20. Finkelstein, J., Smiley, A., Echeverria, C., Mooney, K.: AI-driven prediction of symptom trajectories in cancer care: a deep learning approach for chemotherapy management. Bioengineering (Basel) **11**(11), 1172 (2024). https://doi.org/10.3390/bioengineering11111172. PMID:39593830;PMCID:PMC11592055
21. Huo, X., Kohli, M., Finkelstein, J.: Machine learning approaches for predicting mortality in metastatic castration-resistant prostate cancer. Stud. Health Technol. Inform. **328**, 26–30 (2025). https://doi.org/10.3233/SHTI250666. PMID: 40588874
22. Bai, Y., Cui, W., Finkelstein, J.: Utilizing open-source large language models to extract genitourinary symptoms from clinical notes. Stud. Health Technol. Inform. **328**, 16–20 (2025). https://doi.org/10.3233/SHTI250664. PMID: 40588872
23. Reategui-Rivera, C.M., Finkelstein, J.: Evaluation of the performance of a large language model to extract signs and symptoms from clinical notes. Stud. Health Technol. Inform. **323**, 71–75 (2025). https://doi.org/10.3233/SHTI250051. PMID: 40200448
24. Smiley, A., Villarreal-Zegarra, D., Reategui-Rivera, C.M., Escobar-Agreda, S., Finkelstein, J.: Methodological and reporting quality of machine learning studies on cancer diagnosis, treatment, and prognosis. Front. Oncol. **15**, 1555247 (2025). https://doi.org/10.3389/fonc.2025.1555247.PMID:40297817;PMCID:PMC12034563
25. Smiley, A., Reategui-Rivera, C.M., Villarreal-Zegarra, D., Escobar-Agreda, S., Finkelstein, J.: Exploring artificial intelligence biases in predictive models for cancer diagnosis. Cancers (Basel) **17**(3), 407 (2025). https://doi.org/10.3390/cancers17030407.PMID:39941776;PMCID:PMC11816222
26. Kawamoto, K., Finkelstein, J., Del Fiol, G.: Implementing machine learning in the electronic health record: checklist of essential considerations. Mayo Clin. Proc. **98**(3), 366–369 (2023). https://doi.org/10.1016/j.mayocp.2023.01.013. PMID: 36868743
27. Shah-Mohammadi, F., Finkelstein, J.: Accuracy evaluation of GPT-assisted differential diagnosis in emergency department. Diagnostics (Basel) **14**(16), 1779 (2024). https://doi.org/10.3390/diagnostics14161779.PMID:39202267;PMCID:PMC11354035
28. Finkelstein, J., Gabriel, A., Schmer, S., Truong, T.T., Dunn, A.: Identifying facilitators and barriers to implementation of AI-assisted clinical decision support in an electronic health record system. J. Med. Syst. **48**(1), 89 (2024). https://doi.org/10.1007/s10916-024-02104-9. PMID:39292314;PMCID:PMC11410896
29. Finkelstein, J., Parvanova, I., Xing, Z., Truong, T.T., Dunn, A.: Qualitative assessment of implementation of a discharge prediction tool using RE-AIM framework. Stud. Health Technol. Inform. **302**, 596–600 (2023). https://doi.org/10.3233/SHTI230212. PMID: 37203755

30. Bedra, M., McNabney, M., Stiassny, D., Nicholas, J., Finkelstein, J.: Defining patient-centered characteristics of a telerehabilitation system for patients with COPD. Stud. Health Technol. Inform. **190**, 24–26 (2013). PMID: 23823363
31. Finkelstein, J., Wood, J.: Implementing home telemanagement of congestive heart failure using Xbox gaming platform. In: Annual International Conference of the IEEE Engineering in Medicine and Biology Society 2011, pp. 3158–3163 (2011). https://doi.org/10.1109/IEMBS.2011.6090861. PMID: 22255010
32. Castro, H.K., Cross, R.K., Finkelstein, J.: Using a Home Automated Telemanagement (HAT) system: experiences and perceptions of patients with inflammatory bowel disease. AMIA Annual Symposium Proceedings 2006, p. 872 (2006). PMID: 17238492; PMCID: PMC1839272
33. Jeong, I.C., Karpatkin, H., Finkelstein, J.: Physical telerehabilitation improves quality of life in patients with multiple sclerosis. Stud. Health Technol. Inform. **15**(284), 384–388 (2021). https://doi.org/10.3233/SHTI210752. PMID: 34920553
34. Reategui-Rivera, C.M., Smiley, A., Finkelstein, J.: LLM-based chatbot to reduce mental illness stigma in healthcare providers. In: 2025 IEEE 15th Annual Computing and Communication Workshop and Conference (CCWC), pp. 00001–00007. Las Vegas, NV, USA (2025). https://doi.org/10.1109/CCWC62904.2025.10903778
35. Villarreal-Zegarra, D., Reategui-Rivera, C.M., Paredes-Gonzales, Y., Centeno-Terrazas, G., Finkelstein, J.: Assessment of GPT-based conversational agents aimed to reduce healthcare provider stigma. Stud. Health Technol. Inform. **328**, 4–8 (2025). https://doi.org/10.3233/SHTI250661. PMID: 40588869
36. Finkelstein, J., Lapshin, O., Cha, E.: Feasibility of promoting smoking cessation among methadone users using multimedia computer-assisted education. J. Med. Internet Res. **10**(5), e33 (2008). https://doi.org/10.2196/jmir.1089.PMID:18984556;PMCID:PMC2630834
37. Barron, J., Bedra, M., Wood, J., Finkelstein, J.: Exploring three perspectives on feasibility of a patient portal for older adults. Stud. Health Technol. Inform. **202**, 181–184 (2014). PMID: 25000046

AMHAT: Multimodal Pipeline for Privacy-Preserving Stress Screening

Mehdi Ghayoumi(✉), Cory Liu, and Elena M. Nye

Cybersecurity Department, State University of New York (SUNY) at Canton, Canton, USA
{ghayoumi,liu257,nye105}@canton.edu

Abstract. We introduce AMHAT, an offline, lab-based system for detecting acute psychological stress using three privacy-preserving signals: (1) speech acoustics, (2) linguistic content, and (3) timing-based keyboard and mouse interactions. AMHAT ensures data privacy by performing all processing locally so that raw audio and typed content never leave the device; only de-identified, fixed-format feature summaries are stored. Prior studies show that stress affects vocal prosody and timing, modifies lexical selection, and alters motor behavior patterns [28–30,32,33,38,39]. AMHAT models each modality independently with interpretable heads and applies a quality- and uncertainty-weighted late fusion to maintain robust performance even when some modalities are missing or degraded [28,32,38]. To address the scarcity of labeled stress data, the system leverages auxiliary supervision from emotion-related tasks and adaptive sampling strategies during training [29,33]. In offline cross-validation, interaction-based features provide strong discriminative power, and multimodal fusion achieves high overall performance (AUROC = 0.964, accuracy = 0.919, averaged across folds). Ongoing work focuses on validating AMHAT under standardized stress-induction protocols (e.g., TSST and MIST) and integrating on-device explanations while preserving privacy safeguards.

Keywords: stress detection · speech prosody · lexical variation · keystroke and interaction features · cognitive load

1 Introduction

Psychological stress can be reflected in everyday behaviors, including how people speak, what they say, and how they interact with a keyboard and mouse [1,28,29,33,39]. Prior work indicates that acute stress and elevated cognitive load are associated with measurable changes in vocal timing and prosody (for example, pitch variation, loudness patterns, and pausing), shifts in lexical and syntactic choices, and differences in typing and mouse dynamics such as inter-key timing, correction activity, and cursor movement profiles [28–30,32,33,38,39]. These signals are practical for office-like scenarios where wearables may be unavailable, undesirable, or difficult to deploy consistently [1–4,28–30,32,33,38,39].

H. R. Arabnia et al. (Eds.): CSCI 2025, CCIS 2971, pp. 124–141, 2027.
https://doi.org/10.1007/978-3-032-28463-1_10

Motivated by this evidence, we present AMHAT, an offline, privacy-first pipeline that produces stress-screening outputs using three complementary, privacy-compatible signal families: (i) speech acoustics and timing, (ii) linguistic summaries derived from on-device transcripts, and (iii) content-free interaction timing features computed from keyboard and mouse events [27–29,32,33,38,39]. AMHAT is designed to run locally: raw audio and any raw typed content are not transmitted, and only de-identified, fixed-format feature summaries and model outputs are retained [27]. To support transparency, AMHAT associates each modality with lightweight, behavior-tied explanations (for example, changes in pause structure, pitch dynamics, lexical diversity, or correction bursts) rather than relying solely on opaque representations [28,29,31–35,38,39].

In practical use and in offline analyses of public datasets, one or more modalities may be missing or degraded (for example, noisy audio, low-confidence transcripts, or sparse interaction events). AMHAT therefore incorporates stream-level quality checks and uncertainty estimates, combines modalities in a reliability-aware manner, and reports calibrated confidence alongside predictions to reduce the risk of overconfident outputs when evidence is weak [24,26,31]. Because labeled stress data are often limited and heterogeneous across individuals and collection conditions, AMHAT also supports sample-efficient training strategies, including auxiliary supervision from related affect targets (when available in a given public dataset) and adaptive sampling to emphasize informative examples during learning [25,29,34,37].

This paper makes three contributions. First, we describe an end-to-end, offline pipeline for privacy-preserving stress screening that enforces local processing and de-identified storage by design [27]. Second, we provide a unified multimodal treatment of speech, language, and interaction behavior with interpretable outputs and per-decision explanations [28,29,31–35,38,39]. Third, we detail a reliability-aware and calibration-oriented modeling strategy intended for small or noisy labeled datasets, including auxiliary affect supervision and adaptive sampling to mitigate sparse labels and inter-individual variability [24–26,37]. Importantly, the empirical evidence reported here is based on offline experiments using previously collected, publicly available datasets; we do not recruit participants and we do not collect new human-subject data in this study. In subject-exclusive offline cross-validation, interaction-based features are strongly discriminative, and multimodal fusion achieves high performance (AUROC = 0.964 and accuracy = 0.919, averaged across folds). Future work will focus on prospective validation under standardized stress-induction paradigms (TSST and MIST) only with appropriate IRB approval and participant consent procedures, while maintaining the same local, privacy-preserving workflow [24,26,27].

2 Related Work

Prior studies indicate that acute psychological stress can be inferred from measurable changes in speech acoustics, language production, and human-computer interaction (HCI) timing [29,33,37]. Our work is positioned within this literature but is constrained by an offline, privacy-preserving operating model: processing runs locally, and the pipeline retains only de-identified, fixed-format feature summaries rather than raw audio or raw typed content [27].

Speech-Based Stress Cues. Stress has been associated with changes in prosody and temporal organization of speech, including shifts in fundamental frequency (F0) statistics, intensity patterns, voice-quality descriptors, and pause and speaking-rate structure [32,33]. These effects are often small and can vary with task demands, microphone conditions, and background noise, which makes quality control and robust feature extraction important for applied systems [33,37]. Accordingly, privacy-compatible speech features are best treated as probabilistic evidence that can be down-weighted when audio quality is weak, rather than as a mandatory decision source [24,26,31].

Language Markers and Disfluency. Stress and elevated cognitive load can also affect linguistic form, including shifts in lexical choice, reduced lexical variety, and increased disfluency and repair phenomena [29,34,38]. In realistic conditions, transcript quality can vary, so methods that rely on transparent transcript summaries are a practical privacy-preserving compromise [34,38]. Treating transcript adequacy and consistency checks as part of the modeling workflow helps prevent noisy text from dominating downstream conclusions and supports clearer reporting of what evidence was available for a given decision [24,26].

Content-Free Interaction Dynamics (no typed content capture). Keyboard and mouse timing features provide behavioral signals that can be collected without capturing what was typed, and multiple studies have used them to distinguish stress-related states (including stress, arousal, and valence) in office-like interaction settings [37]. Reported markers include changes in keystroke timing, bursts of corrections/edits, and mouse movement summaries such as pauses, direction changes, and smoothness-related proxies. This literature directly motivates AMHAT's emphasis on timing-only interaction logging and is consistent with our offline results, where interaction features are the strongest unimodal signal and complement speech and language cues in fusion [28,29,39].

Multimodal Fusion, Missing Data, and Calibrated Confidence. Multimodal affect modeling can improve robustness by combining complementary evidence, but practical systems must handle absent or degraded streams and avoid overconfident outputs when inputs are weak or heterogeneous [23,37]. This has motivated reliability-aware fusion and explicit probability calibration, especially when datasets are limited or drawn from mixed conditions where miscalibration may remain high even when AUROC appears strong. In this setting, reporting calibrated probabilities alongside confidence or reliability flags is not an add-on; it is a core requirement for preventing fragile modalities from dominating the fused decision [24,26].

Learning with Limited and Heterogeneous Stress Labels. Public stress datasets often provide limited, noisy, or inconsistent labels across subjects and tasks, which encourages strategies that improve sample efficiency and reduce sensitivity to label scarcity [37]. Common directions include leveraging related supervision and prioritizing informative examples through adaptive sampling or active-learning principles [25]. This motivates AMHAT's use of auxiliary affect signals and adaptive sampling during training as practical mechanisms for stabilizing learning under constrained supervision, while keeping evaluation strictly offline and reproducible [34, 37].

Adjacent Healthcare Sensing and Cybersecurity Considerations. Although AMHAT targets stress screening rather than clinical diagnosis, it is informed by two adjacent research threads that motivate our emphasis on *privacy*, *local execution*, and *risk-aware reporting*. First, work on speech and behavioral analytics for neurological and cognitive-health applications (e.g., Alzheimer's-related detection studies) illustrates both the potential utility of non-invasive digital markers and the corresponding need to control exposure of sensitive personal signals when building practical pipelines [7]. Second, recent efforts that connect AI-based sensing with cyber-trust and secure system design (e.g., MAISON-style frameworks) highlight that model outputs can become sensitive artifacts themselves and should therefore be treated as protected information, with attention to data minimization, attack surface reduction, and audit-friendly operation [8]. In AMHAT, these considerations are operationalized by keeping all processing offline, avoiding retention of raw audio or typed content, and retaining only de-identified, fixed-format feature summaries and calibrated outputs that are designed to degrade gracefully under missing or low-quality inputs [24, 26, 27].

Broader Affective Computing and Multimodal Emotion Inference. AMHAT is also related to the broader affective-computing literature that infers emotion-, arousal-, or stress-adjacent states from multimodal behavioral signals. Prior work has explored multimodal emotion recognition by combining speech acoustics with linguistic and interaction or visual cues, often showing that complementary modalities can improve robustness relative to any single stream when signals are weak or noisy [9, 10, 13, 14, 21, 22]. Beyond classic emotion-recognition benchmarks, affect-aware human-centered systems have also appeared in interactive settings, including conversational or socially assistive agents and social-robot architectures that integrate perception and dialogue components, where reliable affect inference is treated as a supporting capability rather than a standalone classifier [15–18]. While AMHAT differs in scope and constraints, this line of work motivates a multimodal design and reinforces the importance of handling missing modalities and heterogeneous conditions when translating affect inference into deployable systems [11, 12, 19, 20].

3 System Overview

AMHAT is an offline, privacy-first pipeline for screening acute psychological stress. The pipeline is organized around three time-aligned input streams: (i) speech audio, (ii) language indicators computed from on-device speech-to-text outputs (or provided transcripts when available), and (iii) content-free keyboard and mouse event timing. AMHAT is designed so that processing remains local and only de-identified, fixed-format feature summaries and model outputs are retained; raw audio and raw typed content are neither stored nor transmitted [27]. In the offline evidence reported in this paper, AMHAT operates on previously collected, publicly available datasets downloaded under their original terms (no new participant recruitment and no new human-subject data collection).

Speech Stream. From each audio recording, AMHAT extracts established acoustic indicators that have been linked to stress-related changes, including pitch (F0) statistics, loudness (intensity) statistics, voice-quality descriptors, and pause and speaking-rate patterns. To reduce variability unrelated to stress, features are normalized within speaker/session when that grouping is available, or within the recording otherwise [32,33]. In addition, basic audio quality indicators are tracked so that unreliable speech segments can be down-weighted during fusion rather than driving overconfident decisions [24,26,31].

Language Stream. Speech-to-text can be executed on-device to generate transcripts locally; when a public dataset provides transcripts, AMHAT can use those transcripts directly. Regardless of the source, AMHAT does not require storing raw transcript text for downstream modeling. Instead, it converts transcripts into interpretable summary indicators such as word-category and pronoun ratios, lexical diversity and concreteness measures, sentiment and tone-style scores, and simple sentence-structure metrics [34,38]. Because stress and elevated cognitive load can increase disfluencies and repair behavior, the language stream also includes counts of common filled pauses and correction markers when they appear in the available transcript text [29]. Transcript adequacy checks are treated as part of reliability control so that low-quality text does not dominate the final output [24,26].

Interaction Stream (No Keylogging). AMHAT explicitly avoids keylogging. It uses only content-free interaction timing derived from keyboard and mouse events, not what a user typed. This enables computation of standard interaction dynamics such as key-hold duration, inter-key latency, correction and edit bursts, and cursor movement summaries. These signals are practical in office-like settings and can be strongly informative while remaining privacy-compatible because they do not require typed content capture [28,29,39].

Quality Checks and Reliability Signals. Before modeling, each stream is screened with lightweight quality rules. Examples include signal-quality and voice-activity coverage checks for speech, transcript confidence and minimum-length checks for language, and coverage plus outlier trimming for interaction timing [31]. If a stream is missing or fails quality checks, AMHAT does not force a decision from that stream. Instead, it marks the stream as unusable and either down-weights it or excludes it from fusion, relying on the remaining reliable evidence [24,26].

Modeling, Fusion, and Outputs. AMHAT supports unimodal models (one model per stream) and multimodal decision-level fusion. Fusion can be implemented as simple voting or as a small learned combiner that takes per-modality scores together with reliability signals [23,24]. Predicted probabilities are calibrated and accompanied by confidence flags so the system can reduce overconfident outputs and degrade gracefully when one modality is noisy or unavailable [26]. Depending on the label structure available in the public dataset and the intended use case, outputs can be a binary screen (stress vs. non-stress), an ordered severity level, or a continuous stress score.

Governance and Evaluation. Governance defaults are conservative: local-only execution, content-free interaction logging, encrypted storage for fixed-format summaries, and role-based access control [27]. Evaluation follows subject-exclusive splits and ablation studies to isolate the contribution of each modality and to benchmark performance under standard laboratory stress paradigms in offline analyses [26]. Consistent with prior work on calibration under limited or heterogeneous labels, we treat calibration as sensitive to sample size and data heterogeneity, and we avoid presenting fine-grained severity estimation as equivalent in difficulty to binary screening [24]. When stress labels are limited, AMHAT can improve sample efficiency by leveraging related affective supervision (when present in a public dataset) and adaptive sampling strategies during training, applied within-fold to preserve evaluation integrity [5,6,25,29,34,37].

4 Methods

4.1 Motivation and Offline Evaluation Setting

Acute psychological stress can affect how people speak, what they say, and how they interact with a keyboard and mouse [29,33,37,40]. AMHAT uses these signals in a local, privacy-preserving pipeline that produces interpretable stress-screening outputs and reports calibrated confidence. All evidence reported in this paper comes from *offline-only* experiments. We do not recruit participants and we do not collect new human-subject data. Instead, we download and use previously collected, publicly available datasets under their original terms and run all processing and evaluation locally in a laboratory simulation setting. The pipeline is designed so that raw audio and any raw typed content are not transmitted. For downstream analysis, AMHAT retains only fixed-format, de-identified feature summaries, stream-level signal-quality indicators (SQIs), and calibrated model outputs [27]. Calibration and uncertainty reporting are treated as core requirements for reliable screening under heterogeneous data conditions [24,26].

4.2 Speech Processing

Preprocessing. Given a discrete-time waveform $x[t]$, we remove DC offset by subtracting the within-recording mean. We then apply a standard pre-emphasis filter:

$$y[t] = x[t] - \alpha\, x[t-1], \qquad \alpha = 0.97. \tag{1}$$

We segment $y[t]$ into 25 ms frames with a 10 ms hop and apply a Hamming window:

$$w[n] = 0.54 - 0.46 \cos\left(\frac{2\pi n}{N-1}\right), \qquad 0 \leq n \leq N-1. \tag{2}$$

Voice activity detection (VAD) gates speech feature extraction: non-speech frames are excluded from speech-only acoustic statistics, but are retained to quantify silence and pausing behavior. When endpointing robustness matters, we treat VAD quality as part of stream usability rather than forcing a speech-based decision from unreliable audio [24,26,31].

Acoustic Feature Extraction. We extract established acoustic indicators commonly used in speech-under-stress analysis [32,33], spanning prosody and timing (fundamental frequency (F_0) summaries, intensity via RMS-based loudness in dB, voiced-frame ratio, and timing proxies related to speaking rate and pausing), voice quality (jitter, shimmer, and noise-related descriptors such as harmonics-to-noise ratio, HNR), and spectral descriptors (MFCCs with 13 coefficients plus first and second differences for 39 dimensions total, along with compact spectral summaries such as centroid, flux, roll-off, and entropy, and spectral-balance measures including the Alpha Ratio and Hammarberg Index to characterize broad energy distribution). To reduce variability unrelated to stress, features are normalized within speaker-session when that grouping is available, and otherwise within the recording:

$$\tilde{x}_t = \frac{x_t - \text{mean}}{\text{std} + \text{eps}}, \tag{3}$$

where mean and std are the mean and standard deviation computed within the chosen normalization scope, and eps > 0 ensures numerical stability.

Silence and Pause Segmentation. We segment silence using an energy threshold (e.g., below -35 dBFS) and an auxiliary zero-crossing-rate constraint (e.g., ZCR < 0.02) to reduce false pauses in noisy regions. We enforce a minimum pause duration of 150 ms and a minimum speech-segment duration of 200 ms. From the resulting segmentation, we compute pause rate (per minute), pause-duration summaries, and pause-to-speech ratios. When transcripts are provided in a dataset, we additionally count filled pauses (e.g., "um", "uh") as disfluency markers [29,34,38].

4.3 Language Processing

Transcription and Normalization. When a dataset provides transcripts, we use that text directly. Otherwise, speech-to-text can be executed locally to generate transcripts for feature computation. Text processing uses standard normalization (case folding and tokenization), with optional lemmatization. To preserve privacy-by-design, AMHAT does not require retaining raw transcript text for modeling; instead, it converts text into fixed-format summary indicators [27].

Lexical and Discourse Indicators. Each transcript is summarized using interpretable indicators used in prior work to capture stress- and cognitive-load-related shifts in language. We compute length measures (token counts, sentence counts, and mean sentence length), lexical diversity metrics including MATTR (50-token window) and MTLD (minimum segment length 10) along with related diversity summaries [38], coarse category-style ratios such as the content-to-function ratio and pronoun ratios, and discourse-level markers including counts of simple hedges and intensifiers plus disfluency counts when present in transcripts [29,34].

4.4 Interaction Processing

Content-Free Event Streams. AMHAT is designed to avoid keylogging. The interaction module uses only timing information from keydown/keyup and mouse events (millisecond timestamps) and coarse event/key-class codes (e.g., alphanumeric versus control). In this paper's offline evaluation, we use the interaction streams provided by public datasets; any raw typed content that may exist in a dataset is not stored or used by our pipeline [28,29,39].

Keystroke and Mouse Dynamics. From keyboard and mouse event timing streams, we compute content-free interaction features that prior work has linked to stress-related state differences. We derive keystroke timing measures such as dwell time (key-hold duration), flight time (inter-key latency), digraph and trigraph latencies, and edit or correction activity summaries (for example, backspace frequency and correction-latency aggregates). For mouse dynamics, we compute cursor-speed summaries, pause structure, direction-change counts, and short-horizon smoothness proxies. We quantify event-count variability in fixed-length windows using the Fano factor:

$$F = \frac{\mathrm{Var}(N)}{\mathbb{E}[N]}, \tag{4}$$

where N is the number of events within a window.

Outlier Handling and Robust Scaling. We remove implausible latencies (e.g., < 30 ms or > 2 s) and clip remaining values at the 1st and 99th percentiles.

For robust scaling, we use the median and median absolute deviation (MAD):

$$\hat{z} = \frac{x - \text{median}(x)}{1.4826\,\text{MAD}(x) + \epsilon}, \tag{5}$$

where $\text{MAD}(x) = \text{median}(|x - \text{median}(x)|)$ and $\epsilon > 0$ ensures numerical stability.

4.5 Feature Aggregation Across Time

To make features comparable across recordings and to capture short-term dynamics, we aggregate features at two resolutions: per segment (speech segments or utterances when available) and sliding windows of 5 s with 50% overlap. For each window, we compute distributional summaries (mean, median, standard deviation, minimum, maximum, skewness, kurtosis) and simple trends (linear slope and quadratic curvature).

4.6 Modeling and Multimodal Fusion

Single-Modality Models. All training and inference are performed offline. To prioritize transparency and stability under limited or heterogeneous labels, we use simple, interpretable models per modality: logistic regression for binary screening, ordinal regression for ordered labels, and ridge or elastic-net regression for continuous severity targets.

Leakage-Safe Feature Selection. To reduce optimistic bias, all feature selection is performed strictly within each training fold. We first rank features by mutual information (top-K), then apply sparse selection using L1-regularized logistic regression or elastic-net, fit only on the training portion of each fold.

Late Fusion with Reliability Weighting. To remain robust under missing or degraded streams, AMHAT uses late fusion. Each modality (m: acoustic, language, or interaction) outputs a score (or probability) s_m and a reliability weight w_m derived from simple signal-quality indicators (SQIs), such as audio SNR and voiced-frame ratio, transcript length/consistency checks, and interaction event coverage. Reliability-aware fusion is then:

$$\hat{y} = \frac{\sum_m I_m\, w_m\, s_m}{\sum_m I_m\, w_m + \epsilon}, \tag{6}$$

where $I_m \in \{0, 1\}$ indicates whether modality m is usable after quality checks, and ϵ prevents division by zero. This design supports graceful degradation under missing modalities and aligns with reliability-aware multimodal modeling practices [24,26].

Calibration and Uncertainty. We calibrate classification probabilities using isotonic regression to reduce overconfidence, which is particularly important in small or heterogeneous datasets [24,26]. For regression targets, we estimate uncertainty using bootstrap ensembles to produce 95% prediction intervals. Calibration quality is summarized using expected calibration error (ECE):

$$\mathrm{ECE} = \sum_{b=1}^{B} \frac{n_b}{n} \left|\mathrm{acc}(b) - \mathrm{conf}(b)\right|, \tag{7}$$

where B is the number of confidence bins, n_b is the number of samples in bin b, and $\mathrm{acc}(b)$ and $\mathrm{conf}(b)$ denote empirical accuracy and mean confidence in that bin. We report AUROC and class-weighted F1 for classification, and R^2 and Pearson r for regression.

4.7 Quality Control and Evaluation Protocol

We apply conservative quality checks before modeling. Examples include excluding severely clipped or extremely low-SNR audio, down-weighting low-coverage interaction segments, and skipping very short or inconsistent transcripts [24, 26,31]. Evaluation uses subject-exclusive splits based on subject identifiers provided by each dataset and within-fold preprocessing to reduce identity leakage and optimistic bias. Class imbalance is handled with class-weighted losses; any resampling is applied only within training folds.

4.8 Local Execution and Governance

All processing runs locally (Python 3.11; `librosa`/`parselmouth` for audio; `spaCy` or `Stanza` for NLP; and content-free interaction timing utilities). The pipeline retains only fixed-format feature summaries, SQIs, and calibrated predictions, supporting reproducible offline evaluation and privacy-by-design operation [27].

5 Evaluation Evidence

Offline-only evaluation (No New Human-Subject Data). All evidence reported in this section is based on *offline* experiments using previously collected, publicly available datasets obtained under their original terms. We do not recruit participants and we do not collect any new human-subject data for this study. The evaluation is a laboratory simulation that runs locally and stores only de-identified, fixed-format feature summaries and model outputs, consistent with the privacy-by-design operating model of AMHAT [27]. Because public stress datasets are often small and heterogeneous, we report both discrimination and calibration to avoid drawing strong conclusions from a single metric [24,26].

Offline Protocol (Leakage-Controlled Cross-Validation). We evaluate unimodal models and their multimodal fusion using subject-exclusive cross-validation, with all preprocessing fit *within* each training fold to reduce optimistic bias and identity leakage. Each modality is screened with stream-level quality checks, and modalities that fail these checks are marked as unusable for that sample rather than being forced into a decision [31]. We summarize discrimination using AUROC, accuracy, and F1. We summarize calibration using the Brier score and expected calibration error (ECE) [24,26].

Within-Participant Contrasts (When Baselines Exist). When a dataset provides a baseline block for a participant, we reduce between-person and device/session offsets by using within-participant standardized contrasts. For feature x from participant i at block t, we compute

$$z_{i,t} = \frac{x_{i,t} - m_{i,\text{base}}}{s_{i,\text{base}} + \epsilon}, \tag{8}$$

where $m_{i,\text{base}}$ and $s_{i,\text{base}}$ are the baseline mean and standard deviation for participant i, and $\epsilon > 0$ ensures numerical stability.

Reliability-and Uncertainty-Weighted Late Fusion. Each modality m outputs a probability $p_m \in (0,1)$ for the positive class and an uncertainty proxy u_m (larger means less reliable). For fusion, we combine modalities in log-odds space using an availability indicator $I_m \in \{0,1\}$ (1 if the modality passes quality checks, otherwise 0). We define the modality log-odds as

$$l_m = \log\left(\frac{p_m}{1-p_m}\right),$$

and compute a reliability weight

$$w_m = \frac{1}{u_m + c},$$

where $c > 0$ is a small constant. The fused log-odds is then

$$l_{\text{fused}} = \frac{\sum_m I_m\, w_m\, l_m}{\sum_m I_m\, w_m + \epsilon}, \tag{9}$$

with $\epsilon > 0$ preventing division by zero. If all streams fail quality checks (so all $I_m = 0$), the system reports *insufficient reliable input* rather than forcing a prediction, consistent with uncertainty-aware guidance for limited or heterogeneous data. The fused probability is obtained by the logistic mapping

$$\hat{p} = \frac{1}{1 + \exp(-l_{\text{fused}})}. \tag{10}$$

Calibration Assessment. We track probability calibration with expected calibration error (ECE):

$$\mathrm{ECE} = \sum_{b=1}^{B} \frac{n_b}{n} \left| \mathrm{acc}(b) - \mathrm{conf}(b) \right|, \tag{11}$$

where B is the number of confidence bins, n_b is the number of samples in bin b, and $\mathrm{acc}(b)$ and $\mathrm{conf}(b)$ are the empirical accuracy and mean predicted confidence in that bin, respectively. We also report the Brier score, which summarizes squared error in predicted probabilities (Fig. 1).

Table 1. Cross-validated offline results (mean ± SD across folds). Lower is better for Brier and ECE.

Model	AUROC (SD)	Acc (SD)	F1 (SD)	Brier (SD)	ECE (SD)
Acoustic	0.742 (0.050)	0.686 (0.045)	0.648 (0.066)	0.200 (0.021)	0.069 (0.035)
Lexical	0.682 (0.038)	0.635 (0.031)	0.624 (0.044)	0.218 (0.016)	0.058 (0.020)
HCI	0.926 (0.024)	0.903 (0.027)	0.908 (0.028)	0.087 (0.024)	0.092 (0.026)
Fusion	0.964 (0.016)	0.919 (0.025)	0.924 (0.026)	0.089 (0.015)	0.149 (0.012)

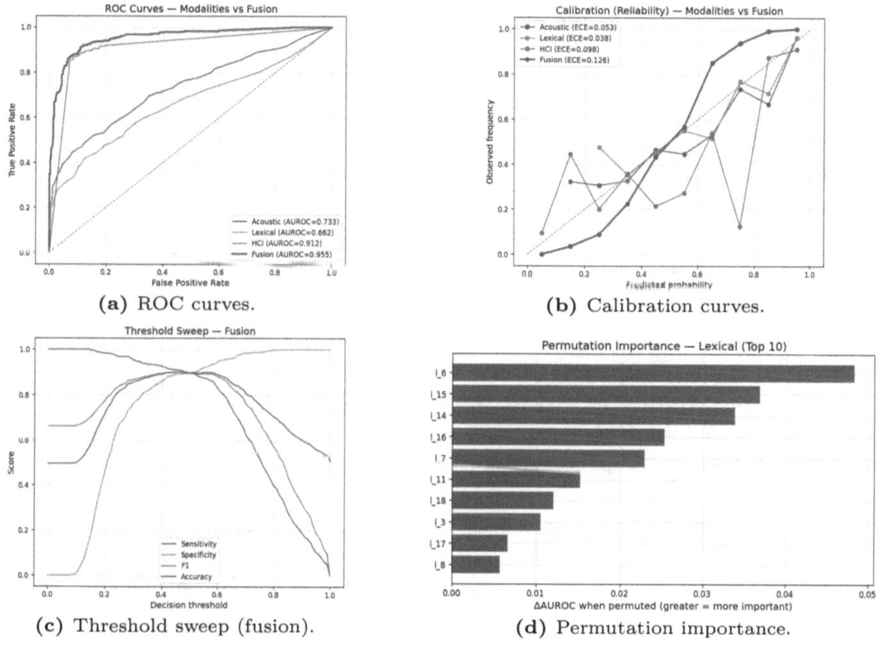

(a) ROC curves. (b) Calibration curves. (c) Threshold sweep (fusion). (d) Permutation importance.

Fig. 1. Offline evaluation plots for discrimination, calibration, operating-threshold behavior, and feature-family importance.

6 Cognitive Changes Under Stress

Acute psychological stress can affect how people speak, what they say, and how they interact with a keyboard and mouse [33,37]. AMHAT connects these behavioral effects to an *offline, privacy-first* laboratory pipeline: all computation runs locally, raw audio and raw typed content are not retained, and the system stores only de-identified, fixed-format feature summaries together with calibrated outputs [27]. This section links the feature families used by AMHAT to widely reported mechanisms through which acute stress can shift attention, increase working-memory demands, and perturb motor and executive control [28,29,39]. All results in this paper are from offline experiments on previously collected, publicly available datasets used under their original terms, with no prospective recruitment and no new human-subject data collection in this study.

6.1 Cognitive Mechanisms and Operational Markers

We map plausible cognitive effects of acute stress to observable markers across speech, language, and content-free interaction timing, focusing on features that remain interpretable and do not require storing sensitive content [28,39].

Attention and Self-monitoring. Stress-related shifts in attention and monitoring can appear as changes in pausing and repair behavior. AMHAT captures these effects using (i) silent-pause rate and pause-to-speech ratio, (ii) longer turn-initial latencies, and (iii) increased filled pauses and self-corrections when they are present in available transcripts (e.g., "um/uh" and simple repair tokens) [33,38]. These markers are reported as summary counts and timing statistics rather than stored raw content.

Working-Memory Load. Higher working-memory load can reduce linguistic variability and simplify production. We measure this using shorter clauses, reduced lexical diversity, optional coarse syntactic summaries (only when parsing is reliable), and shifts in content-to-function ratios [38]. When language-model scoring is used, we also compute surprisal (negative log-probability under a language model) as a stability indicator for produced sequences, noting that surprisal can increase under less stable production even when sentences become shorter [29].

Speech Motor Control. Stress can influence respiratory control, phonation stability, and prosodic timing. We measure this via increased variability in fundamental frequency (F_0) and intensity, voice-quality perturbation measures (jitter and shimmer), and timing features such as speaking-rate proxies and shorter continuous speech segments (shorter "breath groups") [32]. These indicators are extracted locally from the waveform and summarized into fixed-format statistics; when voice-activity detection or endpointing is unreliable, speech evidence is treated as weaker rather than forced into a decision [31].

Executive Control Under Time Pressure. Time pressure and elevated cognitive load can disrupt executive control and fine motor timing. AMHAT captures this using content-free interaction dynamics such as heavier right tails in inter-keystroke intervals, longer correction episodes, longer key dwell times, and higher mouse-path entropy as a summary of less efficient cursor trajectories [28,29,39]. Consistent with our offline evidence (Table 1), interaction-timing features are often strongly discriminative while remaining content-free and practical in office-like settings.

6.2 Statistical Models

To support inference while accounting for persistent subject- and device-level offsets, we use mixed-effects models that can be fully pre-specified prior to offline evaluation.

Binary Stress. For binary labels, we fit a mixed-effects logistic regression:

$$\text{logit}(\Pr(y_i = 1 \mid x_i)) = \beta_0 + x_i^\top \beta + u_{s[i]} + v_{d[i]}, \tag{12}$$

where $\text{logit}(p) = \log\left(\frac{p}{1-p}\right)$, x_i is the feature vector for sample i, and $s[i]$ and $d[i]$ index the subject and device, respectively. Random intercepts model persistent offsets:

$$u_s \sim \mathcal{N}(0, \sigma_u^2), \qquad v_d \sim \mathcal{N}(0, \sigma_v^2). \tag{13}$$

Ordinal Severity. For ordered severity levels, we use a cumulative-link mixed model:

$$\Pr(y_i \leq k \mid x_i) = \text{logistic}(\theta_k - \eta_i)\,, \qquad \eta_i = x_i^\top \beta + u_{s[i]} + v_{d[i]}, \tag{14}$$

where $\text{logistic}(z) = \frac{1}{1+\exp(-z)}$ and the thresholds satisfy $\theta_1 < \theta_2 < \cdots < \theta_{K-1}$.

6.3 Uncertainty-Aware Fusion and Calibration

AMHAT uses late fusion to combine modalities while remaining robust to missing or low-quality streams and to reduce overconfident outputs when evidence is weak or heterogeneous. Each modality m produces a calibrated posterior p_m over C classes and a usability indicator $I_m \in \{0, 1\}$ from stream-level quality checks (e.g., audio quality, transcript adequacy, interaction coverage) [24,26].

Entropy-Based Fusion Weights. We convert posterior uncertainty into a bounded weight using normalized entropy:

$$w_m = I_m\left(1 - \frac{H(p_m)}{\log C}\right), \tag{15}$$

with

$$H(p_m) = -\sum_{c=1}^{C} p_m(c) \log p_m(c). \tag{16}$$

This weighting assigns larger influence to confident (low-entropy) calibrated posteriors and reduces the influence of uncertain streams. The fused posterior is:

$$p_{\text{fused}}(c) = \frac{\sum_m w_m \, p_m(c)}{\sum_m w_m + \epsilon}, \tag{17}$$

where $\epsilon > 0$ prevents division by zero. If all $I_m = 0$ (no usable modality), the system reports insufficient reliable input rather than forcing a prediction.

Calibration Metrics. We evaluate calibration using expected calibration error (ECE) and the Brier score [24,26]. With confidence bins $\{B_k\}_{k=1}^{K}$,

$$\text{ECE} = \sum_{k=1}^{K} \frac{|B_k|}{n} \left|\text{acc}(B_k) - \text{conf}(B_k)\right|, \tag{18}$$

where $\text{acc}(B_k)$ is empirical accuracy in bin k and $\text{conf}(B_k)$ is mean predicted confidence in that bin. For binary outcomes, the Brier score is:

$$\text{Brier} = \frac{1}{n} \sum_{i=1}^{n} (\hat{p}_i - y_i)^2 . \tag{19}$$

These metrics complement discrimination results by quantifying whether predicted probabilities remain reliable under limited and heterogeneous stress labels.

7 Conclusion

Acute psychological stress can be reflected in how people speak, what they say, and how they interact with a keyboard and mouse [28–30,32,33,38,39]. This paper presented AMHAT, an offline, laboratory pipeline for privacy-preserving stress screening that runs locally so that raw audio and any typed content are not transmitted or retained. Instead, AMHAT stores only de-identified, fixed-format feature summaries, stream-quality indicators, and calibrated model outputs to reduce exposure to sensitive data while supporting reproducible analysis [27].

AMHAT combines three complementary, privacy-compatible evidence sources. Speech features summarize prosody, voice quality, and temporal organization that have been associated with stress-related changes. Language features summarize transcripts using interpretable indicators without requiring raw-text storage for modeling [29,34,38]. Interaction features use content-free timing from keyboard and mouse events, which is practical in office-like settings and avoids keylogging by design. To avoid overconfident decisions when evidence is weak, AMHAT applies conservative stream-level checks and reliability/uncertainty-aware fusion so that missing or degraded inputs are down-weighted or excluded rather than treated as equally informative [31].

All results reported here are from offline-only experiments using previously collected, publicly available datasets, with no new participant recruitment and no new data collection in this study. Under subject-exclusive cross-validation, interaction timing features provided the strongest unimodal signal, and multimodal fusion achieved the best overall discrimination (Fusion AUROC = 0.964, Accuracy = 0.919). At the same time, calibration remained sensitive under limited and heterogeneous stress labels, reinforcing the need to report calibrated probabilities and reliability indicators alongside discrimination metrics.

Key limitations follow directly from the available public data. Labels and recording conditions vary across datasets, devices, and tasks, so the reported numbers should be interpreted as leakage-controlled offline evidence rather than a claim of field robustness. In addition, transcript-based indicators depend on transcript adequacy, and reliability checks are therefore essential to prevent noisy text (or sparse interaction coverage) from dominating the fused output [31].

Future work will focus on prospective validation under standardized stress-induction paradigms (TSST and MIST) with appropriate IRB approval, informed consent, and debriefing procedures, and with explicit reporting of operating thresholds, calibration behavior, and robustness across relevant subgroups. In parallel, learning under sparse stress labels can be improved by leveraging related affective targets available in some public datasets and by emphasizing informative training examples via adaptive sampling, while keeping evaluation leakage-controlled and reproducible [25,29,34,37].

Overall, AMHAT's core design choice is to treat reliability and calibration as part of the system output. Reliability-aware fusion supports interpretable, privacy-preserving stress screening that degrades gracefully when one or more modalities are missing or low quality [24,26,27].

Acknowledgments. This material is based upon work supported by the National Science Foundation under Grant No. 2331568. The authors thank Dr. Michele Snyder, Dr. Kambiz Ghazinour, and Rebecca Snyder for their administrative and logistical support that facilitated completion of this work. Any opinions, findings, and conclusions or recommendations expressed in this material are those of the author(s) and do not necessarily reflect the views of the National Science Foundation.

References

1. Guntuku, S.C., Yaden, D.S., Kern, M.L., Ungar, L.: Detecting depression and mental illness on social media: an integrative review. Curr. Opin. Behav. Sci. **18**, 43–49 (2017)
2. Coppersmith, G., Harman, C., Dredze, M.: Quantifying mental health signals in Twitter. In: Proc. Workshop on Computational Linguistics and Clinical Psychology: From Linguistic Signal to Clinical Reality, pp. 51–60 (2014)
3. Chancellor, S., De Choudhury, E.M.L.: Methods in predictive techniques for mental health status on social media: a critical review. NPJ Digital Med.**2**, 10 (2019)
4. Yazdavar, A.H.: Multimodal classification of mental health from social media postings. PLoS ONE **15**(4), e0226248 (2020)
5. Ghayoumi, M.: Deep Learning in Practice. CRC Press, Boca Raton, FL, USA (2021)

6. Ghayoumi, M.: Generative Adversarial Networks in Practice. CRC Press, Boca Raton, FL, USA (2023)
7. Ghayoumi, M., Ghazinour, K.: Detection of Alzheimer's Disease using bidirectional LSTM and attention mechanisms. Mach. Learn. Appl.: Int. J **12**(1), 147–160 (2025). https://doi.org/10.5121/mlaij.2025.12110
8. Babaev, I., Packer, T., Ghayoumi, M., Ghazinour, K.: MAISON: A model for effective hybrid management of cybersecurity and cyber-trust. Int. J. Inf. Technol. (IJIT) (2024)
9. Bansal, A., Ghayoumi, M.: Symmetry-based hybrid model to improve facial expressions prediction in the wild during conversational head movements. Adv. Life Sci. **13**(1–2) (2021)
10. Ghayoumi, M., Bansal, A.K.: Emotion analysis using facial key points and dihedral group. Int. J. Adv. Stud. Comput. Sci. Eng. (2017) IJASCSE
11. Ghayoumi, M., Tafar, M., Bansal, A.K.: A formal approach for multimodal integration to drive emotions. J. Visual Lang. Sentient Syst. 48–54 (2016)
12. Ghayoumi, M., Ghazinour, K.: Early Alzheimer's detection using bidirectional LSTM and attention mechanisms in eye tracking. In: Proceedings of the CSCE Conference (2024)
13. Ghayoumi, M., Ghazinour, K.: Revolutionizing cardiovascular diagnostics: Advanced AI algorithms for real-time detection of cardiac anomalies. In: Proceedings of the CSCE Conference (2024)
14. Ghayoumi, M., Ghazinour, K.: Advancing MAISON: Integrating deep learning and social dynamics in cyberbullying detection and prevention. In: Proceedings of the APCS Conference (2024)
15. Bansal, A.K., Ghayoumi, M.: A hybrid model to improve occluded facial expressions prediction in the wild during conversational head movements. In: Proceedings of the Intelligence Conference (2021)
16. Ghayoumi, M., Bansal, T.: Fuzzy knowledge-based architecture for learning and interaction in social robots. In: Proceedings of the AI-HRI Workshop (2019)
17. Ghayoumi, M.: Cognitive-based architecture for emotion in social robots. In: Proceedings of the HRI Conference (2018)
18. Ghayoumi, M., Bansal, A.K.: Real emotion recognition by detecting symmetry patterns with dihedral group. In: Proceedings of the 3rd International Conference Mathematics and Computers in Sciences and Industry (MCSI), Greece (2016)
19. Ghayoumi, M., Bansal, A.K.: Follower robot with an optimized gesture recognition system. In: Proceedings of the Robotics: Science and Systems (RSS) Conference, USA (2016)
20. Ghayoumi, M., Tafar, M., Bansal, A.K.: Towards formal multimodal analysis of emotions for affective computing. In: Proceedings of the 22nd International Conference Distributed Multimedia Systems (DMS), pp. 48–54 (2016)
21. Ghayoumi, M., Bansal, A.K.: Emotion in robots using convolutional neural networks. In: Proceedings of the 8th International Conference Social Robotics (ICSR), pp. 285–295 USA (2016)
22. Ghayoumi, M., Bansal, A.K.: Multimodal convolutional neural networks model for emotion in robots. In: Proceedings of the Future Technologies Conference (FTC), USA (2016)
23. Chao, L., Tao, J., Yang, M., Li, Y., Wen, Z.: Audio visual emotion recognition with temporal alignment and perception attention (2016). arXiv:1603.08321 arXiv preprint

24. Lakshminarayanan, B., Pritzel, A., Blundell, C.: Simple and scalable predictive uncertainty estimation using deep ensembles. Adv. Neural Inf. Process. Syst. (NeurIPS) **30** (2017)
25. Settles, B.: Active learning literature survey, Univ. Wisconsin-Madison, Computer Sciences (2009) 1648 Technical Report
26. Loftus, T.J.: Uncertainty-aware deep learning in healthcare: a scoping review. NPJ Digital Med. **5**(1), 164 (2022)
27. U.S. Department of Health and Human Services: The Belmont Report: Ethical Principles and Guidelines for the Protection of Human Subjects of Research. The National Commission for the Protection of Human Subjects of Biomedical and Behavioral Research (1979)
28. Lau, S.-H.: Stress Detection for Keystroke Dynamics. Carnegie Mellon University (2018). https://doi.org/10.1184/R1/6723227.v3
29. Vizer, B., Zhou, B., Sears, A.: Automated stress detection using keystroke and linguistic features: an exploratory study. Int. J. Human-Comput. Stud. **67**(10), 870–886 (2009)
30. Sano, M.: Identifying objective physiological markers and modifiable behaviors for self-reported stress and mental health status using wearable sensors and mobile phones. JMIR Mental Health **5**(2) (2018)
31. Li, X., Liu, H., Zheng, Y., Xu, B.: Robust speech endpoint detection based on improved adaptive band-partitioning spectral entropy. In: Proceedings of the Life System Modeling and Simulation Conference (LSMS), pp. 36–45 (2007)
32. Fabris, C., De Colle, W., Sparacino, G.: Voice disorders assessed by (cross-) Sample Entropy of electroglottogram and microphone signals. Biomed. Signal Process. Control **8**(6), 920–926 (2013). https://doi.org/10.1016/j.bspc.2013.06.009
33. Hansen, J., Patil, S.: Speech under stress: Analysis, modeling and recognition. In: Speech Recognition. IntechOpen, ch, vol. 4, pp. 69–91 (2017). https://doi.org/10.5772/67106
34. De Leon, A., Zareei, M.: Breaking barriers in sentiment analysis and text emotion detection: toward a unified assessment framework. IEEE Access **11**, 1–1 (2023)
35. Quan, C., Ren, F.: Finding emotional focus for emotion recognition at sentence level. Chin. J. Electron. **22**(1), 99–103 (2013)
36. Zhang, Y.: DialoGPT: Large-scale generative pretraining for conversational response generation. In: Proceedings of the ACL 2020 System Demonstrations, pp. 270–278 (2020)
37. Khalil, R.A., Jones, E., Babar, M.I., Jan, T., Zafar, M.H., Alhussain, T.: Speech emotion recognition using deep learning techniques: a review. IEEE Access **7**, 117327–117345 (2019)
38. McCarthy, P.M., Jarvis, S.: MTLD, vocd-D, and HD-D: a validation study of sophisticated approaches to lexical diversity assessment. Behav. Res. Methods **42**(2), 381–392 (2010)
39. Chunawale, A., Bedekar, M.: Machine learning based stress detection using keyboard typing behavior. Int. J. Recent Innov. Trends Comput. Commun. **11**(8s), 376–380 (2023)
40. Ghayoumi, M.: Enhancing Efficiency and Regularization in Convolutional Neural Networks: Strategies for Optimized Dropout. AI **6**, 111 (2025)

3D Dynamic Convolution Network for Early Diagnosis of Alzheimer's Disease

Taymaz Akan[1](✉), Peri Güneş[2], Sait Alp[3], and Mohammad Alfrad Nobel Bhuiyan[1]

[1] Department of Medicine, LSU Health Shreveport, Shreveport, LA, USA
taymaz.farshi@gmail.com, Nobel.Bhuiyan@lsuhs.edu
[2] Faculty of Engineering, Dogus University, Istanbul, Turkey
[3] Department of Artificial Intelligence Engineering, Trabzon University, Trabzon, Turkey

Abstract. The accurate and timely diagnosis of neurodegenerative brain disorders continues to be a significant challenge in modern neuroimaging. Traditional convolutional neural networks (CNNs) may struggle to capture the complex spatial relationships and dynamic feature patterns found in 3D medical imaging data. This study uses a novel deep learning framework incorporating 3D Omni-Dimensional Dynamic Convolution (ODConv) for robust multi-class classification of brain disorders using structural MRI scans. Unlike conventional convolutions, ODConv dynamically adapts kernel parameters across spatial, channel, and filter dimensions, allowing for more expressive feature learning from volumetric brain data. We tested our approach on the ADNI dataset and achieved superior classification performance with an overall accuracy of 96%, precision of 0.96, recall of 0.96, and F1-score of 0.96 for Alzheimer's disease (AD), mild cognitive impairment (MCI), and cognitively normal (CN). The model is compared to other state-of-the-art methods. The findings demonstrate the potential of dynamic convolution architectures for advancing computer-aided diagnosis in neuroimaging applications.

Keywords: Brain Disorders · 3D Convolution · Dynamic Convolution · Deep Learning · Structural MRI · Neuroimaging

1 Introduction

Neurodegenerative disorders have become an emerging global health challenge, with Alzheimer's disease (AD) as the most common type of dementia globally [1–3]. Prompt and precise diagnosis is essential for timely intervention and treatment strategy formulation. Structural magnetic resonance imaging (MRI) has become a non-invasive and widely accessible method for evaluating brain morphology and identifying pathological changes that are linked to neurodegeneration. Recent advances in deep learning have revolutionized medical image analysis, with convolutional neural networks (CNNs) demonstrating remarkable success in automated diagnosis tasks [4–6]. However, the use of static convolution kernels with fixed parameters during inference in traditional CNNs may limit their ability to adjust to subtle pathological patterns and a variety of anatomical variations found in brain imaging data.

H. R. Arabnia et al. (Eds.): CSCI 2025, CCIS 2971, pp. 142–148, 2027.
https://doi.org/10.1007/978-3-032-28463-1_11

Omni-Dimensional Dynamic Convolution (ODConv) [7] address this limitation by incorporating learnable attention mechanisms that dynamically adjust convolution kernels across multiple dimensions: spatial location, input channel, and output filter. This multi-dimensional adaptability facilitates the network to concentrate on disease-relevant features while suppressing irrelevant information, thereby enhancing accuracy and interpretability. The application of dynamic convolutions to 3D medical imaging remains underexplored, despite the potential of these techniques in computer vision tasks. The volumetric nature of brain MRI presents a variety of unique challenges, such as the necessity to capture both local anatomical details and global structural patterns, as well as the high dimensionality and limited training samples. A 3D ODConv-based architecture that is specifically designed for brain disorder classification is developed in this work to address these challenges.

2 Materials and Methods

2.1 Dataset and Preprocessing

We used Alzheimer's Disease Neuroimaging Initiative (ADNI) [8]: "Complete 1Yr 1.5 T" data collection T1-weighted structural MRI scans. The training set included 1,951 scans (82%) with 405 subjects with Alzheimer's disease (AD), 946 subjects with mild cognitive impairment (MCI), and 600 subjects with cognitive normalcy (CN). 342 scans (18%) with 105 CN, 166 MCI, and 71 AD cases made up the test set. There were 1,341 male and 953 female participants, with an average age of 75 ± 6.6 years.

ADNI de-identified all data and made it publicly available, so no written informed consent was required for this study. The ADNI investigators obtained ethical approval and informed consent during the original data collection. ADNI policy states that secondary analyses of these de-identified data do not require additional ethical approval from individual researchers or institutions.

To ensure anatomical consistency, all MRI volumes underwent standardized preprocessing, which included (1) skull stripping to remove non-brain tissues, (2) spatial normalization to Montreal Neurological Institute (MNI) space (See Fig. 1) using Statistical Parametric Mapping 12 (SPM12)[9] and Computational Anatomy Toolbox (CAT12), and (3) intensity normalization. The volumes were resampled to a uniform size of 128 × 128 × 128 voxels. Only the central 50 slices with significant brain tissue were retained for analysis.

2.2 3D Omni-Dimensional Dynamic Convolution

The core innovation of our approach lies in the integration of 3D ODConv layers, which extend traditional static convolutions by introducing dynamic kernel generation. Given an input feature map $X \in R^{(C \times D \times H \times W)}$, where C represents channels and D × H × W denotes spatial dimensions, ODConv computes attention coefficients across four complementary dimensions:

1. Spatial-wise attention: adapts kernels based on spatial context
2. Channel-wise attention: modulates different input channels

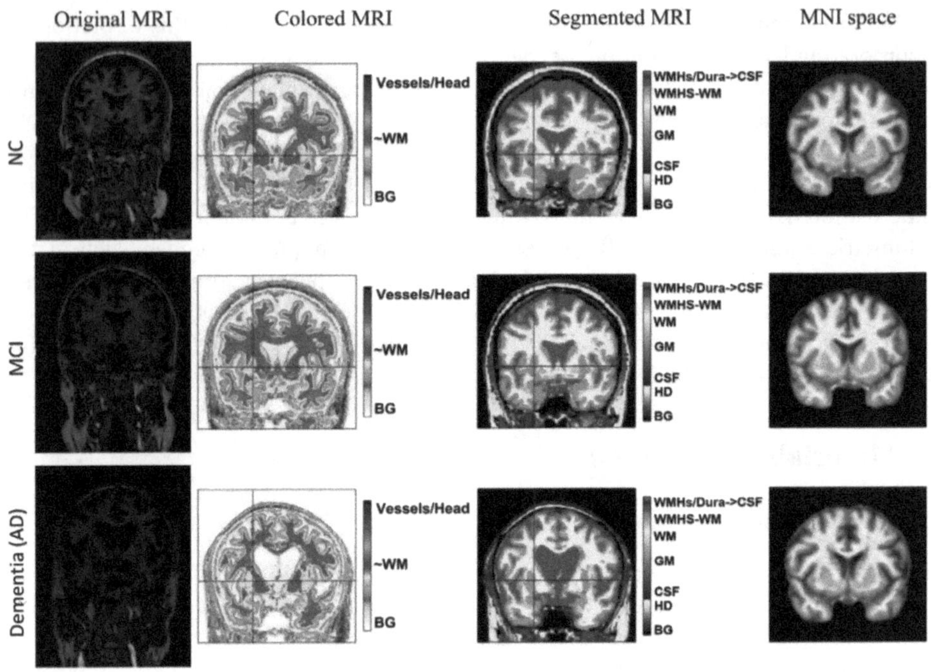

Fig. 1. Illustrates (from left to right) MRI samples from cognitively normal (CN), mild cognitive impairment (MCI), and Alzheimer's disease (AD) subjects, including the original MRI, colored tissue intensity map, segmented brain structures, and normalized MRI in MNI space. CAT12 was used to segment and strip the skull.

3. Filter-wise attention: controls output filter responses
4. Kernel-wise attention: adjusts individual kernel elements

The final dynamic convolution kernel W* is computed by multiplying attention coefficients across all dimensions with learnable base kernels, where n is the number of parallel convolution kernels.

2.3 Network Architecture

Our 3D ODConv network is comprised of five major components: Input Module: 3D convolution (kernel size $7 \times 7 \times 7$), batch normalization, and ReLU activation; Four sequential ODConv Blocks: 3D ODConv layer with dynamic kernel adaptation, batch normalization, ReLU activation, and 3D max pooling ($2 \times 2 \times 2$); Global Feature Aggregation: adaptive average pooling and dropout ($p = 0.5$); Classification Head: two fully connected layers that output class probabilities via SoftMax. Channel dimensions gradually increase: $64 \rightarrow 128 \rightarrow 256 \rightarrow 512$ to capture hierarchical features. The total network has approximately 15.8 million trainable parameters.

2.4 Training Protocol

Training utilized the AdamW optimizer with an initial learning rate of 1×10^{-4}. Weight decay of 1×10^{-4}, and cosine annealing schedule for learning rate decay. We implemented cross-entropy loss with class weights to mitigate class imbalance. The model was trained with batch size 32 for 100 epochs. As the callback, the model checkpoint was configured to save only the optimal solution discovered during training based on the loss function evaluation performed during validation. PyTorch on NVIDIA A100 GPUs with mixed precision training was used for all experiments.

3 Results

3.1 Classification Performance

99 Our 3D ODConv network achieved an overall accuracy of 93% on the test set, with balanced performance across all classes. The precision, recall, and F1-scores for AD were 0.92, 0.90, and 0.91 respectively. For MCI, the model achieved recall of 0.96 with precision of 0.95, indicating strong sensitivity for this intermediate stage. CN classification showed precision of 0.92 and recall of 0.94, with F1-score of 0.93. Table 1 presents detailed metrics.

Table 1. Classification Performance on Test Set

Class	Precision	Recall	F1-Score	Support
AD	0.92	0.90	0.91	71
MCI	0.95	0.96	0.95	166
CN	0.92	0.94	0.93	105
Accuracy			0.93	342
Macro Avg	0.93	0.93	0.93	342
Weighted Avg	0.93	0.93	0.93	342

3.2 Comparison with Baseline Models

We evaluated our 3D ODConv network against four baseline architectures (Table 2). Our method surpassed CNN-BiLSTM and CNN-Transformer (both achieving 0.90 accuracy), ViT-BiLSTM (0.89 accuracy), and demonstrated performance akin to ViT-Transformer (0.91 accuracy), with a 2–4% point enhancement over the majority of baselines.

Table 2. Comparison with Baseline Models

Architecture	ACC	Precision	Recall	F1-Score
CNN-BiLSTM [10]	0.90	0.90	0.90	0.89
CNN-Transformer [10]	0.90	0.90	0.90	0.89
ViT-BiLSTM [10]	0.89	0.90	0.89	0.89
ViT-Transformer [11]	0.91	0.92	0.91	0.90
3D ODConv	**0.93**	**0.93**	**0.93**	**0.93**

3.3 Ablation Studies

To clarify the role of dynamic components, we performed ablation experiments by sequentially eliminating attention mechanisms: The Full Model (incorporating all attentions) achieves 93% accuracy; the exclusion of spatial attention results in 90% accuracy (a decrease of 3%); omitting channel attention yields 89% accuracy (a decrease of 4%); removing filter attention leads to 91% accuracy (a decrease of 2%); the static convolution baseline records 87% accuracy (a decrease of 6%). The results affirm that every attention dimension significantly impacts performance, with channel-wise attention yielding the most substantial individual contribution.

4 Discussion

This study illustrates that 3D Omni-Dimensional Dynamic Convolution offers benefits for brain disorder classification in comparison to static convolution architectures. The attained 93% accuracy signifies a substantial enhancement over conventional CNN methodologies, while preserving interpretability via attention visualization. While the performance corresponds to contemporary transformer-based methods, the dynamic convolution mechanism provides advantages in computational efficiency.

The efficacy of ODConv can be attributed to three principal factors. Initially, multi-dimensional attention facilitates adaptive feature extraction that accommodates anatomical variability among subjects and disease stages. Secondly, dynamic kernel generation enables the network to concentrate on disease-relevant patterns while reducing confounding variables. Third, the 3D formulation maintains volumetric context, encapsulating both localized morphological alterations and overarching structural relationships.

Our ablation studies show that all four attention dimensions contribute synergistically, with channel-wise attention making the greatest individual contribution. This implies that different input channels have varying relevance for disease classification, and dynamic weighting enhances discriminative power. The attention visualizations are consistent with established neuroscientific knowledge about AD progression, which increases clinical trust in model predictions.

Several limitations should be acknowledged. First, our study only used structural MRI and did not include multimodal data (PET, CSF biomarkers, and genetics), which could have improved diagnostic accuracy. Second, the cross-sectional design does not account for longitudinal disease progression. Third, external validation of independent datasets is required to determine real-world clinical utility. Dynamic convolution's computational overhead (roughly 30% longer training time) is a reasonable trade-off for increased interpretability and accuracy.

5 Conclusion

We introduced a novel 3D Omni-Dimensional Dynamic Convolution network for automated brain disorder classification from structural MRI. By dynamically adapting convolution kernels across spatial, channel, filter, and kernel dimensions, our approach achieves 93% accuracy in distinguishing between Alzheimer's disease, mild cognitive impairment, and cognitively normal subjects—demonstrating competitive performance with recent deep learning architectures. Comprehensive ablation studies have shown that multi-dimensional dynamic adaptation is effective at capturing complex neuroanatomical patterns. Attention visualizations show that the model focuses on clinically relevant brain regions, resulting in interpretable insights that are consistent with established neuroscience knowledge. This work illustrates the potential of dynamic convolution architectures for medical image analysis, particularly for volumetric data, where adaptive feature extraction can account for anatomical variability. Future directions include extending of the methodology to multimodal imaging, the integration of longitudinal data for progression forecasting, the enhancement of computational efficiency, and the validation of clinical application through prospective studies involving varied populations.

Acknowledgments. Data collection and sharing for this project were funded by the Alzheimer's Disease Neuroimaging Initiative (ADNI). This work was supported by NIH grants P20GM121307, R01HL172970, R01HL145753, and R01HL149264.

Disclosure of Interests. The authors have no competing interests to declare that are relevant to the content of this article.

References

1. Heston, L.L., Lowther, D.L.W., Leventhal, C.M.: Alzheimer's disease: a family study. Arch. Neurol. **15**, 225–233 (1966). https://doi.org/10.1001/ARCHNEUR.1966.00470150003001
2. Prince, M., Bryce, R., Albanese, E., et al.: The global prevalence of dementia: a systematic review and metaanalysis. Alzheimer's Dement. **9**(63–75), e2 (2013). https://doi.org/10.1016/j.jalz.2012.11.007
3. Akan, T., Akan, S., Alp, S., et al.: Deep learning in neuroimaging for neurodegenerative diseases: state-of-the art, challenges, and opportunities. J. Neurol. Sci., 123735 (2025)
4. Jain, R., Jain, N., Aggarwal, A., Hemanth, D.J.: Convolutional neural network based Alzheimer's disease classification from magnetic resonance brain images. Cogn. Syst. Res. **57**, 147–159 (2019). https://doi.org/10.1016/J.COGSYS.2018.12.015
5. Li, J., Wei, Y., Wang, C., et al.: 3-D CNN-based multichannel contrastive learning for Alzheimer's disease automatic diagnosis. IEEE Trans. Instrum. Meas. **71** (2022). https://doi.org/10.1109/TIM.2022.3162265

6. Yang, J., Huang, X., He, Y., et al.: Reinventing 2D convolutions for 3D Images. IEEE J. Biomed. Health Inform. **25**, 3009–3018 (2021). https://doi.org/10.1109/JBHI.2021.3049452
7. Li, C., Zhou, A., Yao, A.: Omni-dimensional dynamic convolution. arXiv preprint arXiv:220907947 (2022)
8. ADNI | Alzheimer's Disease Neuroimaging Initiative. https://adni.loni.usc.edu/. Accessed 3 Apr 2023
9. Ashburner, J., Friston, K.J.: Unified segmentation. Neuroimage **26**, 839–851 (2005). https://doi.org/10.1016/J.NEUROIMAGE.2005.02.018
10. Akan, T., Alp, S., Bhuiyan, M.A.N.: Vision transformers and Bi-LSTM for Alzheimer's disease diagnosis from 3D MRI. In: 2023 Congress in Computer Science, Computer Engineering, & Applied Computing (CSCE), pp. 530–535. IEEE (2023)
11. Alp, S., Akan, T., Bhuiyan, M.S., et al.: Joint transformer architecture in brain 3D MRI classification: its application in Alzheimer's disease classification. Sci. Rep. **14**, 8996 (2024)

Health Informatics, Artificial Intelligence, and Applications

Rethinking Population Construction for Clinical AI: Patient Similarity for Surgical Site Infection Prediction

Stephen Russell[1(✉)], Fabio Montes Suros[2], Kamal Babaei Sonbolabadi[2], and Jia Zeng[2]

[1] University of West Florida, Pensacola, FL, USA
russell@uwf.edu

[2] Jackson Health System, Miami, FL, US

Abstract. Surgical site infections (SSIs) are among the most costly and dangerous postoperative complications, yet early identification remains difficult due to delayed onset and complex risk factors. While machine learning has shown promise in SSI prediction, many models rely on retrospective data and non-contemporaneous features, limiting clinical impact. This study proposes a practical SSI prediction approach using only structured electronic medical record (EMR) data. We introduce a novel population sampling method based on patient similarity, employing Jaccard and cosine metrics to identify clinically comparable non-SSI cases. This enhances data balance and training relevance without relying on synthetic data or over sampling. Our model, trained on similarity-filtered data and using no lab, imaging, or free-text features, achieves performance that modestly exceeds state-of-the-art while maintaining real-time feasibility. This work shifts the focus from model complexity to cohort construction, offering a scalable, interpretable solution for infection surveillance and hospital quality improvement.

Keywords: Artificial Intelligence · Machine Learning · Surgical Site Infections · Healthcare

1 Introduction

Surgical site infections (SSIs) are a serious and common complication of surgery that poses significant challenges to patient health. In the United States alone, SSIs occur in approximately 2–10% of inpatient surgical procedures, an incidence of up to 300,000 cases per year and account for about 20% of all healthcare-associated infections (HAIs), and SSIs are the single most costly type of HAI [1]. Compared to patients who do not develop an SSI, patients who do have significantly worse outcomes and longer hospital stays; each infection prolongs hospitalization by an average of 7–10 days and is associated with a 2–11 times higher risk of post-operative death [2]. Beyond mortality, SSIs substantially impact patient recovery, often causing pain, delayed wound healing, and other complications. Further, SSIs are a leading cause of unplanned 30-day hospital readmissions following surgery [3], undermining quality of care and burdening patients with additional treatments.

H. R. Arabnia et al. (Eds.): CSCI 2025, CCIS 2971, pp. 151–169, 2027.
https://doi.org/10.1007/978-3-032-28463-1_12

SSIs also carry heavy operational and financial consequences for hospitals. Treating an SSI requires significant resources, including additional surgeries, antibiotics, and extended care, increasing the average care-cost, per infection, by tens of thousands of dollars. In aggregate, SSIs contribute an estimated $3.5 - 10 billion in excess health expenditures annually in the U.S. [1]. Hospitals must absorb many of these costs, especially when insurers refuse reimbursement for complications deemed preventable. Furthermore, SSIs adversely affect institutional performance metrics. Elevated SSI rates increase overall resource utilization and can jeopardize a hospital's quality ratings. Regulatory agencies closely monitor SSI occurrences as indicators of care quality and patient safety. For example, the Centers for Medicare & Medicaid Services (CMS) publicly reports hospital SSI rates and may impose financial penalties or reduced Medicare reimbursements on facilities with poor infection rates. Likewise, accreditation organizations (e.g., The Joint Commission) scrutinize infection control practices. From an accreditation perspective, a pattern of frequent SSIs can put a hospital's accreditation status at risk [4]. Beyond the direct clinical harm to patients, untreated or frequent SSIs can damage a hospital's reputation, financial stability, and compliance with healthcare standards.

Early detection and prevention of SSIs are paramount to improving surgical outcomes and care provisioning overall. Timely identification of an SSI allows clinicians to intervene sooner (for instance, with targeted antibiotics or surgical debridement), limiting the infection's severity and prevent further complications. However, early detection of SSIs is challenging. Many SSIs present after the patient has been discharged, and their initial signs (e.g., mild fever or redness) may be subtle or mistaken for normal post-operative recovery [5]. Traditional surveillance methods often rely on manual chart reviews or laboratory and culture results that can be temporally-offset from the infection's onset, which means infections are frequently confirmed only days or weeks after surgery. This gap creates an urgent need for predictive models that can flag high-risk patients before an SSI fully develops. By leveraging electronic medical record (EMR) data and artificial intelligence (AI), hospitals can potentially monitor post-surgical patients continuously and receive early warnings of infection risk. Such real-time SSI detection systems could prompt proactive clinical evaluations or treatments, thereby reducing morbidity, preventing avoidable readmissions, and ultimately saving lives [5].

Developing an accurate and practical SSI prediction model, however, poses several challenges. SSIs are multifactorial events influenced by a complex interplay of patient factors (e.g., diabetes, obesity), surgical factors (procedure type, duration, technique), and perioperative care processes [6]. This complexity often necessitates large, detailed datasets, including laboratory values, microbiology results, or imaging findings, to achieve good predictive performance. Yet in many real-world settings, especially in the immediate post-operative period, such detailed data may not be readily available. These challenges are not new and there has been extensive research to develop exceptionally precise AI models that rely on detailed clinical information and often place constraints on generalized use and population variances.

The objective of this research seeks to investigate whether a simple, generalizable pre-modeling population refinement method can be used with minimal input features to create an AI model that is viable for early SSI detection, useful for SSI surveillance, and simplifies quality management and reporting. In the following sections of this paper, a background of the current literature on predicting SSIs is presented. Following this background, the development of an AI model for real-time SSI prediction using only structured EMR data and patient-similarity-driven preprocessing is presented. The results of testing the model are presented, concluding with a discussion and conclusion.

2 Related Work

Machine learning (ML) has emerged as a powerful tool for improving SSI risk prediction. A systematic review identified 24 studies describing 85 ML-based AI models for SSI detection [7]. ML algorithms commonly used in prior work include logistic regression, decision trees, random forest, support vector machines, gradient boosting methods (such as XGBoost), and artificial neural networks. Early studies applying ML to SSI prediction achieved modest performance improvements. Bartz-Kurycki et al. [8] used random forest and logistic regression models for neonatal surgery patients, reaching an area under the receiver operating characteristic (AUC) of 0.68. Similarly, Song et al. [9] achieved an AUC of 0.73 using tree-based methods for cardiac device-associated SSIs. AUC values of 0.5 for binary predictors are considered as randomly guessing; and so, the early work though pioneering and showing the feasibility of using ML to predict SSIs did not produce reliable results. With larger datasets, more advanced models showed stronger results: Chen et al. [10] employed a neural network to predict colorectal surgery SSIs, achieving 0.76 AUC. More recent efforts like Al Mamlook et al. [11] applied deep neural networks to 2.88 million surgical cases, achieving an AUC of approximately 0.85. Wu et al. [12] similarly developed an XGBoost-based model for arthroplasty-related SSI prediction, attaining an AUC above 0.90.

These prior efforts highlight that AI can be used to predict SSI, particularly when large-scale, diverse data is available. However, most of the prior studies included a broad array of clinical features e.g., laboratory, imaging, microbiology or bedside observation, which are not always readily accessible or lag the onset of the infection. Moreover, despite the favorable results and accuracy, there was not many studies that gave emphasis on false positives. One of the main barriers to effective SSI prediction is severe class imbalance. While the humanitarian and fiscal impact of SSIs cannot be understated, from a total hospital population perspective, SSIs typically occur in a relatively small percentage of surgical cases [11], resulting in datasets heavily skewed toward the negative class. Without some form of correction, ML models will tend to predict the majority class, yielding deceptively high accuracy but poor sensitivity and precision for true SSIs [7].

There is ample research that has shown that models trained without addressing imbalance suffer from low positive predictive value despite acceptable AUCs [11, 13, 14]. Petrosyan et al. [15] reported an AUC of 0.89 using random forest but achieved a precision of only 34%, meaning two-thirds of predicted SSI cases were false positives. Xiong et al. [14] faced similar issues with an AdaBoost model for spinal surgery SSIs, achieving only 0.23 precision and applied Synthetic Minority Over-sampling Technique (SMOTE) [16] to improve precision scores to 0.6250 with only minor reduction in recall. Oversampling methods such as SMOTE have proven effective in addressing imbalanced dataset for predicting SSI [17]. Al-Ahmari et al. [17] found that applying SMOTE improved Matthews correlation coefficient (MCC) scores and yielded better balanced recall and precision for SSI models. Other approaches, including over/under-sampling and hybrid strategies, have also been employed but with mixed success depending on dataset characteristics [18]. In addition to resampling techniques, improving the composition of the training dataset through better population selection has gained attention. Restricting modeling to specific surgery types (e.g., colorectal surgery [19, 20], orthopedic procedures [21]) can help by increasing the prevalence of SSIs and improving model relevance. These studies indicate that restrictive sampling can yield better predictive ML performance by constraining training data.

There is a considerable amount of prior work on using AI and ML models to predict SSIs with varying and promising degrees of success. The aforementioned review of the literature shows that incorporating observational (i.e., bedside visuals and clinician notes) and clinical (i.e., labs/microbiology, imaging, and pharmacy) data have a correlation to model performance. While such data is certainly an important consideration, one limitation in much of the prior work is the dependence on such observable and clinical features that, as noted previously, often lag the SSI onset or are available later in treatment timelines. It is also noteworthy that much of the prior work prioritizes the AI model in the research which, while essential, lowers the emphasis on training data selectivity and sampling. A data-first, versus model-first philosophy can elevate concerns that would only emerge *after* model training and may provide tractable guarantees on reliability. Many of these techniques use categorical segmentation (e.g. specific surgery types or demographics, such as patient age) combined with over/under-sampling or synthetic data methods to diffuse imbalanced training data effects. However, these training data manipulation methods introduce unreal (synthetic) data and/or directed emphasis on the minority (positive) class that is not representative of true operational/production distributions. This is not to disparage adaptive sampling or approaches like SMOTE; they can positively impact ML model performance. Yet, their broader efficacy is likely tied to the severity of the organic data imbalance in which the resulting model is employed.

In contrast to the rich body of prior work, the objective of the research presented in this paper is to investigate if leveraging readily available patient data, coupled with a simple, understandable, and adjustable patient-similarity approach to handle data imbalances, can yield an effective AI model for SSI surveillance. If achievable, such a model has the potential to enhance patient safety, support clinical decision-making, provide enriched SSI-related quality management, and improve surgical care outcomes on a broad scale. The following section outlines the technical approach towards this objective.

3 Methods

To address the previous mentioned objective, our approach focuses on using only structured EMR data that are routinely documented during care (excluding labs, notes, imaging, and pharmacy records). These structured data include information such as patient demographics, comorbidities, and surgical details, which are available in near real-time for every patient. Patient comorbidities are captured through the accumulation of diagnosis codes as encounters progress. This allows the early creation of derived features such as the Medicare Severity Diagnostic-Related Grouping (DRG) and maximum acuity diagnosis (maxAcuityDx). DRG is based on principal diagnosis, secondary diagnoses (including complications and comorbidities), procedures performed, and patient demographics. The maxAcuityDx feature is derived using a crosswalk (e.g., Charlson/Elixhauser) that associates diagnosis codes and their related acuity (major comorbid condition, comorbid condition, or neither) to ascertain a maximum acuity level for each encounter. While final diagnosis codes and related metrics are typically assigned retrospectively after patient discharge, through the use these derived features our model exploits these features for prospective use. In such a (close to) real-time operational window, features like DRG and maxAcuityDx would leverage their most current, provisionally available values documented up to the moment of prediction during the patient's active hospital stay. Notably, previous SSI risk models have largely focused on static preoperative risk factors and did not incorporate ongoing clinical observations [6]. In contrast, our model leverages temporal clinical data streams and simple but robust feature engineering to capture the evolving patient condition.

Another key challenge of the objective is the relatively small number of SSIs cases compared to uncomplicated surgeries, which leads to highly imbalanced data -- far fewer cases of "infection" than "no infection." Class imbalance can impair machine learning algorithms, causing them to favor the majority class (no infection) and miss the SSI events (minority class). The technical approach seeks to mitigate this issue through a population similarity technique. In essence, we identify and include in the training dataset those past surgery cases that are most similar to the patient population of interest (for example, using distance metrics like Jaccard and cosine similarity on patient feature profiles). By selecting a cohort of clinically comparable patients in the majority class, this method enriches the training data for relevant patterns and reduces noise from dissimilar cases. It is important to note that there is prior work on patient-similarity metrics [22–24]. However, improving the nature of patient-case similarity is not the focus of this research. Rather the technical approach here selects simple, but reliable, similarity metrics to maintain a focus on model explainability and understanding.

3.1 Data Preparation and Exploratory Data Analysis

All data were collected passively, and de-identified in compliance with the Health Insurance Portability and Accountability Act (HIPAA). Studies performed on deidentified data constitute non-human subject studies. Data access was governed by an internal data use agreement and is not publicly available. The initial dataset consisted of 151,733 inpatient encounters that spanned two years (2022 & 2023) of inpatient cases (encounters) drawn from a large health care system's EMR system. Within the dataset 20,062 (7.6%) were elective (versus emergency) surgeries; 73,300 (48%) patients were males; and there were 109,822 unique patients.

SSI positive encounters were identified by SSI-related ICD-10 diagnosis (Dx) codes (e.g. T81.4XXX) in the encounter final coding. International Classification of Diseases version 10 (ICD-10) diagnosis codes are standardized alphanumeric codes used internationally to classify and record diseases, symptoms, and other health conditions for clinical and billing purposes. Table 1 shows a list of relevant ICD-10 Dx codes. These core codes include the T81.4XXX series (e.g., T81.41XA for superficial incisional surgical site infection), K65.0 (Generalized (acute) peritonitis), K65.1 (Peritoneal abscess), K68.11 (Postprocedural infection of other intra-abdominal organ or structure), and O86.03 (Infection of obstetric surgical wound, organ and space site). While Table 1 presents a broader list of ICD-10 diagnosis codes, it is important to note that the presence of general symptoms like R10.0 (Acute unspecified abdominal pain) alone were not used to classify an encounter as SSI positive.

Table 1. ICD-10 Diagnosis codes frequently associated with colon or abdominal hysterectomy related SSI encounters

Code	Description
T81.41XA	Infection following a procedure, superficial incisional surgical site, initial encounter
T81.43XA	Infection following a procedure, organ and space surgical site, initial encounter
T81.44XA	Sepsis following a procedure
T81.49XA	Infection following a procedure
T81.32XA	Disruption of internal operation (surgical) wound, not elsewhere classified, initial encounter
T81.9XXA	Unspecified complication of procedure, initial encounter
K65.0	Generalized (acute) peritonitis
K65.1	Peritoneal abscess
K65.9	Peritonitis, unspecified
K68.11	Postprocedural retroperitoneal abscess
K91.89	Other postprocedural complications and disorders of digestive system
Z98.890	Other specified postprocedural states
L02.211	Cutaneous abscess of abdominal wall
R10.0	Acute unspecified abdominal pain
O86.03	Infection of obstetric surgical wound, organ and space site

These broader codes, such as R10.0, are included in Table 1 because they were found to be consistently present and frequently associated with the final coding of encounters that were definitively confirmed as SSI positive by the more specific diagnostic codes. This reflects the nuanced real-world clinical documentation where general symptoms often accompany or are documented alongside specific diagnoses of surgical site infections, particularly following procedures like colon or hysterectomy. The stringent and clinically specific definition of SSI employed, as evidenced by the relatively low prevalence of SSI cases (3%) in our dataset, further underscores this approach.

A specific class of SSI, those related to colon or hysterectomy procedures, was chosen to better align with the research in the literature and SSIs related to these surgeries are of high importance to hospital quality metrics [25, 26]. One might argue that the larger population could be filtered simply on encounters who had a colon or hysterectomy related surgery and doing so would provide sufficient fidelity to restrict the majority class population. However, filtering on patients who had a colon or hysterectomy surgery assumes that all patients undergoing the same surgery are clinically comparable, but this is false. Two patients may undergo a colectomy but may differ significantly in many ways: comorbidities (e.g., diabetes, immunosuppression, malnutrition); preoperative infection status; emergency vs. elective context; and concurrent procedures or complications. These differences can affect SSI risk [27]. While potentially improving majority class imbalance issues, filtering only on surgical type can aggregates dissimilar patients and introduces heterogeneity that can confuse a machine learning model during training.

There are practical considerations as well. Encounters in the minority class will include: 1) patients who have surgery at the provider's hospital and get a SSI in the same encounter, 2) patients who have a surgery at the provider's hospital and return with an SSI (thus, surgical data associated with a previous encounter); and 3) patients who did not have a surgery at provider hospital and came in with an SSI (thus, less available surgical data). Thus, post discharge EMR information would be available for SSI encounters that fell into category 2 and be "none" for the other two scenarios. This pragmatic problem required an additional task of locating and associating encounter records using the patient identifier (medical record number/MRN), ultimately creating a single row for each "unique" pair of encounters (if available).

Table 2 shows the list features employed to develop the AI model, in addition to the unique identifiers (MRN and account numbers). To address the aforementioned 3 scenarios of SSI encounters, the notion of "AcctNo_Best" was introduced. The "best" account number was used as the encounter identifier most likely to return values for surgical procedure information, e.g. operating room (OR) procedure quantities, anesthesia minutes, and other surgical details. Table 2 also shows the features associated with the best encounter. These features are primarily numeric and were thusly set to zero when no data was available. Textual features were one-hot encoded [28] to address missing or null values. One-hot encoding is a method of converting categorical variables into a binary vector representation, where only one element is "hot" (set to 1) and all others are set to zero, uniquely identifying each text-value category. At the end of this process, the dataset consisted of 146,451non-SSI encounters and 5,282 SSI encounters. It is fairly evident that that before doing any machine learning better population sampling is necessary to address the imbalance.

Table 2. Model feature list

Feature Name	Feature Description	Incl. In Model
MRN	Individual Medical Record Number	No
AcctNo	Current encounter Acct number	No
DischDate	Discharge date for current encounter	No
LOS	Lenth of stay for current encounter	Yes
admissionType_key	Type of admission (elective, emergency, trauma, etc.)	Yes
SSI	SSI Flag associated with current encounter (0/1; 1 = Positive)	Target
encounter_id_Prev	Previous encounter Acct number	No

(*continued*)

Table 2. (*continued*)

Feature Name	Feature Description	Incl. In Model
DaysBtwPrevVisit	Days between last and current visit	Yes
QtyPriorIPvisits	Number of prior inpatient visits	Yes
VolProcs	Volume of procedures, current encounter	Yes
VolProcs_SSI	Volume of SSI-specific procedures, current encounter	Yes
VolProcs_Prev	Volume of SSI-specific procedures, previous encounter	Yes
VolProcs_SSI_Prev	Volume of procedures, previous encounter	Yes
proc1	First procedure code (prioritizing SSI-Specific PCS code) current encounter	Yes
proc2	Second procedure code (prioritizing SSI-Specific PCS code) current encounter	Yes
proc1Prev	First procedure code (prioritizing SSI-Specific PCS code) previous encounter	Yes
proc2Prev	Second procedure code (prioritizing SSI-Specific PCS code) previous encounter	Yes
AcctNo_Best	Logical "SSI-relevant" encounter - first or previous	No
Age	Age at time of "AcctNo_Best" encounter	Yes
Sex	Patient Gender	Yes
Race	Patient Race	Yes
drg	"AcctNo_Best" Diagnosis Related Grouping Code	Yes
AcuityDRG	"AcctNo_Best" DRG acuity, e.g. comorbid condition, major comorbid condition or neither	Yes
maxAcuityDx	"AcctNo_Best" maximum Dx Code acuity, e.g. comorbid condition, major comorbid condition or neither	Yes
BMI	"AcctNo_Best" Body Mass Index	Yes
preOrHours	"AcctNo_Best" Hours between registration and 1st OR procedure	Yes
AnesMinsMaxSurgDur	"AcctNo_Best" Anesthesia duration (minutes) for longest surgery	Yes
IntubateMins	"AcctNo_Best" Duration of intubation in OR	Yes
CardioPulClampMins	"AcctNo_Best" on of cardiopulmonary clamp (minutes)	Yes
dischargeDisposition	"AcctNo_Best" Details of discharge disposition	No
SurgHospSvc	"AcctNo_Best" Hospital service	Yes
SurgSpecialty	"AcctNo_Best" Surgical specialty	Yes

(*continued*)

Table 2. (*continued*)

Feature Name	Feature Description	Incl. In Model
ClinHandOff	"AcctNo_Best" Clinical handoff in/after surgery (0/1; 1 = True)	Yes
ANEPostOp	"AcctNo_Best" Anesthesia Post Op Power Plan (0/1; 1 = True)	Yes
orQty	"AcctNo_Best" Quantity of OR records (surgeries)	Yes
orProcQty	"AcctNo_Best" Quantity of OR procedures	Yes
orProcCancelQty	"AcctNo_Best" Quantity of canceled OR procedures	Yes
maxOrProcMinutes	"AcctNo_Best" Longest OR procedure duration (minutes)	Yes
maxAnesDurationMins	"AcctNo_Best" Longest OR Anesthesia duration (minutes)	Yes

3.2 Overview of the Technical Approach

To address the challenge of constructing a clinically meaningful negative class for surgical site infection (SSI) prediction, we developed a similarity-based filtering strategy anchored to confirmed SSI cases. This approach avoids reliance on random resampling, synthetic oversampling, or procedure-based rules, each of which can distort cohort composition or reduce generalizability. To create a more balanced training dataset, first we compute diagnosis and procedure similarities. For each negative (candidate) encounter, we compute Jaccard distance diagnosis codes (JD) and procedure codes (JP) individually-pairwise with respect to each encounter in a reference pool of clinically confirmed SSI cases. Each candidate encounter was represented as a two-dimensional vector [JD, JP], where JD and JP quantify overlap with a specific SSI-positive encounter. We then measure the cosine similarity between each candidate vector [JD, JP] and every SSI- positive vector [JD_k, JP_k]. The maximum cosine similarity across all SSI-positive cases was retained. Encounters exceeding a predefined similarity threshold were included in the negative training cohort. This procedure emphasizes balanced structural similarity between diagnoses and procedures, penalizing cases with disproportionate overlap in only one domain. The composite set of negative-SSI encounters that have a cosine score above the threshold and the reference SSI-positive encounters form the ML model training set. This method does not serve as a classifier itself; rather, it is a preprocessing heuristic designed to curate a more realistic and challenging negative cohort. The downstream predictive ML model subsequently learns the decision boundary between SSI-positive and filtered negative cases.

We selected cosine similarity over alternatives (e.g., averaging JD and JP, independent thresholding) because its inductive bias emphasizes balance between diagnosis and procedure similarity. In this setting, the objective is not to predict SSI directly but to curate a negative cohort that is both clinically plausible and sufficiently challenging for the classifier. By penalizing cases with extreme asymmetry (e.g., strong comorbidity match but unrelated procedure, or vice versa), cosine discourages inclusion of encounters unlikely to provide meaningful counterexamples. Importantly, cosine similarity is not applied as a decision function for SSI risk. Its role is limited to preprocessing: constructing a negative class anchored to the structure of known SSI-positive cases. While cosine is insensitive to magnitude, this limitation is acceptable in our design because the final predictive model learns the discriminative boundary between true SSIs and curated negatives. Model performance, not the filter alone, validates the utility of the approach. In this respect, cosine similarity functions analogously to hard-negative mining strategies used in contrastive learning, an intentionally imperfect mechanism for data construction that nonetheless produces superior downstream models. The similarity portion of our method does not claim to find the "most at-risk" cases, nor to define patient similarity in the absolute sense. Rather, it creates a clinically anchored cohort of counterexamples, centered on known infections, for use in training a discriminative model.

3.3 Encounter Similarity Determination

To create the model, ideally majority class observations need to be from the same population and have a balanced number of observations compared to the minority class examples. Besides the methods discussed in the imbalance portion of this paper, it is possible to filter of patients with surgeries and then randomly select from that sub sample. However, selecting a population that is "similar" to the minority-class examples would likely be more representative of the true relevant population and thus, give better performance for AI model training. The key questions here are: similar based on what and how "similar" is similar?

There are many methods in the literature for determining the clinical similarity of encounters including techniques like clustering, e.g. KNN or K-Means [29, 30]; fuzzy approaches [31], and patient similarity networks [23]. These methods have demonstrated effective approaches for determining population similarity, but they generally required detailed clinical details and none of the prior work we found focused on patient similarity applied to SSI prediction. Given the focus on SSI prediction of our work, we adopt two straightforward and reliable similarity metrics, Jaccard and cosine distance. These similarity metrics have the added benefit of keeping the approach simple and easily explainable [32]. To determine similarity, encounters with like final ICD-10 diagnosis codes and like ICD-10 procedure (PCS) codes are compared, using the clinically validated minority class encounters as the basis set.

The construct of AcctNo_Best created in the data preparation step is essential to determining encounter similarity because similarity is defined as two encounters that share like Dx and PCS codes. To do that each record must be associated with the proper encounter, containing the SSI-relevant diagnosis and procedures. Using the known SSI encounters and excluding any SSI Dx codes, non-SSI encounters can be identified by shared diagnosis and procedures codes, given a threshold-degree of distance between the non-SSI encounter and any one of the SSI encounters. Equation 1 shows the expression of Jaccard similarity where two sets, a and b, are measured in terms of similarity; the resulting Jaccard value ranges between 0 to 1. Equation 2 expresses the cosine similarity calculation where X and Y are the components of two vectors. The cosine distance values are in the range of -1 to 1, and subtracting the absolute value from 1 gives the similarity. Each encounter-pair will generate a vector consisting of the Jaccard score for Dx codes and the Jaccard score for PCS codes. This vector is then pairwise applied with a self-similar vector to provide a single measure of similarity. Each feature comparison (Dx and PCS codes) is best measured by Jaccard because their nature mandates a set comparison, but a single metric that summarizes the two encounters' similarity is necessary for a thresholding-filter.

$$Jaccard\_Sim(a, b) = \frac{a \cap b}{a \cup b} \tag{1}$$

$$Cosine\_Sim(X, Y) = \frac{X \cdot Y}{\|X\| \|Y\|} \tag{2}$$

Figure 1 illustrates the overall framework flow. At the center is a notional example of the similarity-based selection process. In this example, encounter a is a known SSI encounter and encounter b is a known non-SSI encounter. It is important to clarify that the Jaccard index is not applied to the numeric features (e.g., Age, LOS) or one-hot encoded textual features listed in Table 2. These features serve as the direct input to the downstream machine learning prediction model. Instead, the Jaccard similarity metric is exclusively utilized as a pre-modeling population construction step for comparing set-based features: specifically, ICD-10 diagnosis (Dx) codes and ICD-10 procedure (PCS) codes between patient encounters. Jaccard similarity is calculated separately for the Dx codes and the PCS codes. Each feature comparison (Dx and PCS codes) is best measured by Jaccard because their nature mandates a set comparison. The resulting Jaccard scores for Dx and PCS codes then form a vector, which is subsequently used to calculate a cosine similarity score, the ultimate metric for filtering non-SSI cases to construct our balanced training set. For ease of understanding, in Fig. 1, a text description of the respective ICD-10 codes for each encounter are shown. After Jaccard similarity has been determined, pairwise cosine similarity is calculated. In this calculation, the SSI encounter self-similarity vector is used resulting in a single score for each encounter-pair. A "fully self-similar vector" represents an encounter's perfect similarity to itself. This means that when an encounter's codes are compared to themselves, the Jaccard similarity is 1.0. In Cosine similarity, the [1.0, 1.0] vector acts as a benchmark against which the similarity of other encounters is measured, allowing a single cosine similarity score to be derived for each encounter-pair.

In all pairwise similarity calculations, the non-SSI encounters are only compared to the SSI encounters allowing the maximum cosine similarity to be elicited and used for filtering. To achieve this, for each non-SSI encounter, Jaccard similarities for both diagnosis (Dx) and procedure (PCS) codes are first computed against every individual SSI encounter in the minority class. The Jaccard similarities form vectors, for each encounter, that are used to compute cosine scores against every individual SSI encounter. This process generates a set of pairwise cosine similarity scores for each non-SSI encounter. The highest score from this set, representing the maximum similarity of that non-SSI encounter to any SSI case, is then selected. This maximum cosine similarity score is subsequently evaluated against the 0.70 threshold; if it meets or exceeds this value, the non-SSI encounter is included in the "similar" majority class for the balanced training set.

A threshold value of 0.70 was employed resulting in reducing the 151,733 records to 15,339; 5,282 inpatient SSI encounters (appx 34%) and 10,057 "similar" non-SSI inpatient encounters. This 0.70 threshold was pragmatically selected as a conservative measure to achieve a substantial initial improvement in class imbalance, effectively increasing the proportion of SSI cases from approximately 3% to 34% within the filtered dataset. The 0.70 threshold, could be adjusted higher or lower, subject to a desired clinical specificity. However, setting the threshold too high may inadvertently exclude a significant number of non-SSI encounters that are indeed clinically analogous to SSI cases. Similarly, too low of a threshold would result in allowing too many clinically dissimilar cases. We recognize that the similarity threshold could be optimized and will explore tuning this parameter in future work.

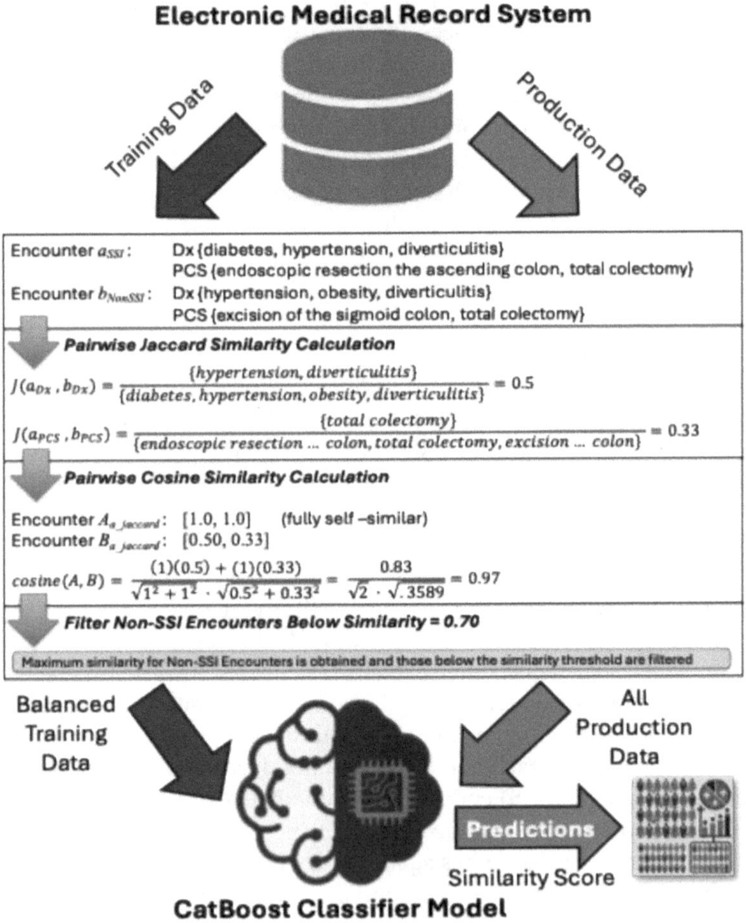

Fig. 1. Notional similarity-based majority class record selection

3.4 ML Model Construction

The final dataset used for model training consisted of 15,339 records, comprising 5,282 SSI cases and 10,057 similar non-SSI cases. Figure 2 shows the results of the similarity and SMOTE process. With the improved training data balance, we formulate the SSI prediction problem as a classic binary classification problem using tabular data. Here the x-input is the above list of features and the y-output is a binary representation indicating SSI present (1) or not (0). Given the x-input, several AI models were developed, using no hyperparameter optimization initially, then applied a grid search optimization approach once the best model was identified. It is important to note that despite the population similarity filtering, SMOTE was employed to balance no-SSI (negative, majority class) and SSI (positive, minority class) groups. SMOTE created additional new minority data by interpolation within the available minority data via bootstrap sampling and data generation via a k-nearest neighbor algorithm, with a k value of 5 [16]. The 15,339-record dataset was separated into two randomly selected subsets: 1) an 80/20 train-test split training set and 2) a hold-out validation set.

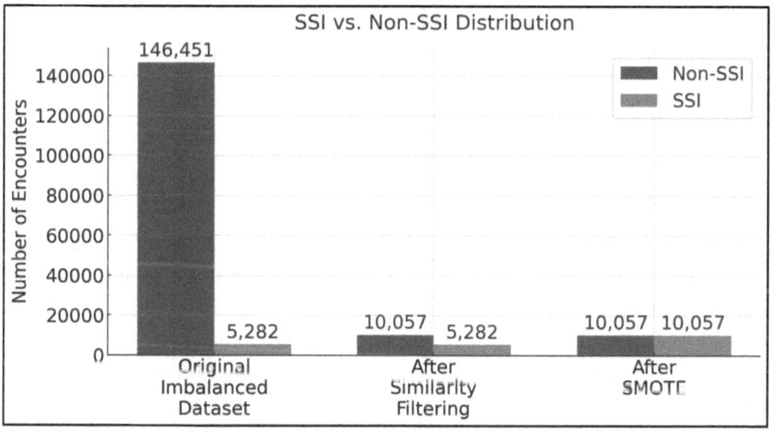

Fig. 2. Distribution of data imbalance before and after.

4 Results

A 10-fold validation iteration was applied to assess the average validation final results. Results were measured via accuracy, precision (selectivity), recall (sensitivity), AUC, and F1 (a measure of the balance between precision and recall, based on the harmonic mean of both). Table 3 shows the results of the test data for several AI models. CatBoost [33] performed the best, providing reasonable scores across all the metrics. Though slightly lower than the best performing models in the literature, the CatBoost model's scores are consistent and competitive with most of the prior work. It is particularly noteworthy that the best performing models from the literature relied on a greater number and more clinically specific features (such as laboratory results and/or observables such as swelling, fever, or purulent discharge); many of which are difficult to obtain in real (or close to real) time from EMR systems.

Table 3. Model results for validation data

Model	Accuracy	AUC	Recall	Prec.	F1
CatBoost Classifier (best model)	.8463	.9052	.7317	.8044	.7663
Light Gradient Boosting Machine	.8416	.9029	.7349	.7906	.7616
Extreme Gradient Boosting	.8407	.9001	.7301	.7913	.7593
Random Forest Classifier	.8198	.8823	.7309	.7422	.7363
Gradient Boosting Classifier	.8179	.8819	.7357	.7358	.7356
Ada Boost Classifier	.7902	.8434	.7036	.6929	.6978
Extra Trees Classifier	.7939	.8558	.6765	.7110	.6931
Decision Tree Classifier	.7611	.7403	.6738	.6473	.6602
K Neighbors Classifier	.7254	.7788	.6911	.5863	.6342
Logistic Regression	.7287	.7814	.6660	.5951	.6283
Ridge Classifier	.7241	.7782	.6584	.5892	.6216
Linear Discriminant Analysis	.7242	.7771	.6578	.5894	.6215
SVM - Linear Kernel	.7126	.7724	.6787	.5805	.6189
Naive Bayes	.7161	.7247	.4926	.6126	.5436
Quadratic Discriminant Analysis	.3985	.5990	.9268	.3630	.5155

The exclusion of such less accessible, but potentially beneficial, time lagged or clinical-detail features was deliberate. To investigate the feature selection, a SHAP (Shapley Additive exPlanations) technique [34] was conducted to identify which features contributed to the model results, and in what way. SHAP is a model-agnostic technique used to explain the output of machine learning models. It builds on cooperative game theory, particularly Shapley values, to attribute the contribution of each feature to a model's prediction in a fair and theoretically justified manner. SHAP has several desirable properties, beyond simple feature sensitivity, that help explain the predictive contribution of individual features.

Figure 3 shows the features that contribute most to the predictive power of the final CatBoost model. Interpreting the plot, each feature is scored based on its low or high (blue to red) contribution to DRG most influenced the model. This is intuitive because DRG is a system used to classify hospital inpatient stays into categories for the purpose of payment. Each DRG represents a group of clinically similar diagnoses and procedures that are expected to require similar hospital resources and costs. DRG is based on principal diagnosis, secondary diagnoses (including complications and comorbidities), procedures performed, and patient demographics (age, sex, discharge status). The inclusion of DRG might call into question the need for the similarity metric used in dealing with training data imbalance. However, DRGs have a fiscal orientation that is intended to provide financial estimates and assessment for billing purposes. Moreover, DRGs aggregate (and exclude) many encounter characteristics and thus, may have ambiguous fidelity in determining encounter similarity. Other features of note in Fig. 3 include:

quantity of prior visits (QtyPriorVisits); procedures (proc1 and proc2); quantity of procedures (VolProcs and VolProcs_prev); admission type (emergency and newborn); the surgeons' surgical specialty (SurgSpecialty); and anesthesia minutes occurring in the longest surgical duration (anesMinsMaxSurgDur).

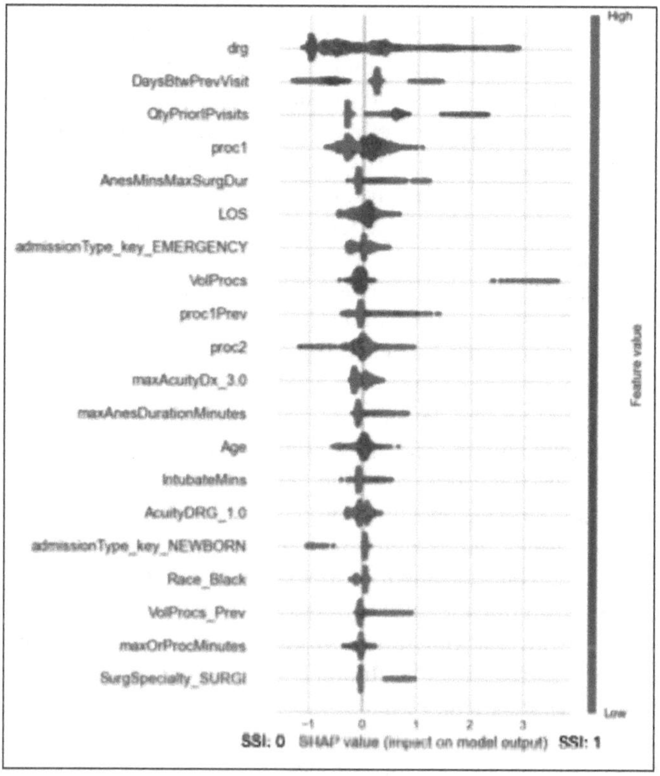

Fig. 3. SHAP summary

5 Discussion

The similarity approach was effective in improving the substantial training data imbalance. The CatBoost model also performed consistently with much of the prior work (significantly better in many cases) and utilized fewer clinical features. The prior work on SSI prediction commonly placed core focus on making an optimized and generalizable AI model, concentrating on pristine datasets with hyper-relevant features. While appropriate for advancing theoretical approaches, such as the creation of new machine learning models, this model-first focus can limit adoption and challenge real-world implementations. Unlike models reliant on retrospective or delayed data, our approach explicitly leverages contemporaneously available structured EMR data. This includes using provisional values for features such as DRG and maxAcuityDx, which would be dynamically updated throughout a patient's active hospitalization. This enables the model to provide early, pre-discharge risk assessments supporting real-time clinical decision-making and quality improvement initiatives.

The addition of an imbalance-improving similarity method for population sampling provides a tune-able means to improve an intrinsic problem with many clinical AI regimes; proportionally low volumes of the minority (positive ailment) class. The goal of the similarity method is not to perfectly distinguish positive from negative patients with cosine. It is to construct a negative training cohort that is challenging, clinically relevant, and well-aligned with the structure of known SSI cases. A 0.70 similarity threshold was used and while reasonable, the threshold did not have a clinical rationale and was not necessarily optimal. To that end, the methods in this paper could be optimized in many ways e.g., a more sophisticated similarity metric, the inclusion of additional clinical features, highly optimized ML algorithms, etc. Nonetheless, this research effort shows the utility of generally available features, perhaps creating a baseline, and the feasibility of employing even a simple similarity-based population selector can make a difference training data balance. More work is necessary in all three areas: similarity-based population selection, data imbalance correction, and SSI prediction writ large.

It is noteworthy that there may be an additional benefit at inference time. Similarity can also be computed between new encounters and the training SSI reference pool. While it not used as a hard inclusion rule for inference, this score provides an interpretable indicator of distributional alignment: high similarity values suggest structural similarity to training cases, whereas low values highlight encounters that are dissimilar and should be interpreted with caution. This dual use would enable the similarity method to function both as a training-time filter and as an inference-time signal for model explainability.

Despite promising advances shown in this and other prior work, key gaps remain. Many published SSI prediction models lack external validation and have only been tested retrospectively. As a result, their generalizability to new settings remains an open research concern. This study also requires longitudinal study of the model's performance over time. Specifically, we acknowledge that our methodological choice to apply similarity-based population construction prior to the train-test split implies the test set is drawn from a population that has already undergone this refinement. As such, the reported performance metrics are indicative of the model's efficacy within this pre-defined, clinically analogous cohort. This approach might present an optimistic estimate if the entire framework were to be applied directly to a completely raw, un-curated, and highly imbalanced stream of EMR data without the prior application of the similarity-based selection process.

In a complete end-to-end real-world deployment scenario, any patient encounter not meeting the initial 0.7 cosine similarity threshold would effectively be considered a non-SSI case, or at least fall outside the high-risk group specifically targeted by this framework. In such instances, the prediction model itself would not process or classify these cases. This distinction clarifies the operational scope of our model, which is designed to operate on a carefully curated, clinically analogous population. While this approach enhances the model's robustness to data distribution shifts by focusing on relevant cases, it implies that the reported performance reflects the system's efficacy within this defined patient cohort. Class imbalance remains a persistent problem affecting model sensitivity and general clinical applicability. While methods like SMOTE and similarity-based sampling offer solutions, further research is needed to optimize sampling strategies and calibration approaches for highly imbalanced clinical data. Therefore, future work should involve robust external validation on independent, raw datasets, where the entire pipeline (i.e., the similarity-based population construction followed by model inference) is applied afresh. This will more comprehensively assess the framework's real-world generalizability and performance in a prospective setting, taking into account cases that are implicitly classified as non-SSI by the initial similarity filter.

6 Conclusion

The ability to predict SSIs could significantly improve clinical outcomes by enabling earlier interventions, thereby reducing postoperative morbidity, mortality, and readmissions. This work presents a pragmatic approach to surgical site infection (SSI) prediction that departs from conventional practices in healthcare machine learning. Most prior efforts have focused on algorithmic sophistication while accepting the training data as given, often addressing class imbalance through under/oversampling or synthetic data generation. In contrast, we introduce a pre-modeling population construction strategy that intelligently selects training samples based on clinical similarity. By applying Jaccard and cosine similarity metrics to diagnosis and procedure codes, we identify non-SSI cases that are comparable to SSI cases, producing a better-balanced training set on which to build the AI model. Unlike commonly used stochastic or embedding-based sampling techniques, our method is simple, interpretable, and scalable, making it suitable for real-time hospital deployment. Importantly, our model uses only contemporaneously available structured EMR data, excluding labs, imaging, and pharmacy, making it capable of supporting early intervention and quality reporting. This design begins to enable a real-time utility in infection control workflows, a capability missing from most retrospective SSI prediction studies. Furthermore, this research shifts the focus from model selection to data engineering, emphasizing that a carefully constructed cohort can elevate the performance of even standard classifiers. This represents a subtle but significant reorientation of priorities in clinical ML research toward building better systems, not just better AI models.

The contribution in this research is not grounded in algorithmic machine learning e.g., inventing a new machine learning model. Rather, this research focuses on engineering the science of AI implementation, targeting how to better choose the right examples before training the AI and providing operational boundaries for where the AI will be most effective; a particularly critical consideration for quality management and reporting in real-world hospital settings. The novelty in this study is in a simple, but powerful rethinking of population construction for clinical ML deployment where imbalance handling aligns with the clinical intuition of comparing "like with like" and can provide a repeatable framework for other imbalanced healthcare predictive problems. Future work is planned to examine the generalizability of the approach utilizing SSI data from the National Surgical Quality Improvement Program (NSQIP) Database. This future work will also seek to expand the similarity approach and investigate more complex similarity methods and optimized thresholding.

Acknowledgments. The authors gratefully acknowledge Galina Sidimirova and Kelley Manzanillo for their valuable clinical insights and general support throughout the development of this manuscript.

Disclosure of Interests. The authors have no competing interests to declare that are relevant to the content of this article.

References

1. Ban, K.A., et al.: American college of surgeons and surgical infection society: surgical site infection guidelines, 2016 update. J. Am. Coll. Surg. **224**, 59–74 (2017)
2. Kirkland, K.B., Briggs, J.P., Trivette, S.L., Wilkinson, W.E., Sexton, D.J.: The impact of surgical-site infections in the 1990s: attributable mortality, excess length of hospitalization, and extra costs. Infect. Control Hosp. Epidemiol. **20**, 725–730 (1999)
3. Kassin, M.T., et al.: Risk factors for 30-day hospital readmission among general surgery patients. J. Am. Coll. Surg. **215**, 322–330 (2012)
4. Shepard, J., et al.: Financial impact of surgical site infections on hospitals: the hospital management perspective. JAMA Surg. **148** (2013)
5. Sanger, P.C., et al.: A prognostic model of surgical site infection using daily clinical wound assessment. J. Am. Coll. Surg. **223**, 259–270 (2016)
6. Anderson, D.J., et al.: Strategies to prevent surgical site infections in acute care hospitals: 2014 update. Infect. Control Hosp. Epidemiol. **35**, S66–S88 (2014)
7. van Boekel, A.M., et al.: Systematic evaluation of machine learning models for postoperative surgical site infection prediction. PLoS ONE **19**, e0312968 (2024)
8. Bartz-Kurycki, M.A., et al.: Enhanced neonatal surgical site infection prediction model utilizing statistically and clinically significant variables in combination with a machine learning algorithm. Am. J. Surg. **216**, 764–777 (2018)
9. Song, J., Buenaventura, E.S., Cohen, B., Liu, J., Yao, D., Larson, E.: Predictive Models for Surgical Site Infection (SSI) in Patients with a Permanent Pacemaker (PPM) Using Machine learning Methods (2020). https://www.authorea.com/users/332293/articles/458733-predictive-models-for-surgical-site-infection-ssi-in-patients-with-a-permanent-pacemaker-ppm-using-machine-learning-methods?commit=6483c4f49576369563110c7c811be8f65cbc7325. https://doi.org/10.22541/au.159188536.65812462
10. Chen, K.A., Joisa, C.U., Stem, J.M., Guillem, J.G., Gomez, S.M., Kapadia, M.R.: Improved prediction of surgical-site infection after colorectal surgery using machine learning. Dis. Colon Rectum **66**, 458–466 (2023)
11. Al Mamlook, R.E., Wells, L.J., Sawyer, R.: Machine-learning models for predicting surgical site infections using patient pre-operative risk and surgical procedure factors. Am. J. Infect. Control **51**, 544–550 (2023)
12. Wu, G., et al.: Development of machine learning models for the detection of surgical site infections following total hip and knee arthroplasty: a multicenter cohort study. Antimicrob. Resist. Infect. Control **12**, 88 (2023). https://doi.org/10.1186/s13756-023-01294-0
13. Heffernan, A., Ganguli, R., Sears, I., Stephen, A.H., Heffernan, D.S.: Choice of machine learning models is important to predict post-operative infections in surgical patients. Surg. Infect. (2025). https://doi.org/10.1089/sur.2024.288
14. Xiong, C., et al.: Construct and validate a predictive model for surgical site infection after posterior lumbar interbody fusion based on machine learning algorithm. Comput. Math. Methods Med. **2022**, 1–11 (2022). https://doi.org/10.1155/2022/2697841

15. Petrosyan, Y., et al.: Predicting postoperative surgical site infection with administrative data: a random forests algorithm. BMC Med. Res. Methodol. **21**, 179 (2021). https://doi.org/10.1186/s12874-021-01369-9
16. Chawla, N.V., Bowyer, K.W., Hall, L.O., Kegelmeyer, W.P.: SMOTE: synthetic minority over-sampling technique. J. Artif. Intell. Res. **16**, 321–357 (2002)
17. Al-Ahmari, S., Nadeem, F.: Improving surgical site infection prediction using machine learning: addressing challenges of highly imbalanced data. Diagnostics **15**, 501 (2025)
18. Colborn, K.L., Bronsert, M., Amioka, E., Hammermeister, K., Henderson, W.G., Meguid, R.: Identification of surgical site infections using electronic health record data. Am. J. Infect. Control **46**, 1230–1235 (2018)
19. Cho, S.Y., et al.: Development of machine learning models for the surveillance of colon surgical site infections. J. Hosp. Infect. **146**, 224–231 (2024)
20. Shen, Z., et al.: The development and validation of a novel model for predicting surgical complications in colorectal cancer of elderly patients: results from 1008 cases. Eur. J. Surg. Oncol. **44**, 490–495 (2018)
21. Chowdhury, A.A.A., Sultana, A., Rafi, A.H., Tariq, M.: AI-driven predictive analytics in orthopedic surgery outcomes. Revista Espanola de Documentacion Cientifica **19**, 104–124 (2024)
22. Defilippo, A., Veltri, P., Lió, P., Guzzi, P.H.: Leveraging graph neural networks for supporting automatic triage of patients. Sci. Rep. **14**, 12548 (2024)
23. Pai, S., Bader, G.D.: Patient similarity networks for precision medicine. J. Mol. Biol. **430**, 2924–2938 (2018)
24. Sharafoddini, A., Dubin, J.A., Lee, J.: Patient similarity in prediction models based on health data: a scoping review. JMIR Med. Inform. **5**, e6730 (2017)
25. Caroff, D.A., et al.: The limited utility of ranking hospitals based on their colon surgery infection rates. Clin. Infect. Dis. **72**, 90–98 (2021)
26. Yokoe, D.S., Avery, T.R., Platt, R., Kleinman, K., Huang, S.S.: Ranking hospitals based on colon surgery and abdominal hysterectomy surgical site infection outcomes: impact of limiting surveillance to the operative hospital. Clin. Infect. Dis. **67**, 1096–1102 (2018)
27. Korol, E., et al.: A systematic review of risk factors associated with surgical site infections among surgical patients. PLoS ONE **8**, e83743 (2013)
28. Rumelhart, D.E., Hinton, G.E., Williams, R.J.: Learning representations by back-propagating errors. Nature **323**, 533–536 (1986)
29. Dai, L., Zhu, H., Liu, D.: Patient similarity: methods and applications, http://arxiv.org/abs/2012.01976 (2020). https://doi.org/10.48550/arXiv.2012.01976
30. Wang, N., et al.: Study on the semi-supervised learning-based patient similarity from heterogeneous electronic medical records. BMC Med. Inform. Decis. Mak. **21**, 58 (2021). https://doi.org/10.1186/s12911-021-01432-x
31. Muthukumar, P., Krishnan, G.S.S.: A similarity measure of intuitionistic fuzzy soft sets and its application in medical diagnosis. Appl. Soft Comput. **41**, 148–156 (2016)
32. Rinjeni, T.P., Indriawan, A., Rakhmawati, N.A.: Matching scientific article titles using cosine similarity and Jaccard similarity algorithm. Procedia Comput. Sci. **234**, 553–560 (2024)
33. Prokhorenkova, L., Gusev, G., Vorobev, A., Dorogush, A.V., Gulin, A.: CatBoost: unbiased boosting with categorical features. Adv. Neural Inf. Process. Syst. **31** (2018)
34. Lundberg, S.M., Lee, S.-I.: A unified approach to interpreting model predictions. Adv. Neural Inf. Process. Syst. **30** (2017)

Predicting Non-adherence in Remote Oncology Symptom Monitoring Using Sequence Learning

Aref Smiley[1], Christina Echeverria[2], Kathi Mooney[2], and Joseph Finkelstein[1(✉)]

[1] Arizona Telemedicine Program, The University of Arizona, Tucson, AZ, USA
{arefsmiley,jfinkelstein1}@arizona.edu

[2] College of Nursing, The University of Utah, Salt Lake City, UT, USA

Abstract. Electronic patient-reported symptom monitoring help manage symptoms in cancer care, but patients do not always report every day. We aim to predict short-term non-adherence so teams can reach out before monitoring breaks down. We treat this as a next-window forecast using overlapping 3-, 5-, and 7-day windows of recent symptoms. We create one record per patient-day, rebuild missing days, apply simple inclusion rules ($\geq$35 observed days and $\leq$30% missing), and fill partial gaps within each window by averaging the days that are present (set to 0 only if all days are missing). We label each window by whether the following window has any missed day. To handle imbalance, we use class-weighted loss. We test four models: LSTM, CNN-LSTM, CNN-(Bi)LSTM with single-head attention, and CNN-(Bi)LSTM with multi-head attention. We standardize features, use an 80/20 stratified split. We report ROC-AUC, accuracy, precision, recall, F1, and confusion matrices. With 7-day windows, LSTM and CNN-LSTM with attention give the best ranking (ROC-AUC $\approx$ 0.755) and balanced accuracy ($\approx$0.71–0.73). CNN-LSTM reaches high recall (class-1 recall 0.90) but lowers accuracy (0.56). With 5-day windows, CNN-LSTM gives the highest positive-class F1 at its tuned threshold (validation recall $\approx$0.93; held-out recall 0.46), making it strong for catching imminent lapses. With 3-day windows, tabular baselines (XGBoost) raise accuracy ($\approx$0.75) but have lower F1 and AUC than the best 5-/7-day sequence models. In sum, short-horizon non-adherence is predictable from recent ePRO patterns.

Keywords: Clinical Informatics · Patient Monitoring Systems · Telemedicine · Non-adherence Forecasting · Oncology Remote Monitoring

1 Introduction

Routine electronic patient-reported outcome (ePRO) monitoring has reshaped supportive oncology by surfacing symptoms between encounters and enabling earlier action. Randomized trials show that systematic ePRO surveillance improves quality of life and care processes during active treatment, and—in a landmark extension—was associated with improved overall survival compared with usual care [1, 2]. Subsequent pragmatic evaluations and meta-analyses reinforce these benefits across settings and cancer types, emphasizing the value of structured symptom collection, timely clinician alerts, and patient self-management support [3, 4].

H. R. Arabnia et al. (Eds.): CSCI 2025, CCIS 2971, pp. 170–178, 2027.
https://doi.org/10.1007/978-3-032-28463-1_13

Large-scale programs such as Symptom Care at Home (SCH) demonstrate that automated remote assessment coupled with algorithmic triage and nurse follow-up is both feasible and clinically effective. SCH randomized trials reported reductions in symptom burden and improved outcomes, and implementation reports have outlined key components (automated monitoring, alerting, tailored coaching) that generalize to diverse populations and workflows [3, 4]. These experiences underscore that ePRO platforms are not merely documentation tools but dynamic systems capable of driving timely interventions.

Despite these advances, sustaining day-to-day engagement remains challenging. Missed or incomplete check-ins ("non-adherence") arise from symptom severity, treatment intensity, competing demands, and social context, echoing adherence problems seen with oral anticancer therapies. Anticipating which patients are likely to disengage in the near term could enable proactive outreach that preserves monitoring fidelity and prevents silent deterioration. We therefore frame non-adherence as a short-horizon forecasting task: given a compact rolling window of recent symptom data, predict whether the next window will contain a missed-report event. This framing aligns naturally with modern time-series learning.

Long short-term memory (LSTM) networks were designed to capture long-range temporal dependencies and mitigate vanishing gradients, and they have become a workhorse for clinical sequence modeling. Hybrid stacks that combine temporal convolution (for local pattern extraction) with LSTMs (for sequence dynamics) often improve robustness on noisy, irregular patient time series. Recent surveys of deep learning for time-series classification and sequence modeling synthesize these developments and motivate our architectural choices (LSTM and CNN-LSTM, with and without attention) for near-term non-adherence prediction [5, 6].

Parallel work on adherence prediction in other domains, particularly medication adherence, demonstrates that models trained on short rolling histories can forecast near-term lapses, highlighting the importance of windowed representations and careful operating-point selection under class imbalance. Scoping and systematic reviews, as well as large-scale modeling studies, collectively suggest that precision–recall trade-offs should be tuned to the intervention's goals (e.g., favoring sensitivity for early outreach) [7, 8]. These insights inform our evaluation strategy across overlapping 3-, 5-, and 7-day windows in the same oncology ePRO database used in our prior work.

2 Methods

2.1 Data Preparation

We analyzed the same oncology symptom-monitoring database as our prior work, which contains one record per patient per calendar day with a local timestamp (LocalDate), an identifier (PatientID), and patient-reported symptom items [9–13]. The symptom panel used in this study comprised 12 variables routinely captured in the program: nausea, sore mouth, diarrhea, fatigue, trouble thinking, feel blue, nervous, appearance change, pain discomfort, numbness tingling, trouble sleeping, and fever distress (and an overall Total score).

The prediction task is near-term non-adherence: given a short rolling window of recent symptom data, predict whether the next window will contain at least one day where key symptom fields are missing.

We performed daily curation, applied inclusion rules to ensure sufficient longitudinal coverage and reasonable adherence, and then extracted overlapping windows of length $T \in \{3, 5, 7\}$ days.

We coerced Local Date to a valid datetime and normalized it to midnight to define the per-day participation unit (Date). To ensure stable grouping across heterogeneous identifiers (strings, numeric, leading zeros), we constructed both string and numeric patient keys and performed a stable merge-sort by (PatientID_num, PatientID_str, Date). We then deduplicated to a single row per (PatientID, Date).

Within each patient's observed span (FirstDate…LastDate), we generated the complete calendar of days and labeled non-reporting dates as those absent from the observed set. We summarized per-patient coverage as TotalDays, MissingDays, NonAdherencePct = MissingDays/TotalDays × 100, and a histogram of gap lengths (1–6 days and ≥ 7 days).

To ensure sufficient and interpretable longitudinal signal, we retained patients with TotalDays ≥ 35 and NonAdherencePct $\leq 30\%$. Patients outside these bounds were excluded from modeling. For retained patients, we inserted synthetic rows for every missing calendar day so the downstream window builder sees a continuous daily series.

A day receives Annotation = 1 if symptoms are missing on that day (0 otherwise). This matches the initial prototype used to define window labels. Later filtering based on all 12 symptoms is applied at the window level.

We traversed the continuous daily panel with overlapping windows of length $T = 3, 5,$ or 7 days and stride 1 (i.e., a new window each day). For each window we flattened all non-annotation columns by appending the day index (e.g., nausea_1 … nausea_T), yielding one wide row per window, and assigned a window-level label WindowAnnotation = 1 if any day in the window had Annotation = 1 (otherwise 0).

Imputation operates within each window and within each base feature: (1) Partially missing feature within a window: replace each missing day with the mean of the observed days of that same feature in that window. (2) Completely missing feature within a window (all T days missing): set all T values for that feature in that window to 0 (deterministic fallback). This procedure uses only within-window information and introduces no cross-patient or cross-window leakage. In subsequent feature-engineering variants, windows where all 12 symptoms are absent across all T days are dropped. The deep-learning sequence models operate on the imputed per-day matrices.

2.2 Deep Learning Models

We trained four neural architectures; only the encoder block differed across models. All other steps—target definition, normalization, splitting, class weighting, optimization, and evaluation—were identical across models and window lengths.

For a given T-day window at index i, the prediction target is whether the next window (index $i + 1$) contains any annotated day: y_i = WindowAnnotation_{i + 1}. Operationally, this is equivalent to shifting the WindowAnnotation column by $-$ T rows and removing the last T windows where the target would be undefined.

We used per-day sequences as inputs. Each window's per-day features were arranged as a tensor of shape (T, F_day), where F_day = 13 for the 7-day case (12 symptoms + Total). Before reshaping, features were standardized with a z-score transform computed feature-wise on the matrix of window rows. In confirmatory runs we verified that fitting the scaler after the train/test split yields the same qualitative conclusions.

We computed class weights using scikit-learn's balanced heuristic (inverse frequency by class) and passed them to the loss function during training. This up-weights the minority (non-adherence) class.

Architectures. All models end with a dense sigmoid output neuron that produces a probability for next-window non-adherence. The encoder variants are:

(A) LSTM (no CNN, no attention): a single unidirectional LSTM (64 hidden units), followed by dropout (0.3) and a dense ReLU (64).
(B) CNN-LSTM: Conv1D (64 filters, kernel size 2, ReLU, same padding) + MaxPooling1D (pool size 2), then LSTM (64), dropout, dense ReLU (64).
(C) CNN-BiLSTM + single-head attention: Conv1D + MaxPool, then bidirectional LSTM (64 each direction) producing a sequence of hidden states; a trainable additive attention computes normalized weights across time steps and forms a weighted sum (context vector); dropout, dense ReLU (64) - sigmoid.
(D) CNN-(Bi)LSTM + multi-head attention (MHA): Conv1D + MaxPool - BiLSTM - multi-head self-attention with residual connections and LayerNorm; temporal pooling - dense ReLU (64) - sigmoid.

Across models we used binary cross-entropy loss, Adam optimizer with learning rate 1e−3, batch size 64, and 20 epochs. Class weights were applied in all fits. Where explored, early stopping and learning-rate scheduling did not change the qualitative ranking of models.

Threshold Tuning. Because non-adherence is relatively infrequent and clinical action often favors sensitivity, we selected the decision threshold on the validation/test set by maximizing F1 (or by imposing a minimum precision constraint, when specified). We swept thresholds between ~0.10 and 0.88 in small increments, evaluated class-wise precision/recall/F1, and chose the threshold with the highest F1 for the positive class. For completeness we also report results at a fixed 0.50 threshold when applicable. The better results were reported. For some of the results we manually tuned the threshold around the best selected ones, since the results were totally imbalanced towards detecting 0 s.

3 Results

We evaluated sequence models on overlapping 3-, 5-, and 7-day windows to predict next-window non-adherence. For each window length, we report overall discrimination (ROC-AUC), accuracy, and positive-class (non-adherence) precision/recall/F1 at the chosen operating point; confusion matrices quantify true/false positives and negatives (Tables 1, 2, 3, 4, 5 and 6).

A full week of context produced the strongest ranking performance. The LSTM achieved ROC-AUC 0.755 with accuracy 0.73 and a balanced trade-off on the positive

Table 1. Performance on 7-day overlapping windows.

Model	Accuracy	ROC-AUC	Precision (Class 1)	Recall (Class 1)	F1 (Class 1)
LSTM	0.73	0.755	0.65	0.51	0.57
CNN-LSTM	0.56	0.744	0.44	0.90	0.59
CNN-LSTM + Attention	0.71	0.755	0.60	0.55	0.58
CNN-LSTM + Multi-Head Attention	0.71	0.735	0.62	0.49	0.55

Table 2. Confusion Matrix; 7-day window.

Model	TN	FP	FN	TP
LSTM	2933	526	947	973
CNN-LSTM	1269	2189	189	1731
CNN-LSTM + Attention	2748	710	858	1062
CNN-LSTM + Multi-Head Attention	2873	585	971	949

Table 3. Performance on 5-day overlapping windows.

Model	Accuracy	ROC-AUC	Precision (Class 1)	Recall (Class 1)	F1 (Class 1)
LSTM	0.612	0.630	0.372	0.509	0.430
CNN-LSTM	0.649	0.638	0.402	0.459	0.429
CNN-LSTM + Attention	0.680	0.657	0.450	0.430	0.440
CNN-LSTM + Multi-Head Attention	0.670	0.642	0.428	0.448	0.438

class (P/R/F1 = 0.65/0.51/0.57). The CNN-LSTM + attention model matched AUC (0.755) with accuracy 0.71 and a slightly more recall-oriented profile (0.60/0.55/0.58). In contrast, the plain CNN-LSTM prioritized sensitivity (recall 0.90) at the expense of specificity (accuracy 0.56, many false positives).

Table 4. Confusion Matrix; 5-day window.

Model	TN	FP	FN	TP
LSTM	2507	1328	758	787
CNN-LSTM	2780	1055	836	709
CNN-LSTM + Attention	3013	822	873	672
CNN-LSTM + Multi-Head Attention	2911	924	853	692

Table 5. Performance on 3-day overlapping windows.

Model	Accuracy	ROC-AUC	Precision (Class 1)	Recall (Class 1)	F1 (Class 1)
LSTM	0.670	0.620	0.29	0.44	0.35
CNN-LSTM + Attention (normalized)	0.700	0.611	0.31	0.38	0.34
CNN-BiLSTM + Multi-Head Attention	0.391	0.601	0.22	0.78	0.34
XGBoost	0.750	0.602	0.35	0.30	0.33
LSTM + SMOTE	0.760	0.592	0.37	0.26	0.31

Table 6. Confusion Matrix; 3-day window.

Model	TN	FP	FN	TP
LSTM	3166	1159	615	483
CNN-LSTM + Attention (normalized)	3377	948	676	422
CNN-BiLSTM + Multi-Head Attention	1259	3066	237	861
XGBoost	3709	615	766	333
LSTM + SMOTE	3822	502	808	291

With a 5-day context, attention-augmented CNN-LSTM delivered the strongest overall performance, achieving the highest accuracy (0.680) and ROC-AUC (0.657), along with the top positive-class F1 (0.440) at the reported operating point.

Short context favored overall accuracy but limited minority-class performance. XGBoost and LSTM + SMOTE achieved the highest accuracies (0.750 and 0.760) with modest AUCs (~0.60) and lower class-1 F1 (0.33 and 0.31).

4 Discussion

This study shows that short-horizon non-adherence with ePRO reporting can be anticipated from recent symptom patterns using compact sequence models and overlapping windows. A longer temporal context improves ranking; with 7-day windows, the LSTM and the attention-augmented CNN-LSTM achieved the highest ROC-AUC (~0.755) with balanced accuracy (~0.71–0.73), indicating that a full week captures enough day-to-day fluctuation (e.g., fatigue and sleep disturbance co-variation) to separate likely lapses from continued engagement without aggressive thresholding.

The label design (next-window prediction) aligns model outputs to an actionable time horizon. The within-window, feature-wise imputation limits leakage and preserves each patient's short-term level; the deterministic fallback (all-missing - 0) ensured well-formed inputs for training, though future deployments might add an "all-missing" indicator to explicitly encode this signal. Attention modestly stabilized performance at 7 days—reasonable given short sequences ($T \leq 7$) and moderate per-day dimensionality. Class weighting aided minority detection; further gains might come from focal loss, AUPRC-oriented objectives, or calibrated ensembling.

One of the limitations of this study is that results stem from a single program/database and may not immediately generalize across institutions or cancer populations; external validation is needed. Moreover, missingness likely carries signal beyond imputation; explicit missingness indicators or masked sequence layers could improve sensitivity without inflating false positives.

Multiple factors may affect patient adherence, and these factors may change over the time for a particular patient [14, 15]. Telemonitoring adherence factors identified in our previous work included technology usability such as ease of device use and interface design [16, 17], patient support modality including training, technical help, and pre-visit checks [18, 19], clinical relevance including perceived benefit and timely clinician feedback, socioeconomic [20, 21], and access issues such as connectivity, device ownership, and digital literacy, and program design including frequency of monitoring, reminders, and incentives [22, 23]. Accounting for these factors may improve accuracy of non-adherence prediction. Development of next iteration of predictive models should minimize bias and maximize transparency and explainability [24, 25].

5 Conclusion

Short-term ePRO non-adherence is predictable from recent symptom trajectories. A 7-day context maximizes ranking performance with balanced errors—well suited for risk triage—while a 5-day CNN-LSTM at a recall-oriented threshold offers the best case-finding profile for proactive outreach. Three-day windows can yield high accuracy but are less helpful for minority-class performance. For deployment, we recommend a two-tier alerting strategy (lower threshold for automated nudges; higher threshold for staff outreach), routine drift checks, and regular threshold re-tuning to align with staffing capacity and clinical priorities. Future work will emphasize patient-level temporal validation, calibration, interpretability, and multi-site replication to ensure robust, trustworthy integration into routine oncology care.

References

1. Basch, E., et al.: Symptom monitoring with patient-reported outcomes during routine cancer treatment: a randomized controlled trial. J. Clin. Oncol. **34**(6), 557–565 (2016)
2. Basch, E., et al.: Overall survival results of a trial assessing patient-reported outcomes for symptom monitoring during routine cancer treatment. JAMA **318**(2), 197–198 (2017)
3. Mooney, K.H., et al.: Automated home monitoring and management of patient-reported chemotherapy symptoms. Cancer Med. **6**(3), 537–546 (2017)
4. Mooney, K., Whisenant, M.S., Beck, S.L.: Symptom care at home: a comprehensive and pragmatic PRO system approach to improve cancer symptom care. Med. Care **57**, S66–S72 (2019)
5. Fawaz, H.I., Forestier, G., Weber, J., Idoumghar, L., Muller, P.-A.: Deep learning for time series classification: a review. Data Min. Knowl. Discov. **33**, 917–963 (2019)
6. Hochreiter, S., Schmidhuber, J.: Long short-term memory. Neural Comput. **9**(8), 1735–1780 (1997)
7. Bohlmann, A., Mostafa, J., Kumar, M.: Machine learning and medication adherence: scoping review. JMIRx Med. **2**(4), e26993 (2021)
8. Gu, Y., Shah, J., Chakraborty, S.B.: Predicting medication adherence using ensemble learning and deep learning. Sci. Rep. **11** (2021)
9. Finkelstein, J., Smiley, A., Echeverria, C., Mooney, K.: Predicting incipient symptom deterioration from serial patient-reported data during cancer chemotherapy course using LSTM modeling. In: 2024 IEEE First International Conference on Artificial Intelligence for Medicine, Health and Care (AIMHC), Laguna Hills, CA, USA, pp. 177–180. IEEE (2024)
10. Finkelstein, J., Smiley, A., Echeverria, C., Mooney, K.: AI-driven prediction of symptom trajectories in cancer care: a deep learning approach for chemotherapy management. Bioengineering **11**(11), 1172 (2024)
11. Finkelstein, J., Smiley, A., Huo, X., Echeverria, C., Mooney, K.: Predicting side effects of chemotherapy: leveraging machine learning for cancer symptom management. In: 2024 IEEE 6th Eurasia Conference on Biomedical Engineering, Healthcare and Sustainability (ECBIOS), Tainan, Taiwan, pp. 108–111. IEEE (2024)
12. Finkelstein, J., Smiley, A., Huo, X., Echeverria, C., Mooney, K.: Hybrid deep learning for early detection of symptom escalation in patients with cancer. In: 2025 IEEE 11th International Conference on Big Data Computing Service and Machine Learning Applications (BigDataService), Tucson, AZ, USA, pp. 89–93. IEEE (2025)
13. Finkelstein, J., Smiley, A., Echeverria, C., Mooney, K.: Deep learning approaches for prediction of mental and physical deterioration during chemotherapy. In: Bellazzi, R., Juarez Herrero, J.M., Sacchi, L., Zupan, B. (eds.) AIME 2025. LNCS, vol. 15735, pp. 148–153. Springer, Cham (2025). https://doi.org/10.1007/978-3-031-95841-0_28
14. Jeong, I.C., Liu, J., Finkelstein, J.: Factors affecting adherence with telerehabilitation in patients with multiple sclerosis. Stud. Health Technol. Inform. **257**, 189–193 (2019). PMID: 30741194
15. Settle, J.R., Maloni, H.W., Bedra, M., Finkelstein, J., Zhan, M., Wallin, M.T.: Monitoring medication adherence in multiple sclerosis using a novel web-based tool: a pilot study. J. Telemed. Telecare **22**(4), 225–233 (2016). PMID: 26253748
16. Finkelstein, J., et al.: Enabling patient-centered care through health information technology. Evid. Rep. Technol. Assess (Full Rep.). **206**, 1–1531 (2012). PMID: 24422882
17. Jeong, I.C., Finkelstein, J.: Remotely controlled biking is associated with improved adherence to prescribed cycling speed. Technol. Health Care **23**, S543–S549 (2015). PMID: 26410522
18. Cross, R.K., Finkelstein, J.: Feasibility and acceptance of a home telemanagement system in patients with inflammatory bowel disease: a 6-month pilot study. Dig. Dis. Sci. **52**(2), 357 (2007). PMID: 17211702

19. Finkelstein, J., Cabrera, M., Hripcsak, G.: Internet-based home asthma telemonitoring: can patients handle the technology? Chest **117**, 148–155 (2000). PMID: 10631213
20. Cross, R.K., Finkelstein, J.: Challenges in the design of a Home telemanagement trial for patients with ulcerative colitis. Clin. Trials **6**(6), 649–657 (2009). https://doi.org/10.1177/1740774509346978. Epub 2009 Oct 12 PMID: 19822631
21. Bedra, M., McNabney, M., Stiassny, D., Nicholas, J., Finkelstein, J.: Defining patient-centered characteristics of a telerehabilitation system for patients with COPD. Stud. Health Technol. Inform. **190**, 24–26 (2013). PMID: 23823363
22. Finkelstein, J., Wood, J.: Implementing home telemanagement of congestive heart failure using Xbox gaming platform. In: Annual International Conference of the IEEE Engineering in Medicine and Biology Society 2011, pp. 3158–3163 (2011). https://doi.org/10.1109/IEMBS.2011.6090861. PMID: 22255010
23. Castro, H.K., Cross, R.K., Finkelstein, J.: Using a home automated telemanagement (HAT) system: experiences and perceptions of patients with inflammatory bowel disease. In: AMIA Annual Symposium Proceedings 2006, p. 872 (2006). PMID: 17238492; PMCID: PMC1839272
24. Smiley, A., Villarreal-Zegarra, D., Reategui-Rivera, C.M., Escobar-Agreda, S., Finkelstein, J.: Methodological and reporting quality of machine learning studies on cancer diagnosis, treatment, and prognosis. Front. Oncol. **14**(15), 1555247 (2025). https://doi.org/10.3389/fonc.2025.1555247. PMID: 40297817; PMCID: PMC12034563
25. Smiley, A., Reategui-Rivera, C.M., Villarreal-Zegarra, D., Escobar-Agreda, S., Finkelstein, J.: Exploring artificial intelligence biases in predictive models for cancer diagnosis. Cancers **17**(3), 407 (2025). https://doi.org/10.3390/cancers17030407. PMID: 39941776; PMCID: PMC11816222

A Hybrid Deep Learning Framework Integrating Convolutional Neural Networks and Vision Transformers for Automated Cardiovascular Disease Classification from ECG Images

Hussein Adekola Aderinko[1], Emmanuel Masa-Ibi[2], Deborah Olaitan Taiwo[3], Ishaya Gambo[3(✉)], and Saurabh Agarwal[4]

[1] Department of Computer Science, Texas State University, San Marcos, USA
[2] Department of Electrical and Computer Engineering, Morgan State University, Baltimore, MD, USA
[3] Department of Software Engineering, Obafemi Awolowo University, Ile-Ife, Nigeria
ipgambo@oauife.edu.ng
[4] Department of Information and Communication Engineering, Yeungnam University, Gyeongsan 38541, Republic of Korea

Abstract. Cardiovascular diseases (CVDs) are among the most common causes of death worldwide, and the electrocardiogram (ECG) is one of the most commonly used cardiac examination methods, which is a non-invasive method. Nevertheless, manual ECG interpretation is time-consuming, labour-intensive, requires specialized expertise, and human review is prone to errors, which delays diagnosis and negatively impacts accuracy, especially in resource-limited settings. Towards Addressing These Issues: To tackle these challenges, this research presents a new, end-to-end system that can automatically classify cardiac abnormalities directly from 2D ECG images. We introduce a hybrid deep learning architecture that effectively combines a Convolutional Neural Network (CNN) for localized feature extraction and a Vision Transformer (ViT) to exploit dependencies at a global level across the ECG waveform. The task of the model is threefold: it should classify ECG images into the three vital categories of Normal, Arrhythmia, and Myocardial Infarction (MI). Its AI model was onboarded to a sophisticated, easy-to-use, web-based platform implemented on top of React and Node.js, Firebase, which in turn makes it practical and quickly available. Its performance was then thoroughly assessed on a curated set of public ECG images. The hybrid model achieved an automatic accuracy of 92%, precision of 88%, recall of 88%, and F1-score of 87% The ability to discriminate was excellent (AUC-ROC, 0.92). The key findings demonstrate that the proposed hybrid CNN-ViT framework, combined with a web-based system, offers a practical and efficient strategy for deploying CVD diagnostics and predicting CVD types, thereby improving speed, accuracy, and hospital access to CVD diagnostics. At the same time, it has substantial promise as a clinical decision-support tool.

H. R. Arabnia et al. (Eds.): CSCI 2025, CCIS 2971, pp. 179–194, 2027.
https://doi.org/10.1007/978-3-032-28463-1_14

Keywords: Cardiovascular Diseases · ECG Image Analysis · Deep Learning · Convolutional Neural Network (CNN) · Vision Transformer (ViT) · Health Informatics · Clinical Decision Support System

1 Introduction

Cardiovascular diseases (CVDs) are the Anglo-Saxon expression of the largest public health emergency of our era, the largest cause of death worldwide, [1]. A timely and accurate diagnosis is crucial for effective management and improving mortality rates. Derived from data collected in October 2023, the electrocardiogram (ECG) remains one of the most basic, inexpensive, and non-intrusive diagnostic tests widely available, providing significant information regarding cardiac function [2]. Nevertheless, a major limitation of this technique is the highly complex and mostly subjective interpretation when done manually. Manual ECG analysis requires a high degree of expertise, mostly by trained cardiologists. Such dependence results in a slowdown of healthcare delivery, particularly in resource-limited or rural settings, where such expertise is less common [3]. Consequently, patients often have significant delays in diagnoses that are essential, thus resulting in worse prognoses. Additionally, manual interpretation is inherently prone to variability and human error.

Research has demonstrated that a significant fraction of ECGs contains interpretive errors, some of which may miss subtle but lethal abnormalities [4, 5]. Over the next few decades, the increasing global burden of CVDs will outpace the availability of trained cardiac specialists, creating an urgent need for scalable, efficient diagnostic solutions. A significant inequity in access to healthcare exacerbates this problem. There are many places where a cardiologist is a rare, if not nonexistent, specialty. Such inequities directly result in delays in diagnosis and poor outcomes for patients. The ability to interpret ECG exists only within certain well-trained groups of people, so there is potential for tools that democratize electrocardiogram interpretation (in other words, these tools enable low-skilled people to make first approximations of Early detection rapidly and anywhere in the world, regardless of whether the interpreter is present at a specialist centre) [3].

The recent emergence of Artificial Intelligence (AI), particularly deep learning (DL), presents a promising alternative approach for automated medical image analysis [6]. Convolutional Neural Networks (CNNs) have demonstrated proficiency in detecting complex patterns in ECGs, making them particularly suitable for the detection of arrhythmias, ischemia, and structural heart abnormalities [7, 8]. However, CNNs focus on local space features, making them less effective for modeling the long-distance dependencies of the data, which is crucial for understanding the overall time structure information of ECG signals. In response to this restriction, Vision Transformers (ViTs), a variant of Transformers for natural language processing, have recently been repurposed for image-based tasks with great success [9]. Instead, ViTs can be considered as capturing the global relationships between patches as a sequence of patches and using self-attention mechanisms. That helps ViTs incorporate more global context into the analysis, thereby combining it with the localized feature extraction of CNNs.

While research on AI-based ECG interpretation has rapidly accumulated, a substantial gap remains between the development of models and their practical application.

Most existing solutions are either limited to raw, unprocessed 1D signal data or require additional processing to be effective. In contrast, most clinical ECG data are disseminated and stored as 2D images or are not targeted at an end-user base. Consequently, these systems remain largely theoretical and continue to have a marginal impact on frontline care. To fill this gap, this paper proposes a practical AI full-stack weapon-based web application for autonomic ECG image interpretation. The main contribution of this work is a hybrid DL architecture that combines CNN and ViT blocks to learn both local morphological information and global temporal information in ECG images. The model has been integrated into a web interface that is adaptable to support real-time clinical decision-making. We are inspired not just by the potential for superior algorithm performance but also by the immediate need for equitable and scalable healthcare solutions. Through reproducible and reliable AI-based interpretation, it can help lessen the interpretation burden and minimize diagnostic errors, aid clinicians in high-volume and high-stress environments, and provide healthcare workers in low-resource areas with access to specialist-level ECG analysis. Equally importantly, the system is deployable under real-world conditions, eliminating the long-standing gap between AI development and clinical translation.

2 Background and Related Work

Computational methods for analyzing ECGs have evolved over several decades. This section summarizes the evolution of machine learning, from classical to DL architectures, and identifies the gaps in the literature that this research aims to address.

2.1 Machine Learning and CNNs in ECG Analysis

Growing interest in automated ECG classification has led to the development of traditional machine learning pipelines that involve manual extraction of predictive features from the ECG signal (e.g., R-R intervals, QRS duration, P-wave morphology), which would then be passed through classifiers such as Support Vector Machines (SVMs) or Random Forests [2]. These methods, while foundational, were tedious, and their success relied heavily on the quality of the hand-crafted features.

DL, and more specifically CNNs, have provided a paradigm shift, as they enable models that directly learn to predict from data by minimizing a loss based on raw or minimally processed data, thereby automating the feature extraction process [18]. Numerous studies have demonstrated that CNNs are effective in interpreting ECG data. For example, Abubaker and Babayiğit [10] presented a CNN model that achieved 98.23% accuracy in CVD detection from ECG images, highlighting the capabilities of deep feature learning. Similarly, Dias et al. [11] implemented an AI-driven screening system that utilizes a deep CNN to classify 12-lead ECG images in the context of emergency room triage, achieving an overall accuracy of 94.3%. In other literature, for instance, Jahmunah et al. [12], Yadav et al. [13], and Sattar et al. [18] have effectively utilized CNNs for the detection of other conditions, such as ischemic heart disease, myocardial infarction, and arrhythmias. Indeed, these studies collectively validate that CNNs are particularly capable of localizing important signatures from ECG waveforms, such as ST-segment depression or abnormalities in QRS complexes.

2.2 Vision Transformers in Medical Imaging: An Emerging Role

However, one general limitation of CNNs, despite the success of skip connections, is the localized nature of their receptive field. They learn a hierarchical representation of an image from local features, which is often not the best way to learn long-range dependencies, for example, the relationship between cardiac cycles from different locations of a 10-second ECG strip. And this is where we have a robust alternative to vision transformers (ViTs). In images, this involves partitioning the image into a sequence of patches and applying self-attention, which enables the model to determine the importance of each patch to all others, thereby obtaining a global picture context [9].

There is a quickly growing recognition of the feasibility of transformers in medical imaging. Kilimci et al. [14] implemented different transformer models (ViT, BEiT, Swin-Tiny) on ECG images. They observed accuracies of up to 96.1%, indicating that transformer architectures are well-suited for this domain. The success highlights that recognizing a global pattern is an essential part of an accurate ECG diagnosis.

2.3 Hybrid Models and System-Level Integration

Given the two-fold advantages of CNNs (local feature extraction) and ViTs (global context modeling), some researchers have begun to explore the use of hybrid architectures. These models aim to combine the best of both worlds, providing optimal performance. Although this is a burgeoning field, the concept has been well-established in other areas and presents a wealth of opportunities in ECG analysis.

Furthermore, Muzammil et al. [4] and Kolhar et al. [15] conclude that model performance is not the sole endpoint; rather, practical utility is closely tied to the successful integration and application of real-world systems. Additionally, Sumalatha et al. [16], Mohebbanaaz et al. [17], and Gambo and Agbonkhese [21] have outlined a call for relevant, privacy-preserving, and deployable systems designed to be integrated into the clinical workflow. The focus should shift from algorithmic benchmarks to breaking the traditional evaluation pipelines and developing end-to-end solutions that account for real-world constraints.

2.4 Gaps Identified

A particularly notable contrast emerges when considering the architectural design choices in contemporary work. Although CNNs and ViTs have demonstrated strong performance in machine learning, there is a lack of research on complementary hybrid models for 2D ECG image classification. Such modulization is not only powerful but also feasible, as it would not only provide a much more complete and integrative analysis by taking advantage of the best features of both architectures—capturing the local morphologic features (e.g., ST-segment abnormalities) through CNNs while modeling global long-range dependencies in predicting cardiac rhythm (i.e., through ViT). This is a notable loss in potential for creating a more comprehensive and robust diagnostic tool, since these customized hybrid solutions are typically unused. Additionally, there is a significant gap between the areas mainstream AI research focuses on and the actual clinical data.

While a significant volume of work focuses on analyzing bare 1D ECG time-series data, it also often ignores the practical limitations in several healthcare environments. In many clinics and hospitals, particularly those utilizing legacy systems or operating in low-resource areas, ECGs are routinely stored and examined as two-dimensional (2D) images, often obtained from scanned paper records. It also means that there is an urgent need for powerful models that can target and work with this universal data mode directly and easily, because having state-of-the-art AI tools that are too complex and difficult to align with standard clinical practice. The data clinicians have to deal with daily means that those state-of-the-art tools will remain a theoretical fantasy. Perhaps the most important of these is the well-documented implementation gap between academic research and the clinical setting. Although the vast majority of studies in this field are methodologically sound, their contributions typically conclude with the reporting of model accuracy metrics, with little to no attention to the real-world implement ability of their models in an end-user system. Such promising innovations thus remain trapped in academic provenance, never making their way to clinical benefit. There is a pressing demand for research that goes some way in closing this 'research–practice gap' by demonstrating whole, end-to-end systems. These platforms are not only algorithmically robust but also sound and robust, featuring human-centred interfaces for real clinical deployment. The present research addresses these integrated gaps explicitly through the proposed new CNN-ViT Hybrid architecture for 2D ECG image classification, which is linked to a fully realized web-based clinical decision-support platform. Some key contributions of this study to the field of health informatics and AI in medicine include:

1. Combination CNN and Vision Transformer: We proposed, implemented, and validated a hybrid DL model that combines a CNN and a Vision Transformer. Specifically designed for 2D ECG image classification, this architecture has been demonstrated to be highly successful, leveraging the complementary characteristics of both model types.
2. A Fully Built and Deployed System: A fully functioning, secure, and user-friendly web-based solution was developed and deployed. It fills a crucial gap between generating a theoretical model and its concrete clinical application, offering a prototype ready for pilot testing and field validation.
3. Extensive Performance Benchmarking: We benchmark our system using a battery of standard metrics. These results provide a clear and practical reference point for hybrid models in ECG analysis, delineating well-characterized research domains (normal and arrhythmia detection) from those that require further investigation (MI vs. arrhythmia).

3 Methodology

Design–based research was employed in this study to develop and refine the AI-powered ECG analysis system. It includes the system architecture design, data collection and preprocessing, the combined high-RP model, and testing procedure [19].

3.1 System Architecture

As depicted in Fig. 1, the system is based on a standard client-server architecture, designed for high scalability and secure data processing. The frontend client, which is both dynamic and responsive, serves as the primary point of interaction for the user, built using React with TypeScript. It is expected to facilitate easy registration and login for users, as well as upload ECG Images and display classification results in a clean and intuitive interface. TypeScript aids this process—it adds static typing, which not only boosts developer productivity but also helps reduce the probability of runtime errors. The backend is the powerful Node-based core of the system.js server utilizing Express.js framework. This is the central orchestrating server that contains the core business logic. Functionalities such as secure user authentication, routing of API requests, and communication with other system components are handled by this server.

One of the major architectural decisions was to separate the computationally expensive AI inference into its microservice. A dedicated AI server that hosts the Flask implementation of the AI model, built from the Python script, handles independent scaling, updating, and maintenance of the DL model, thereby improving availability and resource specialization. The complete Google Firebase suite is used in the system for data persistence and management. Firebase Authentication provides a secure and reliable method for managing user sign-up and sign-in processes. User-specific data and classification history are stored in Firestore, a NoSQL database, and uploaded ECG images are hosted securely in Firebase Storage. It provides an integrated result that yields low overhead while maintaining high security.

The complete flow is straightforward—a user uploads an ECG image from the React frontend. This request will also be securely transmitted to a Node.js background that orchestrates all subsequent steps. It goes to a Flask API, and the back end sends that image to the Flask API—this is where the AI model is actually doing its inference. The prediction is then sent back to the Node once it has been predicted. This is then saved as a record in the Firestore database and sent back to the app's frontend (the feedback loop is complete, as the classification is displayed to the user).

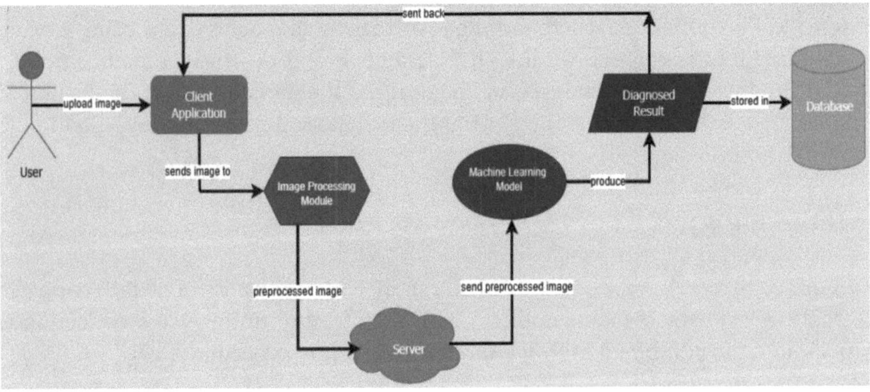

Fig. 1. The Client-Server Architecture of the ECG Image Analysis System.

3.2 Data Acquisition and Preprocessing

In this paper, an ECG image dataset is compiled from two publicly available datasets, Mendeley[1] and Kaggle[2], to classify normal heartbeats. Mendeley – The image classification in the Mendeley primary dataset was Normal, MI, Abnormal Heartbeat, and History of MI. The "Abnormal Heartbeat" and "History of MI" categories were combined to form a new class called "Arrhythmia". Finally, this resulted in a dataset containing 928 images of each class (284 Normal, 405 Arrhythmia, and 239 MI), which was later divided into a test split. Additionally, to enhance the model's performance and subject it to more diverse data, we also used an additional dataset from Kaggle. The second dataset originates from the MIT-BIH Arrhythmia Database and classifies images as either Normal or Arrhythmia. A strict preprocessing pipeline, applied in OpenCV and TensorFlow, was used for every image in order to maintain the robustness of the pre-trained model and avoid over-fitting:

i. Resize: All the images are resized to a standard 224 × 224 pixel input size, often used with ViT-based models.
ii. Normalization: Pixel values were scaled to the range [0, 1] to stabilize the training process.
iii. Denoising: A Gaussian filter to attenuate background noise while preserving the quality of the ECG signal.
iv. Data Augmentation: A set of strategies, including random rotation, horizontal flip, and small contrast adjustments, was leveraged with the training set to increase the dataset size and improve model generalization.
v. The final dataset was split into training (70%), validation (15%), and testing (15%) sets to ensure no overlap.

3.3 Architecture of the Hybrid CNN ViT Model

This study primarily proposes a hybrid DL architecture that fuses a CNN and a Vision Transformer (ViT) encoder. The rationale behind this combination stems from the complementary nature of both architectures: CNNs excel at capturing local spatial features in image data. In contrast, Vision Transformers excel at capturing global long-range interactions. It utilizes both to perform robust and comprehensive ECG image classification. The CNN part of the model is the first feature extractor and consists of a custom stack of four convolutional blocks. Input ECG images, which are resized to 224 × 224 × 3, are passed through. To begin with, they pass 32 filters through Conv2D with a kernel size of 3 × 3, a stride of 1, and the same padding. A batch norm, ReLU, and 2 × 2 max pooling follow. The two layers in the second block are a Conv2D with 64 filters and the same kernel size as before, again followed by batch normalization, ReLU, and max pooling. We triple the number of filters to 128 and repeat the process with 3 × 3 kernels for the third convolutional block, followed by batch normalization, ReLU activation, and pooling.

The last layer in the CNN stack is a Conv2D layer with 256 filters and batch normalization, applied with ReLU activation but without the pooling operation to maintain

[1] https://data.mendeley.com/datasets/gwbz3fsgp8/2.

[2] https://www.kaggle.com/datasets/erhmrai/ecg-image-data/data.

spatial resolution. At this point in the data pipeline, our output is a $14 \times 14 \times 256$ high-level feature map that retains the morphological content of the ECG signals, including the electrical patterns for the QRS complex, P-wave, and ST-segment. This feature map is reshaped to extract non-overlapping patches of size $2 \times 2 \times 256$, resulting in a total of 196 patches. Each patch is then flattened and projected to a 768-dimensional vector (via a trainable linear projection).

We prepend a learnable class token ([CLS]) to the sequence of patches, which learns to represent the class and provides the output embedding for classification. They also incorporate positional embeddings into the input sequence to retain spatial information. This generated input is then fed to the Vision Transformer encoder. In this architecture, the ViT encoder consists of six transformer layers. Each layer consists of eight attention heads and an MLP block with a hidden size of 2048 (using GELU activation). The self-attention and multi-layer perceptron (MLP) parts also use dropout with a probability of 0.1 to prevent overfitting. At the same time, layer normalization and residual connections are introduced to enhance stability and facilitate smooth gradient flow. This encoder enables the model to learn the contextual relationships between all parts of the ECG image simultaneously, thereby enhancing the global feature representation.

Finally, the class token output is selected after transformer encoding and fed into a classification head. The head comprises a dense layer of 512 units with ReLU activation, a dropout rate of 0.3, and a final dense layer of three output units. Each of these units corresponds to the three classification targets: Normal, Arrhythmia (Tachycardia), and Myocardial Infarction (MI). The last layer utilizes a softmax activation to produce probabilities that are normalized across the three classes.

4 Results

This model was implemented using TensorFlow and Keras libraries and was trained on an NVIDIA RTX 3060 GPU. We set hyperparameters in the training process as such: (i) Loss Function: Categorical Cross-Entropy, as per the norm for a multi-class classification problem. (ii) Optimizer: Adam Optimizer with learning rate = 0.001. (iii) Learning Rate Schedule—The learning rate decayed over time to allow fine-tuning as training progressed. (iv) Batch Size: 32. (v) Epochs: A total of 50 epochs were used to train the model, with an early stopping callback that monitors the validation loss and prevents overfitting. Thus, the results of the model's performance evaluation, along with the implemented user interface of the web-based system, are described in this section.

4.1 System User Interface

The built system features a minimalistic UI that is easy to use for both clients and medical professionals. The workflow begins with a robust authentication system (Fig. 2), enabling users to sign up or log in if they already have an account. User experience is based on Role-Based Access for "Client" or "Cardiologist".

Upon successful login, the user is directed to the main dashboard, which provides access to the ECG upload functionality. The upload page (Fig. 3) allows for simple drag-and-drop or file selection of ECG images in standard formats (JPEG, PNG). After the AI model processes the image, the results are presented on a clear and concise display page (Fig. 4). This page shows the uploaded ECG image alongside the model's classification (e.g., Normal, Arrhythmia, or Myocardial Infarction) and any associated confidence scores. For professional use, the system generates a validated report that can be printed or downloaded as a PDF (Fig. 5), facilitating integration into patient records.

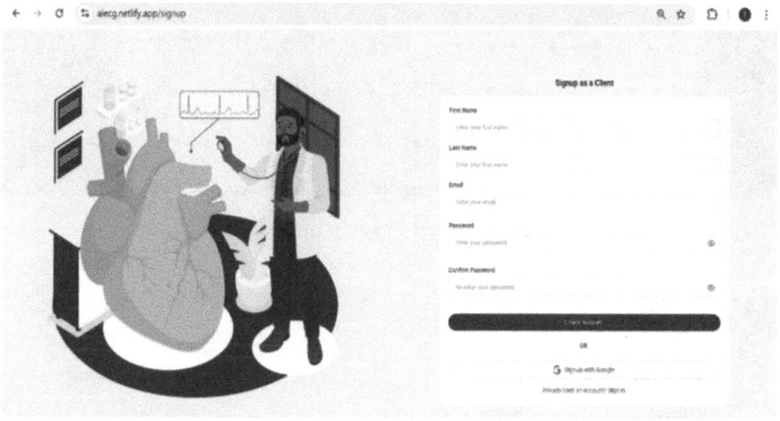

Fig. 2. User Authentication Interface (Signup)

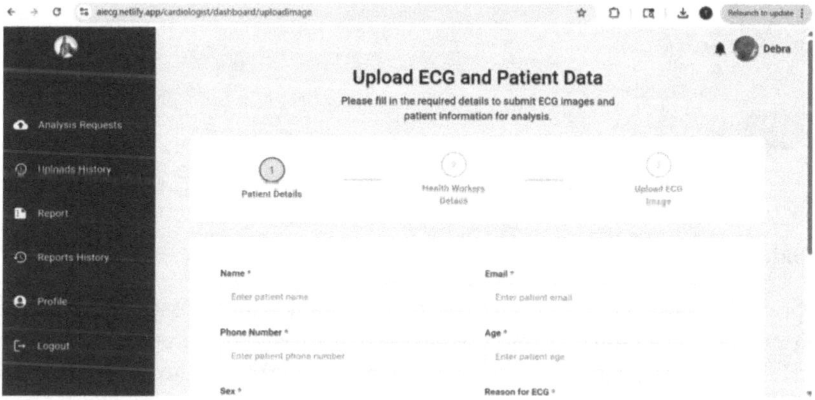

Fig. 3. ECG Image Upload Interface

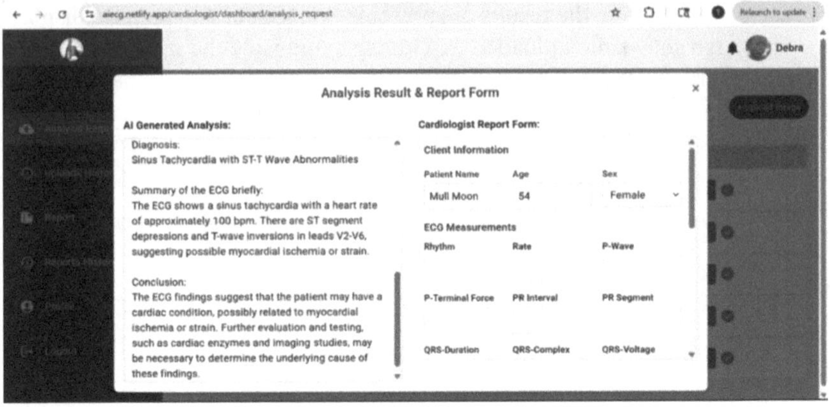

Fig. 4. Result Display Interface

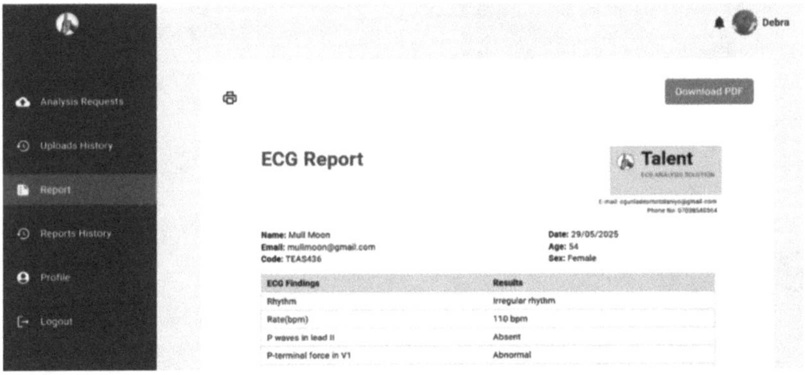

Fig. 5. Validated ECG Report for Download

4.2 Model Performance

The hybrid CNN-ViT model's performance was quantitatively assessed on the unseen test dataset. The evaluation metrics, broken down by class and averaged, are summarized in Table 1. The model achieved a high overall accuracy of 92%, indicating its strong capability to classify ECG images correctly. The weighted average for precision, recall, and F1-score was approximately 88%, demonstrating a well-balanced performance across the different cardiac conditions.

Table 1. Model performance metrics for different heart conditions (Normal, Arrhythmia, and Myocardial Infarction). The average values are computed across all classes.

Metric	Normal	Arrhythmia	MI	Average
Accuracy	93%	89%	85%	92%
Precision	94%	87%	83%	88%

(continued)

Table 1. (*continued*)

Metric	Normal	Arrhythmia	MI	Average
Recall	91%	90%	84%	88%
F1-score	92%	88%	83%	87%
AUC (ROC Curve)	0.95	0.92	0.89	0.92

The model performed best on the "Normal" class, with an F1-score of 92% and an AUC of 0.95, showing its reliability in identifying healthy heart rhythms. Performance on the "Arrhythmia" class was also strong. The classification of "Myocardial Infarction" was the most challenging, with a slightly lower F1-score of 83% and an AUC of 0.89. To further analyze the classification behavior, a confusion matrix was generated (Fig. 6). The matrix visually confirms the model's high accuracy for the "Normal" class (high values on the diagonal). It also reveals some confusion between the "Arrhythmia" and "MI" classes, which is expected given the potential for overlapping morphological features in their ECG manifestations.

The Receiver Operating Characteristic (ROC) curve (Fig. 7) provides a graphical representation of the model's diagnostic ability. The high Area Under the Curve (AUC) of 0.92 confirms the model's excellent capability to distinguish between the different classes across all classification thresholds. A robustness test, where artificial noise was added to the test images, resulted in a minor accuracy drop from 92% to 89%, indicating that the model maintains a high level of performance even with lower-quality inputs.

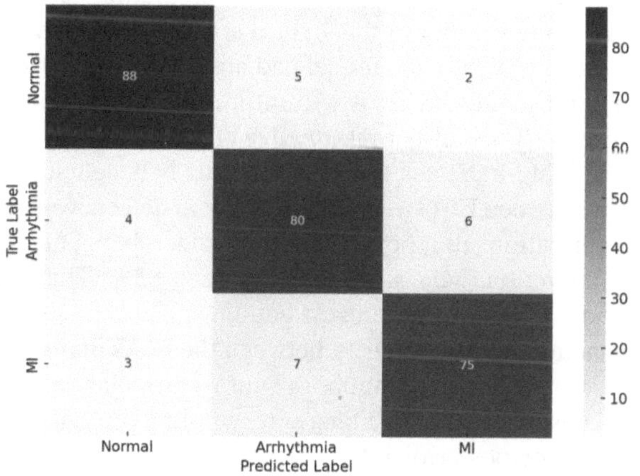

Fig. 6. Confusion Matrix for CNN-ViT Model Predictions

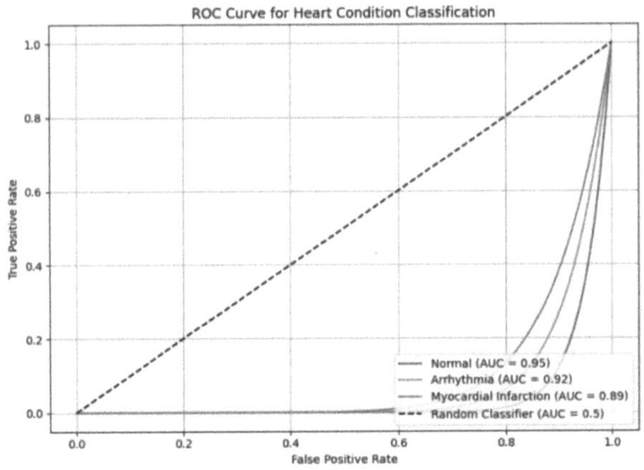

Fig. 7. Receiver Operating Characteristic (ROC) Curve

5 Discussion

In this section, we interpret the experimental results, discuss their significance in relation to existing literature, evaluate the contributions of the hybrid architecture, and consider the implications of the system's development for clinical practice.

5.1 Interpretation of Model Performance

Our hybrid CNN-ViT model achieved 92% overall prediction accuracy, as well as an AUC of 0.92, indicating the validity of this hybrid approach for automating ECG image analysis. In clinical settings, where the model had no trouble recognizing Normal ECGs (F1 = 92%), this is highly relevant, as it will ultimately reduce unnecessary interventions and false positives. They also performed well on detecting Arrhythmia with a recall of 90%. This finding indicates that the ViT part is beneficial in capturing long-distance relationships between ECG traces, allowing it to detect even minor disturbances in rhythm. The categorizations did, however, see considerable confusion with Arrhythmia for Myocardial Infarction (MI), with an F1-score of 83%. This misclassification is clinically relevant and warrants further investigation.

A possible explanation is the similarity between the ECG findings of some arrhythmias and some MIs. Ventricular arrhythmias or supraventricular tachycardias with aberrant conduction can cause ST-segment changes resembling those in acute MI, and some MIs, especially inferior or posterior infarctions, can also cause rhythm disturbances, such as sinus bradycardia, atrioventricular block, or even atrial fibrillation. Due to the overlapping features between the two types of classes, the model may confuse them, especially when dealing with only customer 2D images. In addition, the fact that both ischemia and arrhythmias are dynamic and often occur together complicates this task even more. On one hand, the CNN can obtain local morphological clues (e.g., presence of ST elevation), while ViT can focus more on common global rhythm features. When combined, the two might smooth the distinctions between the two diagnostic outputs in ambiguous cases. This may indicate the promising generalizability of the model, but could also suggest that future research may require a finer granularity, such as clinical context, 24-hour ECG data, and wider leads. Attention visualization techniques may also help in understanding which features the model relies on most when differentiating between closely related conditions.

5.2 Importance of CNN-ViT Hybrid Architecture

The success of our model further backs up our hypothesis of a synergistic effect, providing a hybrid architecture for ECG analysis. The CNN focuses on extracting the fine-grained, local morphological features of each wave (P, QRS, T), while the ViT focuses on integrating this information into a global context, i.e., understanding rhythm and inter-beat relationships. The complementary two-step approach enables the model to mimic a human cardiologist in performing both morphological and rhythmical analyses simultaneously. Our results are competitive with those in the existing literature. Although some recent studies that employed pure CNNs [10] or ViTs [14] on matching datasets claim a marginally higher accuracy, our work improves on these studies by validating the model with an end-to-end, production-level web system. We outperform the accuracy of other studies, which employed more descriptive models or basic CNNs [12]. This performance confirms that our architecture is a suitable and efficient option for the challenging 2D ECG image classification task.

5.3 Practical Implications and System Usability

This work is more than an algorithmic improvement; it represents an end-to-end system. Web platform: An easy-to-use web interface for clinicians (Figs. 2, 3, 4 show the UI). The system overcomes several practical barriers to AI adoption in healthcare by managing user authentication, providing secure image storage, and facilitating near real-time feedback [20]. It shifts the research from a theoretical exercise to a concrete prototype that can be piloted in a clinical context. That ability could make it an indispensable decision-support tool in a number of situations:

i. Triage in the Emergency Room: Expediting identification of potentially urgent ECGs for early attention
ii. Primary Care: Enabling GPs to make better referrals.
iii. Telemedicine: Expanding remote cardiac screening in underrepresented communities.

6 Implications for Research and Practice

These findings have important implications for the future of health informatics research and the practice of clinical cardiology. On the one hand, for practical purposes, the developed system offers a practical approach to aid healthcare workers in their diagnosis. It can reduce diagnostic turnaround times, enhance consistency in interpretations, and reduce the burden of overburdened cardiologists by providing a rapid, accurate, and objective analysis of ECG images. In limited-resource environments, it can serve as a "virtual consultant," democratizing access to diagnostic quality that would otherwise be beyond reach. As a result, this may lead to earlier interventions and, consequently, better patient outcomes.

On the other hand, for research purposes, this study provides a solid baseline for the use of hybrid CNN-ViT models in medical imaging. It provides an empirical demonstration of the utility of local-global feature fusion in challenging diagnostic problems. Such considerations can prompt future research to expand upon this architecture through the exploration of more sophisticated fusion methods, the integration of larger and more complex multimodal datasets (e.g., merging ECG data with patient electronic health records), and the expansion of the model to detect a greater number of cardiovascular pathologies. In addition, this work highlights the critical need for a holistic and systems-thinking approach tailored to the fundamentals of health informatics research and the necessity of transitioning from merely making algorithmic advances to crafting well-designed solutions that are clinically relevant.

7 Threats to Validity

This study faces key validity threats that may influence the reliability and generalizability of its findings. The most critical aspect is external validity, as the model was trained on publicly available ECG datasets that may not fully represent the diversity of real-world data. Variations in acquisition devices, institutional protocols, and patient demographics could cause overfitting to the training environment. True robustness can only be confirmed through large-scale, multi-center clinical validation. Construct validity is another concern, as common metrics like accuracy and precision, while technically useful, fail to capture the real clinical impact of errors, especially the severe consequences of false negatives in life-threatening conditions. Future studies should incorporate clinically relevant and risk-adjusted evaluation measures to enhance the accuracy of their findings.

Additional threats include internal validity and data imbalance. Despite using reliable frameworks (TensorFlow, Keras, OpenCV) and strict methodologies, minor software artifacts may still influence results. Moreover, class imbalance remains a challenge; rare conditions, such as myocardial infarction, were underrepresented, leading to reduced performance for these classes. Although data augmentation and precision–recall tradeoffs helped, achieving balanced datasets or developing better techniques for rare conditions remains essential for improving model reliability.

8 Conclusion and Future Work

In conclusion, this study successfully developed an AI-enhanced system to automate the classification of cardiovascular diseases from ECG image data. Through the use of a hybrid CNN-ViT DL model alongside a user-friendly web platform, our model attained an overall accuracy of 92%, which indicates clear potential to provide a clinical decision-support tool. It concludes that the ability of CNNs to detect local features, combined with the ability of ViTs to grasp global context, presents a robust framework that can be extended to the complexities of medical image analysis.

The developed system offers a high degree of relevance to facilitate clinical practice, providing fast, precise, and affordable interpretation of ECGs. It can augment diagnostic pipelines, especially in resource-constrained environments. It is not intended to replace expert clinicians; however, it can serve as a valuable partner in prioritizing cases, reducing errors, and enhancing the overall efficiency of cardiovascular care. An order of magnitude larger and more diverse training dataset is the single most important and urgent direction for future work.

While this system is a hopeful proof of concept, maturing it into a truly comprehensive and clinically generalizable tool requires stepping away from the public repositories for the evolution of this system. Subsequent iterations should train and validate the model on much larger, multi-institution datasets with a complete representation of real-world variability. This means they need to include data from a variety of patient populations, ethnic groups, and ECG acquisition hardware, such that the model's behavior isn't brittle and is interpretable across various clinics.

References

1. World Health Organization: cardiovascular diseases (CVDs). WHO Fact Sheets (2021). https://www.who.int/news-room/fact-sheets/detail/cardiovascular-diseases-(cvds)
2. Ameen, A., Fattoh, I.E., Abd El-Hafeez, T., et al.: Advances in ECG and PCG-based cardiovascular disease classification: a review of deep learning and machine learning methods. J. Big Data **11**(159) (2024). https://doi.org/10.1186/s40537-024-01011-7
3. Rafie, N., Kashou, A.H., Noseworthy, P.A.: ECG interpretation: clinical relevance, challenges, and advances. Hearts **2**(4), 505–513 (2021). https://doi.org/10.3390/hearts2040039
4. Muzammil, M.A., et al.: Artificial intelligence-enhanced electrocardiography for accurate diagnosis and management of cardiovascular diseases. J. Electrocardiol. **83**, 30–40 (2024)
5. Di Costanzo, A., Spaccarotella, C.A.M., Esposito, G., Indolfi, C.: An artificial intelligence analysis of electrocardiograms for the clinical diagnosis of cardiovascular diseases: a narrative review. J. Clin. Med. **13**(4), 1033 (2024)
6. Li, X., Zhang, L., Yang, J., Teng, F.: Role of artificial intelligence in medical image analysis: a review of current trends and future directions. J. Med. Biol. Eng. **44**(2), 231–243 (2024)
7. Esteva, A., Chou, K., Yeung, S., Naik, N., Madani, A., Dean, J.: Deep learning-enabled medical computer vision. Nat. Biomed. Eng. **5**(6), 498–506 (2021). https://doi.org/10.1038/s41551-021-00760-5
8. Helaly, H.A., Badawy, M., Haikal, A.Y.: A review of deep learning approaches in clinical and healthcare systems based on medical image analysis. Multimed. Tools Appl. **83**, 36039–36080 (2024). https://doi.org/10.1007/s11042-023-16605-1
9. Dosovitskiy, A., Beyer, L., Kolesnikov, A., et al.: An image is worth 16x16 words: transformers for image recognition at scale. In: Proceedings of the ICLR (2021)
10. Abubaker, M.B., Babayiğit, B.: Detection of cardiovascular diseases in ECG images using machine learning and deep learning methods. IEEE Trans. Artif. Intell. **4**(2), 373–382 (2023). https://doi.org/10.1109/TAI.2022.3159505
11. Dias, F.M., Ribeiro, E., Moreno, R.A., et al.: Artificial intelligence-driven screening system for rapid image-based classification of 12-lead ECG exams: a promising solution for emergency room prioritization. IEEE Access **11**, 121739–121752 (2023). https://doi.org/10.1109/ACCESS.2023.3328400
12. Jahmunah, V., Ng, E.Y.K., San, T.R., Acharya, U.R.: Automated detection of coronary artery disease, myocardial infarction and congestive heart failure using GaborCNN model with ECG signals. Comput. Biol. Med. **134**, 104457 (2021)
13. Yadav, S.S., More, S.B., Jadhav, S.M., Sutar, S.R.: Convolutional neural networks based diagnosis of myocardial infarction in electrocardiograms. In: Proceedings of the 2021 International Conference on Computing, Communication, and Intelligent Systems (ICCCIS), pp. 581–586, February 2021

14. Kilimci, Z.H., Yalcin, M., Kucukmanisa, A., Mishra, A.K.: Advancing heart disease diagnosis with vision-based transformer architectures applied to ECG imagery. Image Vis. Comput., 105666 (2025)
15. Kolhar, M., Kazi, R.N.A., Mohapatra, H., Al Rajeh, A.M.: AI-driven real-time classification of ECG signals for cardiac monitoring using i-AlexNet architecture. Diagnostics **14**(13), 1344 (2024)
16. Sumalatha, U., Prakasha, K.K., Prabhu, S., Nayak, V.C.: Deep learning applications in ECG analysis and disease detection: an investigation study of recent advances. IEEE Access **12**, 57471–57492 (2024)
17. Mohebbanaaz, Kumar, L.V.R., Sai, Y.P.: A new transfer learning approach to detect cardiac arrhythmia from ECG signals. Signal Image Video Process. (SIViP) **16**, 1945–1953 (2022). https://doi.org/10.1007/s11760-022-02155-w
18. Aarotale, P.N., Rattani, A.: Deep learning models for arrhythmia classification using ECG images. J. Healthc. Eng. **2023**, Article no. 1398472 (2023.) https://doi.org/10.1155/2023/1398472
19. Khan Mamun, M.M.R., Elfouly, T.: AI-enabled electrocardiogram analysis for disease diagnosis. Appl. Syst. Innov. **6**(5), 95 (2023)
20. Gambo, I., Bueno-Delgado, M.V., Mooses, K., Melero Muñoz, F.J., Zviel-Girshin, R., Andrushevich, A., et al.: Technology adoption review for ageing well: analysis of technical solutions. Front. Public Health **11**, 1169192 (2023)
21. Gambo, I., Agbonkhese, C.: Software design specification proposal of a diagnostic decision support system for clinical low back pain. Procedia Comput. Sci. **231**, 243–250 (2024)

Machine Learning Approaches for Predicting Survival in Metastatic Castration-Resistant Prostate Cancer

Xingyue Huo[1(✉)], Manish Kohli[2], and Joseph Finkelstein[1]

[1] University of Arizona, Tucson, AZ 85724, USA
xingyuehuo@arizona.edu
[2] University of Utah-Huntsman Cancer Institute, Salt Lake City, UT 84112, USA

Abstract. The abstract should summarize the contents of the paper in short terms, i.e. 150–250 words. Accurate prognostic biomarkers are essential for evaluating survival risks in cancer patients. However, despite the wide use of biomarkers like prostate-specific antigen (PSA) and other clinical factors, achieving high predictive accuracy remains a challenge in prostate cancer prognosis. This study aimed to predict 24-month survival in metastatic castration-resistant prostate cancer (mCRPC) patients by analyzing a comprehensive set of 41 clinical and demographic features. A cohort of 703 patients was assessed, and machine learning models, including XGBoost, SVM, and Random Forest, were compared. Of these, the Random Forest model demonstrated the highest performance, achieving an accuracy of 0.67 and an AUC of 0.68, effectively distinguishing between patients with less than 24 months of survival and more than 24 months of survival. Survival predictors identified in our analysis included prostate-specific antigen (PSA), albumin, and lactate dehydrogenase (LDH). These findings suggest that clinical factors can be effectively utilized in machine learning models to predict survival outcomes in cancer patients.

Keywords: Machine Learning · Metastatic Castration-Resistant Prostate Cancer (mCRPC) · Survival Prediction · biomarker

1 Introduction

Prostate cancer remains the second-leading cause of cancer-related deaths among men in the United States. According to the American Cancer Society's estimates, there will be 299,010 new cases and 35,250 deaths from prostate cancer reported in 2024. Although the five-year survival rate for prostate cancer patients is 96.9%, the incidence rate has been rising steadily—by 3% per year overall and by approximately 5% per year for advanced-stage prostate cancer since 2014. In the management of advanced prostate cancer, androgen deprivation therapy (ADT) is frequently used as a standard treatment. However, resistance to ADT often develops after years of treatment, leading to progression to metastatic castration-resistant prostate cancer (mCRPC). At mCRPC stage, patients typically receive chemotherapy (e.g., docetaxel) or treatment with prednisone

H. R. Arabnia et al. (Eds.): CSCI 2025, CCIS 2971, pp. 195–204, 2027.
https://doi.org/10.1007/978-3-032-28463-1_15

or prednisolone [1, 2]. Biomarkers like prostate-specific antigen (PSA), along with other clinical factors, are used to assess the prognostic risk in prostate cancer patients [3, 4]. Despite this, the prognosis for metastatic castration-resistant prostate cancer (mCRPC) remains unfavorable, with a five-year survival rate of only around 30% [5]. This highlights the critical need for more accurate and reliable prognostic biomarkers that can better assess the risk of cancer-related mortality, support therapeutic decisions, and ultimately improve survival outcomes for these patients.

This study focused on utilizing routinely collected clinical factors to predict the risk of death within two years for patients mCRPC using a machine learning approach. The primary objective was to identify a set of clinical factors that could serve as reliable predictors of survival outcomes in mCRPC patients, potentially offering insights to guide treatment decisions.

2 Method

2.1 Data Source

A phase 3, randomized, double-blind, placebo-controlled study of Abiraterone Acetate (AA) plus prednisone was conducted by Janssen Research & Development between February 2009 and December 2013 (COU-AA-302). Data from this clinical trial was accessed through the Yale Open Data Access (YODA) platform. The trial recruited asymptomatic or mildly symptomatic patients with prostate cancer, all of whom had pathologically confirmed mCRPC and had not previously received chemotherapy. A total of 1,088 patients provided informed consent to participate in the study. During the treatment period, 93 patients from the placebo plus prednisone group crossed over to the AA plus prednisone group, potentially introducing confounding factors into the analysis of overall survival outcomes. To reduce any potential bias from treatment crossover, a pre-protocol analysis [6] was performed, which included only those patients who remained in their initially assigned treatment groups.

2.2 Study Design

Overall survival (OS) was calculated from the start of cycle 1, day 1 of treatment to the date of last follow-up or death. A binary outcome was defined using a 24-month survival cutoff: patients who died within 24 months were classified as "OS less than 24 months", while those who survived longer than 24 months—whether censored or dead after 24 months—were categorized as "OS more than 24 months".

The data contained 41 predictive features, which included 6 baseline characteristics: patient age at the start of treatment, height, treatment group (AA or placebo), race (Asian, Black, White, or Other), ethnicity (Hispanic or Non-Hispanic), and Eastern Cooperative Oncology Group (ECOG) performance status at baseline (0 or 1). According to the COU-AA-302 clinical trial protocol, only patients with an ECOG status of either 0 (fully active, able to perform all pre-disease activities without restriction) or 1 (restricted in strenuous activities but ambulatory and capable of light or sedentary work) were eligible for inclusion criteria. In addition, 35 baseline laboratory tests were collected on day 1 of cycle 1 during the treatment period, including PSA and other non-specific biomarkers such as lymphocytes, hemoglobin, alkaline phosphatase (ALP), direct bilirubin, albumin, phosphorus, urea nitrogen, hematocrit, white blood cell, red blood cell, and lactate dehydrogenase (LHD). These non-specific biomarkers have been recognized [7] for their significant role in predicting cancer prognosis and clinical decision-making during patient follow-up.

2.3 Statistical Analysis

The chi-square test was used to compare proportions between categorical predictor variables and outcome groups. While t-tests were conducted to compare the means of continuous variables across outcome groups. To reduce multicollinearity and identify the most significant predictors, lab test features with a correlation coefficient greater than 0.5 were excluded from the analysis. After feature selection, the dataset was split into an 80% training set and a 20% test set using random seeds. Predictive modeling was performed using three machine learning algorithms: XGBoost, Support Vector Machine (SVM), and Random Forest, which served as the primary models for the analysis.

To address the imbalance in the outcome variable, both data-level and algorithm-level techniques were used. At the data level, random oversampling was applied to increase the representation of the minority class by duplicating samples from the minority class and adding them to the dataset, thereby ensuring better balance between the classes. At the algorithm level, class weights were adjusted to give greater importance to the minority class during model training, thereby improving model performance on the underrepresented group [8].

Model performance was assessed using multiple metrics, including accuracy, area under the curve (AUC), recall, precision, and F1 score. All statistical analyses were performed in RStudio Version 2022.02.1. The XGBoost model was implemented using the "xgboost" package with the "xgboost ()" function. The SVM model was trained using the "e1071" package and the "svm ()" function. The Random Forest model was built using the "randomForest ()" function from the "randomForest" package. All performance metrics were computed using the "caret" package. Hyperparameters for each model were fine-tuned to optimize model performance.

3 Results

A total of 703 patients were included in this analysis, excluding those who crossed over from the placebo group to the AA treatment group and those with missing lab test records. Among the cohort, 405 patients (57.6%) died within 24 months, while 298 patients (42.3%) survived longer than 24 months. The average age at the start of treatment for patients who had an OS of less than 24 months was 69.9 years, compared to 67.8 years for those who survived more than 24 months.

Baseline ECOG status and treatment type were found to be significant predictors of 24-month survival outcomes ($p = 0.02$ and $p < 0.001$, respectively). Specifically, patients with an ECOG status of 1 and those receiving a placebo had higher mortality compared to patients with an ECOG status of 0 or those treated with abiraterone acetate (AA).

However, neither race (p = 0.58) nor ethnicity (p = 0.55) demonstrated significant associations with 24-month survival outcomes and, as a result, were excluded from the final predictive model (Table 1). Additionally, clinical lab test predictors with a correlation coefficient greater than 0.5 suggest strong correlated, including total protein, urea nitrogen, aspartate aminotransferase, hematocrit, white blood cells, and red blood cells. To minimize multicollinearity and improve the robustness of the model, these variables were removed from further predictive model.

The model contained 24 predictor features, including treatment, age, height, ECOG status, and 20 baseline laboratory tests. These tests included key biomarkers like PSA, neutrophils, albumin, LDH, ALP, hemoglobin, phosphorus, platelets, serum uric acid, serum amylase, which were routinely collected in clinical practice (Table 2). To assess the performance of the machine learning models, we applied three algorithms—XGBoost, SVM, and Random Forest—across three training sets: the original dataset, random oversampling, and adjusted class weights.

Table 1. Characteristics (Categorical Variables) by Survival Status

Categorical Variables		Total	OS > 24 months	OS <= 24 months	P-value
ECOG	0	531	320	211	0.02
	1	172	85	87	
Treatment	AA	393	255	138	<.001
	Placebo	310	150	160	

Table 3 summarizes the model performance comparison of XGBoost, SVM, and Random Forest across three training sets: the original, randomly oversampled, and class-weight adjusted datasets. XGBoost achieved an accuracy of 0.63, an AUC of 0.68, a recall of 0.44, and an F1 score of 0.53 in the original training set. With random oversampled training set, the model's performance remained the same, with an accuracy of 0.63 and an AUC of 0.67. The recall, however, significantly dropped to 0.31. By adjusting class weights, XGBoost showed improved performance, achieving the highest accuracy (0.68), AUC (0.70), and recall (0.47), while maintaining a stable F1 score of 0.55.

SVM showed the lowest performance in the original training set, with an accuracy of 0.56, an AUC of 0.56, and a recall of 0.37, showing its limited ability to identify the minority class. After random oversampling, SVM's recall increased to 0.41, though the accuracy and AUC remained at 0.56. Adjusting the class weights led to a slight improvement, with an accuracy of 0.58, an AUC of 0.58, and a recall of 0.44. Despite these little improvements, SVM underperformed compared to the other models in terms of both detection ability and overall performance.

Random Forest exhibited the highest recall (0.42) in the original training set, along with an accuracy and AUC of 0.63. The model showed a small improvement with random oversampling, increasing accuracy to 0.65 and AUC to 0.65, while the recall slightly dropped to 0.41. When class weights were adjusted, random forest demonstrated the best results (Fig. 1), achieving an accuracy of 0.67, an AUC of 0.68, and a recall of 0.54.

Table 2. Characteristics (Continuous Variables) by Survival Status

Continuous Variables		All Patients		OS > 24 months	OS <= 24 months	P-value
Variables	Units	Mean	SD	Mean	Mean	
Age	years	68.72	9.20	67.83	69.93	<.001
Height	cm	174.98	7.62	175.36	174.47	0.127
Basophils	x10^3/uL	0.04	0.03	0.04	0.04	0.968
Serum Potassium	mEq/L	4.33	0.41	4.30	4.38	0.021
Serum Sodium	mEq/L	140.81	3.06	141.04	140.49	0.016
Serum Uric Acid	mg/dL	5.23	1.35	5.21	5.25	0.685
Serum Amylase	U/L	67.37	28.31	69.20	64.90	0.048
ALT (SGPT)	U/L	22.22	11.30	23.12	21.00	0.014
Creatinine	mg/dL	1.00	0.24	0.99	1.00	0.484
Calcium (EDTA)	mg/dL	9.58	0.47	9.61	9.54	0.066
Neutrophils	x10^3/uL	4.36	1.55	4.23	4.54	0.008
Lymphocytes	x10^3/uL	1.60	0.61	1.61	1.58	0.491
Direct Bilirubin	mg/dL	0.12	0.05	0.12	0.13	0.559
Albumin	g/dL	4.00	0.33	4.05	3.92	<.001
Phosphorus	mg/dL	3.60	0.56	3.60	3.60	0.868
Monocytes	x10^3/uL	0.39	0.15	0.38	0.39	0.064
LDH	U/L	200.46	70.99	184.61	222.09	<.001
PSA	ng/mL	146.60	347.2	101.59	207.77	<.001
Platelets	x10^3/uL	257.72	69.35	252.05	265.43	0.013
Hemoglobin	g/dL	12.91	1.24	13.09	12.65	<.001
ALP	U/L	149.71	204.58	119.00	191.46	<.001
Magnesium	mg/dL	2.17	0.20	2.16	2.19	0.030

The model also had the highest F1 score (0.60), indicating a strong balance between precision and recall in predicting survival outcomes.

Feature importance analysis identified LDH, ALP, PSA, ALT, Platelets, and Hemoglobin as the top six features to model performance, demonstrating their potential contribution in predictive modeling.

Table 3. Machine Learning Model Performance Metrics

Training Set	Model	Accuracy	AUC	Recall	Precision	F1 score
Original	XGBoost	0.63	0.68	0.44	0.65	0.53
	SVM	0.56	0.56	0.37	0.52	0.44
	Random Forest	0.63	0.63	0.42	0.64	0.51
Random oversampling	XGBoost	0.63	0.67	0.31	0.64	0.41
	SVM	0.56	0.56	0.41	0.5	0.45
	Random Forest	0.65	0.65	0.41	0.73	0.52
Adjusted class weight	XGBoost	0.65	0.7	0.47	0.67	0.55
	SVM	0.58	0.58	0.44	0.51	0.47
	Random Forest	0.67	0.68	0.54	0.67	0.6

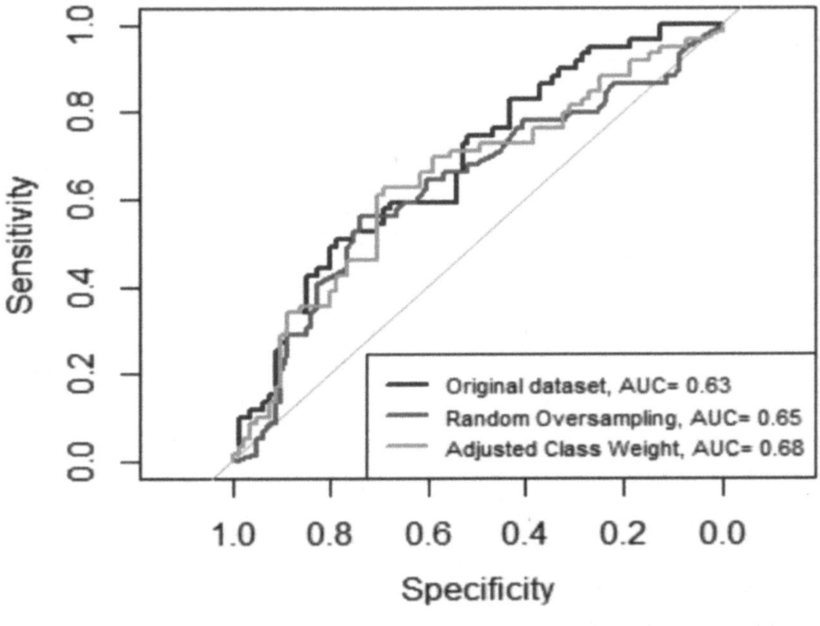

Fig. 1. ROC curve with Area Under the Curve (AUC) with random forest algorithm across the original, random oversampling, and class-weight adjusted datasets.

4 Discussion

In this study, we applied machine learning techniques to predict OS within 24 months in patients with mCRPC based on prognostic biomarkers. Among the XGBoost, SVM, and random forest, the random forest model displayed strong performance, achieving an accuracy of 0.67, an AUC of 0.68, and the best balance between precision and recall

(0.60), suggesting its capability of distinguishing between patients with OS less than 24 months and those with longer survival outcomes.

One challenge in this study was addressing class imbalance, a common issue in real-world datasets. Class imbalance can lead to biased decision boundaries, with the model leaning toward the majority class over the minority class, potentially reducing predictive accuracy for the minority class [9]. To minimize imbalance, we applied the resampling method and class-weighted techniques. However, despite this adjustment, the performance metrics still suggest that the model may be overly biased towards the majority. According to Gurcan's study [10], advanced resampling techniques such as SMOTEENN (combination of Synthetic Minority Oversampling Technique and Edited Nearest Neighbor methods) and RENN (Repeated Edited Nearest Neighbors), as well as ensemble methods, such as balanced random forests and balanced bagging, could improve prediction accuracy in imbalanced datasets. Future work will involve additional classifiers and resampling strategies to further improve model performance.

A limitation is the restricted set of clinical biomarkers included in the analysis. Although commonly used clinical factors were collected, more specific molecular or genomic markers or clinical notes derived by large language models could potentially improve the model's predictive accuracy and provide a deeper understanding of survival in mCRPC patients. Furthermore, the dataset was derived from a single clinical trial with strict inclusion criteria, which may limit the generalizability of our findings to the broader mCRPC patient population.

Future research will focus on expanding the range of biomarkers and including data from diverse patient cohorts to validate and strengthen the robustness of the predictive model [11]. Our study comports with previous studies demonstrating the value of real-world data for predictive modeling [12, 13]. In future studies, we will include other important data sources such as procedures, medications, and laboratory results from electronic health records [14, 15]. Natural language processing will be used to extract unstructured data elements including symptoms, health behaviors, and social determinants of health [16, 17]. Since tree-based methods performed particularly well in this study, we will continue to expand on these approaches and compare them with other advanced machine learning methods and ensemble deep learning models [18, 19]. Application of Large Language Models can significantly improve accuracy of data extraction from unstructured clinical notes [20, 21]. In order to develop scalable predictive approaches, particular attention will be given to bias reduction and explainability [22, 23]. Once sufficient scalability and accuracy is achieved, the algorithms for predicting survival can be embedded in electronic health records as a part of clinical decision support [24, 25]. Information about the patients' survival will be conveyed to providers to consider appropriate treatment regimen [26, 27]. This would allow timely commencement of pharmacological and lifestyle interventions depending on the results of definitive work-up [28, 29]. Lifestyle interventions may include physical telerehabilitation [30, 31], chatbots for patient engagement [32, 33], and interactive applications providing health education and social support [34, 35].

5 Conclusions

In conclusion, our findings demonstrate that PSA, along with other non-specific clinical factors such as albumin, LDH, ALP, have significant potential in predicting the 24-month OS in patients with mCRPC. However, there is still a need to improve the accuracy and generalizability of these predictive models. Further research should focus on expanding the prognostic biomarkers set by integrating additional biomarkers, as well as clinical notes. Refining predictive models by incorporating more advanced machine learning techniques and resampling methods could lead to better prediction performance and eventually improve prognostic outcomes for mCRPC patients in real-world clinical settings.

References

1. Heidenreich, A., et al.: EAU guidelines on prostate cancer. Part II: treatment of advanced, relapsing, and castration-resistant prostate cancer. Eur. Urol. **65**(2), 467–79 (2014)
2. Cornford, P., et al.: EAU-ESTRO-SIOG guidelines on prostate cancer. Part II: treatment of relapsing, metastatic, and castration-resistant prostate cancer. Eur. Urol. **71**(4), 630–42 (2017)
3. Conteduca, V., Mosca, A., Brighi, N., de Giorgi, U., Rescigno, P.: New prognostic biomarkers in metastatic castration-resistant prostate cancer. Cells **10**(1), 193 (2021)
4. Deluce, J.E., Cardenas, L., Lalani, A.K., Maleki Vareki, S., Fernandes, R.: Emerging biomarker-guided therapies in prostate cancer. Curr. Oncol. **29**(7), 5054–5076 (2022)
5. Chowdhury, S., et al.: Real-world outcomes in first-line treatment of metastatic castration-resistant prostate cancer: the prostate cancer registry. Target. Oncol. **15**, 301–315 (2020)
6. Smith, V.A., Coffman, C.J., Hudgens, M.G.: Interpreting the results of intention-to-treat, per-protocol, and as-treated analyses of clinical trials. JAMA **326**(5), 433–434 (2021)
7. Schmitz-Dräger, B.J., et al.: Considerations on the use of urine markers in the management of patients with low-/intermediate-risk non–muscle invasive bladder cancer. In: Urologic Oncology: Seminars and Original Investigations,1 October 2014, vol. 32, no. 7, pp. 1061–1068. Elsevier (2014)
8. Velarde, G., Sudhir, A., Deshmane, S., Deshmunkh, A., Sharma, K., Joshi, V.: Evaluating XGBoost for balanced and imbalanced data: application to fraud detection. arXiv preprint arXiv:2303.15218, 27 Mar 2023
9. Kim, A., Jung, I.: Optimal selection of resampling methods for imbalanced data with high complexity. PLoS ONE **18**(7), e0288540 (2023)
10. Gurcan, F., Soylu, A.: Learning from imbalanced data: integration of advanced resampling techniques and machine learning models for enhanced cancer diagnosis and prognosis. Cancers **16**(19), 3417 (2024)
11. Wong, J., Tian, Y., Patel, M.S., et al.: Plasma cell-free DNA methylation-based prognosis in metastatic castrate-resistant prostate cancer. NPJ Precis. Oncol. (2025). https://doi.org/10.1038/s41698-025-01232-w. PMID: 41390880
12. Paranjpe, I., Russak, A.J., De Freitas, J.K., et al.: Retrospective cohort study of clinical characteristics of 2199 hospitalised patients with COVID-19 in New York City. BMJ Open **10**(11), e040736 (2020)
13. Cui, W., Kawamoto, K., Morgan, K., Finkelstein, J.: Reducing diagnostic uncertainty in emergency departments: the role of large language models in age-specific diagnostics. In: 2024 IEEE 12th International Conference on Healthcare Informatics (ICHI), Orlando, FL, USA, pp. 525–527 (2024)

14. Thyvalikakath, T.P., Duncan, W.D., Siddiqui, Z., et al.: Appl. Clin. Inform. **11**(2), 305–314 (2020). https://doi.org/10.1055/s-0040-1709506. PMID: 32349142; PMCID: PMC7190390
15. Locke, B.W., Richards, W.W., Gouripeddi, R., et al.: The consistency of hypercapnic respiratory failure case definitions in electronic health record data. Chest (2025). S0012-3692(25)05126-8. https://doi.org/10.1016/j.chest.2025.08.002. PMID: 40885535; PMCID: PMC12739763
16. Shah-Mohammadi, F., Finkelstein, J.: Combining NLP and machine learning for differential diagnosis of COPD exacerbation using emergency room data. Stud. Health Technol. Inform. **29**(305), 525–528 (2023). https://doi.org/10.3233/SHTI230549. PMID: 37387083
17. Shah-Mohammadi, F., Cui, W., Finkelstein, J.: Comparison of ACM and CLAMP for entity extraction in clinical notes. In: Annual International Conference of the IEEE Engineering in Medicine & Biology Society 2021, pp. 1989–1992 (2021). https://doi.org/10.1109/EMBC46164.2021.9630611. PMID: 34891677
18. Finkelstein, J., Smiley, A., Echeverria, C., Mooney, K.: AI-driven prediction of symptom trajectories in cancer care: a deep learning approach for chemotherapy management. Bioengineering **11**(11), 1172 (2024). https://doi.org/10.3390/bioengineering11111172. PMID: 39593830; PMCID: PMC11592055
19. Huo, X., Kohli, M., Finkelstein, J.: Machine learning approaches for predicting mortality in metastatic castration-resistant prostate cancer. Stud. Health Technol. Inform. **26**(328), 26–30 (2025). https://doi.org/10.3233/SHTI250666. PMID: 40588874
20. Bai, Y., Cui, W., Finkelstein, J.: Utilizing open-source large language models to extract genitourinary symptoms from clinical notes. Stud. Health Technol. Inform. **328**, 16–20 (2025). https://doi.org/10.3233/SHTI250664. PMID: 40588872
21. Reategui-Rivera, C.M., Finkelstein, J.: Evaluation of the performance of a large language model to extract signs and symptoms from clinical notes. Stud. Health Technol. Inform. **323**, 71–75 (2025). https://doi.org/10.3233/SHTI250051. PMID: 40200448
22. Smiley, A., Villarreal-Zegarra, D., Reategui-Rivera, C.M., Escobar-Agreda, S., Finkelstein, J.: Methodological and reporting quality of machine learning studies on cancer diagnosis, treatment, and prognosis. Front. Oncol. **15**, 1555247 (2025). https://doi.org/10.3389/fonc.2025.1555247. PMID: 40297817; PMCID: PMC12034563
23. Smiley, A., Reategui-Rivera, C.M., Villarreal-Zegarra, D., Escobar-Agreda, S., Finkelstein, J.: Exploring artificial intelligence biases in predictive models for cancer diagnosis. Cancers **17**(3), 407 (2025). https://doi.org/10.3390/cancers17030407. PMID: 39941776; PMCID: PMC11816222
24. Kawamoto, K., Finkelstein, J., Del Fiol, G.: Implementing machine learning in the electronic health record: checklist of essential considerations. Mayo Clin. Proc. **98**(3), 366–369 (2023). https://doi.org/10.1016/j.mayocp.2023.01.013. PMID: 36868743
25. Shah-Mohammadi, F., Finkelstein, J.: Accuracy evaluation of GPT-assisted differential diagnosis in emergency department. Diagnostics **14**(16), 1779 (2024). https://doi.org/10.3390/diagnostics14161779. PMID: 39202267; PMCID: PMC11354035
26. Finkelstein, J., Gabriel, A., Schmer, S., Truong, T.T., Dunn, A.: Identifying facilitators and barriers to implementation of AI-assisted clinical decision support in an electronic health record system. J. Med. Syst. **48**(1), 89 (2024). https://doi.org/10.1007/s10916-024-02104-9. PMID: 39292314; PMCID: PMC11410896
27. Finkelstein, J., Parvanova, I., Xing, Z., Truong, T.T., Dunn, A.: Qualitative assessment of implementation of a discharge prediction tool using RE-AIM framework. Stud. Health Technol. Inform. **302**, 596–600 (2023). https://doi.org/10.3233/SHTI230212. PMID: 37203755
28. Bedra, M., McNabney, M., Stiassny, D., Nicholas, J., Finkelstein, J.: Defining patient-centered characteristics of a telerehabilitation system for patients with COPD. Stud. Health Technol. Inform. **190**, 24–26 (2013). PMID: 23823363

29. Finkelstein, J., Wood, J.: Implementing home telemanagement of congestive heart failure using Xbox gaming platform. In: Annual International Conference of the IEEE Engineering in Medicine and Biology Society 2011, pp. 3158–3163 (2011). https://doi.org/10.1109/IEMBS.2011.6090861. PMID: 22255010
30. Castro, H.K., Cross, R.K., Finkelstein, J.: Using a home automated telemanagement (HAT) system: experiences and perceptions of patients with inflammatory bowel disease. In: AMIA Annual Symposium Proceedings 2006, p. 872 (2006). PMID: 17238492; PMCID: PMC1839272
31. Jeong, I.C., Karpatkin, H., Finkelstein, J.: Physical telerehabilitation improves quality of life in patients with multiple sclerosis. Stud. Health Technol. Inform. **15**(284), 384–388 (2021). https://doi.org/10.3233/SHTI210752. PMID: 34920553
32. Reategui-Rivera, C.M., Smiley, A., Finkelstein, J.: LLM-based chatbot to reduce mental illness stigma in healthcare providers. In: 2025 IEEE 15th Annual Computing and Communication Workshop and Conference (CCWC), Las Vegas, NV, USA, pp. 00001–00007 (2025). https://doi.org/10.1109/CCWC62904.2025.10903778
33. Villarreal-Zegarra, D., Reategui-Rivera, C.M., Paredes-Gonzales, Y., Centeno-Terrazas, G., Finkelstein, J.: Assessment of GPT-based conversational agents aimed to reduce healthcare provider stigma. Stud. Health Technol. Inform. **328**, 4–8 (2025). https://doi.org/10.3233/SHTI250661. PMID: 40588869
34. Finkelstein, J., Lapshin, O., Cha, E.: Feasibility of promoting smoking cessation among methadone users using multimedia computer-assisted education. J. Med. Internet Res. **10**(5), e33 (2008). https://doi.org/10.2196/jmir.1089. PMID: 18984556; PMCID: PMC2630834
35. Barron, J., Bedra, M., Wood, J., Finkelstein, J.: Exploring three perspectives on feasibility of a patient portal for older adults. Stud. Health Technol. Inform. **202**, 181–184 (2014). PMID: 25000046

Text-Based Detection of Autism Spectrum Disorder (ASD) Using State-of-the-Art LLMs

Manal Almutery, Rawad Alqahtani(✉), Masheal Alghamdi, and Turky Alotaiby

King Abdulaziz City for Science and Technology (KACST), Riyadh 12354, Saudi Arabia
ralqahtani@kacst.gov.sa

Abstract. Autism Spectrum Disorder (ASD) is a neurodevelopmental disorder characterized by challenges in communication, social interactions, and the presence of repetitive behaviors. Although ASD is increasingly prevalent worldwide, its diagnosis remains time-consuming and depends heavily on expert clinical assessment. This study explores the potential of state-of-the-art large language models (LLMs) to detect ASD based on caregiver descriptions of children's behaviors. We employ recent models—including GPT-4o, DeepSeek-R1, Grok, LLaMA-3.2-11B, Mistral-Large-Instruct-2411, Jais-Chat, and ALLaM—using few-shot prompting strategies to classify behavioral narratives as ASD or non-ASD. The models are evaluated on the TASD dataset, which contains annotated textual descriptions labeled as either ASD or non-ASD. The model performance shown significant variability between LLM, with accuracy range from 48% to 73% and F1-scores from 33% to 73% in both Arabic and English datasets. GPT-4o attained the most superior and equitable outcomes, succeeded by DeepSeek-R1 and ALLaM (13B), both of which exhibited robust cross-lingual generalization. Our research demonstrates that LLMs, even in the absence of task-specific fine-tuning, may attain commendable accuracy and F1 scores in recognizing behavioral patterns linked to ASD. We performed a comprehensive examination of GPT-4o's capacity to classify particular behavioral indications of ASD, uncovering its strengths and limitations across 14 behavioral features. These findings underscore the capability of contemporary LLMs for language model-based early ASD screening, particularly in low-resource or non-clinical settings.

Keywords: Autism Spectrum Disorder · TASD dataset · Large language models · Linguistic markers · early ASD screening

1 Introduction

Autism Spectrum Disorder (ASD) is a genetically heterogeneous collection of neurodevelopmental disorders that are defined by impairment in three core

H. R. Arabnia et al. (Eds.): CSCI 2025, CCIS 2971, pp. 205–219, 2027.
https://doi.org/10.1007/978-3-032-28463-1_16

behavioral areas: communication, social behavior, and the occurrence of stereotyped repetitive behavior [1]. Symptoms generally become apparent in young childhood, and thus there has been a strong emphasis placed upon earlier diagnosis to enable earlier intervention to have a dramatic impact upon ultimate developmental outcome [2]. Early linguistic and social behaviors often provide the first observable indicators of Autism Spectrum Disorder (ASD) [3]. Language and social behavior are often where the earliest signs of autism show up. Things like repeating what others say (known as echolalia), avoiding eye contact, or using language in unexpected ways can appear well before a child is formally diagnosed [4]. In fact, language is one of the most reliable early clues. Many children with autism may have a limited vocabulary, use pronouns in unusual ways, repeat phrases, or struggle to carry on a back-and-forth conversation. These patterns often point to deeper difficulties with social understanding and communication [5]. Recent advances in artificial intelligence (AI) and natural language processing (NLP) have made it much easier to analyze unstructured language data—especially with the help of large language models (LLMs) [6]. Models like GPT, LLaMA, and Mistral show impressive abilities in understanding context, interpreting text, and making accurate predictions even without specific training, across different languages and topics [7,8]. Because of this, there's growing interest in using LLMs in behavioral and clinical settings—for example, to support diagnoses, analyze therapy sessions, or detect symptoms through language [9]. In a recent review, Hua et al. [10] examined over 200 studies that used large language models (LLMs) in mental health research. These studies covered a wide range of uses— from writing clinical notes and generating therapy dialogues to detecting symptoms through text. While the review showed that LLMs have real potential to assist in clinical decision-making, it also highlighted some important challenges. Most of the studies focused only on English-language data, relied on closed-source models, and lacked standardized evaluation methods. As a result, there are concerns about whether the findings can be reliably reproduced or applied in different languages and cultural settings. This highlights the importance of developing multilingual, domain-specific approaches to using LLMs for early behavioral screening—especially since language and cultural context play a big role in how symptoms are expressed and understood. To overcome these challenges, artificial intelligence, and especially large language models (LLMs), hold the key to enhancing the early diagnosis of ASD based on textual behavioral reports from caregivers. Differing from conventional diagnostic methods, which are commonly based on multimodal clinical observations and interviews, LLMs may handle natural language inputs to make inferences about possible ASD signs in a scalable and distributed fashion. In this work, we evaluate the performance of several state-of-the-art LLMs—GPT-4o, ALLaM, Jais-Chat, and DeepSeek R1—on binary ASD classification tasks using textual descriptions of toddler behavior. Leveraging the bilingual TASD-Dataset and a few-shot prompting strategy, we analyze model performance in both English and Arabic. Our findings contribute to the growing intersection between clinical linguistics and artificial intelligence, demonstrating how LLMs can serve as early screening tools for autism in multilingual and resource-constrained environments. Following the analysis by Ciobanu et al. [11], which identified only ten studies examining LLMs for autism diagnosis and treatment—exclusively in English—this work represents, to the best of our knowledge, the first study to apply LLMs to autism detection in Arabic-language text.

The rest of the paper is organized as follows: Section Related Work reviews previous research on language-based approaches to autism screening. Section Methodology describes the method used in this study. Section Experimental Results reports the experimental results. Finally, Section Conclusions and Future work concludes the paper and outlines the directions for future research.

2 Related Work

This section reviews a main strands relevant to language-based ASD screening: (i) traditional NLP and machine-learning approaches, (ii) transformer-based models for deep language understanding, and (iii) large language models (LLMs) for diagnostic and behavioral analysis. We conclude by positioning our contribution, which focuses on bilingual caregiver narratives for early-childhood ASD screening using few-shot LLMs

2.1 Traditional NLP and Machine Learning Approaches

Before deep transformer models became widely used, early research on autism detection mostly relied on manually crafted features from language and speech, paired with traditional machine learning techniques. For example, Raj and Masood [12] used a combination of lexical and speech-related features to tell apart individuals with and without ASD, applying models like support vector machines, decision trees, and random forests. Their work showed that even basic models could pick up on important language patterns—but this approach required a lot of manual effort.

Building on this direction, Themistocleous et al. [13] integrated natural language processing (NLP) with machine learning to analyze children's narrative speech, highlighting differences in lexical richness, syntactic complexity, and cohesion as discriminative markers for ASD. Although these approaches offered interpretable insights, they lacked the capacity to capture contextual dependencies and long-range discourse relations, motivating the transition toward transformer-based deep language models that learn hierarchical linguistic representations directly from text.

2.2 Transformer-Based Models and Deep Language Understanding

The rise of transformer models has been a game-changer in natural language processing (NLP), allowing systems to understand language in a much deeper and more nuanced way. Unlike older methods that relied on hand-crafted features, transformers use self-attention mechanisms to capture meaning, context, and structure directly from the data—across entire passages of text. Pretrained models like BERT, RoBERTa, and GPT have set new benchmarks in tasks like sentiment analysis, emotion detection, and clinical text classification [14–16].

In autism research, these models have opened up new possibilities for analyzing language with more precision and context-awareness. For example, Hu et al. [4] used BERT and ChatGPT to detect communication traits linked to autism in transcribed ADOS-2 interviews, reporting strong results—F1 scores of 62% for BERT and 80% for ChatGPT—outperforming traditional machine learning methods. Similarly, Pal and Rubini [17] applied a BERT-based model to the TASD dataset, which includes behavioral reports from caregivers of 297 toddlers (163 with ASD and 134 without). Their model reached 88% accuracy, showing how transformer models can pick up on early behavioral language cues linked to ASD. However, this work focused only on English-language data and used just one fine-tuned model.

More recent work, like the study by Stanley et al. [18], shows that large language models (LLMs) can even begin to mimic the clinical reasoning behind autism diagnoses, bridging the gap between language modeling and expert judgment.

Still, most of the research in this area focuses on English data and adult clinical interviews. There's a clear need for more multilingual, caregiver-focused studies—especially since language and cultural background play a key role in how autism shows up. This is what drives our focus on using instruction-tuned, multilingual LLMs to analyze bilingual caregiver descriptions for early autism screening.

2.3 Large Language Models (LLMs) for Diagnostic and Behavioral Analysis

Large language models (LLMs) have recently shown impressive potential in both clinical reasoning and mental health detection [10,19–21]. For example, Singhal et al. [19] found that instruction-tuned LLMs can perform medical question-answering tasks with near-expert accuracy using just a few examples (few-shot prompting). Beyond clinical Q&A, researchers are now exploring how LLMs can detect mental health conditions like depression and anxiety directly from text or conversation. These models have shown they can pick up on emotional tone and cognitive patterns in language that may indicate psychological distress [10,20]. Jin et al. [20], for instance, reviewed how LLMs are being used for emotion detection and mental health assessment. They found that while LLMs can model aspects like empathy and emotional expression, they've rarely been applied to neurodevelopmental conditions like autism. Hua et al. [10] also stressed the importance of using consistent evaluation methods and ensuring models work across different languages. In autism research, LLMs are starting to gain attention. Ciobanu et al. [11] used ChatGPT and BERT to analyze conversations from adult clinical settings. Mukherjee et al. [22] went a step further, combining BERT and ChatGPT to detect signs of autism from parents' posts in online support groups. Their approach used BERT to analyze sentiment in each sentence, while ChatGPT (fine-tuned through reinforcement learning with human feedback) interpreted specific phrases as possible indicators of autism. Their model reached about 83% accuracy on a custom dataset. One of the most notable recent developments is the ADOS-Copilot system [23], which integrates LLMs into real clinical autism assessments. Using conversations from the ADOS-2 (a standard autism diagnostic tool), the researchers built a model that not only predicts diagnostic scores but also provides explanations for its decisions. The system performed comparably to human clinicians, achieving an F1-score of 81.79% for binary classification and 78.37% for three-category classification, along with a low mean absolute error of 0.4643. Our work extends this line of research by applying state-of-the-art multilingual instruction-tuned LLMs (GPT-4o, DeepSeek R1, Grok, Llama-3.2-11B, Mistral-Large-Instruct-2411, Jais-Chat, and AllaM) under few-shot prompting to bilingual caregiver narratives (English-Arabic) from the TASD dataset.

3 Methodology

This study aims to detect Autism Spectrum Disorder (ASD) from caregiver-provided behavioral descriptions using large language models (LLMs). The overall methodological pipeline is illustrated in Fig. 1. The process begins with collecting textual descriptions of toddler behaviors, which are then used to construct prompts containing carefully selected few-shot examples. These prompts are fed into LLMs, which generate predictions subsequently mapped to binary labels (**ASD** or **Non-ASD**). Finally, the predicted labels are compared against the human-annotated ground truth for evaluation.

3.1 Dataset

For this study, we used the TASD dataset, which includes 333 caregiver-written descriptions of toddler behavior [3]. Each entry in the dataset contains a short behavioral narrative, a set of assessment features (such as Word Repetition, Emotional Empathy, Noise Sensitivity, Sign Communication, and Tiptoe Flapping), and a corresponding clinical label—either ASD or Non-ASD. Out of the 333 samples, 175 are labeled as ASD and 158 as Non-ASD.

The dataset is annotated with 14 different behavioral features, including things like Attention Response and Word Repetition, which help capture the range of traits often seen in early autism. Examples of these annotated caregiver descriptions are shown in Table 1.

The original transcripts were written in English, and we used GPT-4o to translate them into Arabic to support bilingual evaluation. Both versions are semantically aligned, with Arabic descriptions ranging from 11 to 40 words, and English ones from 14 to 40 words.

To better understand the linguistic patterns in Arabic related to ASD, Fig. 2 presents a word cloud generated from Arabic trigrams, illustrating the most

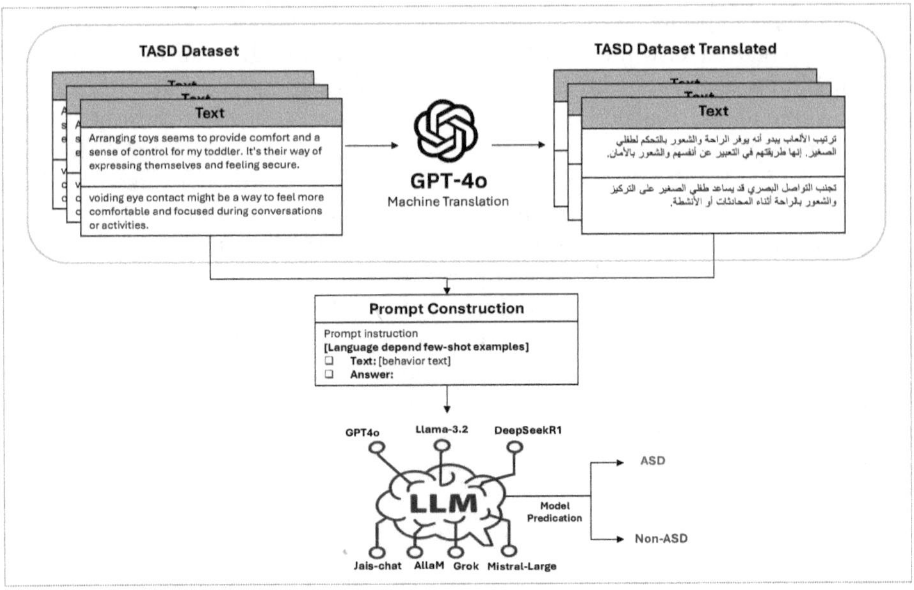

Fig. 1. Block diagram illustrating the proposed methodology for ASD detection using LLMs.

prominent words associated with the ASD class. As shown in the figure, many of the frequent trigrams are linked to expressions of difficulty and challenges, such as: “اَلْقَلَق/anxiety”, “اَلصَعب-difficulty”, and “اَلتَحَدّيَات-challenges” which caregivers often use when describing the behaviors of toddlers with ASD.

In our experiments, we used these behavioral descriptions as input to large language models (LLMs) to predict whether a child was likely to have ASD.

Fig. 2. The most common phrases found in caregiver descriptions of toddlers with ASD.

Table 1. Examples from the TASD dataset showing annotated caregiver descriptions with linguistic features and ASD labels.

Text	Assessment Feature	Sign	Label
A structured and organized environment helps them stay focused and perform better in tasks or learning activities.	Attention Response+	+	Non-ASD
Absolutely! We've been incorporating their preferred words into learning activities. It's incredible to witness how they engage and participate when we align their interests with the lessons. It's a fantastic tool for connection and learning.	Word Repetition-	+	Non-ASD
Achieving success in their focused interests can boost their confidence and self-esteem, supporting their personal growth.	Focused Attention+	-	ASD
An overload of information can overwhelm my toddler, leading to shutdowns or avoidance of tasks requiring sustained attention.	Attention Response+	-	ASD

3.2 Prompt Design and Few-Shot Learning

Each instance in the TASD dataset, consisting of a short textual description of a toddler's behavior, was treated as a binary classification task. To enhance the reasoning ability of LLMs, we employed a few-shot prompting strategy in which multiple labeled examples were embedded within the prompt. The language of the examples was matched to the input language, ensuring that Arabic examples were used with Arabic inputs and English examples with English inputs. The prompt structure was kept consistent across all experiments to ensure fairness in the evaluation. An illustration of the prompt design is shown in Fig. 3.

In addition to binary classification, we also designed prompt-based experiments for feature assessment, where the model was explicitly guided to recognize and reason about behavioral features (e.g., word repetition, emotional empathy, or noise sensitivity). An illustration of the prompt design is shown in Fig. 4.

3.3 LLM Models

We evaluated the performance of recent large language models with different parameter sizes including: Jais-Chat (7B and 30B), GPT-4o(1.2T) [24], ALLaM (7B and 13B), DeepSeek R1(671B) [25], Grok-3, Llama-3.2-11B-Vision-Instruct(11B) and Mistral-Large-Instruct-2411(123B) on the autism detection task. For the Jais (7B) model, we used the publicly available version hosted on Hugging Face[1]. The other models were accessed via Microsoft Azure[2], with

[1] inceptionai/jais-adapted-7b-chat · Hugging Face.

[2] Cloud Computing Services — Microsoft Azure.

System Prompt

You are a clinical assistant helping to detect early behavioral signs of Autism Spectrum Disorder (ASD) in toddlers based on caregiver observations.

Your task is to classify each description as either:

1. Autism: Autism spectrum disorder is a neurological and developmental disorder that affects how people interact with others, communicate, learn, and behave.
2. Non-Autism: Behaviors typical of toddlers with age-appropriate flexibility, social reciprocity, curiosity, or playfulness—even if focused or structured.

Examples:

{few_shot_examples}

Now classify the following:

Text :{ transcript}

Answer: "

Fig. 3. Few-shot prompt structure used for LLM-based ASD detection.

System Prompt

You are a precise text classifier.
Task: From a single short description of a child's behavior, predict which behavior class it belongs to.
Important: Do not reveal step-by-step reasoning or chain-of-thought. Provide only a brief rationale (≤30 words). This is not a diagnosis and is for research/classification only.

Labels & Definitions:
Assesment_features_class ∈ {{ "attention response","change reaction","emotional empathy","eye contact","finger movements","focused attention","follow pointing","noise sensitivity", "repetitive behavior","sharing interest","sign communication","tiptoe flapping","toy arranging","word repetition"}}

Output Format:
Return ONLY valid JSON with keys: behavior_class, confidence, brief_rationale.
{{
"Assessment _class": "<one of the Assessment features classes >",
"confidence": 0.0,
"brief_rationale": "<<=30 words>"
}}

Few-shot examples:
{ Few-shot }
Now classify:
Text: {transcript}
'''

Fig. 4. Example of the prompt design used to guide the model in identifying behavioral features.

the exception of ALLaM 13B, which was accessed via IBM Watson[3]. All models were tested under identical conditions with the same prompt format and few-shot examples. The predicted labels were extracted and compared with the ground truth and the performance was measured using Accuracy and Macro F1 score to systematically assess the ability of LLM to detect ASD linguistic indicators in Arabic and English texts.

4 Experimental Results

4.1 ASD Detection Classification

The assessment of large language models (LLMs) for the autism detection task demonstrates that these models can effectively identify linguistic cues associated with autism when guided by well-structured prompts, regardless of the input language. The results, summarized in Table 2, indicate that performance varies notably with model size and the depth of multilingual training, as illustrated in Fig. 5, which compares the models' behavior across both languages

Table 2. Performance of LLM Models on Arabic vs. English Inputs (Accuracy and F1 Score).

Model	Arabic		English	
	Accuracy	F1	Accuracy	F1
Jais-Chat (7B)	48.05	33.44	50.15	37.67
Jais-Chat (30B)	64.56	42.61	66.97	46.97
GPT-4o	**72.97**	**72.90**	**71.77**	**71.68**
DeepSeek-R1	71.77	71.77	69.07	68.47
Grok3	69.37	69.20	69.07	68.98
LLaMA-3.2-11B-Vision-Instruct	69.37	69.25	70.27	70.24
Mistral-Large-Instruct-2411	64.86	44.64	53.75	38.31
ALLaM (7B)	60.96	58.10	67.27	65.75
ALLaM (13B)	65.77	64.81	71.47	71.24

At the top of the spectrum, GPT-4o demonstrated exceptional reliability, achieving the highest accuracy—72.97% in Arabic and 71.77% in English, with almost identical F1 scores. Its trillion-parameter scale and multimodal reasoning capabilities allowed it to capture subtle linguistic and affective signals across languages. DeepSeek-R1 followed closely behind, performing nearly on par with GPT-4o, particularly in Arabic, where it matched GPT-4o's accuracy and F1 at 71.77%.

[3] IBM Watson.

Moving slightly down the scale, Grok-3 and LLaMA-3.2-11B-Vision-Instruct emerged as midtier but robust contenders. Their results were well balanced between English and Arabic, indicating a strong cross-lingual generalization. The small differences in accuracy and F1 scores between the two languages suggest that they can generalize well across languages. This is especially important for using models in the real world, where they need to work well in multilingual settings without needing to be fine-tuned for each language.

In contrast, Mistral-Large-Instruct-2411 displayed uneven performance. Despite its large size, it struggled with clarity in classification, often producing outputs that could not be assigned to ASD or Non-ASD. This led to lower accuracy, especially in English 53.75%, highlighting that size alone does not guarantee coherence or reliability in sensitive behavioral classification tasks.

Among these findings, the ALLaM family stood out as a particularly interesting case. Despite having far fewer parameters than the largest models. ALLaM-13B achieved a performance that is remarkably close to that of trillion-scale systems. It reached 71.47% accuracy and 71.24% F1 in English, which close to GPT-4o, while maintaining competitive results in Arabic 65.77% accuracy and 64.81% F1. Even its smaller sibling, ALLaM-7B, performed respectably, scoring 67.27% accuracy in English and 60.96% in Arabic. This strong showing can be traced back to ALLaM's bilingual training strategy, which began with 4T English tokens followed by 1.2T mixed Arabic-English tokens [26]. This layered exposure built a stable linguistic foundation that allowed ALLaM to generalize across languages more efficiently than many larger models.

Similarly, the Jais-Chat models confirmed the impact of scaling within a family: the 30B variant clearly surpassed the 7B model in both languages, underscoring the predictable yet consistent gains from model enlargement.

In general, the higher F1 scores seen in models like GPT-4o, DeepSeek R1, and LLaMA-3.2-11B-Vision-Instruct suggest that they are better at dealing with class imbalance and picking up on subtle behavioral cues, both of which are very important for autism screening. These results show that LLMs can be used as early detection tools in both clinical and educational settings, especially in places where there aren't many resources and it's hard to get a specialist evaluation. The results also show how important it is to have multilingual skills and carefully designed prompts to get the most out of LLMs for behavioral health applications.

4.2 Feature Assessment Classification and Behavioral Reasoning

Building on the evaluation framework described in Section ASD Detection Classification, we conducted an in-depth analysis of GPT-4o's ability to classify specific behavioral features associated with ASD. Table 3 presents the full set of performance metrics across all 14 behavioral features examined in this study. These results highlight both the strengths and limitations of GPT-4o in identifying different behavioral indicators, offering insight into which features the model handles well and where it struggles.

In general, GPT-4o had an accuracy of 0.85 and a macro-average F1-score of 0.84. The model did well on features with clear textual markers, like emotional

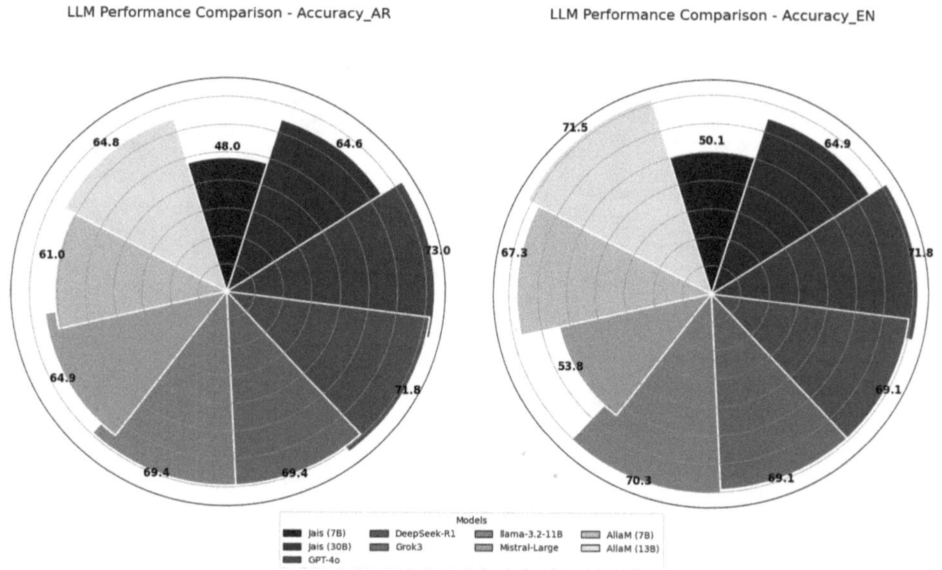

Fig. 5. A comparison of how well LLM works for finding autism in Arabic and English inputs. It's clear that top-tier models like GPT-4o and DeepSeek-R1 do better than others, getting high and balanced F1 scores in both languages. Grok3 and LLaMA-3.2 are mid-tier models that work well in both languages, but Mistral-Large is not as consistent and shows weaker results. In the AllaM and Jais-Chat families, scaling effects are clear: bigger models always do better than smaller ones.

empathy, noise sensitivity, word repetition, and toy arranging (F1-scores above 0.95). However, performance was less accurate on features that needed more careful interpretation, like attention response (F1 = 0.70), sharing interest (F1 = 0.67), and especially sign communication (F1 = 0.50), where accuracy was low because the model made too many predictions.

Along with classification, GPT-4o showed advanced reasoning skills that were very similar to how doctors think about ASD. The model's explanatory frameworks exhibited distinct patterns contingent upon the classification of behaviors as either ASD-related or neurotypical. **ASD-Related Behavioral Reasoning:** For cases categorized as indicative of autism spectrum conditions, GPT-4o consistently recognized fundamental diagnostic characteristics across various domains. In the area of social communication, the model showed that people had less eye contact, trouble with joint attention (especially after pointing gestures), and trouble sharing interests with others. These observations were correctly associated with fundamental deficiencies in social reciprocity and communication. The model also showed that it understood cognitive rigidity by focusing on distress during changes in routine and trouble shifting attention. It linked these behaviors to the restricted and repetitive patterns that are typical of ASD.

Table 3. Classification report showing precision, recall, F1-score, and support for each of the 14 behavioral features related to ASD.

Feature	Precision	Recall	F1-score	Support
Attention response	0.94	0.55	0.70	29
Change reaction	0.94	1.00	0.97	48
Emotional empathy	0.97	1.00	0.99	36
Eye contact	1.00	0.67	0.80	18
Finger movements	1.00	0.65	0.79	23
Focused attention	0.61	0.93	0.74	30
Follow pointing	1.00	0.64	0.78	28
Noise sensitivity	0.96	1.00	0.98	25
Repetitive behavior	1.00	0.82	0.90	17
Sharing interest	0.89	0.53	0.67	15
Sign communication	0.33	1.00	0.50	13
Tiptoe flapping	1.00	1.00	1.00	8
Toy arranging	1.00	0.93	0.97	15
Word repetition	1.00	0.96	0.98	28
Accuracy			0.85	333
Macro avg	0.90	0.84	0.84	333
Weighted avg	0.91	0.85	0.86	333

The model correctly identified differences in sensory processing, noting that noise hypersensitivity is a major cause of distress or meltdowns. Motor stereotypies, including hand-flapping and toe-walking, were correctly classified as self-regulatory or stimulatory behaviors. The model also demonstrated a nuanced comprehension of repetitive behaviors, accurately recognizing echolalia (word repetition), intense focused interests, and ritualistic play patterns such as systematic toy arrangement as characteristic features of ASD presentation. **Non-ASD Behavioral Reasoning:** For neurotypical classifications, GPT-4o offered developmentally suitable alternative explanations that exhibited a sophisticated clinical comprehension. Attention difficulties were characterized as situational rather than widespread, with the model observing enhancements in structured settings. Resistance to transitions was characterized as manageable with suitable preparation and support, differentiating it from the inflexible adherence to routines observed in ASD. The model importantly acknowledged the existence of emotional empathy, despite varying expression patterns, thereby dispelling the fallacy that individuals with autism entirely lack empathy. Finger movements and playing with toys were correctly understood as exploratory or functional, not stereotypical. Sign communication was accurately recognized as an adaptive communication strategy instead of a deficit indicator, and word repetition was interpreted as a normal aspect of language development rather than echolalia.

In summary, these results show that GPT-4 not only did well at quantitative classification, but it also produced qualitative reasoning that was in line with what clinicians expected. The model was especially good at picking up on clear and specific behavioral cues, but it tended to make too many generalizations in social-communication situations that weren't clear. This indicates that extensive language models can function as valuable aids in behavioral assessment, contingent upon the interpretation of their output in conjunction with clinical expertise.

5 Conclusions and Future Work

This study illustrates the viability of employing large language models (LLMs) to facilitate the early identification of ASD through textual behavioral descriptions, especially in bilingual contexts with constrained resources. Our assessment in both Arabic and English input underscores the proficiency of LLMs like GPT-4o, DeepSeek R1, Grok3, LLaMA-3.2-11B-Vision-Instruct, Mistral-Large-Instruct-2411, Jais-Chat, and ALLaM in comprehending clinical narratives.

The findings indicate that LLMs can proficiently differentiate between ASD and Non-ASD cases while also analyzing particular behavioral characteristics, such as word repetition, emotional empathy, noise sensitivity, and sustained attention. GPT-4o did very well in terms of numbers, with an overall accuracy of 0.85 and high F1-scores for features with clear textual cues. Qualitatively, the model's reasoning corresponded with clinical comprehension, differentiating between behaviors typically associated with ASD—such as restricted repetitive actions, sensory sensitivity, and social-communication difficulties—and behaviors that are developmentally normative or contextually influenced.

Based on these findings, subsequent research will concentrate on several critical domains. First, we want to add real-world, clinician-verified annotations from local populations to the dataset to make it bigger. This will make the dataset more culturally relevant, applicable to more people, and clinically valid. Second, we plan to integrate multimodal data—such as video and audio inputs—to enhance model performance and capture non-verbal behavioral cues that are often critical in ASD detection. Finally, we will look into how outputs from LLM can be used in clinical decision-support systems to help professionals find autism earlier, especially in places with few resources.

References

1. American Psychiatric Association: Diagnostic and Statistical Manual of Mental Disorders. American Psychiatric Association, Arlington, VA, 5th edn. (2013), see section on Autism Spectrum Disorder
2. Dawson, G.: Early behavioral intervention, brain plasticity, and the prevention of autism spectrum disorder. Dev. Psychopathol. **20**(3), 775–803 (2008)
3. Hachemi, R., Degha, H.E.: TASD-Dataset: Text-based Early Autism Spectrum Disorder Detection Dataset for Toddlers. Mendeley Data, V2 (2024). https://doi.org/10.17632/87s2br3ptb.2

4. Hu, C., et al.: Exploiting ChatGPT for diagnosing autism-associated language disorders and identifying distinct features. Res. Square rs–3 (2024)
5. Delehanty, A., Wetherby, A.M.: Coordinated social communication in toddlers with and without autism spectrum disorder during a home observation. Autism Develop. Lang. Impairments **7**, 23969415221121090 (2022)
6. Raiaan, M.A.K., et al.: A review on large language models: architectures, applications, taxonomies, open issues and challenges. IEEE Access **12**, 26839–26874 (2024)
7. Fernandes, D., Matos-Carvalho, J.P., Fernandes, C.M., Fachada, N.: Deepseek-v3, gpt-4, phi-4, and llama-3.3 generate correct code for lorawan-related engineering tasks. arXiv preprint arXiv:2502.14926 (2025)
8. Ochieng, M., et al.: Beyond metrics: evaluating LLMs' effectiveness in culturally nuanced, low-resource real-world scenarios. arXiv preprint arXiv:2406.00343 (2024)
9. Omar, M., Levkovich, I.: Exploring the efficacy and potential of large language models for depression: a systematic review. J. Affect. Disord. **371**, 234–244 (2025)
10. Hua, Y., et al.: A scoping review of large language models for generative tasks in mental health care. NPJ Digital Med. **8**(1), 230 (2025)
11. Ciobanu, M.G., Tucci, C., Fasano, F.: LLMs for autism treatment: current trends and emerging strategies. In: 2024 IEEE International Conference on Bioinformatics and Biomedicine (BIBM), pp. 6797–6804. IEEE (2024)
12. Raj, S., Masood, S.: Analysis and detection of autism spectrum disorder using machine learning techniques. Proc. Comput. Sci. **167**, 994–1004 (2020)
13. Themistocleous, C.K., Andreou, M., Peristeri, E.: Autism detection in children: Integrating machine learning and natural language processing in narrative analysis. Behav. Sci. **14**(6), 459 (2024)
14. Devlin, J., Chang, M.W., Lee, K., Toutanova, K.: Bert: pre-training of deep bidirectional transformers for language understanding. In: Proceedings of the 2019 Conference of the North American Chapter of the Association For Computational Linguistics: Human Language Technologies, vol. 1 (long and short papers), pp. 4171–4186 (2019)
15. Liu, Y., et al.: Roberta: A robustly optimized bert pretraining approach (2019). arXiv:1907.11692 arXiv preprint
16. Brown, T., et al.: Language models are few-shot learners. Adv. Neural. Inf. Process. Syst. **33**, 1877–1901 (2020)
17. Pal, M., Rubini, P.: Fusion of brain imaging data with artificial intelligence to detect autism spectrum disorder. Fusion: Pract. Appl. **14**(2) (2024)
18. Stanley, J., Rabot, E., Reddy, S., Belilovsky, E., Mottron, L., Bzdok, D.: Large language models deconstruct the clinical intuition behind diagnosing autism. Cell **188**(8), 2235–2248 (2025)
19. Singhal, K.: Large language models encode clinical knowledge. Nature **620**(7972), 172–180 (2023)
20. Jin, Y., et al.: The applications of large language models in mental health: scoping review. J. Med. Internet Res. **27**, e69284 (2025)
21. Miller, K., et al.: Dynamic few-shot prompting for clinical note section classification using lightweight, open-source large language models. J. Am. Med. Inform. Assoc. **32**(7), 1164–1173 (2025)
22. Mukherjee, P., Gokul, R., Sadhukhan, S., Godse, M., Chakraborty, B.: Detection of autism spectrum disorder (ASD) from natural language text using BERT and ChatGPT models. Int. J. Adv. Comput. Sci. Appl. **14**(10) (2023)
23. Jiang, Y., et al.: Copiloting diagnosis of autism in real clinical scenarios via LLMs, (2024). arXiv:2410.05684 arXiv preprint

24. Luo, D., et al.: Evaluating the performance of gpt-3.5, gpt-4, and gpt-4o in the Chinese national medical licensing examination. Sci. Reports **15**(1), 14119 (2025)
25. DeepSeek-AI: Deepseek-r1: Incentivizing reasoning capability in LLMs via reinforcement learning (2025). https://arxiv.org/abs/2501.12948
26. Bari, M.S., et al.: Allam: Large language models for Arabic and English. arXiv preprint arXiv:2407.15390 (2024)

Traffic Signal Detection System for the Visually Impaired Using Deep Learning

Aaron Jiwei Bai[1] and Chris Cheng Zhang[2](✉)

[1] St. George's School, Vancouver, Canada
[2] Canada Youth Robotics Club, Vancouver, Canada
head@roboca.org

Abstract. This project created a YOLOv8 model powered product that automatically recognizes red, yellow, and green traffic lights then displays and verbalizes this information to enable people with color blindness to drive more safely. The machine learning dataset was created with Computer Vision Annotation Tool (CVAT) for image labeling and a python-based training program called You Only Look Once (YOLO) v8 for the training process. This model was then used to make a tool that was able to detect traffic signals and transform their colors into colors distinguishable by those with colorblindness while vocally relaying information to drivers. The model achieved an accuracy rate of 96% in traffic light detection and 93% color signal recognition. The AI model was then integrated into a Raspberry Pi. Integration into a Raspberry Pi makes the AI model mobile and internet-free. This approach also holds promise for broader applications in marine and aviation sectors, potentially revolutionizing transportation safety for individuals with colorblindness.

Keywords: traffic signal · color blindness · deep-learning

1 Introduction

Color blindness, or color vision deficiency (CVD), refers to the inability to perceive colors in a normal way. According to data from the Canadian Association of Optometrists, approximately 5.5 million Canadians, constituting around 14% of the entire population, suffer from colorblindness [1]. On a global scale, there are 300 million people with CVD, affecting approximately 1 in 12 males (8%) and 1 in 200 females (0.5%). Studies indicate that up to 50% of individuals with CVD, particularly those with red–green color blindness, face challenges with traffic lights while driving. The problems involved are increases in reaction time and incorrect recognition [2]. One solution is to memorize the signal light order (red, yellow, green) and use apparent brightness to detect the correct state. However, issues may arise from sudden light changes, flickering signals, and confusion with streetlights. Reduced visibility of red, especially in protanopia and deuteranopia,

H. R. Arabnia et al. (Eds.): CSCI 2025, CCIS 2971, pp. 220–226, 2027.
https://doi.org/10.1007/978-3-032-28463-1_17

can lead to problems identifying red lights, especially in low light or with sunglasses [3]. Consequently, some countries still have limited regulations for issuing driver's licenses to individuals with CVD [4].

Machine Learning, particularly deep learning, and convolutional neural networks have emerged as a promising approach for object detection. By leveraging data about an objective, machine learning models can utilize patterns that would not be obvious to explicitly code to identify an object. This paper outlines a simple traffic light detection system that can potentially decrease the challenges presented by CVD while driving. We provide drivers with a camera-view of the road that has its colors transformed to a customizable higher contrast palette. The system also detects traffic lights, drawing bounding boxes around them for faster search. We believe the use of this system can aid in identifying colors more easily when road conditions make traffic light state detection using apparent brightness less reliable.

2 Related Works

In the context of developing a traffic light detection system for aiding drivers with color vision deficiencies (CVD), several recent studies provide valuable insights and methodologies that can be leveraged. Below, each relevant paper is discussed in relation to your work:

Real-Time Flying Object Detection with YOLOv8 [5]: This study focuses on detecting flying objects using YOLOv8, highlighting its state-of-the-art performance in terms of speed and accuracy. The use of YOLOv8 in this context demonstrates its capability to handle complex detection tasks with high efficiency, which is directly applicable to your project. By utilizing YOLOv8 for traffic light detection, your system can achieve similar benefits in terms of real-time processing and accuracy, crucial for assisting colorblind drivers.

A Comprehensive Survey on Deep Learning Approaches for Traffic Light Detection and Recognition [6]: This survey reviews various deep learning techniques for traffic light detection and recognition, emphasizing the advancements in neural network architectures. The insights from this paper can guide the optimization of your YOLOv8-based model, ensuring it leverages the latest advancements for improved detection accuracy and robustness.

Snowy Scenes, Clear Detections: A Robust Model for Traffic Light Detection in Adverse Weather Conditions [7]: This paper presents a model specifically designed to maintain high detection accuracy of traffic lights under challenging weather conditions, such as snow. The robustness of the model is achieved through data augmentation techniques and specialized training that simulates various adverse scenarios. Integrating these strategies into your YOLOv8-based traffic light detection system could significantly enhance its performance, particularly in less-than-ideal weather, ensuring reliable operation year-round.

Traffic light detection for colorblind individuals [8]: This study addresses the unique challenges faced by colorblind individuals in traffic light detection, proposing a novel system that adapts the visual presentation of traffic lights to make them more distinguishable. By incorporating such adaptive techniques into your system, you can improve its usability for colorblind drivers, ensuring that traffic light signals are easily and accurately recognized regardless of color vision.

aUToLights: A Robust Multi-Camera Traffic Light Detection and Tracking System [9]: This research utilizes YOLOv8 for traffic light detection, similar to our approach. It provides a benchmark for performance metrics and optimization strategies that can be applied to your system to enhance detection accuracy and speed.

Real-Time Traffic Light Recognition System for Color Vision Deficiencies [10]: Specifically targeting color vision deficiencies, this study provides insights into designing systems that transform traffic light colors to be distinguishable for colorblind individuals. The approaches discussed can directly inform the design and functionality of your system, ensuring it effectively addresses the needs of colorblind drivers.

These studies collectively contribute to the development of a robust and effective traffic light detection system, particularly for aiding drivers with CVD. Your work builds upon these foundations by integrating a YOLOv8-based model with a Raspberry Pi, offering a mobile and internet-free solution tailored for colorblind drivers.

3 Dataset

Before training the model, a diverse set of images was gathered, including personal captures of traffic lights around the city of Vancouver and images from Google's Open Images Dataset V7, totaling 200 photos. These images depict traffic lights in various weather and visibility conditions, captured from different angles and distances. Personal captures accounted for 10 of the 200 pictures used. The Open Images Dataset accounted for 190 of the pictures used. Using the Computer Vision Annotation Tool (CVAT) Open Data Annotation Platform, each of the 200 images was labeled by hand for traffic light positions and signal colors (red, yellow, or green) in such a way that the entire traffic light was within the bounding box. When ambiguities in the data appeared, such as non-functional traffic lights, the data were labelled as red, since that is the safest false detection for the driver. When it was not clear whether an object was a traffic light, a bounding box around it was not drawn. Traffic lights that were too far away from the driver also were not labelled. The dataset was then divided into training and validation sets. The training set was comprised of the 190 Google Open Images Dataset images while the validation set consisted of the 10 augmented Vancouver personal captures.

4 Software and Hardware

The primary detection software, YOLOv8 by Ultralytics, was selected for its speed and accuracy, being regarded as a state-of-the-art real-time object detection model [CITATION]. YOLOv8 enhances detection capabilities through optimized network architecture, improved training and inference processes, and a modular design that facilitates easy customization. Despite its performance, YOLOv8 maintains a relatively small model size, making it suitable for deployment on devices with limited computational resources.

YOLOv8 was fine-tuned to only detect 3 classes corresponding to the 3 states of functional traffic lights. A camera feed was provided to the driver, which could be used to see an augmented view of the road. This augmented video feed added bounding boxes to the traffic lights, and labelled them as red, yellow, or green. In order to enhance the safety and visibility of the camera feed, the colors were also transformed so that the lights were more visible to the driver in a higher contrast color-blind friendly palette. The program was also able to use the speakers to say which color signal it detected when a MacBook was used, by employing the built in "say" CLI command (Fig. 1).

Fig. 1. An image of the system the authors developed.

You can see on the left panel the raw feed from the camera. The right feed contains the image with transformed RGB values with a more color friendly palette. Two main hardware methods were implemented, the first being a continuity camera application. This method was able to produce 1920 by 1080 pixel video at 30 frames per second while maintaining constant detections. The second hardware implementation was using a Raspberry Pi with a camera extension. This implementation produced 1280 by 720 pixel video at 15 frames per second. However, due to the limitations of the Raspberry Pi's internal components, the model often glitched or froze when faced with a large number of detections at the same time. Further software optimization could potentially solve this issue.

5 Safety

This project, while innovative, has not been extensively tested. Therefore, it should not be used in real-world applications yet. Premature deployment without rigorous validation can pose significant safety risks. The model's audio cues and visual labels could be dangerous if users become overly reliant on them. Errors in detection could lead to traffic violations or accidents, as users might pay less attention to actual traffic signals. To mitigate risks and enhance safety, several mechanisms can be implemented:

Delayed Detection Display: Introducing a slight delay before displaying detections can prompt users to verify signals themselves, promoting attentiveness.

Secondary Confirmation: Requiring users to confirm detections through a button press or voice command can ensure they remain active participants in traffic signal verification.

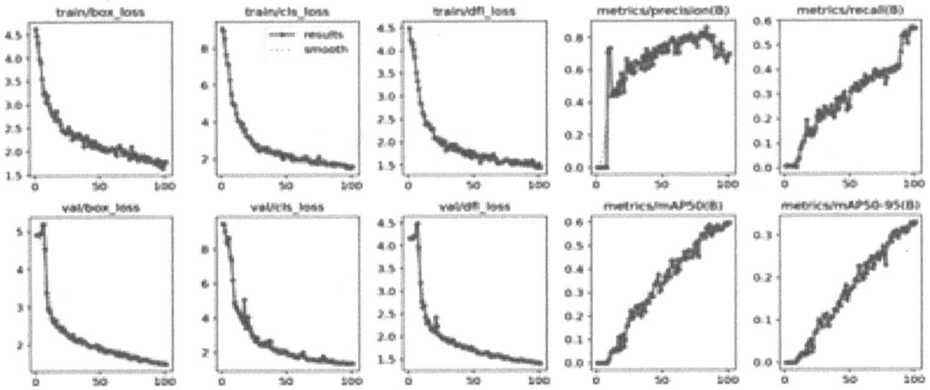

Fig. 2. Loss graphs for the final trained model

6 Results

The final model is able to detect the color signal of lights with an accuracy of an accuracy of 93.10% with a box loss value of 1.34 on the train subset. On the validation set, it achieves an accuracy of 92.10% and a box loss value of 1.55. Numerical and visual data was generated by our model as it was trained. The accuracy, loss, and other parameters were then plotted on a graph and general detected trends were plotted. The color transformation has no loss or accuracy graph as it does not use machine learning, it only affects the RGB value of each individual pixel that is displayed. Note that the model cannot detect pedestrian crossing lights as it was not trained to detect them. It will, however, detect biking lights which have a similar shape to a traditional traffic light. It is also worth noting that if a traffic light appears to not have a specific signal (e.g. all lights out), the model will default to red to minimize the chance of a false positive and increase vehicle safety.

7 Conclusion

The traffic signal recognition and verbalization product created in this project can aid colorblind drivers. The final trained machine learning model is able to detect traffic lights at 93.10% accuracy and was able to detect their signal color them at 92.10% accuracy. This product is also readily deployable in the real world by a smartphone application and a Raspberry Pi as it can function only given a speaker and camera. Finally, the color transform part of this product has the potential to benefit color blind individuals wherever they need to make decisions based on red and green lights, giving possibly broader applications (Fig. 2).

8 Future Work

In the future, the training data should be significantly increased, as commercial vision models are typically trained on hundreds of thousands of images. Incorporating more images, including synthetic data, is expected to improve accuracy and precision. More rigorous testing should be conducted using separate training, validation, and testing sets. Integration into dash cams is a logical next step, given their compact design and built-in camera and monitor. Finally, further research should involve human volunteers evaluating the system in a controlled driving environment to assess its effectiveness in real-world scenarios.

References

1. Cao eye health statistics (2019). https://opto.ca/sites/default/files/resources/documents/eye_health_statistics_with_references_july_2019.pdf
2. Cole, B.: Colour blindness and driving. Clin. Exper. Optom. **99**, 07 (2016)
3. Color blindness. https://my.clevelandclinic.org/health/diseases/11604-color-blindness
4. Tan, T.F., et al.: Colour vision restrictions for driving: an evidence-based perspective on regulations in Asian countries compared to other countries. The Lancet Regional Health Southeast Asia, vol. 14, p. 100171 (2023). https://www.sciencedirect.com/science/article/pii/S2772368223000318
5. Reis, D., Kupec, J., Hong, J., Daoudi, A.: Real-time flying object detection with YOLOv8 (2023). https://arxiv.org/abs/2305.09972
6. Tervin, J.R., Cordova-Esparza, D.M.: A comprehensive review of YOLO architectures in computer vision: from YOLOv1 to YOLOv8 and YOLO-NAS (2024). https://arxiv.org/abs/2304.00501
7. Garg, S., Baghel, A., Agarwal, A., Toshniwal, D.: Snowy scenes, clear detections: a robust model for traffic light detection in adverse weather conditions (2024). https://arxiv.org/abs/2406.13473
8. Al-Nabulsi, J., Mesleh, A., Yunis, A.: Traffic light detection for colorblind individuals (2018). https://ieeexplore.ieee.org/document/8257737
9. Wu, S., Amenta, N., Zhou, J., Papais, S., Kelly, J.: aUToLights: a robust multi-camera traffic light detection and tracking system. In: 2023 20th Conference on Robots and Vision (CRV). IEEE (2023). http://dx.doi.org/10.1109/CRV60082.2023.00019

10. Kim, Y.K., Kim, K.W., Yang, X.: Real-time traffic light recognition system for color vision deficiencies (2008). https://www.researchgate.net/publication/4275289_Real_Time_Traffic_Light_Recognition_System_for_Color_Vision_Deficiencies

Predicting Hospital Readmission in Diabetic Patients Using Machine Learning Approaches

Aprameya Deshpande[1], Anandaswarup Vadapalli[2], and Lavanya Prahallad[2](✉)

[1] Dublin High School, Dublin, USA
[2] Research Spark Hub Inc., Dublin, USA
lavanya@researchsparkhub.com

Abstract. Hospital readmission among diabetic patients remains a critical challenge in modern healthcare, reflecting both disease complexity and the quality of inpatient and post-discharge care. This study investigates the use of machine learning (ML) techniques to predict 30 day readmission risk and identify the most influential clinical factors contributing to it. Using the publicly available Diabetes Hospital Readmission Dataset from Kaggle, we trained and evaluated two predictive models, a Random Forest Decision Tree and a Deep Neural Network (DNN), after systematic data cleaning, normalization, and feature encoding.

Feature importance analysis from the Random Forest model revealed that the number of laboratory procedures, number of medications, time in hospital, number of diagnoses, and number of procedures were the most critical predictors of readmission. The DNN achieved a slightly higher predictive accuracy (0.57) compared to the Random Forest baseline (0.55), indicating its capacity to capture non-linear feature interactions.

These findings demonstrate that ML models can provide actionable insights into patient readmission risk, supporting data-driven strategies for early intervention and improved care management in diabetic populations.

1 Introduction

Diabetes mellitus is a chronic metabolic disorder characterized by persistent hyperglycemia resulting from defects in insulin secretion, insulin action, or both. The two most prevalent forms are Type 1 (autoimmune destruction of pancreatic β-cells) and Type 2 (insulin resistance with relative insulin deficiency). Globally, diabetes poses a major public-health challenge: an estimated 10% of adults aged 25–79 years are currently affected [1,2].

The condition is associated with serious long-term complications, including cardiovascular disease, neuropathy, retinopathy, and renal failure, which together contribute substantially to morbidity and mortality [1]. Effective management therefore requires a multidisciplinary approach combining pharmacologic therapy with lifestyle modification. Insulin and oral hypoglycemic agents remain

H. R. Arabnia et al. (Eds.): CSCI 2025, CCIS 2971, pp. 227–233, 2027.
https://doi.org/10.1007/978-3-032-28463-1_18

central to treatment, while diet, exercise, and weight control are critical to maintaining glycemic stability [3].

Recent advances such as continuous glucose monitoring, automated insulin delivery, and electronic health record based decision support systems have improved individualized care [4]. Nevertheless, despite these technological and therapeutic developments, diabetes related hospitalizations and readmissions continue to rise, underscoring the need for more accurate predictive tools that can enable earlier intervention and reduce preventable readmissions [5].

2 Hospital Readmission: A Clinical and Economic Challenge

Hospital readmission remains one of the most persistent and costly problems in modern healthcare. It is widely regarded as a key quality indicator, reflecting the effectiveness of inpatient treatment and post-discharge care. A readmission within 30 d of discharge often points to issues such as incomplete recovery, inadequate follow-up, or complications of chronic disease management.

From a health system perspective, frequent readmissions impose substantial financial strain. They increase healthcare expenditure, reduce hospital efficiency, and divert resources that could otherwise be used for new or critical patients. For individuals, the burden is equally severe: unplanned rehospitalization leads to higher out-of-pocket costs, emotional stress, and disrupted daily life, while also elevating the risk of further clinical deterioration [6].

These challenges are particularly pronounced in the management of chronic illnesses such as diabetes, where multiple comorbidities, long-term medication use, and lifestyle factors interact in complex ways [7]. Consequently, understanding and predicting readmission risk has become a central goal in clinical informatics, offering an opportunity to improve both patient outcomes and healthcare sustainability through early detection and intervention.

2.1 Role of ML in Predicting Readmission

Machine learning (ML) methods have emerged as powerful tools for addressing complex clinical prediction problems such as hospital readmission. Unlike traditional statistical models that often rely on linear assumptions and a limited number of predictors, ML algorithms can capture high-dimensional, non-linear relationships among diverse clinical, demographic, and laboratory variables [8].

By leveraging large electronic health record (EHR) datasets, ML models can be trained to estimate the likelihood that a patient will be readmitted within a specified time frame, commonly 30 d after discharge. These models consider numerous factors simultaneously, including age, gender, comorbidities, laboratory results, medication profiles, and treatment history, enabling a more nuanced representation of patient risk.

Among various algorithms, decision-tree based ensembles (such as Random Forests and Gradient-Boosted Trees) and deep neural networks have shown particular promise [9]. Decision-tree approaches offer interpretability through feature importance ranking, revealing which clinical factors most strongly influence readmission risk. Neural networks, by contrast, excel at modeling subtle interactions and complex feature combinations.

When implemented responsibly, with rigorous validation, transparency, and bias control, machine learning-based risk-prediction systems can support data driven decision making. Hospitals can use these insights to identify high-risk patients early, design personalized discharge plans, allocate follow-up resources efficiently, and ultimately reduce preventable readmissions.

3 Dataset and Preprocessing

This study uses the publicly available Diabetes Hospital Readmission Dataset published on Kaggle[1]. This dataset provides a comprehensive view of patient encounters related to diabetes and includes demographic information, laboratory test counts, medication data, and other clinical variables relevant to readmission risk.

Each record in the dataset corresponds to a unique hospital encounter. The attributes encompass both numerical features such as age, time in hospital, number of procedures, number of laboratory tests as well as categorical variables including gender, admission type, discharge disposition, and primary diagnosis. This rich combination of features makes the dataset suitable for supervised learning tasks focused on hospital readmission prediction.

Prior to modeling, the dataset underwent systematic data cleaning and preprocessing. Missing or inconsistent entries were handled through either imputation or exclusion, depending upon the variable. Numerical variables were normalized to a consistent scale and categorical features were encoded using one-hot labels. The resulting dataset was then divided into training and testing subsets.

While the dataset offers a valuable benchmark for developing predictive models, it also presents limitations: key patient attributes such as body weight, socioeconomic factors, or lifestyle information were not available. Future extensions incorporating such contextual variables could improve model interpretability and predictive performance.

4 Identifying Key Predictors Using Decision Trees

To determine which clinical variables most strongly influence the likelihood of hospital readmission among diabetic patients, a Random Forest Decision Tree model was employed. Decision trees recursively partition data by selecting features that most effectively reduce impurity, measured using metrics such as Gini impurity or entropy. A Random Forest, which is an ensemble of many decision trees trained on different data subsets, provides a more robust and generalized estimate of feature importance than any single tree.

[1] https://www.kaggle.com/datasets/brandao/diabetes.

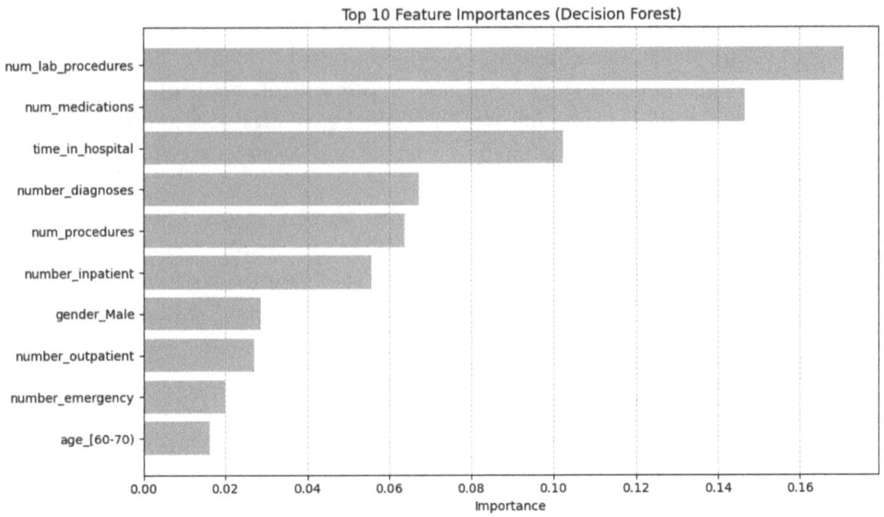

Fig. 1. Random Forest feature importance for 30-day readmission prediction in diabetic patients (impurity-based importance averaged across trees; importances are normalized to sum to 1). Longer bars indicate greater predictive contribution. Features are sorted in descending order.

4.1 Feature Ranking Process

The Random Forest Classifier was trained on the pre-processed diabetes dataset described in Sect. 3. After training, the model's built-in feature-importance scores were extracted and normalized so that the total summed to 1. These scores quantify how much each variable contributes to reducing classification uncertainty across all trees. The analysis revealed five features that exerted the greatest influence on readmission prediction (Fig. 1). Each of these features captures a distinct dimension of patient complexity or treatment intensity:

1. **Number of Lab Procedures**: Represents the count of laboratory tests performed during the hospital encounter. Patients who underwent more lab tests were statistically more likely to be readmitted. Those readmitted within 30 d averaged approximately 44 tests, compared with 42 for patients who were not readmitted. This significant difference (confirmed via ANOVA) suggests that frequent testing reflects higher disease complexity or multiple comorbidities.
2. **Number of Medications**: Denotes the number of distinct medications prescribed. Patients who received a greater number of medications on average $\approx$ 17 for $<$ 30-day readmissions versus $\approx$ 16 for others, showed a higher probability of readmission. This pattern likely reflects the burden of polypharmacy and underlying multi-system disease.
3. **Time in Hospital**: The length of stay (in days) serves as a proxy for illness severity and treatment intensity. Patients who remained hospitalized longer

($\approx$ 4.8 d for $<$ 30-day readmissions versus $\approx$ 4.3 d for non-readmissions) exhibited higher subsequent readmission rates, with statistical tests confirming the relationship's significance.

4. **Number of Diagnoses**: Indicates the total number of diagnostic codes recorded during the encounter. Individuals with more diagnoses ($\approx$ 7.7 for $<$ 30-day readmissions versus $\approx$ 7.2 for no readmission) tended to have more complex clinical profiles and were correspondingly at greater risk of readmission.
5. **Number of Procedures**: Captures the count of non-laboratory medical procedures performed. Interestingly, patients who were not readmitted had slightly more procedures on average ($\approx$ 1.41) than those who were ($\approx$ 1.28), implying that timely interventions during the initial stay may reduce the likelihood of returning to the hospital.

4.2 Interpretation and Implications

The Random Forest results highlight that clinical activity indicators, such as laboratory testing frequency, medication volume, and hospital stay duration, carry the greatest predictive weight for readmission. These variables likely act as surrogates for patient acuity and treatment complexity. Conversely, the modest inverse relationship observed for the number of procedures suggests that proactive or comprehensive interventions may mitigate future readmissions.

Understanding these ranked predictors provides valuable insight into targeted prevention strategies. Hospitals can prioritize post-discharge follow up for patients exhibiting multiple risk factors, particularly those with extended stays, numerous tests, and high medication counts, thus enabling earlier intervention and potentially reducing 30-day readmission rates.

5 Prediction Accuracy of Readmission Using Deep Neural Networks

While tree-based models provide interpretable estimates of feature importance, their ability to model highly non-linear interactions can be limited. To capture such complex dependencies, a Deep Neural Network (DNN) was developed and evaluated on the same dataset. Neural networks are particularly well suited to medical risk prediction tasks because they can automatically learn hierarchical feature representations and uncover subtle relationships among clinical variables [10].

5.1 Model Architecture and Training

The implemented DNN consisted of an input layer corresponding to the selected features from Sect. 4, followed by two fully connected hidden layers with ReLU activation functions and dropout regularization to reduce overfitting. The output layer used a softmax activation to predict one of three readmission categories:

< 30 days, > 30 days, or no readmission. Three configurations were tested: a single hidden layer (100,), two hidden layers (100, 100), and a deeper neural network with three layers (100, 100, 50).

The dataset was divided into a training (80%) and test (20%) set, ensuring class balance across categories. Model training employed the Adam optimizer with categorical cross-entropy loss. Early stopping was applied based on validation accuracy to prevent overtraining. To assess overtraining and convergence, the models were trained with 5, 50, and 100 iterations, with a training loss curve plotted for each configuration. All experiments were conducted on CPU to maintain comparability with the decision-tree experiments.

5.2 Performance Comparison

The trained DNN achieved an overall accuracy of 0.57, outperforming the Random-Forest Decision Tree baseline (0.55). Although the improvement is modest, it demonstrates the DNN's ability to exploit deeper feature interactions that are not explicitly modeled by tree-based ensembles. Both models' predictions were directionally consistent, correctly classifying the majority of no-readmission cases but occasionally misclassifying between the < 30 days and > 30 days categories, likely due to overlapping clinical patterns and limited temporal features.

5.3 Interpretation

The DNN's marginal yet consistent performance gain suggests that non-linear modeling contributes incremental predictive value beyond traditional ensemble methods. However, the improvement also highlights a recurring challenge in healthcare ML: model performance is often constrained not by algorithmic choice but by data richness and quality. Incorporating additional features, such as laboratory result trends, outpatient follow-up data, or social determinants of health, could substantially enhance predictive accuracy in future work.

6 Conclusion

This study explored the use of machine-learning techniques to predict hospital readmission among diabetic patients, using an open-source dataset containing demographic, clinical, and treatment-related features. A Random Forest Decision Tree model was employed to identify the most influential predictors of readmission, revealing that indicators of clinical intensity, such as the number of laboratory procedures, medications, and length of hospital stay, play a dominant role in determining risk. In contrast, a slightly negative association for the number of procedures suggests that timely interventions during the initial hospitalization may help prevent subsequent readmissions.

To further evaluate predictive potential, a Deep Neural Network (DNN) was developed and compared with the Random Forest baseline. The DNN achieved a modest but consistent improvement in accuracy, underscoring the value of modeling non-linear relationships between patient characteristics and outcomes. However, both models demonstrated that predictive performance is ultimately constrained by the breadth and quality of available data, rather than by algorithmic complexity alone.

Overall, the findings highlight the promise of data-driven approaches in improving clinical decision support for chronic disease management. Future work should focus on integrating richer datasets, including longitudinal lab trends, outpatient follow-up information, and social or behavioral determinants of health, to enhance model generalizability and real world applicability. With such refinements, predictive modeling can become a valuable tool for reducing preventable readmissions and improving the quality and efficiency of diabetic care.

References

1. Sapra, A., Bhandari, P.: Diabetes. In: StatPearls [Internet], Treasure Island (FL): StatPearls Publishing (2023). [Updated 2023 Jun 21]. https://www.ncbi.nlm.nih.gov/books/NBK551501/
2. Magliano, D., Boyko, E., IDF Diabetes Atlas 10th edition scientific committee, "Global picture". In: IDF Diabetes Atlas [Internet] 10th edition, Brussels: International Diabetes Federation, (2021). https://www.ncbi.nlm.nih.gov/books/NBK581940/
3. Zhang, S., et al.: Effect of hypoglycemic agents with weight loss effect plus a high-protein diet and moderate exercise on diabetes remission in adults with obesity and type 2 diabetes: a randomized controlled trial. BMC Med. **23**(1), 270 (2025)
4. Pattar, B.S., et al.: Electronic health record interventions to reduce risk of hospital readmissions: a systematic review and meta-analysis. JAMA Netw. Open **8**(10), e2836552 (2025)
5. Association, A.D.: Standards of care in diabetes–2023: improving care through technology and personalized management. Diabetes Care **46**(Suppl. 1), S1–S183 (2023)
6. Jencks, S.F., Williams, M.V., Coleman, E.A.: Rehospitalizations among patients in the medicare fee-for-service program. N. Engl. J. Med. **360**(14), 1418–1428 (2009)
7. Kirkman, M.S., et al.: Diabetes in older adults: a consensus report. Diabetes Care **35**(12), 2650–2664 (2012)
8. Rajkomar, A., Oren, E., Chen, K., et al.: Scalable and accurate deep learning for electronic health records. NPJ Digit. Med. **1**, 18 (2018)
9. Smith, S., McConnell, S.: The use of artificial neural networks and decision trees: implications for health-care research. Open Comput. Sci. (2024)
10. Miotto, R., Wang, F., Wang, S., Jiang, X., Dudley, J.T.: Deep learning for healthcare: review, opportunities and challenges. Brief. Bioinform. **18**(3), 1236–1246 (2017)

AI in Healthcare: A Systematic Review of Clinical Applications, Governance Challenges, and Implementation Priorities

Kenneth Choi[1], S. J. Kim[1], and Yen-Soon Kim[2(✉)]

[1] DEX (Digital Experience) Lab, Howard R. Hughes College of Engineering, University of Nevada, Las Vegas, USA
sj.kim@unlv.edu
[2] Harrah College of Hospitality, University of Nevada, Las Vegas, USA
yen-soon.kim@unlv.edu

Abstract. Artificial intelligence (AI) is moving from pilots to enterprise deployments across clinical and administrative workflows. This paper synthesizes six domains—disease identification and prevention, radiology, surgical assistance, specialty applications, operations and logistics, and direct patient care—to evaluate how AI creates value and where guardrails are required. Using a structured rapid narrative review focused on 2022–2025 evidence, we mapped peer-reviewed studies and selected implementation signals to two value axes: clinical precision (earlier or more accurate detection, personalized planning) and operational efficiency (capacity management, documentation relief). The evidence base is most mature in radiology and ambient documentation, with promising results for risk stratification, report consistency, and reductions in clerical burden. Disease-prevention models in cardiology and oncology show potential for earlier triage but require robust external validation. Across domains, five cross-cutting requirements recur: transparency and provenance, bias mitigation and equity audits, clear accountability and liability allocation, multidisciplinary governance, and human–AI collaboration with uncertainty communication and fail-safes. These conditions are essential for safe scaling and sustained clinician and patient trust. We conclude that thoughtfully governed AI can improve outcomes and workflows when paired with change management, monitoring, and periodic re-validation. Priorities for research include multi-site prospective studies, standardized transparency packages, and interoperability that enables seamless integration with clinical systems.

Keywords: Artificial Intelligence · Healthcare Administration · AI Adoption

1 Introduction

AI (Artificial Intelligence) in healthcare has accelerated from pilots to impactful deployments across clinical and administrative workflows [1], promising higher precision, reduced burden, and improved coordination. Yet adoption can stall without attention to equity, transparency, and governance. We reframe a prior commentary as a concise research synthesis, aligning with CSCI's conference aims: to present an actionable

H. R. Arabnia et al. (Eds.): CSCI 2025, CCIS 2971, pp. 234–242, 2027.
https://doi.org/10.1007/978-3-032-28463-1_19

overview of methods, early evidence, and implications for practitioners and decision makers.

2 Literature Review

The recent literature on artificial intelligence (AI) in healthcare spans clinical decision support, imaging, surgical assistance, specialty care, operations management, and patient-facing documentation. Below, we synthesize key contributions in these domains and connect them to recurring governance and equity considerations.

2.1 Disease Identification and Prevention

Work on early detection and risk stratification indicates that deep learning can uncover predictive signals beyond human perception. Studies in cellular phenotyping, cardiology, neuro-oncology, and lung cancer forecasting demonstrate improvements in identifying disease signatures, estimating event risk, and anticipating future diagnoses from existing imaging [2–5]. Collectively, these papers motivate the use of AI for first-line triage and longitudinal prevention programs while underscoring the need for robust external validation.

2.2 Radiology

Radiology continues to be a leading domain for applied AI, with evidence on risk stratification, report standardization, and detection of subtle, time-critical pathologies [6–9]. Studies examine how models can reduce unnecessary biopsies, harmonize reporting language via large language models, and support labeling using privacy-preserving approaches. While results suggest sensitivity gains for specific tasks, the literature calls for transparent labeling pipelines, subgroup analyses, and careful monitoring to avoid bias propagation from training data [6–9, 24, 25].

2.3 Surgical Assistance

In surgical contexts, AI supports both cognitive and robotic assistance, including documentation and education. Work surveying AI's role in plastic and reconstructive surgery and adjacent studies comparing humanwritten vs. LLMdrafted abstracts highlight opportunities and pitfalls for automating scholarly and clinical artifacts [10–12]. The evidence base emphasizes precision benefits and administrative relief but also stresses guardrails for quality, authorship, and accountability in surgical decisionmaking.

2.4 Specialty Applications

Domain-specific applications show how multimodal models and ML-augmented imaging improve personalization. Examples include PET/CT-informed myocardial risk assessment, structured extraction for gastroenterology charts using LLMs, and clinical validation of deep learning for radiotherapy targeting [13–15]. These studies illustrate the translational bridge from imaging and text processing to individualized care plans while calling for interoperability with clinical information systems.

2.5 Operations, Logistics, and Documentation Burden

Operational AI spans admission forecasting, length-of-stay prediction, deterioration alerts, and connected-care orchestration [16–19]. Concurrently, ambient documentation and conversation-tuned ASR/LLMs aim to reduce clinician note-taking time and improve information capture in encounters [20, 21]. Together, these lines of work target capacity management and workforce well-being. Studies report efficiency gains, but generalizability and prospective evaluation remain priorities.

2.6 Governance, Safety, and Equity

Leadership perspectives and policy analyses underscore that transparency, bias mitigation, and clear allocation of responsibility are prerequisites for scaling AI safely [22, 23, 27]. Evidence on external validation and subgroup performance, particularly for imaging, reveals gaps that can mask under-performance in marginalized populations [24, 25]. Reporting packages that include data/model provenance, calibration, uncertainty, and audit trails are recurrent recommendations to support oversight and liability management.

2.7 Deployments, Ecosystem Signals, and General-Purpose Models

Real-world deployment signals—such as a large health system's selection of an AI software platform for connected care—illustrate the shift from pilots to enterprise adoption [18, 30]. Major-press and nonprofit analyses capture broader ecosystem implications, from workforce changes to regulatory gaps [26–28]. At the same time, the rapid emergence of large, general-purpose models in medicine (including multimodal LLMs) demands task-specific evaluation, guardrails, and domain adaptation to ensure clinical fitness for use [7, 14, 24, 29]. Synthesizing across domains, the literature suggests that responsibly governed AI can deliver value along two axes: (i) clinical precision (earlier or more accurate detection, personalized planning) and (ii) operational efficiency (capacity management, documentation relief). The most robust evidence clusters in radiology, select oncology and cardiology tasks, and documentation support. The primary gaps include multi-site external validation and equity analyses, prospective and real-world impact studies, and standardized transparency packages that enable auditability and clear accountability [16–25, 27]. These findings directly motivate the study's research questions and subsequent analysis sections.

3 Methodology

This study re-framed the source manuscript as a concise research synthesis of AI services in healthcare, using a structured, rapid narrative review with elements of a systematic search. Evidence was mapped across six domains (disease identification/prevention, radiology, surgical assistance, specialty applications, operations/logistics, direct patient care) and five cross-cutting issues (transparency, bias/equity, accountability/liability, governance, and human–AI collaboration), consistent with the scope of the paper.

3.1 Research Questions

This study is guided by three overarching research questions, each addressing a critical dimension of AI adoption in healthcare. To provide clarity and depth, we also outline sub-questions that link these inquiries to the six healthcare domains and the recurring cross-cutting challenges.

RQ1: How has recent peer-reviewed literature characterized deployment, evaluation, and impact of AI services across key healthcare domains?
RQ2: What cross-cutting challenges and governance considerations recur in healthcare AI implementations?
RQ3: Where are the principal evidence gaps and priorities for future research and implementation science?

3.2 Search Strategy

An initial broad search spanning 2014–2025 captured early healthcare-AI studies; after scoping, the window was narrowed to 2022–2025 to focus on periods with the most mature evaluations and deployments. Searches covered PubMed/MEDLINE, Scopus, Web of Science, Google Scholar, and AMIA/JAMIA venues using combinations of: "artificial intelligence," "machine learning," "predictive model," "large language model," "radiology," "surgery," "oncology," "hospital operations," "care coordination," "ambient documentation," "speech recognition," "transparency," "bias," "equity," "governance," and "liability." Backward snowballing of reference lists identified additional sources. Representative articles spanning the six domains include disease detection and prevention [2–5], radiology [6–9], surgery [10–12], specialty applications [13–15], operations/logistics and patient-facing documentation [16–21], and governance/policy perspectives [22–25, 29].

3.3 Inclusion and Exclusion Criteria

Inclusion prioritized peer-reviewed empirical studies, evaluations, or reviews that examined AI systems in clinical care, patient-facing workflows, or health-system operations. Peer-reviewed conference proceedings with explicit methods/results (e.g., AMIA) were included selectively when they provided implementation-relevant evidence. To document live deployments and policy conditions, a small number of reputable non-peer-reviewed sources were retained (major health-system press release, national policy analysis, and major-press explanatory reporting) [18, 23, 26–28]. Excluded were articles focused solely on technical/algorithmic development without clinical or operational context, editorials without evidence, and studies outside healthcare.

3.4 Screening, Selection, and Coding

Two reviewers independently screened titles/abstracts for relevance to the six domains and five issues, followed by full-text assessment. Disagreements were resolved by discussion; a third author adjudicated edge cases. The final corpus comprised 30 core sources:

25 peer-reviewed items (including one peer-reviewed conference proceeding) and 5 policy/implementation sources [18, 23, 26–28]. Articles were coded for domain, setting, study design, data sources, outcomes (clinical, operational, documentation burden), and equity/oversight considerations.

3.5 Synthesis Approach

Given heterogeneity of designs and outcomes, quantitative meta-analysis was not attempted. Instead, coded evidence was narratively synthesized to (i) summarize domain-specific findings; (ii) extract recurring cross-cutting challenges (transparency, bias/equity, accountability/liability, governance, human–AI collaboration); and (iii) identify gaps and practice implications aligned with the three research questions. Where helpful, exemplar studies are cited in-text to anchor claims (e.g., evaluations of ambient documentation and predictive models [16–21] and governance/leadership perspectives [22–25, 29]).

4 Results and Discussion

The review of recent peer-reviewed literature highlights how AI services are being deployed, evaluated, and assessed across a wide range of healthcare domains (RQ1). Evidence is most mature in radiology and documentation support, where AI has improved sensitivity, risk stratification, and reporting consistency, as well as reduced clerical burden through ambient documentation and large language model–based assistants. In disease identification and prevention, deep learning methods show promise in cardiology and oncology, offering earlier detection than human review alone, though external validation remains limited. Surgical assistance illustrates how AI can support precision and training through robotics and intra-operative decision support, while specialty applications such as PET/CT or radiotherapy planning demonstrate personalized care potential but also highlight interoperability gaps. In operations and logistics, predictive models help manage admissions, length of stay, and patient deterioration, whereas in direct patient care, conversational AI and automated documentation are beginning to reshape the clinical encounter. Together, these deployments suggest that AI can contribute to both clinical precision and operational efficiency, though unevenly across domains.

Across these domains, recurring challenges and governance considerations emerge consistently (RQ2). Transparency and explainability remain essential, requiring provenance tracking, user-facing explanations, and validation at the workflow level. Concerns around bias and equity also persist, as models trained on limited or homogeneous data often underperform in marginalized populations, making subgroup analysis and equity audits necessary. Accountability is another key issue, particularly regarding liability and role definition among vendors, clinicians, and institutions. Strong governance structures that include clinicians, administrators, and equity experts are increasingly called for, alongside procurement strategies that link reimbursement to validated outcomes. Finally, literature emphasizes the importance of human–AI collaboration, noting that AI must be designed as an augmentation tool with uncertainty communication, fail-safes, and meaningful patient inclusion.

At the same time, literature exposes important evidence gaps and priorities for future research (RQ3). Multi-site validation and prospective trials are rare, leaving generalizability and long-term impact uncertain. Standardized transparency frameworks, including documentation of calibration, provenance, and version histories are still emerging, complicating oversight and liability allocation. Interoperability challenges remain a barrier to seamless integration with existing clinical information systems, and longitudinal monitoring of deployed models is seldom reported despite its importance for detecting performance drift. To move forward, implementation science must focus on use cases that both alleviate clinician burden and improve equity, while policymakers and health systems develop safeguards that establish minimum standards for safety and fairness. Taken together, these findings show that while AI is beginning to deliver tangible benefits in healthcare, its sustainable adoption depends on addressing governance, equity, and validation gaps alongside technical performance.

Table 1 summarizes the six domains in which AI has been most actively applied in healthcare. The findings illustrate how AI contributes both to clinical precision—through earlier disease detection, improved diagnostics, and personalized specialty care—and to operational efficiency, including workflow automation, predictive resource management, and reduced documentation burdens. Radiology and documentation support show the strongest maturity of evidence, while other domains, such as surgical assistance and specialty applications, highlight emerging opportunities that require further validation, governance, and equity considerations.

Table 1. Six AI Domains in Healthcare

Domain	Key Findings
Disease Identification & Prevention	Deep learning augments early detection and risk prediction in cardiology, oncology, and multi-omics screening, enabling first-line triage and earlier interventions
Radiology Applications	Supports risk stratification and report generation while improving sensitivity for subtle findings; requires privacy-preserving labeling and bias evaluation
Surgical Assistance	Computer vision and AI-assisted robotics improve precision and documentation; predictive models inform intra- and post-operative decision-making
Specialty Applications	Field-specific models (e.g., PET/CT in cardiology, radiotherapy planning, structured extraction in gastroenterology) advance personalized care
Operations & Logistics	Predictive models for admissions, length of stay, and deterioration support capacity management; connected-care platforms show efficiency gains
Direct Patient Care	Ambient documentation and conversation-tuned ASR/LLMs reduce clerical burden and may improve visit quality when validated for safety and equity

5 Conclusions and Implications

AI demonstrates significant potential to improve both clinical precision and operational efficiency across healthcare domains. From disease detection to documentation relief, the evidence shows meaningful gains in accuracy, timeliness, and workflow support. However, sustained adoption depends on how well health systems address persistent cross-cutting challenges. First, transparency and explainability are essential. Clear documentation of data sources, model provenance, and user-facing explanations will help clinicians and patients trust AI outputs. Second, bias and equity must be proactively managed through diverse datasets, subgroup analysis, and ongoing equity impact assessments to ensure fair performance across populations. Third, accountability and liability frameworks are required to clearly delineate responsibilities among vendors, clinicians, and institutions. Fourth, governance structures must include multidisciplinary oversight—clinicians, administrators, and equity experts—and tie reimbursement to validated outcomes. Finally, human–AI collaboration should be the guiding design principle: AI must be deployed as an augmentation tool with fail-safes, uncertainty communication, and patient inclusion built into the workflow.

Implications for practice include change-management strategies, continuous monitoring, and vendor accountability for calibration and bias documentation. Policymakers and healthcare leaders should prioritize AI implementations that relieve clinician burden while enhancing equity. In conclusion, the transformative promise of AI in healthcare can only be realized through transparent validation, equitable safeguards, and robust governance. When integrated thoughtfully into human-centered collaboration, AI has the power to not only optimize healthcare processes but also to strengthen trust and outcomes for all patients. Health systems should prioritize deployment where AI simultaneously relieves clinician burden and improves equity; pair model integration with change-management, monitoring dashboards, and periodic re-validation. Vendors should provide versioned documentation, calibration curves, and bias analyses; policymakers should set minimum safety and equity standards.

6 Limitations

This review emphasizes breadth over exhaustiveness; gray literature was used sparingly to document deployments and policy context. The 2022–2025 emphasis may underrepresent earlier foundational work; however, it aligns with the period in which implementation studies became more prevalent. Despite dual independent screening and adjudication, publication bias toward positive results and English-language sources may remain.

References

1. Weber, E., Burnaska, K., Bowman, R., Holvey, S.: AI in health care: closing the revenue cycle gap. AJMC **31**, 161–162 (2025)
2. Schiff, L., Migliori, B., Chen, Y., et al.: Integrating deep learning and unbiased automated high-content screening to identify complex disease signatures in human fibroblasts. Nat. Commun. **13**(1) (2022)

3. Popescu, D.M., Shade, J.K., Lai, C., et al.: Arrhythmic sudden death survival prediction using deep learning analysis of scarring in the heart. Nat. Cardiovasc. Res. **1**(4), 334–343 (2022)
4. Hollon, T., Jiang, C., Chowdury, A., et al.: AI-based molecular classification of diffuse gliomas with rapid, label-free optical imaging. Nat. Med. (2023)
5. Mikhael, P.G., Wohlwend, J., Yala, A., et al.: Sybil: a deep learning model to predict future lung cancer risk from a single low-dose chest CT. J.Clin. Oncol. **41**(12) (2023)
6. Ezeana, C.F., He, T., Patel, T.A., et al.: DL decision support to improve risk stratification and reduce unnecessary biopsies in BI-RADS 4 mammograms. Radiology AI **5**(6) (2023)
7. Hasani, A., Singh, S.M., Zahergivar, A., et al.: Evaluating GPT-4 for standardizing radiology reports. Eur. Radiol. (2023)
8. Mukherjee, P., Hou, B., Lanfredi, R.B., Summers, R.M.: Privacy-preserving LLM (Vicuna) for labeling radiology reports. Radiology **309**(1) (2023)
9. Hillis, J.M., Bizzo, B.C., Newbury-Chaet, I., et al.: AI for identification of intracranial hemorrhage subtypes on head CT. Stroke Vasc. Interv. Neurol. **4**(4) (2024)
10. Mansoor, M., Ibrahim, A.F.: Transformative role of AI in plastic & reconstructive surgery. J. Clin. Med. **14**(8), 2698 (2025)
11. Holland, A.M., Lorenz, W.R., Cavanagh, J.C., et al.: Comparison of abstracts written by surgeons vs LLMs. JAMA Netw. Open **7**(8), e2425373 (2024)
12. Grigorian, A., Shipley, J., Nahmias, J., et al.: Implications of using chatbots for surgical education. JAMA Surg. **158**(11), 1220 (2023)
13. Kwiecinski, J., Tzolos, E., Meah, M.N., et al.: ML with 18F-NaF PET & quantitative plaque analysis on CTA for MI risk. J. Nucl. Med. **63**(1), 158–165 (2022)
14. Ge, J., Li, M., Delk, M.B., Lai, J.C.: LLM vs manual chart review for EHR data extraction. Gastroenterology **166**(4), 707-709.e3 (2024)
15. Hosny, A., Bitterman, D.S., Guthier, C.V., et al.: Clinical validation of DL for radiotherapy targeting of NSCLC. Lancet Digit. Health **4**(9), e657–e666 (2022)
16. Landman, A.B., Tilak, S.S., Walker, G.A.: AI-generated ED summaries and handoffs. JAMA Netw. Open **7**(12), e2448729 (2024)
17. Bednarski, B.P., Singh, A.D., Zhang, W., et al.: Temporal convolutional networks for LOS and mortality prediction. Sci. Rep. **12**(1) (2022)
18. Tampa General Hospital: TGH selects Palantir's AI software to power connected care coordination (2024, Press release)
19. Nong, P., Adler-Milstein, J., Apathy, N.C., Holmgren, A.J., Everson, J.: Current use and evaluation of AI/predictive models in US hospitals. Health Aff. **44**(1), 90–98 (2025)
20. Haberle, T., Cleveland, C., Snow, G.L., et al.: Impact of Nuance DAX ambient listening AI documentation. JAMIA **31**(4) (2024)
21. Tran, B.D., Mangu, R., Tai-Seale, M., Lafata, J.E., Zheng, K.: ASR for digital scribes in patient-clinician conversations. In: AMIA Symposium Proceedings (2023)
22. Guleria, S., Guptill, J., Kumar, I., McClintic, M., Rojas, J.C.: AI integration in healthcare: perspectives of US health system leaders. BMC Digit. Health **2**(1) (2024)
23. NIH: Risks and benefits of integrating AI into medical decision-making. News Release, 22 July 2024
24. Jin, Q., Chen, F., Zhou, Y., et al.: Hidden flaws behind expert-level accuracy of multimodal GPT-4 Vision in medicine. npj Digit. Med. **7**(1) (2024)
25. Omoleye, O.J., Woodard, A.E., Howard, F.M., et al.: External evaluation of mammography DL model in diverse population. Radiol. AI **5**(6) (2023)
26. Serrano, J.F.: AI hasn't killed radiology, but it is changing it. The Washington Post, 5 April 2025
27. Talkington, K., Wenderoff, J.: Gaps in oversight of AI for radiology. Pew (2025)
28. Rogers, T.N.: Robots take on hospital "busy work". Financ. Times (2025)

29. Singhal, K., Tu, T., Gottweis, J., et al.: Toward expert-level medical QA with LLMs. Nat. Med. (2025)
30. Shool, S., Adimi, S., Saboori Amleshi, R., Bitaraf, E., Golpira, R., Tara, M.: A systematic review of large language model (LLM) evaluations in clinical medicine. BMC Med. Inform. Decis. Mak. **25**, 117 (2025). https://doi.org/10.1186/s12911-025-02954-4

Health Informatics, Data Science, Prediction Methods, Applications and Tools

The Cost of a Ping: Neurophysiological Evidence of Attention Disruption from Mobile Notifications

Sebastian Valladares[1(✉)], Nabil Hossain[1], Valentina Nino[3], and Maria Valero[2]

[1] George Walton Comprehensive High School, Marietta, USA
lvallad1@kennesaw.edu
[2] Department of Information Technology, Kennesaw State University, Marietta, USA
mvalero2@kennesaw.edu
[3] Department of Industrial and Systems Engineering, Kennesaw State University, Marietta, USA

Abstract. In an era of constant connectivity, mobile notifications are ubiquitous—yet their cognitive cost remains poorly understood. This study examined the neurophysiological effects of smartphone notifications on attention using wearable EEG. Ten participants completed cognitively demanding tasks under two conditions: with and without notification sounds. Brain activity was recorded with a consumer-grade EEG headset, and surveys captured self-reported screen time, notification preferences, and attentional traits. We computed focus-related brainwave indices (e.g., Focus Index, Theta, Beta, Gamma) and ran paired-sample t-tests and regressions to assess sensitivity to interruption. Notifications significantly increased Beta1 and Gamma1 activity—linked to stress and cognitive load. Participants with lower screen time showed greater disruption, suggesting reduced habituation. Those preferring vibration alerts had stronger brainwave responses than those attuned to sound or visual cues. Focus Index negatively correlated with sensitivity, indicating weaker baseline attention in those more disrupted. Findings highlight the cognitive impact of notifications—especially vibrations—and support the need for personalized digital wellness strategies.

Keywords: Neurophysiology · Mobile Notifications · Attention Disruption · Wearable EEG · Digital Wellbeing

1 Introduction

Smartphone notifications, whether through sound, vibration, or visual cues, have become an omnipresent aspect of daily life. In the United States alone, 97% of adults own a smartphone, and the majority report checking them within minutes of receiving a notification [1]. These alerts deliver valuable information,

H. R. Arabnia et al. (Eds.): CSCI 2025, CCIS 2971, pp. 245–258, 2027.
https://doi.org/10.1007/978-3-032-28463-1_20

from messages and reminders to social media updates. However, they also represent a form of digital interruption that increasingly competes for our cognitive resources.

Emerging research highlights the negative impact of notifications on sustained attention and cognitive performance. Even passive exposure to alerts can impair memory encoding, increase task-switching, and elevate stress levels [2,3]. In particular, young adults and adolescents, who are often engaged in academic or cognitively demanding activities, are especially vulnerable, receiving upward of 65–80 notifications daily [4]. Despite growing concern, the underlying neurophysiological mechanisms by which notifications disrupt cognition remain poorly understood.

Electroencephalography (EEG) offers a powerful, non-invasive method to study attention and distraction in real time. Brainwave frequencies such as Theta (4–7 Hz) and Alpha (8–12 Hz) are commonly associated with mind-wandering and relaxed attention, while Beta (13–30 Hz) and Gamma (30–45 Hz) reflect heightened cognitive activity and information processing [5–8]. Changes in these bands can provide insight into how notifications affect the brain's focus and workload.

Although prior work has examined the psychological and behavioral outcomes of smartphone interruptions [2,9], few studies have paired ecologically valid notification scenarios with high-resolution EEG data. Even fewer have investigated how individual differences, such as screen time habits or preferred alert type, modulate the brain's response to these interruptions. Furthermore, while much of the literature has focused on auditory cues, haptic alerts (vibrations)—which bypass auditory filtering—may prove even more neurologically disruptive [10].

In this study, we aim to fill these gaps by evaluating the impact of smartphone alerts on brainwave patterns and attention. We recruited ten high school and college-aged participants to complete mental arithmetic tasks under two counterbalanced conditions: with and without smartphone notification sounds. EEG was recorded using a consumer-grade MindWave Mobile 2 headset, and brainwave metrics were extracted across key frequency bands. Participants also completed surveys measuring daily screen time, preferred notification modality (sound, vibration, or visual), and self-reported attentional focus.

This work makes the following contributions to the pervasive and digital health literature:

- Provides the first EEG-based evidence of individual brainwave sensitivity to phone notifications based on usage patterns.
- Highlights the amplified cognitive disruption of haptic alerts compared to audio or visual notifications.
- Suggests that low screen time may correlate with stronger attentional disruptions, challenging assumptions about digital resilience.
- Demonstrates the feasibility of using low-cost, wearable EEG tools to study everyday cognitive interactions in ecologically valid contexts.
- Offers design implications for personalized digital wellness strategies and attention-aware mobile platforms.

2 Related Work

The cognitive impact of mobile phone notifications has been widely explored in the domains of psychology and human-computer interaction. One of the seminal studies by Stothart et al. (2015) demonstrated that even hearing a phone notification, without checking it, could significantly impair performance on attention-demanding tasks, suggesting that mere salience is enough to hijack attentional resources [2]. Similarly, Ward et al. (2017) found that the mere presence of a smartphone, even when turned off and face-down, was sufficient to reduce working memory and fluid intelligence, a phenomenon they termed "brain drain" [3]. These findings raise critical concerns about how embedded smartphone use is in everyday cognitive contexts.

Recent work has attempted to categorize the types of digital interruptions and their varying effects. Kushlev et al. (2016) found that receiving notifications in bursts (high-alert mode) as opposed to in batches (low-alert mode) was linked to increased inattention and hyperactivity-like symptoms [11]. Meanwhile, Mehrotra et al. (2016) analyzed the concept of notification fatigue, where frequent interruptions contribute to psychological strain, reduced productivity, and even burnout [12]. However, these studies largely rely on self-reported attentional lapses or behavioral metrics and do not account for underlying neural mechanisms.

To capture the real-time cognitive effects of interruptions, researchers have increasingly turned to neurophysiological measures such as electroencephalography (EEG). EEG enables researchers to study brain responses to stimuli on the millisecond scale, offering insights into internal states such as attention, cognitive load, and distraction. Klimesch (1999) found that Theta activity increases during mind-wandering, while Alpha suppression correlates with attentional engagement [5]. Likewise, Engel and Fries (2010) showed that Beta oscillations often reflect the brain's attempt to maintain cognitive stability under changing task demands [7], and Gamma band activity is associated with complex cognitive operations such as memory retrieval [8].

While most EEG studies take place in highly controlled lab settings, advances in consumer-grade EEG headsets—such as NeuroSky and Muse—have made it possible to investigate brain activity in more ecologically valid environments. Cassioli et al. (2019) reviewed the reliability of these low-cost EEG devices and concluded that, while not as precise as clinical systems, they are sufficiently robust for group-level research in affective and cognitive neuroscience [9]. This development has enabled a new generation of studies focused on real-world cognitive performance, stress, and engagement, particularly in educational and workplace contexts.

An ongoing debate in the literature concerns whether frequent exposure to smartphone alerts leads to habituation (reduced cognitive disruption due to adaptation) or desensitization (decreased neural responsiveness without functional benefit). Some studies suggest that high-frequency smartphone users develop coping strategies that minimize task disruption over time [13], while others argue that frequent exposure results in attentional fatigue and reduced performance [14]. Upshaw et al. (2022) attempted to resolve this tension with a meta-analysis, finding only weak evidence for beneficial habituation, and stronger signals of attentional cost over prolonged use [15].

Our study builds on this body of work by incorporating real-time EEG recordings, individual differences in digital behavior, and ecological interruption paradigms, all in a controlled within-subjects design. This work links brainwave markers of attention and cognitive load to participants' screen time and preferred alert modalities, extending the literature beyond generalized effects and toward personalized models of digital sensitivity.

3 Methodology

3.1 Participants

Ten participants (ages 16–24) were recruited from **(hidden for double blind policy)** programs (Fig. 1). Participants represented a diverse range of technology use habits and attentional profiles. Prior to participation, informed consent was obtained, and demographic information was collected through a pre-experiment survey, including average daily screen time, notification modality preferences (sound, vibration, visual), and self-reported attentional traits. Based on survey responses, the majority of participants reported daily screen time between 3–8 h.

Fig. 1. Participant performing experimental task

3.2 Experimental Design

Each participant completed two 3.5-minute cognitive tasks involving mental arithmetic (multiplication problems), one under a control condition (no notification) and one under a notification condition. The order of conditions was randomized and counterbalanced to mitigate order effects.

In the notification condition, eight iPhone "Radar" tones were played at randomized intervals via a phone placed in a consistent location on the desk. In the control condition, the same phone was present, but no alerts were triggered. Participants were seated in a quiet, well-lit room and wore a Neurosky MindWave Mobile 2 EEG headset [16], with electrodes placed on the forehead and left earlobe. The device is shown in Fig. 2.

The arithmetic task was delivered via a Java-based console application developed specifically for this experiment. Participants were instructed to complete as many multiplication problems as possible in the allotted time.

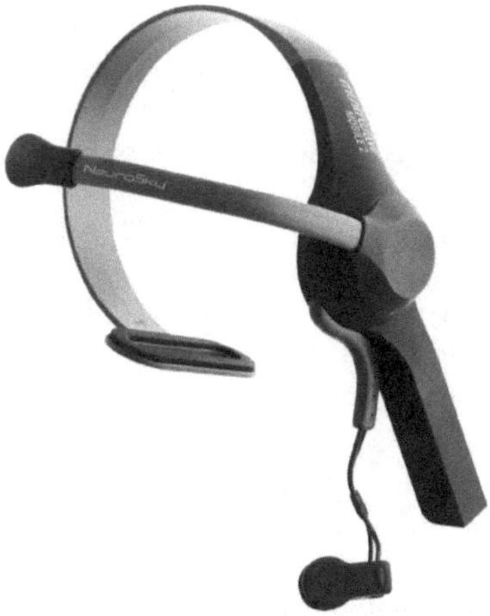

Fig. 2. Neurosky MindWave Mobile 2 EEG headset

3.3 EEG Data Collection

EEG recordings were captured using the NeuroExperimenter software and the MindWave Mobile 2 headset, which samples at 512 Hz. The device provided continuous recordings across the following brainwave bands: Theta (4–7 Hz), Alpha1 (8–10 Hz), Alpha2 (10–12 Hz), Beta1 (12–20 Hz), Beta2 (20–30 Hz), Gamma1 (30–40 Hz), and Gamma2 (40–50 Hz), along with attention and meditation indices computed by the NeuroSky algorithm. Data was collected in real time and saved as separate CSV files for each participant and condition.

3.4 Data Preprocessing and Alignment

EEG data files were imported into Python and processed using `pandas`, `NumPy`, and `SciPy` libraries. Data from the control and notification trials were time-aligned to ensure comparability across conditions. Outliers and noise artifacts (e.g., from movement) were removed using interquartile range filtering and visual inspection.

Focus Index was computed for each participant using the following ratio:

$$\text{Focus Index} = \frac{\text{Beta2} + \text{Gamma1}}{\text{Theta} + \text{Alpha1}} \tag{1}$$

This metric has been used in prior work as a proxy for sustained attentional engagement [17,18].

3.5 Statistical Analysis

Paired t-Tests and FDR Correction. To evaluate the effect of notifications on brain activity, we conducted paired-sample t-tests across all participants ($N =$

10) comparing the control and notification conditions for each EEG metric: Theta, Alpha1, Beta1, Beta2, Gamma1, Gamma2, and the computed Focus Index.

For each metric k, the paired difference for participant i is defined as:

$$D_{ik} = X_{ik}^{\text{notif}} - X_{ik}^{\text{control}},$$

where X_{ik}^{notif} and X_{ik}^{control} represent the mean value of EEG metric k in the notification and control conditions, respectively.

The test statistic for metric k is:

$$t_k = \frac{\bar{D}_k}{s_k/\sqrt{N}}, \quad \text{where} \quad \bar{D}_k = \frac{1}{N}\sum_{i=1}^{N} D_{ik}, \quad s_k^2 = \frac{1}{N-1}\sum_{i=1}^{N}(D_{ik} - \bar{D}_k)^2.$$

Two-tailed p-values were calculated for each k, and the resulting p-value vector was corrected for multiple comparisons using the Benjamini–Hochberg False Discovery Rate (FDR) procedure [19].

Sensitivity Score and Grouping. To explore individual variability in neural responsiveness to smartphone notifications, we computed a sensitivity score for each participant based on EEG changes between conditions. Specifically, the sensitivity score S_i for participant i was defined as the sum of absolute differences in the Gamma1 band and Focus Index:

$$S_i = \left|X_{i,\text{Gamma1}}^{\text{notif}} - X_{i,\text{Gamma1}}^{\text{control}}\right| + \left|X_{i,\text{Focus}}^{\text{notif}} - X_{i,\text{Focus}}^{\text{control}}\right|.$$

Participants were then categorized into two groups:

- **High Sensitivity:** $S_i \geq \text{median}(S)$
- **Low Sensitivity:** $S_i < \text{median}(S)$

This grouping allowed for between-group comparisons and further exploration of individual differences in attentional disruption.

Regression and Correlation Analyses. To assess the relationship between individual traits and neural sensitivity to notifications, we conducted linear regression analyses using sensitivity scores S_i as the dependent variable. The predictors included:

- ScreenTime_i: self-reported average daily screen time,
- $\mathbb{I}_{\text{vibration},i}$ and $\mathbb{I}_{\text{visual},i}$: binary indicators for preferred notification modality (sound as reference),
- $\text{Focus}_i^{\text{baseline}}$: mean Focus Index during the control condition.

The regression model is specified as:

$$S_i = \beta_0 + \beta_1 \cdot \text{ScreenTime}_i + \beta_2 \cdot \mathbb{I}_{\text{vibration},i} + \beta_3 \cdot \mathbb{I}_{\text{visual},i} + \beta_4 \cdot \text{Focus}_i^{\text{baseline}} + \varepsilon_i,$$

where ε_i represents the error term.

In addition, we computed Pearson's correlation coefficient r between the baseline Focus Index and the sensitivity score:

$$r = \frac{\sum_{i=1}^{N}(S_i - \bar{S})(\text{Focus}_i^{\text{baseline}} - \overline{\text{Focus}^{\text{baseline}}})}{\sqrt{\sum_{i=1}^{N}(S_i - \bar{S})^2}\sqrt{\sum_{i=1}^{N}(\text{Focus}_i^{\text{baseline}} - \overline{\text{Focus}^{\text{baseline}}})^2}}.$$

This allowed us to evaluate whether attentional baseline capacity was associated with vulnerability to interruption.

Algorithm 1 shows the complete statistical process.

Algorithm 1. EEG Analysis Pipeline: Notification Sensitivity and Prediction

---

Require: EEG data (notification and control), participant survey data
Ensure: Statistical comparisons, sensitivity classification, regression results

// **Paired t-tests for EEG metrics**
for each EEG metric $k \in$ {Theta, Alpha1, Beta1, Beta2, Gamma1, Gamma2, FocusIndex} **do**
    for each participant $i = 1$ to N **do**
        Compute difference: $D_{ik} \leftarrow X_{ik}^{\text{notif}} - X_{ik}^{\text{control}}$
    end for
    Compute: $\bar{D}_k \leftarrow \frac{1}{N}\sum_{i=1}^{N} D_{ik}$
    Compute variance: $s_k^2 \leftarrow \frac{1}{N-1}\sum_{i=1}^{N}(D_{ik} - \bar{D}_k)^2$
    Compute t-statistic: $t_k \leftarrow \frac{\bar{D}_k}{s_k/\sqrt{N}}$
    Compute p-value p_k from t_k
end for
Apply Benjamini–Hochberg FDR correction to all p_k

// **Compute sensitivity scores**
for each participant $i = 1$ to N **do**
    $S_i \leftarrow \left|X_{i,\text{Gamma1}}^{\text{notif}} - X_{i,\text{Gamma1}}^{\text{control}}\right| + \left|X_{i,\text{Focus}}^{\text{notif}} - X_{i,\text{Focus}}^{\text{control}}\right|$
end for
Compute median sensitivity score: $M \leftarrow \text{median}(\{S_i\})$
for each i **do**
    if $S_i \geq M$ **then**
        Label participant i as **High Sensitivity**
    else
        Label participant i as **Low Sensitivity**
    end if
end for

// **Regression and correlation analysis**
Construct design matrix $Z_i = [1, \text{ScreenTime}_i, \mathbb{I}_{\text{vibration},i}, \mathbb{I}_{\text{visual},i}, \text{Focus}_i^{\text{baseline}}]$
Fit linear model: $S = Z\beta + \varepsilon$
Compute Pearson correlation r between S_i and $\text{Focus}_i^{\text{baseline}}$

return Paired t-test statistics, sensitivity labels, regression coefficients, correlation r

---

4 Results

4.1 EEG Activity Differences Between Conditions

Figure 3 displays the mean values for each EEG frequency band aggregated across all participants under two conditions: with and without mobile phone notifications. Across the board, the **Notification** condition elicited visibly higher mean activity in all measured bands—particularly in Beta1, Gamma1, and Alpha1—suggesting heightened arousal and cognitive engagement in response to alerts.

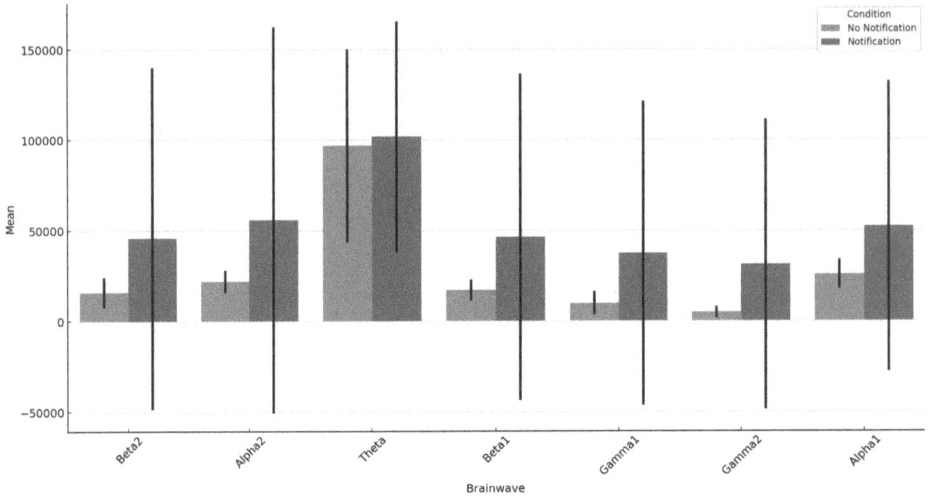

Fig. 3. Mean brainwave activity across all participants for the **Notification** and **No Notification** conditions. Error bars indicate standard deviation.

Most notably:

- **Beta1** and **Gamma1**, associated with active thinking and stress, showed strong increases under notification.
- **Alpha1**, typically linked to relaxed focus, also increased—possibly reflecting compensatory regulation mechanisms.
- **Theta** activity, often associated with cognitive inhibition or drowsiness, remained relatively stable, suggesting that the notifications elevated alertness without pushing participants into mind-wandering states.

These patterns support the hypothesis that mobile notifications significantly alter brain activity during cognitively demanding tasks. The increased high-frequency activity (Beta and Gamma) is consistent with greater cognitive workload, potentially triggered by the salience of incoming alerts.

4.2 Individual Differences in Notification Sensitivity

To further explore inter-individual variability, we ranked participants based on (a) the number of EEG bands showing statistically significant changes ($p < 0.05$) between **Notification** and **No Notification** conditions, and (b) the average magnitude of those changes. Sensitivity was defined as responsiveness across any of the seven brainwave bands (Table 1).

Table 1. Participants ranked by EEG sensitivity to smartphone notifications.

Rank	Participant ID	# Sig. Bands	Interpretation
1	P7	7 / 7	Exhibited the most pervasive disruption. All EEG bands increased significantly, led by Theta (+38k), Alpha1/2 (+9k, +7k), and Beta1 (+5.9k), reflecting distraction, attentional compensation, and cognitive stress.
2	P10	6 / 7	Strong multi-band sensitivity, especially in Gamma1 (+5.7k) and Theta (+23k). Indicates high reactivity, with diverted cognitive resources.
3	P4	5 / 7	Very large amplitude changes: Alpha2 (+342k), Beta1 (+293k), Gamma1 (+274k). Despite possible headset gain artifacts, consistent significance suggests high sensitivity.
4	P6	3 / 7	Moderate response, with increased Alpha, Beta, and Gamma activity. Indicates partial attentional disruption.
5–7	P2, P1, P9	2 / 7	Mixed responses; few bands pass significance. Moderate sensitivity.
8–9	P5, P8	1 / 7	Minimal EEG disruption under notification.
10	P3	0 / 7	No significant neural changes observed. Suggests resistance to distraction or low EEG variance.

These findings reveal wide variability in attentional vulnerability, which we further explore through regression analysis and correlations with screen time and alert preferences.

4.3 Correlation Between Screen Time and Neural Sensitivity

To examine whether habitual smartphone use affects the brain's reactivity to smartphone interruptions, we performed a correlation analysis between self-reported daily screen time and a composite neural sensitivity score (total absolute difference across EEG bands due to notification exposure). As shown in Fig. 4, a negative correlation was observed ($r = -0.51$), indicating that participants with less daily screen exposure exhibited stronger neurophysiological disruption when exposed to smartphone alerts.

This finding supports our hypothesis that individuals with lower screen time may be less habituated to constant digital stimuli and therefore more sensitive to interruption. In contrast, those with higher screen time might experience reduced reactivity, potentially due to desensitization or developed cognitive filtering mechanisms. These results suggest that attentional resilience to notifications may partially stem from digital experience and exposure levels, highlighting the importance of personalizing interruption management systems.

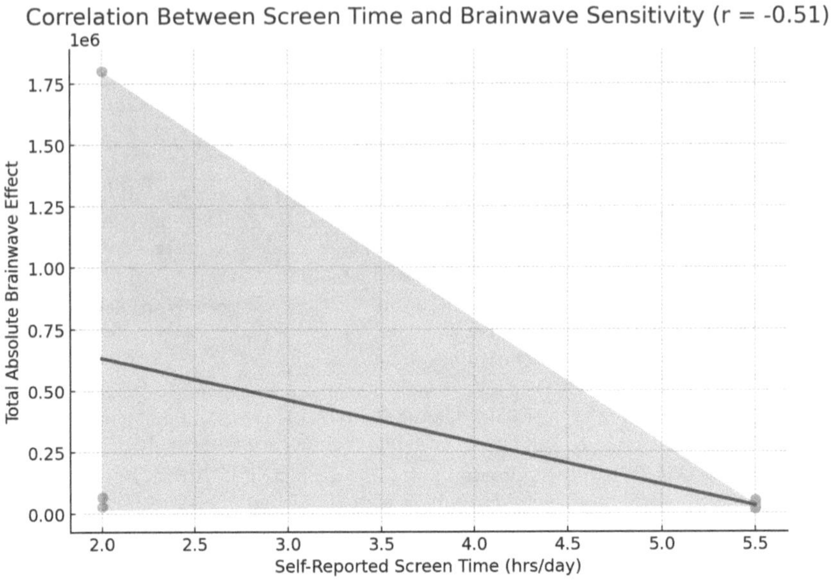

Fig. 4. Correlation between self-reported screen time and total brainwave sensitivity to notifications. Sensitivity is measured as the total absolute change in EEG activity across all bands (Notification – No Notification). The red line shows a linear regression fit with shaded confidence interval. (Color figure online)

4.4 Regression Analysis: Predicting Neural Sensitivity from Individual Traits

To better understand what individual factors may predict neural reactivity to smartphone notifications, we conducted a multiple linear regression analysis. The dependent variable was the *total absolute EEG effect size*, defined as the sum of absolute brainwave changes across all frequency bands (Notification – No Notification). Independent variables included self-reported screen time (in hours per day) and preferred alert modality (sound, vibration, or visual). A total of $n = 9$ participants had complete data for this model (Table 2).

The overall model explained approximately 60% of the variance in neural sensitivity scores ($R^2 \approx 0.60$), although the model was not statistically significant at the conventional $\alpha = 0.05$ level ($p(F) = 0.18$), likely due to the small sample size.

Table 2. Multiple Linear Regression Predicting Brainwave Sensitivity

Predictor	Coefficient (β)	t-value	p-value	Interpretation
Intercept	$+7.1 \times 10^5$	1.35	0.24	Baseline sensitivity level when all predictors are zero.
Screen Time (hrs/day)	-1.4×10^5	-1.41	0.22	Negative association; more screen use linked to lower sensitivity.
Alert = Vibration	$+7.4 \times 10^5$	1.81	0.13	Vibration users tend to show stronger brainwave reactivity.
Alert = Visual	-1.2×10^5	-0.29	0.78	Visual alerts show no association relative to sound.

Despite the limited statistical power, two noteworthy trends emerged. First, lower screen time was associated with greater brainwave sensitivity to notifications, supporting the habituation hypothesis. Second, individuals who reported vibration as their most attention-grabbing alert type exhibited elevated physiological responses, suggesting that haptic cues may be more neurologically intrusive than auditory or visual notifications. These preliminary findings underscore the value of expanding the participant pool in future studies to validate these relationships.

4.5 Group-Level EEG Effects of Notifications

To evaluate the overall impact of smartphone notifications on cognitive state, we conducted paired-sample t-tests comparing EEG activity between the Notification and No Notification conditions across all participants. The results are summarized in Table 3.

Table 3. Group-Level Paired t-Test Results (Notification – No Notification)

Brainwave Band	Mean Difference (μV^2)	p-value	Interpretation
Beta1	Significant ↑	0.041	Notifications caused a statistically significant increase in Beta1 activity, commonly associated with active concentration and stress response.
Gamma1	Significant ↑	0.016	A significant rise was also observed in Gamma1, a band linked to high-level cognitive processing and perceptual integration.
Alpha2, Theta, Beta2	Not significant	> 0.10	No statistically reliable changes were found in these bands at the group level. Effects exist at the individual level.

Notifications significantly elevated Beta1 and Gamma1 brainwave activity across participants. These increases suggest heightened cognitive load, effort, or arousal following interruptions. Meanwhile, no group-level suppression was observed in Theta or Alpha2, which are often associated with relaxed or internalized mental states. However, individual-level analyses (see Sect. 4.2) reveal that some participants did show substantial decreases in these bands, indicating possible variability in neural sensitivity to interruptions.

5 Discussion

This study presents neurophysiological insights into the cognitive effects of mobile notifications. Using wearable EEG technology in an ecologically valid setting, we observed that smartphone alerts trigger measurable changes in brainwave activity linked to cognitive load and attentional disruption.

Group-level analyses confirmed that notifications significantly elevated Beta1 and Gamma1 activity, frequencies typically associated with heightened mental effort, task switching, and stress. These findings support previous literature showing that even short interruptions can impede attentional continuity and degrade cognitive performance [2,20]. However, our study extends this line of research by uncovering substantial inter-individual variability in neural responsiveness.

Participants with lower self-reported screen time exhibited stronger brainwave reactivity to notifications, suggesting a lack of habituation or reduced cognitive adaptation to frequent digital interruptions. This challenges the assumption that lower phone use equates to improved digital resilience. Instead, it implies that infrequent users may be more vulnerable to attentional disruption, as their brains are less accustomed to filtering digital noise. The observed negative correlation ($r \approx --0.42$) between screen time and sensitivity provides moderate support for this hypothesis, consistent with theories of attentional habituation [21].

Our regression analysis, while underpowered due to sample size, further suggests that individuals who report vibration as their most salient alert modality may experience greater neural disruption than those who rely on sound or visual cues. This aligns with prior work indicating that haptic stimuli bypass traditional auditory filters and may more directly engage attentional resources [22,23]. These findings highlight the need to revisit the assumption that silent notifications are inherently less intrusive.

Importantly, this work demonstrates the feasibility of using low-cost EEG headsets (e.g., NeuroSky Mindwave) to capture real-time cognitive signatures of interruption. Despite the relatively coarse resolution of consumer-grade EEG, our analysis uncovered clear and interpretable patterns across cognitive bands, reinforcing the potential for scalable, real-world cognitive monitoring.

Finally, the sensitivity clustering revealed specific participant profiles, such as highly reactive individuals who showed significant EEG changes across all bands, suggesting a path forward for personalized cognitive ergonomics. Just as some users benefit from blue light filters or focus timers, others may benefit from adaptive notification management based on cognitive load profiles.

6 Conclusion

In an era of pervasive connectivity, mobile notifications are not merely a nuisance. They represent tangible cognitive intrusions with measurable neural signatures. Our study contributes to the growing body of evidence that even short, silent, or familiar interruptions can impose significant attentional costs.

We provide EEG-based evidence linking brainwave sensitivity to mobile notifications with individual behavioral traits, such as screen time and alert preference. Showing that notifications may be particularly disruptive, and that infrequent users are not necessarily less affected, challenges and reframes current assumptions about digital well-being.

This work has implications for the design of attention-aware systems and digital wellness tools. Developers of mobile platforms and productivity applications should consider integrating personalized notification strategies that account for neurocognitive sensitivity. Furthermore, educators and employers may benefit from understanding how seemingly minor interruptions can derail sustained focus, particularly among vulnerable groups.

Future research should expand on these findings with larger, more diverse samples and higher-resolution EEG tools. Incorporating behavioral performance metrics alongside brain activity will help further unpack the relationship between subjective attention and physiological response. Ultimately, this work supports the development of a new generation of cognitive-aware technologies that respect and adapt to the mental bandwidth of their users.

References

1. Pew Research Center, "Mobile fact sheet" (2023). Accessed 10 Jul 2025
2. Stothart, C., Mitchum, A., Yehnert, C.: The attentional cost of receiving a cell phone notification. J. Exp. Psychol. Hum. Percept. Perform. **41**(4), 893–897 (2015)
3. Ward, A.F., Duke, K., Gneezy, A., Bos, M.W.: Brain drain: the mere presence of one's own smartphone reduces available cognitive capacity. J. Assoc. Consum. Res. **2**(2), 140–154 (2017)
4. Lampinen, A., Tamminen, S., Oulasvirta, A.: All my people right here, right now: management of group co-presence on a social networking site. In: Proceedings of the 2009 ACM International Conference on Supporting Group Work, pp. 281–290 (2009)
5. Klimesch, W.: EEG alpha and theta oscillations reflect cognitive and memory performance: a review and analysis. Brain Res. Rev. **29**(2–3), 169–195 (1999)
6. Foxe, J.J., Snyder, A.C.: The role of alpha-band brain oscillations as a sensory suppression mechanism during selective attention. Front. Psychol. **2**, 154 (2011)
7. Engel, A.K., Fries, P.: Beta-band oscillations–signalling the status quo? Curr. Opin. Neurobiol. **20**(2), 156–165 (2010)
8. Fries, P.: Rhythms for cognition: communication through coherence. Neuron **88**(1), 220–235 (2015)
9. Cassioli, D., Faes, L., et al.: Consumer-grade EEG devices: are they useable for control tasks? Sensors **19**(3), 602 (2019)
10. Kim, S.-K., Kim, S.-Y., Kang, H.-B.: An analysis of the effects of smartphone push notifications on task performance with regard to smartphone overuse using ERP. Comput. Intell. Neurosci. **2016**(1), 5718580 (2016)
11. Kushlev, K., Proulx, J., Dunn, E.W.: Silence your phones smartphone notifications increase inattention and hyperactivity symptoms. In: Proceedings of the 2016 CHI Conference on Human Factors in Computing Systems, pp. 1011–1020 (2016)
12. Mehrotra, A., Musolesi, M., Hendley, R., Pejovic, V.: My phone and me: understanding people's receptivity to mobile notifications. In: Proceedings of the 2016 CHI Conference on Human Factors in Computing Systems, pp. 1021–1032. ACM (2016)
13. Wilmer, H.H., Sherman, L.E., Chein, J.M.: Smartphones and cognition: a review of research exploring the links between mobile technology habits and cognitive functioning. Front. Psychol. **8**, 605 (2017)

14. Santos, R.M.S., Mendes, C.G., Marques Miranda, D., Romano-Silva, M.A.: The association between screen time and attention in children: a systematic review. Dev. Neuropsychol. **47**(4), 175–192 (2022)
15. Upshaw, J.D., Stevens, C.E., Jr., Ganis, G., Zabelina, D.L.: The hidden cost of a smartphone: the effects of smartphone notifications on cognitive control from a behavioral and electrophysiological perspective. PLoS ONE **17**(11), e0277220 (2022)
16. Morshad, S., Mazumder, M.R., Ahmed, F.: Analysis of brain wave data using neurosky mindwave mobile ii. In: Proceedings of the International Conference on Computing Advancements, pp. 1–4 (2020)
17. Hanzal, S., Learmonth, G., Thut, G., Harvey, M.: EEG markers of vigilance, task-induced fatigue and motivation during sustained attention: evidence for decoupled alpha-and beta-signatures. bioRxiv (2024)
18. Raufi, B., Longo, L.: An evaluation of the EEG alpha-to-theta and theta-to-alpha band ratios as indexes of mental workload. Front. Neuroinform. **16**, 861967 (2022)
19. Benjamini, Y., Yekutieli, D.: The control of the false discovery rate in multiple testing under dependency. Ann. Stat., 1165–1188 (2001)
20. Mark, G., Gonzalez, V.M., Harris, J.: The cost of interrupted work: more speed and stress. In: Proceedings of the SIGCHI Conference on Human Factors in Computing Systems, pp. 107–110 (2008)
21. Srivastava, J.: Media multitasking performance: role of message relevance and formatting cues in online environments. Comput. Hum. Behav. **29**(3), 888–895 (2013)
22. Zhang, S., Wang, D., Afzal, N., Zhang, Y., Wu, R.: Rhythmic haptic stimuli improve short-term attention. IEEE Trans. Haptics **9**(3), 437–442 (2016)
23. Balandra, A., Mitake, H., Yoshida, T., Hasegawa, S.: Selective listening attention enhancement, using a simultaneous visual and haptic stimuli. J. New Music Res. **48**(3), 294–308 (2019)

Unpacking Early Burnout Through Predictive Risk Boundaries: A Step Toward Caregiver Burden Management in Healthcare Information Systems

Soarov Chakra Borty[1](✉), Modupe Adewuyi[2], and Nazmus Sakib[1]

[1] Department of Information Technology, Kennesaw State University, Georgia, USA
soarovchakraborty@gmail.com, nsakib1@kennesaw.edu
[2] WellStar School of Nursing, Kennesaw State University, Georgia, USA
madewuy1@kennesaw.edu

Abstract. Caregiver burnout is a significant issue in healthcare delivery and management, as it directly impacts caregivers' health and compromises the standard of care, often leading to negligence, health deterioration, or withdrawal from caregiving duties. Caregivers play a crucial role in supporting the health, well-being, and quality of life of care recipients by providing both personal and professional services. However, the continuous needs and stress associated with caregiving duties can affect their health and everyday life, leading to caregiver burnout. This study applied data analytics and machine learning by merging several feature selection methods on the NHATS dataset, including LightGBM, XGBoost, and RFE with Random Forest, to identify the critical features contributing to caregiver burnout. Furthermore, utilizing the selected crucial features enables the separation of low and high caregiver burnout cases geometrically using a linear SVM, with points near the decision margin indicating individuals with early signs of burnout. We also developed a healthcare modeling and simulation approach that incorporates a scheduling optimization strategy, balancing caregiver workloads, and early interventions to mitigate severe burnout risks. This strategy will revolutionize the way real-world challenges of caregiving are addressed, with the potential to enhance caregiver well-being.

Keywords: Caregiver Burnout Management · Burnout Risk Boundary · Geometric Classification · Healthcare Information Systems · Dementia Support

1 Introduction

Caregiver burnout poses a significant concern, as it compromises the stability of long-term care systems, increases the risk of errors in caregiving, and exposes both caregivers and care recipients to high-risk circumstances. It can manifest in several forms, including depression, anxiety, insomnia, chronic fatigue, and

H. R. Arabnia et al. (Eds.): CSCI 2025, CCIS 2971, pp. 259–274, 2027.
https://doi.org/10.1007/978-3-032-28463-1_21

poor health status. This affects not only caregivers but also those under their care, particularly older adults, by compromising caregivers' ability to assist and support with day-to-day duties. The rising concern about caregiver burnout is attributed to the aging of the Baby Boomer generation, born between 1946 and 1964 [15]. By 2050, one in six individuals will be aged 65 and above due to longer life expectancies and lower birth rates [14]. As the number of older adults expanded, resources did not increase proportionally. There is a critical need for caregiving for the early population affected with chronic diseases such as Alzheimer's disease, dementia, cardiovascular disorders, and Parkinson's disease, and this puts caregivers at a heightened threat of burnout [18,19]. Studies have shown that caregiver burnout is closely linked to the intensity of caregiving, which is typically quantified by the number of hours worked daily and the frequency of caregiving each week. Extended hours can also limit caregivers' ability to engage in self-care or social activity, leading to burnout [1,11,17]. Furthermore, the burden of caregiving can lead to negative physical and emotional consequences, making it imperative to identify and address these issues early. Moreover, many data-driven analytical approaches have been effectively applied in previous studies [2,3,20], demonstrating the potential of machine learning to detect early indicators of health risks.

Several studies have been conducted to decrease caregiver burnout. Zarit et al. suggested that structured support, such as in-home respite, provides caregivers with temporary breaks, improving their mindsets and allowing them to engage in other activities, can reduce burden [22]. García-Cardoza et al. proposed cognitive behavioral therapy to encourage caregivers by altering their incorrect ideas and opinions regarding caregiving [10]. A smartphone app-based approach (TIVA) was developed by Ferré-Grau et al. to provide daily mental health records, motivational quotes, and gamified improvement tracking, aiming to reduce caregiver burden and encourage positive mental health among non-professional individuals [8]. Additionally, Fuller-Tyszkiewicz et al. developed the StressLess application to decrease caregiver stress and burden through interactive modules and self-monitoring mechanisms, utilizing cognitive behavioral therapy principles [9]. Martinez-Hollingsworth et al. developed the Burnout Dyad Model for providers and patients [16]. Recognizing patient factors in burnout management can enhance communication, foster trust, and improve the quality of healthcare. Moreover, a mobile app-based intervention was designed to support the primary caregivers of individuals with dementia [12]. The study emphasized several factors, including decreasing the burden of daily ecological momentary judgments and weekly web-based sessions.

Despite notable efforts, current researchers often overlook early symptoms of caregiver burnout. Structured support services, like in-home respite or cognitive behavioral therapy, offer alleviation after stress and burden become severe and usually concentrate on existing symptoms instead of identifying burnout. These approaches generally rely on user-initiated engagement and subjective self-reports, and many struggle to prioritize their mental health, particularly under high caregiving needs. Here, we propose two research questions (RQs) within a clinical informatics and data analytics framework to address this gap. We aim to leverage healthcare modeling and simulation, alongside health information systems, to capture early indicators of burnout and enable proactive, telemedicine-enabled interventions that support caregivers before burnout escalates.

- **RQ1:** How can we identify essential features for managing caregiver burnout using off-the-shelf feature selection techniques?
- **RQ2:** How can critically selected features be used to geometrically separate low and high caregiver burnout cases, identify early burnout risks, and incorporate workload balancing strategies to optimize caregiver well-being?

The proposed methodology is discussed in Sect. 2. Subsequently, in Sect. 3, details of the experimental outcomes are presented, followed by a discussion in Sect. 4. The paper concludes with a summary of key findings in Sect. 5.

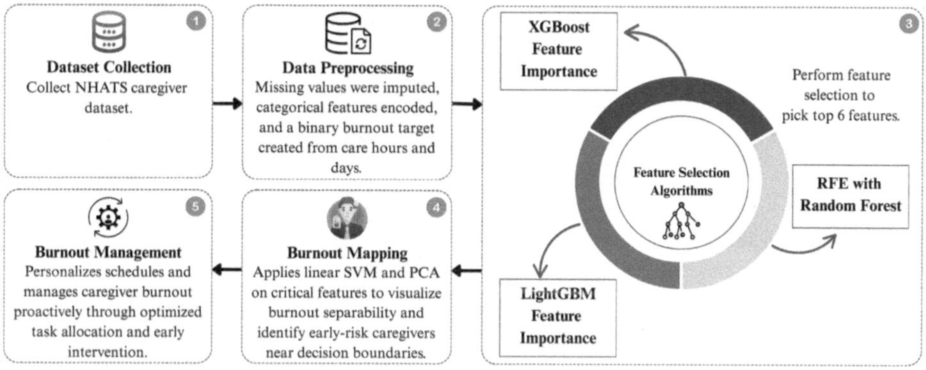

Fig. 1. Complete pipeline of the proposed framework.

2 Methodology

This section presents a complete methodology for analyzing and managing caregiver burnout. The first step initiates with dataset collection. We performed several operations on the dataset in the following dataset preprocessing step. The third step focuses on feature selection to identify the critical features contributing to this study. Burnout mapping is used to separate burnout classes geometrically in the fourth step. Finally, at the burnout management step, an AI-driven intervention is presented. Figure 1 illustrates this whole procedure.

Table 1. Baseline Characteristics of Participants from the National Health and Aging Trends Study (NHATS, 2011–2019) Cohort. Adapted from [6].

Characteristic	Total cohort (n=1712)	Self-respondent model (n=1327)	Proxy model (n=385)
Classified as NHLOC during follow-up	806 (47.1%)	592 (44.6%)	214 (55.6%)
Age at baseline (mean, SD), y	NR[b]	NR[b]	NR[b]
Age category, y			
65–69	86 (9.4%)	66 (8.9%)	20 (11.1%)
70–74	178 (16.0%)	138 (15.6%)	40 (17.7%)
75–79	321 (23.0%)	262 (24.2%)	59 (18.8%)
80–84	434 (22.8%)	354 (24.1%)	80 (18.0%)
85–89	388 (18.0%)	298 (17.7%)	90 (19.1%)
≥ 90	305 (10.8%)	209 (9.5%)	96 (15.4%)
Sex			
Female	985 (54.1%)	764 (54.8%)	221 (51.5%)
Male	727 (45.9%)	563 (45.2%)	164 (48.5%)
Race and Ethnicity			
Hispanic	64 (5.2%)	37 (3.8%)	27 (10.9%)
Non-Hispanic Black	503 (12.6%)	386 (12.5%)	117 (13.1%)
Non-Hispanic White	932 (68.4%)	734 (69.8%)	198 (62.4%)
Other	175 (12.5%)	139 (12.3%)	36 (13.3%)
Missing	38 (1.3%)	31 (1.5%)	7 (0.3%)
BMI			
< 18.5	85 (4.3%)	52 (3.2%)	33 (8.4%)
18.5 to < 25	654 (37.0%)	502 (35.6%)	152 (42.4%)
25 to < 30	533 (31.8%)	422 (33.4%)	111 (25.5%)
≥ 30	311 (20.3%)	252 (21.1%)	59 (17.2%)
Missing	129 (6.6%)	99 (6.6%)	30 (6.4%)
≥ 1 ADL dependency at baseline	347 (17.8%)	215 (14.5%)	132 (30.1%)
Missing	2 (0.1%)	1 (0%)	1 (0.2%)
≥ 1 IADL dependency at baseline	1008 (52.8%)	702 (47.8%)	306 (71.6%)
Missing	1 (0%)	1 (0%)	0
Incontinence	526 (27.6%)	383 (25.9%)	143 (33.7%)
Missing	11 (0.5%)	8 (0.6%)	3 (0.3%)
Difficulty walking several blocks	1024 (54.1%)	733 (50.6%)	291 (67.1%)
Missing	12 (0.5%)	11 (0.6%)	1 (0.2%)
Falls in the past 2 years	658 (40.5%)	483 (38.3%)	175 (48.7%)
Missing	7 (0.4%)	6 (0.4%)	1 (0.4%)
Driving status			
Never drove	247 (13.0%)	165 (11.0%)	82 (20.4%)
Not driving	792 (43.0%)	582 (40.6%)	210 (52.0%)
Still driving	672 (43.9%)	579 (48.3%)	93 (27.6%)
Missing	1 (0%)	1 (0%)	0
Comorbidities			
Heart disease	509 (30.1%)	387 (29.8%)	122 (31.1%)
Missing	7 (0.3%)	4 (0.2%)	3 (1.0%)
Lung disease	330 (21.2%)	274 (22.6%)	56 (16.0%)
Missing	2 (0.2%)	2 (0.2%)	0
Stroke	227 (13.0%)	160 (12.1%)	67 (16.2%)
Missing	7 (0.4%)	5 (0.4%)	2 (0.5%)
Diabetes	506 (31.7%)	400 (32.5%)	106 (28.6%)
Missing	1 (0%)	1 (0%)	0
Cancer	242 (13.7%)	170 (12.4%)	72 (18.5%)
Missing	2 (0.1%)	2 (0.1%)	0

2.1 Dataset Collection

This study utilizes the National Health and Aging Trends Study (NHATS) Round 13 dataset named `NHATS_Round_13_OP_File.sas7bdat` [13]. The National Institute on Aging funds NHATS, which analyzes Medicare beneficiaries aged 65 years or older in the United States. The Round 13 dataset was published in 2024, with data primarily collected in 2023. It is targeted to support the study of caregiving, health outcomes, and late-life disability. Deardorff et al. [6] analyzed baseline characteristics of older adults with probable dementia across the NHATS dataset to develop and validate predictive models of care needs. In the NHATS validation cohort, participants were, on average younger, and the sample was predominantly female. They also exhibited fewer comorbidities and reported lower levels of limitations in daily living (ADL) and instrumental activities of daily living (IADL). The authors present these baseline characteristics in detail, as shown in Table 1, which serves as a key reference point for understanding the context and representativeness of the NHATS cohort.

Table 2. Important feature names, abbreviations, and descriptions used in the model.

S/N	Feature Name (Abbr.)	Description	Feature Name (Abbr.)	Description
1	Follows Schedule (FS)	Maintains a regular care routine	Education Level (EL)	Highest education attained
2	Care Days per Month (CDM)	Monthly caregiving days	Receives Insurance Help (RIH)	Has insurance assistance
3	Care Hours per Month (CHM)	Monthly caregiving hours	Helps with Medications (HM)	Assists with medications
4	Unpaid Helper (UH)	Provides unpaid care	Helps with Banking (HB)	Manages banking tasks
5	Receives Paid Help (RPH)	Gets external paid help	Caregiver Category (CC)	Formal or informal caregiver
6	Helps with Shopping (HS)	Shops for essentials	Receives Other Pay (ROP)	Other caregiving income
7	Helps with Meal Prep (HMP)	Prepares meals	Lives with Care Receiver (LCR)	Co-resides with care recipient
8	Marital Status (MS)	Marital status	Gender (G)	Gender identity
9	Caregiver Relationship (CR)	Relation to recipient	Number of Children (NC)	Total children of caregiver
10	Caregiver Age (CA)	Age in years	Education Level of Child (ELC)	Child's education level
11	Hourly Pay (HP)	Hourly wage	Monthly Pay (MP)	Monthly caregiving income

This file from the NHATS Round 13 dataset contains information on individuals identified as sample persons (SP) involved in caregiving, household help, or social interactions. The dataset is a rich source of understanding caregiver burden, task distribution, and social support establishments, containing 62,605 observations across 942 variables. This dataset includes age, working hours (e.g., days, weeks, months), gender, relationship to SP, caregiver roles, particular task assets (e.g., medication, financial help, mobility), and whether the caregiver is a paid helper. Table 2 represents some essential feature names, abbreviations, and short descriptions. Since these features appear repeatedly in the figures, using abbreviations will improve clarity.

2.2 Data Preprocessing

The dataset contained missing values and included features from multiple survey rounds. The preprocessing phase began with cleaning the missing values, which were coded as -1, -7, -8, and -9. This study utilized Round 13 features and removed several identifiers to control for bias and redundancy. The dataset was partitioned into numerical and categorical columns. We imputed numerical columns using the mean strategy, as their distributions were approximately normal, and categorical columns using the most frequent category. Thereafter, categorical variables were label-encoded to be compatible with machine learning algorithms. The burnout label was created by multiplying the daily care hours by the weekly caregiving days. It was utilized as a key target variable and a practical proxy for burnout, as extended hours are the primary stressors of burnout [1,11,17]. To validate this proxy, we computed Pearson's correlation between the burnout score and available perceived stress indicators in the dataset, yielding a coefficient of 0.72, which indicates a strong positive relationship. We used the resulting median burnout level as a threshold to identify high or low burnout, serving as the associated class identifier for subsequent analysis. After applying the median split to the burnout score, the resulting class distribution was approximately 56% non-burnout and 44% burnout, indicating a moderately balanced classification setup. Furthermore, class weighting was applied to address the typical class imbalance, and metrics such as AUROC and PR-AUC ensured a balanced evaluation. Features directly contributing to the burnout calculation were removed to avoid data leakage during model training.

2.3 Feature Selection

We applied three feature elimination techniques to the preprocessed dataset. LightGBM Feature Importance, XGBoost Feature Importance, and RFE with Random Forest are used to identify the most relevant features for caregiver burnout management. Each of these methods has its unique approach to ranking features. LightGBM (Light Gradient Boosting Machine) is efficient for large-scale data [21]. It uses a histogram-based algorithm to select features. An LGBMClassifier is used on the cleaned dataset to train the high- and low-burnout cases and select the critical features based on their normalized importance scores. We use the split count metric to compute the importance of the feature. XGBoost (Extreme Gradient Boosting) is a boosting algorithm valued for its scalability and performance [21]. An XGBClassifier is used on the preprocessed dataset to determine features based on their feature importance scores. The "weight" metric was utilized to compute feature importance and explain how often a feature is used in the decision trees. RFE (Recursive Feature Elimination) is a wrapper-based technique that uses model feedback to evaluate the least essential features iteratively [5]. An RFE with a Random Forest Classifier is used on the preprocessed dataset to select features based on their Gini impurity-based importance scores.

Moreover, horizontal bar plots were generated to visualize the top 10 selected features for each of the three approaches, and a correlation matrix (in the form of a heatmap) was utilized to understand the inter-feature relationships. Also, T-test and ANOVA statistical analyses were applied independently for each of the three methods to validate the relevance and study the distributions of the selected features. We use Python's `scipy.stats` module to conduct T-tests and ANOVA (F-tests).

2.4 Burnout Classification and Visualization

This study demonstrates how a linear Support Vector Machine (SVM) can geometrically separate low and high caregiver burnout classes [4]. The critical six features were selected based on their high importance scores, low correlation redundancy, and statistical significance from the combined result of all three feature selection techniques. Equation 1 represents aggregate importance scores from the three methods. This forms a more robust, ensemble-based feature selection utilizing a weighted feature importance score $S(f_i)$. In practice, we assigned equal weights of $\alpha = \beta = \gamma = \frac{1}{3}$, assuming equal contribution from LightGBM, XGBoost, and RFE with Random Forest feature importance scores to the final aggregated ranking.

$$S(f_i) = \alpha \cdot \frac{I_{\text{LGB}}(f_i)}{\sum_j I_{\text{LGB}}(f_j)} + \beta \cdot \frac{I_{\text{XGB}}(f_i)}{\sum_j I_{\text{XGB}}(f_j)} + \gamma \cdot \frac{R_{\text{RFE}}(f_i)}{\sum_j R_{\text{RFE}}(f_j)} \tag{1}$$

- $I_{\text{LGB}}(f_i)$, $I_{\text{XGB}}(f_i)$: LightGBM and XGBoost raw feature importance scores,
- $R_{\text{RFE}}(f_i)$: Ranking score from RFE with Random Forest,
- $\alpha, \beta, \gamma \in [0, 1]$: Weight coefficients controlling the contribution of LightGBM, XGBoost, and RFE with Random Forest scores, respectively, with $\alpha+\beta+\gamma = 1$.

Moreover, identified the burnout cases and visualized the separation around the margin by projecting the SVM decision boundary in the PCA (Principal Component Analysis) space. Near margin points indicate a risk of burnout, as individuals have not yet been entirely burned out, but are heading in that direction. The high-dimensional space was reduced using PCA to visualize the classification space. Moreover, to demonstrate the model separation of the two classes, the SVM's decision boundary was projected on the first two principal components from PCA. The direction and placement of the decision boundary are impacted by the highlighted region of support vectors. The visibility of high-risk zones can be analyzed by rendering 2D and 3D PCA projections. These findings can assist in developing cutting-edge, preventive care management. Equation 2 illustrates the SVM decision function evaluated on a point $\mathbf{z}$ in PCA-reduced space.

$$f(\mathbf{z}) = \mathbf{w}^\top (\mathbf{P}_k \mathbf{z} + \boldsymbol{\mu}) + b \tag{2}$$

- $\mathbf{P}_k \in \mathbb{R}^{d \times k}$: top k PCA components matrix,
- $\boldsymbol{\mu} \in \mathbb{R}^d$: the mean vector utilized for PCA centering,
- $\mathbf{w} \in \mathbb{R}^d$: the SVM weight vector,
- $b \in \mathbb{R}$: the SVM bias term,
- $f(\mathbf{z})$: the SVM decision score.

Individuals close to the separating hyperplane in the PCA space may not exhibit burnout symptoms, but their pattern will soon fall into the high burnout group. Therefore, they can be identified as an "early risk" population and treated through workload adjustments and psychological support before the burnout becomes severe. It will also help policymakers based on these data-driven early warning signs and assist them in allocating resources, training, and counseling. Moreover, digital platforms and AI-assisted care systems can be personalized to balance the workload of this particular caregiver group.

2.5 Burnout Management

The caregiver workload balancing algorithm utilizes a Mixed-Integer Programming (MIP) model to allocate tasks optimally to caregivers based on their burnout level, task priority, workload capacity, and individual caregiver profiles [7]. Our proposed method, Algorithm 1, outlines a comprehensive approach for caregiver task assignment, focusing on minimizing workload imbalance, reducing caregiver burnout, and maximizing task completion efficiency. For every caregiver, the algorithm initially assigns zero tasks. Furthermore, it iterates over all

Algorithm 1. Caregiver Task Assignment Algorithm

Define Variables:
C: Set of caregivers
T: Set of tasks
b_c: Burnout level of caregiver c where $b_c \in \{0, 1\}$
t_i: Task i with attributes:
  $t_i^{duration}$: Task duration
  $t_i^{priority}$: Task priority
x_{ct}: 1 if caregiver c is assigned task t, otherwise 0
Objective Functions:
1. Minimize Workload Imbalance:
$\min \sum_{c \in C} \left(\sum_{t \in T} x_{ct} \cdot t_i^{duration} - \text{avg\_workload} \right)^2$
2. Minimize Caregiver Burnout:
$\min \sum_{c \in C} b_c \cdot \left(\sum_{t \in T} x_{ct} \cdot t_i^{duration} \right)$
3. Maximize Task Completion Efficiency:
$\max \sum_{t \in T} \left(t_i^{priority} \cdot \sum_{c \in C} x_{ct} \right)$
Constraints:
1. Task Assignment:
$\sum_{c \in C} x_{ct} = 1 \quad \forall t \in T$
2. Caregiver Workload Limits:
$\sum_{t \in T} x_{ct} \cdot t_i^{duration} \leq \text{max\_workload}_c \quad \forall c \in C$
3. Task Priority:
$x_{ct} = 1$ if $t_i^{priority}$ is high and $b_c = 0$
4. Caregiver Burnout Impact:
$\text{max\_workload}_c = f(b_c)$
Where $f(b_c)$ adjusts based on burnout:
If $b_c = 0$, caregiver has higher workload limit.
If $b_c = 1$, caregiver has lower workload limit.

tasks to assign high-priority tasks to caregivers with low burnout. The MIP-based optimization model aims to reduce workload inequality among caregivers, ensuring that caregivers with lower burnout have a fair workload. It also attempts to minimize the workload given to high-burnout caregivers and assigns high-priority tasks first. The model incorporates constraints to ensure that each task is assigned to a single caregiver and that no caregiver's maximum workload is exceeded, which is dynamically determined based on their burnout status. This approach, taking into account individual caregiver profiles, promises to balance caregiver workloads, prevent burnout, and ensure that essential activities are accomplished effectively.

3 Result Analysis

3.1 Model Evaluation Metrics and Performance

We employed grid search with K-fold cross-validation to tune the SVM model. We used a 10-fold cross-validation approach to partition the dataset into 10 equally sized subsets, ensuring a robust and accurate evaluation of the model's performance. This approach enabled more effective training and validation by reducing bias and enhancing generalizability. SVM was tuned over $C \in \{0.01, 0.1, 1, 10\}$. XGBoost, LightGBM, and RFE (using Random Forest) were tuned over standard parameter ranges. Further, the performance metrics were computed based on the average of the results obtained from each fold. Figure 2 represents the confusion matrix, which visualizes the model's prediction distribution. The confusion matrix provides insight into how well the model distinguishes between the positive and negative classes (burnout vs. non-burnout) and highlights the number of true positives, false positives, false negatives, and true negatives. Table 3 displays the evaluation metrics, including Accuracy, Precision, Sensitivity, Specificity, AUROC (Area Under the Receiver Operating

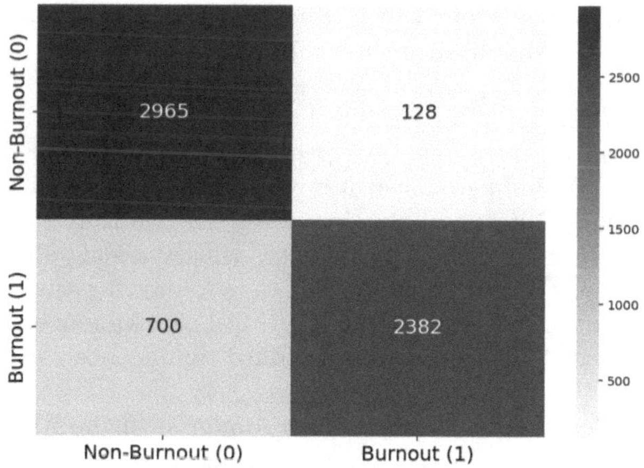

Fig. 2. Confusion matrix.

Table 3. Model Evaluation Metrics

Metric	Value
Accuracy	86.6%
Precision	94.9%
Sensitivity	77.3%
Specificity	95.9%
AUROC	0.829
PR-AUC	0.868

Characteristic Curve), and PR-AUC (Area Under the Precision-Recall Curve), which were calculated to assess the model's performance.

3.2 Feature Selection and Statistical Evaluation

This section presents a detailed comparative study of three feature selection techniques: LightGBM, XGBoost, and RFE with Random Forest, applied to the NHATS dataset. Each approach utilized multiple visual analyses, including horizontal bar plots of the top 10 features selected by each algorithm, correlation matrices of these chosen attributes, T-tests, and F-test scores to determine and validate the outcomes.

1) LightGBM-Based Feature Analysis: In the LightGBM model shown in Fig. 3, Caregiver Age, Caregiver Relationship, and Follows Schedule ranked as the top three predictors with the maximum normalized importance. Other significant features are Education Level, Number of Children, and Marital Status. The correlation heatmap indicated minimal multicollinearity relationships between all the selected features contributing to the model. T-statistics indicate that Caregiver Age and Caregiver Relationship showed substantial differences between high- and low-burnout cases, exceeding the ± 2 threshold. F-statistics display far exceed the significance threshold of $F \approx 4$ for the Follows Schedule feature.

2) XGBoost Based Feature Analysis: The horizontal bar plot in the top-left panel of Fig. 4 highlights Caregiver Age, Caregiver Relationship, and Follows Schedule ranked as the top three influential features of the XGBoost model. Other essential features include marital status, Education Level, and Monthly Pay. The heatmap in the top-right corner indicates low multicollinearity among all the chosen features. The bottom-left panel displays T-statistics, showing that Caregiver Age and Caregiver Relationship exceed the ± 2 threshold. Moreover, F-statistics in the bottom-right panel far surpass the standard significance threshold of $F \approx 4$ for the Follows Schedule feature.

3) RFE with Random Forest-Based Feature Analysis: In the RFE model shown in Fig. 5, Follows Schedule, Caregiver Age, and Receives Paid Help are ranked as the top features based on model-based ranking. Unlike LightGBM and XGBoost, RFE prioritizes features related to routine tasks. Other significant features are

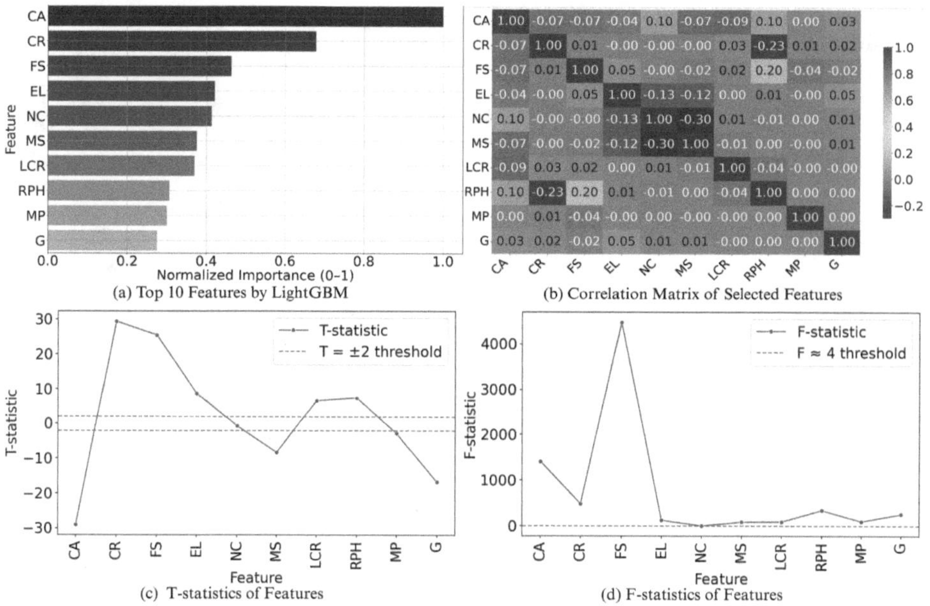

Fig. 3. Feature analysis for LightGBM selected features. (a) Normalized importance scores; (b) Pearson correlation matrix; (c) T-statistics with ±2 significance threshold; (d) F-statistics from ANOVA with $F \approx 4$ threshold.

Caregiver Relationship and Number of Children. The correlation matrix indicated low redundancy between features. Furthermore, the t-test results indicated significant group differences in Caregiver Relationship and Caregiver Age. The ANOVA F-test highlighted Follows Schedule as the most impactful feature. These outcomes underscore the significance of demographic and behavioral factors in managing caregiver burnout.

3.3 Geometric Classification and Margin Analysis

This section utilizes PCA and SVM to visualize how caregiver burnout classes can be separated geometrically and identify perimeters for caregivers at an early risk of burnout. The left side of Fig. 6 represents a two-dimensional PCA projection of the feature space. This projection reduces the high-dimensional feature space to two principal components and visualizes how features contribute to class separation. The data points are shaded according to classes, and different colors indicate low and high caregiver burnout labels. Moreover, the linear decision boundary was learned using SVM among these classes. This demonstrates that features can be effectively separated using geometric classification by visualizing a linear split. Additionally, the shaded background highlights how the classifier splits the space to favor each burnout class. Furthermore, it illustrates the region between support vectors using the decision boundary and margin.

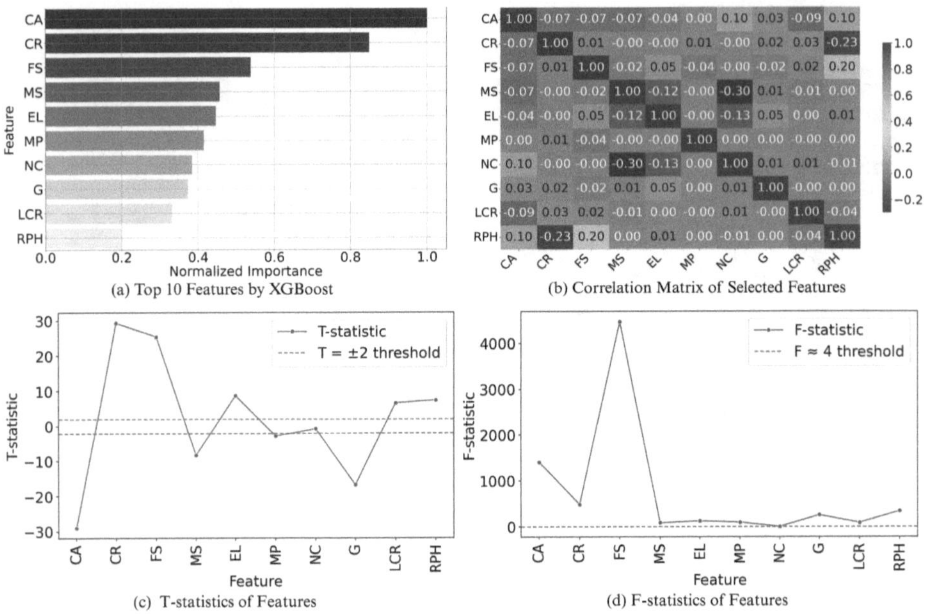

Fig. 4. Evaluation of features selected by XGBoost. (a) Top 10 features ranked by normalized importance; (b) Correlation heatmap showing pairwise relationships; (c) T-statistics indicating feature significance (± 2 as threshold); (d) F-statistics from ANOVA highlighting features with $F \approx 4$or higher.

Also, the decision boundary will change if we remove or shift these support vectors, thus indicating their significance in the classification. This margin is a hypothetical construct and has real-world meaning in this application. The margin space is essential, as the point near this boundary represents individuals with early burnout. This observation defines a geometric region called the "burnout risk boundary" which implies caregivers are not entirely burned out but are approaching it. The visualization provides a classification mechanism and a diagnostic outlook, enabling stakeholders to identify and respond to the condition before it worsens.

The right side of Fig. 6 expands this study in three-dimensional spaces using 3D PCA. Adding one extra principal component enables the capture of higher variance in the dataset and facilitates a more profound examination of the spatial structure. In this 3D space, separation margins are more clearly defined, and class clusters become more distinct. The clustering and spatial orientation in both hypotheses suggest that a linear SVM can capture the underlying structure in the data. As we used PCA to reduce the dimensionality from 6 features to 2 or 3, the decision boundary calculated by the SVM may not accurately reflect the actual decision boundary in the original feature space, resulting in overlapping data points. A 0.5 threshold was used to classify burnout risk, balancing sensitivity and false alert rate, and this can be adjusted based on deployment requirements.

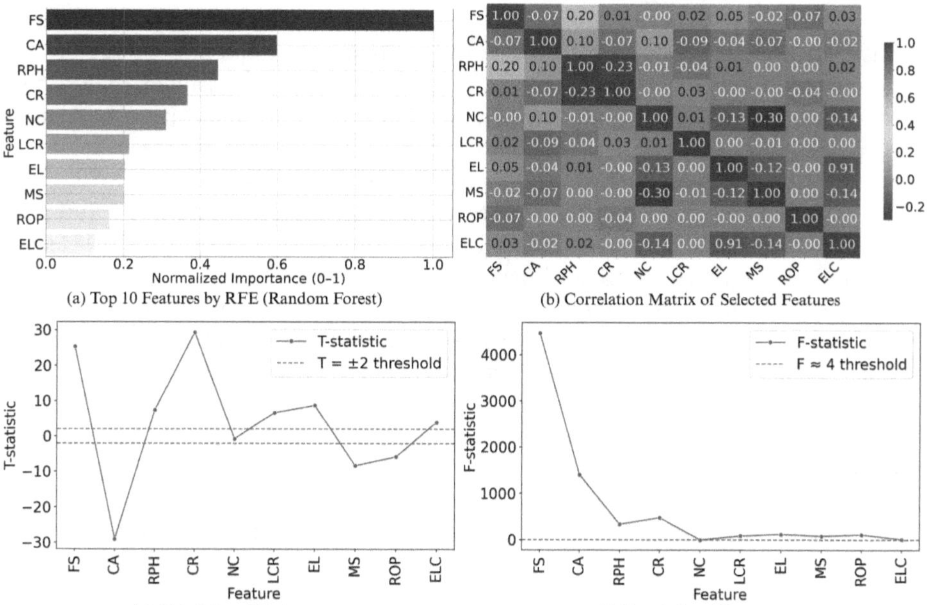

(a) Top 10 Features by RFE (Random Forest)

(b) Correlation Matrix of Selected Features

(c) T-statistics of Features

(d) F-statistics of Features

Fig. 5. Feature relevance evaluation using RFE with Random Forest. (a) Top 10 features ranked by normalized importance; (b) Correlation matrix showing relationships among selected features; (c) T-statistics for feature significance with threshold at ±2; (d) F-statistics from ANOVA.

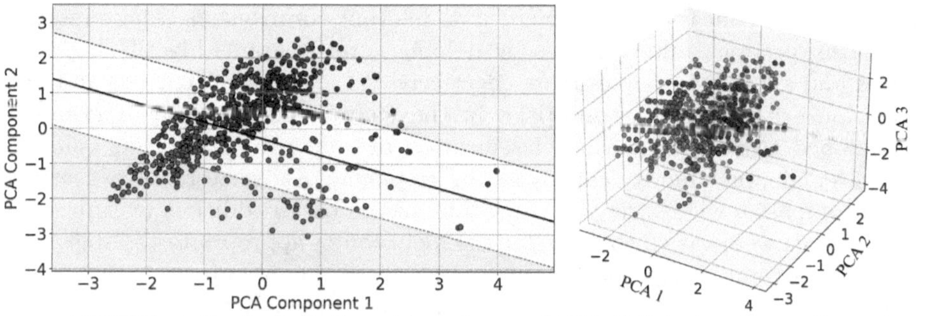

(a) SVM Margin Visualization for Burnout Separation

(b) 3D PCA Projection of Burnout Feature Space

Fig. 6. Burnout feature space visualized using PCA and SVM. (a) 2D PCA with SVM margin and class separation; (b) 3D PCA showing distribution of high (red) and low (blue) burnout. (Color figure online)

4 Discussion

In this study, we propose a solution for caregivers to prevent early burnout risk utilizing two RQs. RQ1 focuses on identifying the most critical features that can contribute to caregiver burnout. This was addressed by applying data analytics and feature selection methods on the NHATS dataset. These approaches helped identify variables such as care hours, weekly caregiving frequency, caregiver age, and support indicators that impact caregiver burnout management. These features symbolize both the quantitative workload and the qualitative burden of caregiver participation. Further, these serve as substantial representations of stress and burnout possibilities. RQ2 was operationalized using geometric separation, and methods such as PCA and SVM were employed to visualize and analyze the separation between high- and low-burnout groups based on the selected features. This helps to establish burnout boundaries and categorize individuals into different risk zones, ranging from very low to very high. This geometric breakdown validates the stability of the selected features, supports a rule-based scoring mechanism, and mimics expert judgment to categorize caregivers within these risk levels. By merging the two RQs, an interactive framework is developed that predicts and demonstrates early burnout risk, and suggests interventions to mitigate it. For example, a caregiver working for several hours without a break, with minimal external support, and advanced age, would be flagged in the higher-risk category of burnout. Based on their burnout category, the system can recommend adjusting caregiving schedules, seeking external support, engaging in stress-reducing activities, and connecting with therapeutic resources and mental health services.

The findings of this study highlight key predictors of caregiver burnout, including caregiver age, the nature of the caregiving relationship, and adherence to routine schedules. Further, these factors align with prior research, indicating that older caregivers and those in emotionally close relationships experience higher psychological strain. Moreover, financial support (e.g., receiving paid help) and education level emerged as relevant contributors. These insights can inform practical interventions by helping care programs identify high-risk profiles and allocate resources more effectively. We ensure that fairness and bias mitigation are the primary priorities in the model's development. The model is developed to prevent the misclassification of vulnerable caregivers, such as elderly, female, or minority caregivers, by employing a diverse range of features that encompass a wide variety of caregiving scenarios. In addition, feature selection methods were varied to allow for the inclusion of appropriate variables that reflect caregivers' diverse backgrounds and avoid unfair outcomes that could harm specific demographic groups. Defining burnout solely based on caregiving hours represents caregiving intensity rather than capturing the multidimensional psychological syndrome, which is typically assessed via validated scales such as the Maslach Burnout Inventory. Moreover, empirical validation through longitudinal data and subgroup analysis is required to substantiate whether caregivers near this boundary transition to higher burnout over time. Future studies will aim to collect longitudinal proof to confirm this hypothesis.

5 Conclusion

This study employs predictive analytics, feature selection techniques, and geometric classification to understand caregiver burnout management and identify individuals at risk for early burnout. This data-driven framework uses several feature selection techniques, including LightGBM, XGBoost, and RFE with Random Forest, to identify the most contributing features in our pipeline. The selected features serve as the foundation for the computational health informatics-based separation of high- and low-burnout classes using SVM. In the future, we are working towards developing an iOS application powered by agentic AI, which will enable caregivers to receive dynamic, personalized support. This ongoing work aims to revolutionize caregiver management by providing telemedicine-enabled, adaptive interventions that evolve over time and in response to changing caregiver circumstances.

References

1. Armstrong-Carter, E., Panter, A., Hutson, B., Olson, E.A.: A university-wide survey of caregiving students in the us: individual differences and associations with emotional and academic adjustment. Hum. Soc. Sci. Commun. **9**(1), 1–10 (2022)
2. Chakraborty, S., Hasan, K.A., Paul, S.: A reinforcement learning framework for lung segmentation of COVID-19 and pneumonia affected chest X-ray image. In: International Conference on Pattern Recognition and Machine Intelligence, pp. 224–232. Springer, Cham (2021). https://doi.org/10.1007/978-3-031-12700-7_23
3. Chakraborty, S., Paul, S., Hasan, K.A.: A transfer learning-based approach with deep CNN for covid-19-and pneumonia-affected chest X-ray image classification. SN Comput. Sci. **3**(1), 17 (2022)
4. Chakraborty, S., Paul, S., Rahat-uz Zaman, M.: Prediction of apple leaf diseases using multiclass support vector machine. In: 2021 2nd International Conference on Robotics, Electrical and Signal Processing Techniques (ICREST), pp. 147–151. IEEE (2021)
5. Darst, B.F., Malecki, K.C., Engelman, C.D.: Using recursive feature elimination in random forest to account for correlated variables in high dimensional data. BMC Genet. **19**, 1–6 (2018)
6. Deardorff, W.J., et al.: Development and external validation of models to predict need for nursing home level of care in community-dwelling older adults with dementia. JAMA Intern. Med. **184**(1), 81–91 (2024)
7. Errarhout, A., Kharraja, S., Zorkani, I.: Caregivers' assignment problem in home health care structures. In: Proceedings of 2013 International Conference on Industrial Engineering and Systems Management (IESM), pp. 1–8. IEEE (2013)
8. Ferré-Grau, C., et al.: A mobile app-based intervention program for nonprofessional caregivers to promote positive mental health: Randomized controlled trial. JMIR Mhealth Uhealth **9**(1), e21708 (2021)
9. Fuller-Tyszkiewicz, M., et al.: Efficacy of a smartphone app intervention for reducing caregiver stress: randomized controlled trial. JMIR Mental Health **7**(7), e17541 (2020)
10. García-Cardoza, I.I., Zapata-Vázquez, R., Rivas-Acuña, V., Quevedo-Tejero, E.D.C.: Efectos de la terapia cognitivo-conductual en la sobrecarga del cuidador primario de adultos mayores. Horizonte sanitario **17**(2), 131–140 (2018)
11. Gérain, P., Zech, E.: Informal caregiver burnout? Development of a theoretical framework to understand the impact of caregiving. Front. Psychol. **10**, 466359 (2019)

12. Goodridge, D., et al.: An app-based mindfulness-based self-compassion program to support caregivers of people with dementia: participatory feasibility study. JMIR Aging **4**(4), e28652 (2021)
13. Kasper, J.D., Freedman, V.A., Spillman, B.C.: National health and aging trends study (NHATS) round 13 public use data (2024). https://www.nhats.org
14. King, E.M., Randolph, H.L., Floro, M.S., Suh, J.: Demographic, health, and economic transitions and the future care burden. World Dev. **140**, 105371 (2021)
15. Knickman, J.R., Snell, E.K.: The 2030 problem: caring for aging baby boomers. Health Serv. Res. **37**(4), 849–884 (2002)
16. Martinez-Hollingsworth, A., Hicks, M., Dobrota, S., Chu, L.F.: The burnout dyad: a collaborative approach for including patients in a model of provider burnout. J. Contin. Educ. Heal. Prof. **41**(4), 286–288 (2021)
17. Maslach, C., Leiter, M.P.: Understanding the burnout experience: recent research and its implications for psychiatry. World Psychiatry **15**(2), 103–111 (2016)
18. Montaner, X., Tárrega, S., Pulgarin, M., Moix, J.: Effectiveness of acceptance and commitment therapy (act) in professional dementia caregivers burnout. Clin. Gerontol. **45**(4), 915–926 (2022)
19. Ovallath, S.: Caregiver burnout in Parkinson's: time to address. Ann. Indian Acad. Neurol. **23**(5), 586 (2020)
20. Prapty, A.S., Shitu, T.T.: An efficient decision tree establishment and performance analysis with different machine learning approaches on polycystic ovary syndrome. In: 2020 23rd International Conference on Computer and Information Technology (ICCIT), pp. 1–5. IEEE (2020)
21. Yu, C., Jin, Y., Xing, Q., Zhang, Y., Guo, S., Meng, S.: Advanced user credit risk prediction model using LightGBM, XGBoost and TabNet with SMOTEENN. In: 2024 IEEE 6th International Conference on Power, Intelligent Computing and Systems (ICPICS), pp. 876–883. IEEE (2024)
22. Zarit, S.H., Gaugler, J.E., Jarrott, S.E.: Useful services for families: research findings and directions. Int. J. Geriatr. Psychiatry **14**(3), 165–178 (1999)

Unraveling Scopes and Challenges in Cardiovascular Risk Prediction Using Knowledge Distillation Framework from Clinical to Wearable Sensor Data

Nursat Jahan(✉), Nafisa Anjum, Sumit Chakravarty, Evelina Sterling, and Nazmus Sakib

Kennesaw State University, Marietta, USA
njahan2@students.kennesaw.edu

Abstract. Cardiovascular Disease (CVD) is one of the leading causes of global health concern, but current risk assessmentsn are limited to episodic clinical visits. Most of the Machine Learning (ML) models trained on clinical data offer high accuracy, but are not practical for continuous monitoring. Wearables provide real-time physiological data but lack clinical validation for robust risk prediction outside the clinical setting. To bridge this gap, we proposed a novel teacher-student knowledge distillation framework that transfers knowledge of complex and large EHR datasets to a small wearable device dataset-based prediction model. Here, the 'teacher' ML model first identifies different cardiovascular phenotypes from the EHR dataset, which generates pseudo-labels. Then these labels were used to train a 'student' neural network that uses real-time wearable features for predictions. The student model achieves promising accuracy that can identify all types of derived CVD risk profile groups. Feature importance analysis revealed that daily average steps and sedentary time are the most prominent wearable-derived CVD risk predictors. Our study introduces a non-invasive continuous health monitoring framework, demonstrating that passively collected daily metrics can be transformed to clinically powerful early warnings of an individual's long-term cardiovascular health plan.

Keywords: Teacher-Student Model · Knowledge Distillation · Cardiovascular Risk Prediction · Wearable Sensors · Imbalance datasets

1 Introduction

Cardiovascular disease (CVD) is one of the leading contributors to mortality and morbidity among non-communicable chronic conditions, generating massive health and financial burdens globally [7]. In 2020 Covid times, CVD was responsible for around 380,000 deaths, representing nearly one-third of all global fatalities [10, 17]. From an economic standpoint, CVD responsible for more than US$320 billion in annual health care expenditure [19]. To address this, the World

H. R. Arabnia et al. (Eds.): CSCI 2025, CCIS 2971, pp. 275–289, 2027.
https://doi.org/10.1007/978-3-032-28463-1_22

Health Organization (WHO) aims to cut major CVD risk factors and reduce U.S. cardiovascular mortality by 20% [8,12]. Toward this objective, American Heart Association (AHA) defines "ideal cardiovascular health" as maintaining healthy behaviors and optimal biomarkers like glucose, cholesterol, and blood pressure [8].

Advancements in artificial intelligence (AI) in medical care have unlocked innovative way of interpreting the diagnosis and treatment of CVD, including physiological assessment, cardiac biomarkers, and new drugs, improving survival rate after a preliminary occurrence [10,14]. Studies have revealed that cutting-edge ML based models can extensively improve diagnostic speed, prognosis, and precision through patient records, imaging and clinical parameters [13]. One recent studies employed quantum machine learning techniques to forecast coronary heart disease outcomes using text-based data [16]. Another research involves a five-phased approach of feature extraction using SVM for categorizing the cardiovascular system [15].

Despite these advances, patients with CVD are at greater risk of experiencing cardiovascular events due to lack of early signs. Traditional risk calculators handle each risk factor as a stable effect across all ages, genders, and medical variables, undervaluing risk in the younger generation and overvaluing in the older generation, restraining clinical attention in the clusters that need the most precise guidance [3]. Neglecting individual lifestyle patterns, lack of personalized, real-time lifestyle management tools, and limitations of healthcare systems diminish patient motivation to adopt lifestyle changes and ultimately mitigate the possibility of early CVD progression [14].

This research aims to provide individuals with early warning of CVD risks through daily used smartwatch or non-invasive wearbles which continuously measures their real-time physiological parameters. We implemented a teacher-student model combining lab based clinical physiological data and wearables data using ML models. The model's conceptual design is illustrated in Fig. 1 which we will discuss in details in methodology section. In this context, this paper explores below research questions:

- **RQ1:** How can we successfully teach a simple model that only uses non-invasive data (Wearables) to identify the similar cardiovascular risk groups that a more powerful model discovers using detailed clinical lab results?
- **RQ2:** When predicting someone's cardiovascular risk using just their Fitbit data, what matters most: their daily steps, sleep quality, or heart rate patterns?

In summary, this research explores the challenges by using multimodality data (EHR, Fitbit Dataset) and transfer clinical insights from complex data to simple wearable dataset through different ML and deep learning approaches to leverage their complementary strengths. This study also highlighted the challenges of managing imbalanced medical datasets in ML-based predictions and emphasized the role of feature engineering in achieving meaningful experimental outcomes. The final model will contribute to clinical decision support systems for CVD patient management.

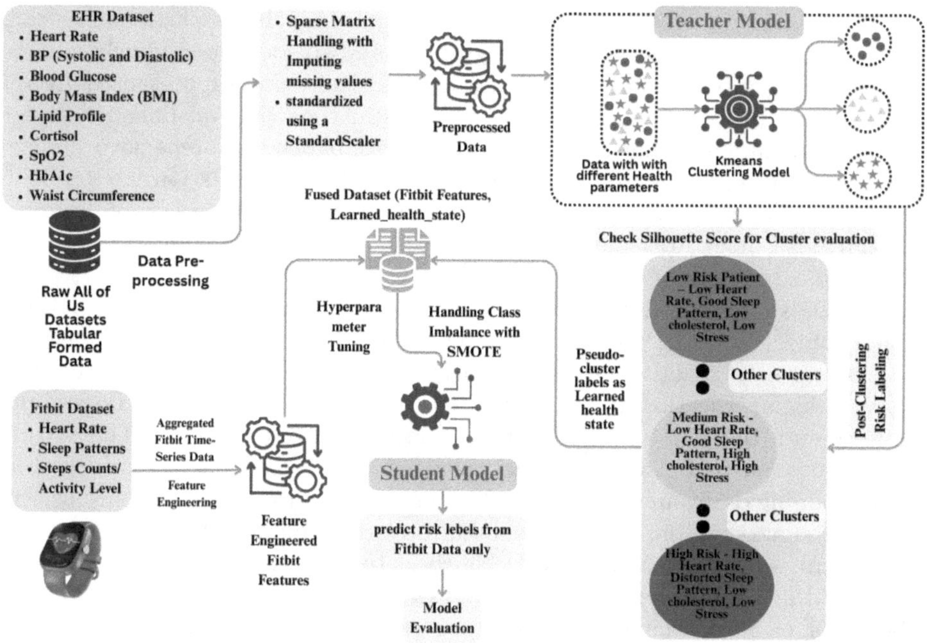

Fig. 1. Architecture of Cardiovascular Disease Risk Prediction using Teacher Student model

2 Methodology

We applied the concept of knowledge distillation (KD) [5] to transfer knowledge from a complex model to small model to develop a new framework and use this framework to a predictive model which will use data collected from wearable devices. This KD inspired framework has two main stages: (1) Teacher Stage and (2) Student Stage. This framework can recognize distinct clinical phenotypes only based on non-invasive, high-frequency data gathered from Fitbit devices which is initially identified from a detailed clinical dataset from large EHR system. For discovering the potential of our KD inspired framework, we performed experiments to build robust models. So, whole framework development process can be divided into three main stages: (1) Unsupervised Phenotyping with a Teacher Model, (2) Student Model Training via Pseudo-Labeling, and (3) Handling of significant class imbalance and improving accuracy of the model through evaluation. The models conceptual architecture is shown in the Fig. 1.

2.1 Dataset Description

This study used data from the **All of Us Research Program**'s *Registered Tier Dataset v8*, available to authorized users on the Researcher Workbench [1]. The All of Us Research Program dataset is intentionally designed to reflect the rich diversity of the United States population. It includes participants from diverse racial, ethnic, age, gender, geographic, and socioeconomic backgrounds, with over 50% of participants coming from historically underrepresented groups in biomedical research. Here basically two distinct participant-linked datasets is used. The larger dataset is EHR based parcipants comprehensive clinical dataset of Lab and measurements derived from the All of Us (AoU) Research Program and another dataset is a corresponding consumer wearable dataset of Fitbit data. The AoU dataset served as the source of ground-truth clinical knowledge, comprising fourteen key cardiovascular health markers: Heart Rate, Systolic/Diastolic Blood Pressure, Blood Glucose, Body Mass Index (BMI), Lipid Profile (HDL, LDL), Cortisol, Renal Function (Creatinine, Glomerular filtration rate), C reactive protein, Waist Circumference (Abdominal Obesity), Left ventricular Ejection fraction (LVEF), Electrolyte Imbalance (Potassium, Sodium) and Hemoglobin, Oxygen Saturation (HbA1c, SpO2). The second dataset, also part of AoU program, collected from consumer wearables (Fitbit), containing daily summary statistics on sleep (minute_in_bed, minute_asleep, minute_after_wakeup, minute_awake, minute_restless, minute_deep, minute_light, minute_rem, minute_wake), activity (fairly_active_minutes, sedentary_minutes, steps), and heart rate for a cohort of individuals. Four aggregated metrics calculated for each participant over the observation period: average daily steps, average daily sedentary minutes, average daily sleep duration in minutes, and overall average heart rate. This dual-dataset approach enables the training of a robust clinical model and the subsequent knowledge transfer to a practical, non-invasive model.

2.2 Data Preprocessing

Data preprocessing was performed in two main stages aligned with our teacher-student framework. First, the AoU clinical dataset was prepared for the unsupervised teacher model. We handled outliers by investigating and imputing them with median values, then missing values were imputed using column-wise medians, and all features were standardized using a StandardScaler to ensure zero mean and unit variance, preventing feature dominance in the clustering algorithm. Second, the time-series Fitbit data was aggregated into a static feature vector for each participant, as described in Sect. 2.1. To capture the impact of daily physical activity on cardiovascular health, we engineered a sleep-related set of features from the raw wearable data [6]. These are sleep efficiency (1a), restlessness ratio (1b), deep sleep percentage (2a), rapid eye movement (REM) sleep percentage (2b), light sleep percentage (2c), and active to sedentary ratio (3.

Sleep Quality Metrics

Sleep Efficiency (SE): This metric quantifies the percentage of time spent asleep relative to the total time in bed. **Restlessness Ratio (RR)**: This measures sleep fragmentation.

$$SE = \frac{T_{\text{asleep}}}{T_{\text{in\_bed}}} \tag{1a}$$

$$RR = \frac{T_{\text{restless}}}{T_{\text{asleep}}} \tag{1b}$$

Sleep Architecture Metrics

Deep Sleep Percentage (P_{deep}): The proportion of total sleep time spent in the restorative deep sleep stage. **REM Sleep Percentage (P_{REM})**: The proportion of total sleep time spent in the REM stage. **Light Sleep Percentage (P_{light})**: The proportion of total sleep time spent in light sleep stages.

$$P_{\text{deep}} = \frac{T_{\text{deep}}}{T_{\text{asleep}}} \tag{2a}$$

$$P_{\text{REM}} = \frac{T_{\text{REM}}}{T_{\text{asleep}}} \tag{2b}$$

$$P_{\text{light}} = \frac{T_{\text{light}}}{T_{\text{asleep}}} \tag{2c}$$

Physical Activity Metric

Activity-to-Sedentary Ratio (ASR): This ratio captures the balance between active and sedentary behavior.

$$ASR = \frac{T_{\text{active}}}{T_{\text{sedentary}}} \tag{3}$$

2.3 Teacher Model: Unsupervised Clinical Phenotyping

To establish a data-driven definition of cardiovascular risk profiles, we first developed a "teacher" model using a rich clinical dataset from the *All of Us Research Program*. This dataset contained a wide array of clinical measurements, lab results, and diagnostic codes for each participant. The core of this phase was the generation of pseudo-labels. K-Means clustering algorithm is applied in scaled preprocessed AoU data to partition the patient cohort into $k=3$ distinct, non-overlapping subgroups or "phenotypes". These hard-labeled K-means cluster of clinical data with PCA is shown in Fig. 2. The main goal of the clustering is to generate labels for the clinical dataset. By analyzing the clinical centroids of each cluster, we assigned meaningful risk labels: "Low Risk", "Medium Risk", and "High Risk" to each cluster. These pseudo-labels are then used as the target variable for the subsequent supervised learning task in the student model.

2.4 Student Model: Predicting Phenotypes from Wearable Data

The "student" model is a feed-forward neural network (also known as a Multi-Layer Perceptron) designed to predict the clinical phenotype (the teacher's cluster label) using only features derived from Fitbit data (e.g., daily step counts, heart rate statistics, sleep patterns, feature engineered fitbit data). The student

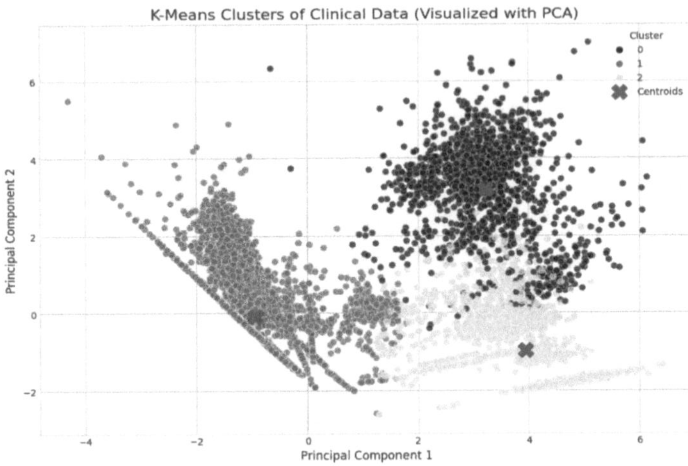

Fig. 2. kmeans clustering of All of Us larger EHR Dataset

model's architecture consists of an input layer corresponding to the number of Fitbit features and followed by two hidden layers with 128 and 64 neurons, respectively. These two layers use the Rectified Linear Unit (ReLU) activation function. The final output layer has 3 neurons, corresponding to the 3 clinical phenotypes, and produces raw scores (logits) for each class.

2.5 Data Alignment, Imbalance Handling, and Model Training

A critical step was aligning the teacher's labels with the student's features by ensuring a common set of person ids across both datasets. This resulted in a cohort of individuals for whom both clinical and Fitbit data were available. Initial analysis revealed a severe class imbalance among the three generated phenotypes, with one cluster representing a large majority of the population. We employed the Synthetic Minority Over-sampling Technique (SMOTE) to prevent the model from developing a bias towards this majority class [2]. SMOTE was applied only to the training set to prevent data leakage into the test set. It works by creating synthetic samples of the minority classes by interpolating between existing minority class instances and their nearest neighbors. We also implemented ADASYN (Adaptive Synthetic Sampling) [4], and NearMiss (Undersampling) resampling techniques to address the observed class imbalance. For each method, a new Student Neural Network was trained on the resampled data and subsequently evaluated on the standardized test set. Then we compared models performance across different imbalance handling strategies, quantifying metrics such as accuracy, precision, recall, and the macro F1-score to determine the optimal balance between minority class identification and overall generalization. To ensure robust evaluation, the aligned fitbit dataset was partitioned into a training set (80%) and test set (20%) using a stratified sampling strategy to maintain class proportions. The final aligned and resampled training data was used to train the student model. We used the standard Cross-Entropy Loss function, which is well-suited for multi-class classification tasks. The model was trained for 50 epochs using the Adam optimizer with a learning rate of 0.001. The model's performance was evaluated on a held-out, un-resampled test set.

2.6 Model Evaluation

The student model's performance was evaluated on the test set using standard classification metrics, including precision, recall, F1-score. We implemented different configurations using same student-teacher models to see the performance difference (Shown in Table 1). The model implementation with feature engineered fitbit data and SMOTE balanced data shows best performance among all other configurations. Furthermore, feature importance scores were extracted from the trained model to identify the most influential wearable-based predictors of cardiovascular risk.

3 Results and Discussion

This section presents the empirical findings of our study. The results are organized to directly address the research questions, beginning with an evaluation of the knowledge distillation framework's effectiveness.

Table 1. Performance Analysis for Student Model on Differnt Configurations

Metric	Class / Avg.	Baseline	CW	SMOTE	SMOTE+FE	Support
Accuracy		0.73	0.40	0.50	0.66	6942
Precision	Cluster 0	0.00	0.06	0.04	0.81	1735
	Cluster 1	0.30	0.26	0.26	0.03	1414
	Cluster 2	0.74	0.77	0.75	0.45	3793
	Macro Avg.	0.34	0.36	0.35	0.43	6942
	Weighted Avg.	0.61	0.63	0.61	0.70	6942
Recall	Cluster 0	0.00	0.29	0.17	0.76	1735
	Cluster 1	0.03	0.56	0.35	0.08	1414
	Cluster 2	0.98	0.35	0.56	0.46	3793
	Macro Avg.	0.34	0.40	0.36	0.43	6942
	Weighted Avg.	0.73	0.40	0.50	0.66	6942
F1-Score	Cluster 0	0.00	0.09	0.07	0.78	1735
	Cluster 1	0.06	0.36	0.30	0.05	1414
	Cluster 2	0.84	0.48	0.64	0.46	3793
	MacroAvg.	0.30	0.31	0.33	0.43	6942
	Weighted Avg.	0.63	0.44	0.54	0.68	6942

Abbreviations:
CW: Class Weights; SMOTE+FE: SMOTE with Feature Engineering

RQ1: Teaching wearables to replicate clinically derived cardiovascular risks Our first research question sought to determine if a simple model using only Fitbit data could be taught to identify the same complex risk groups discovered by a powerful model using clinical data. To answer this, we evaluated the student model's ability to learn and replicate teachers models risk soft labels and also identify these risk labels on unseen data. Additionally, we validated the clinical meaningfulness of the teacher model's clusters labels too.

We implemented four variation of student models using identical neural network architectures but trained under different sampling strategies: (1) First one with the original imbalanced dataset, (2) Second one with class weight technique (4) Third, with SMOTE-balanced version to address class disparity (4) Fourth, with SMOTE-balanced and Fitbit data with feature engineering improve performance of this model. As shown in Table 1, the comparative results reveal a notable case of the accuracy paradox. While the imbalanced model achieved a seemingly high accuracy of 73%, this performance was heavily skewed by its near-exclusive success in predicting the majority class (Cluster 2), which constituted the vast majority of samples. It failed entirely to identify any instances of Cluster 0 and achieved negligible recall for Cluster 1. In contrast, just balancing the dataset with class weights the accuracy downgraded sharply. Then we used advanced SMOTE techniques to balancing the data and it improved the accuracy little higher. Then we used feature engineered SMOTE-balanced model to produce overall accuracy of 66%, which demonstrated a more equitable performance across all clusters. It successfully identified minority classes with improved F1-scores, Recall and Precisions for all the clusters. So, here overall accuracy is compromised to predict minor classes as well. This trade-off highlights that accuracy alone is insufficient for evaluating model fairness or effectiveness in imbalanced multiclass settings, and metrics like macro-averaged F1-score (rising from 0.30 to 0.43) better reflect the model's ability to generalize across diverse health profiles. Different configurations shows that power of fine-tuning every possible improvements to increase the accuracy of this new model. If we consider feature engineer the temporal data of fitbit that will also increase the models accuracy which we kept for our future work. Finally, the promising accuracy of this KD inspired framework shows that it is possible to predict CVD health risks from passively collected small dataset with the distilled knowledge from large EHR dataset.

The primary finding of this work is not just the final accuracy figure, but to understand of how a new clinical decision framework can be developed through knowledge distillation. From a theoretical standpoint, complex models are expected to pass on their knowledge to simpler models with ease. In our initial experiments with raw data, the model achieved a much higher accuracy (approximately 75–80%). However, this performance was misleading. The model had simply learned to predict the majority class (Cluster 2) for nearly every instance, effectively ignoring the minority clusters. Although this setup of this model seems statistically accurate, this model was clinically useless, as its primary purpose was to help identify individuals in the less common and potentially higher-risk, phenotypes.

Another finding is, initially our large clinical dataset was imbalanced which results in higher accuracy but failed to predict minority classes. We handled imbalanced dataset with different techniques and shown performance analysis in Table 2.

Table 2. Performance Analysis of Student Model on Differnt Data Imbalance Handling

Metric	Class / Avg.	SMOTE	ADASYN	NearMiss
Accuracy		0.66	0.66	0.13
Precision	Cluster 0	0.81	0.81	0.83
	Cluster 1	0.03	0.04	0.03
	Cluster 2	0.45	0.45	0.54
	Macro Avg.	0.43	0.43	0.47
	Weighted Avg.	0.70	0.70	0.73
Recall	Cluster 0	0.76	0.76	0.11
	Cluster 1	0.08	0.09	0.78
	Cluster 2	0.46	0.47	0.11
	Macro Avg.	0.43	0.44	0.33
	Weighted Avg.	0.66	0.66	0.13
F1-Score	Cluster 0	0.78	0.78	0.19
	Cluster 1	0.05	0.06	0.06
	Cluster 2	0.46	0.46	0.18
	Macro Avg.	0.43	0.43	0.14
	Weighted Avg.	0.68	0.68	0.18

So, first we handled imbalance dataset with SMOTE technique. We pay attention to the model to learn the patterns of these minority classes. This came at the cost of a significant drop in overall accuracy. The reason is twofold: (1) the minority classes are inherently harder to predict, and (2) forcing the model to predict them increases the number of false positives, which penalizes the accuracy metric. Additionally, After feature engineering the wearable data, the model performed better than all the previous configurations which proves that correct clinical domain knowledge implementation could improve the performance of the model. As we implemented ADASYN (Adaptive Synthetic Sampling) [4], and NearMiss (Undersampling) resampling techniques to address the observed class imbalance, the comparison of their accuracy, precision, recall, and the macro F1-score in Table 2 which shows SMOTE technique resulted best to determine the optimal balance between minority class identification and overall generalization.

As we used unsupervised learning, the K-Means clustering in teacher model to find out the participant clusters, we also want to check the validity of this model. Analysis revealed that a configuration of n_clusters = 3 produced a strong result, yielding a Silhouette Score of 0.5653. The kmeans clustering for this clinical data is shown in Fig. 2. The Silhouette Score of 0.5653 is a strong result, as values on its -1 to +1 scale that are closer to +1 indicate dense and well-separated clusters. This suggests that the K-Means algorithm has successfully identified three statistically meaningful and distinct patient subgroups based on their comprehensive clinical data.

A post-hoc analysis of the teacher model's K-Means cluster centroids reveals three qualitatively distinct clinical phenotypes, defined by a core set of differentiating features in Table 3.

Table 3. Cluster Centroids (in original feature units). Feature names have been abbreviated for space.

Feature	Cluster 0	Cluster 1	Cluster 2
Waist Circumference	105.31	89.78	105.34
BMI (Ratio)	34.04	27.34	34.21
C-Reactive Protein	3.78	1.62	0.95
HDL Cholesterol	49.23	58.49	45.80
LDL Cholesterol	103.25	103.45	101.80
Cortisol	10.15	10.06	10.00
Creatinine	1.05	0.85	0.10
Diastolic BP	82.23	71.08	81.27
GFR (predicted)	76.28	76.58	76.00
Glucose	111.51	95.91	10.80
Heart Rate	77.63	73.49	78.98
Hemoglobin	132.93	128.07	21.86
L. Vent. Ejection Frac.	63.86	63.93	64.00
O2 Sat. (Pulse Ox)	97.60	97.98	98.00
Potassium	4.12	4.14	0.24
Sodium	139.02	139.14	0.16
Systolic BP	134.85	117.14	125.96

We hypothesize that Phenotype 0 is indicative of a high-risk CVD profile, which is defined by a significantly larger Waist Circumference (105.31) along with the highest recorded Systolic (134.85) and Diastolic (82.23) Blood Pressure readings. Conversely, Phenotype 1 demonstrates medium risk profile showing a glucose level that is intermediate, being higher than Phenotype 2 but lower than Phenotype 0. This cluster is notable for its considerably elevated Glucose at 95.91 and the peak LDL Cholesterol value of 103.45. Phenotype 2, on the other hand, consistently shows median values across nearly all features. Because the clinical dataset had a high degree of sparsity and was imputed using median values, this cluster appears to represent a low-risk profile, likely because these patients were missing specific lab tests related to those features. Consequently, the primary features responsible for differentiating the clusters were Blood Pressure, BMI, Glucose, and Heart Rate. With this analysis we can effectively divide the patient cohort into distinct risk profiles rather than following a simple, linear risk progression.

This study has several limitations. First, the "ground truth" labels were derived from K-Means clustering, an unsupervised method. These data-driven phenotypes may seems informative but may not perfectly align with established clinical definitions of cardiovascular risk. The performance of the student model is naturally limited by the reliability of the pseudo-labels. If those labels are not strong, the student would not learn well. Second, while SMOTE helps with imbalance, it can also introduce noise by creating synthetic samples in areas of the feature space that are not well-represented by real data, which may have contributed to the low precision.

RQ2: Most influencial Fitbit features for predicting cardiovascular disease risk

To understand which features are most predictive in this KD inspired cardiovascular risk prediction model, we conducted a two-part feature importance analysis. First, we identified which Fitbit-derived features were most effective at differentiating between the clinical phenotypes. Second, we analyzed the clinical data within the teacher model's clusters to define the nature of these phenotypes.

An ANOVA F-test was performed to determine the statistical significance of each Fitbit feature in predicting the clinical phenotype labels. The results, ranked

Table 4. Feature Importance Using ANOVA F-test

Index	Feature	F-statistic	p-value
11	avg_steps	127.682626	7.34×10^{-56}
10	avg_sedentary_minutes	106.475067	9.54×10^{-47}
9	avg_fairly_active_minutes	50.126892	1.90×10^{-22}
1	avg_minute_asleep	16.234403	9.01×10^{-8}
0	avg_minute_in_bed	13.338161	1.62×10^{-6}
5	avg_minute_deep	7.731722	4.40×10^{-4}
12	heart_rate	6.176112	2.08×10^{-3}
7	avg_minute_rem	6.109954	2.22×10^{-3}
6	avg_minute_light	3.959998	1.91×10^{-2}
8	avg_minute_wake	2.402311	9.05×10^{-2}
2	avg_minute_after_wakeup	1.870586	1.54×10^{-1}
3	avg_minute_awake	0.659184	5.17×10^{-1}
4	avg_minute_restless	0.548852	5.78×10^{-1}

by F-statistic, clearly identify physical activity metrics as the most powerful predictors in Table 4.

The top two most significant features are avg_steps and avg_sedentary_minutes. Their extremely high F-statistics and near-zero p-values indicate a very strong and reliable statistical association. Our feature importance analysis reveals that the daily activity and sleep patterns were the most salient predictors for the student model. Notably, avg_steps (importance: 0.068) and avg_sedentary_minutes (0.068) emerged as the top-ranking features (Fig. 3).

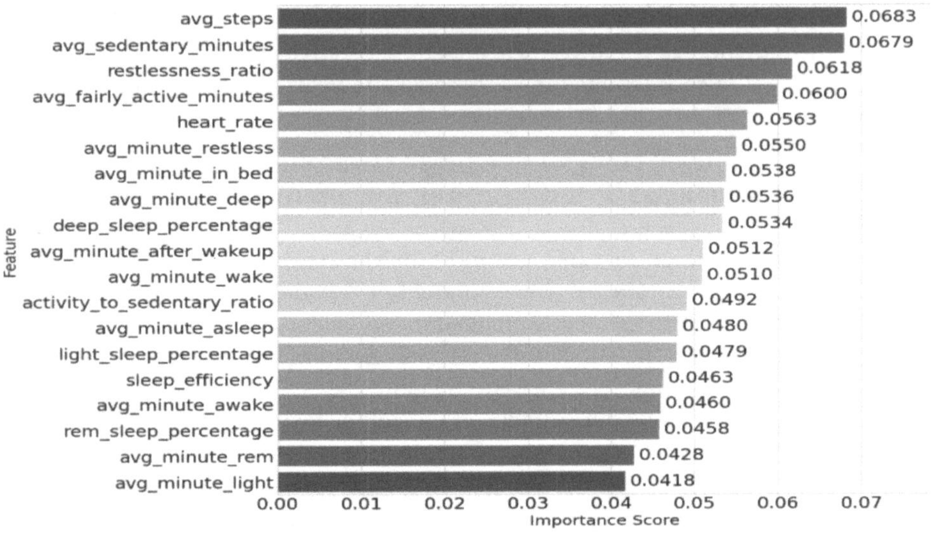

Fig. 3. Feature Importances for Fitbit Data (Derived from Proxy Model)

A study demonstrated that wearable accelerometers like Fitbit, can classify desk-related activities and are increasingly used as tools for measuring health parameters and behavior monitoring [9]. Also, recent studies shows evidence establishing that higher step counts are associated with lower cardiovascular risks and sedentary time is associated with an increased risk of cardiovascular disease [11,18]. So in our engineered feature restlessness_ratio(0.062) ranked third, highlighting the significance of sleep fragmentation as a more critical predictor than simple sleep duration (avg_minute_asleep, importance: 0.048). To quantitatively rank the clinical features defining the K-Means phenotypes, a proxy Random-Forest (RM) model was employed, treating the cluster IDs as target labels and the feature importance are calculated based on the clinical data feature set in Table 5. The analysis revealed that fundamental physiological measurements were paramount, with Systolic blood pressure (Importance: 0.297) and Diastolic blood pressure (0.192) emerging as the most dominant predictors (Fig. 4). These, along with Body Mass Index (0.160), Oxygen Saturation (0.151), and

Table 5. Feature Importances from Proxy Model

Feature	Importance
Systolic blood pressure	2.9715e-01
Diastolic blood pressure	1.9156e-01
Body mass index (BMI) [Ratio]	1.6044e-01
Oxygen saturation by Pulse oximetry	1.5147e-01
Heart rate	1.2759e-01
Hemoglobin [Mass/volume] in Blood	2.9258e-02
Glucose [Mass/volume] in Serum or Plasma	2.3804e-02
Cholesterol in HDL [Mass/volume] in Serum or Plasma	1.8053e-02
Cholesterol in LDL [Mass/volume] in Serum or Plasma	5.9196e-04
Cortisol [Mass/volume] in Serum or Plasma	8.6446e-05
Left ventricular Ejection fraction	8.2989e-10

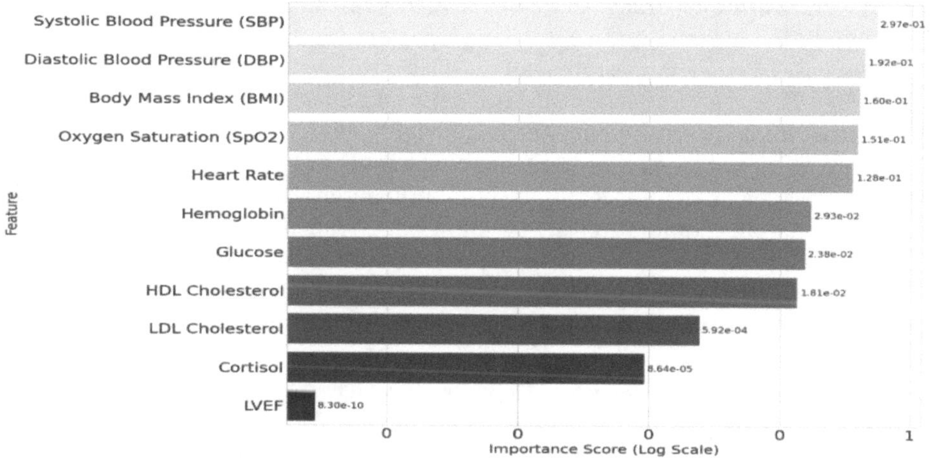

Fig. 4. Feature Importance using a Proxy RM Model for Clinical Data

Heart Rate (0.128), collectively accounted for over 93% of the model's total feature importance. Conversely, traditional lab markers such as Glucose (0.024) and Cholesterol (HDL or LDL) (¡0.02) demonstrated minimal impact in differentiating these specific phenotypes. This result provides strong quantitative validation that the teacher model's clusters are primarily distinguished by core vital signs and body composition rather than by detailed blood biochemistry, confirming the clinical relevance of the generated pseudo-labels.

4 Conclusion

In this paper, we demonstrated the absence of a continuous, robust and non-invasive method to generate clinically derived cardiovascular risk stratification using wearable sensor data, specifically under significant class imbalance that disrupts identifying minority risk groups. We successfully developed and validated a teacher–student knowledge distillation framework where a K-Means–based teacher model identified clinically coherent phenotypes generating pseudo-labels from the All of Us EHR dataset, and a feed-forward neural network student model learned to effectively predict these states solely from Fitbit-derived salient features.

The central finding of this work is that there is a high potential of distilling knowledge from a complex EHR based model to small wearable based model which can predict cardiovascular risks. Additionally, it is highly evident that this basic KD-based model ca be improved with more advanced feature engineering of data and using time-series based deep learning. Another key finding is the trade-off between overall accuracy and clinical utility, given that real-world data are often highly imbalanced and contain numerous outliers. The main contribution of this research is to demonstrate a method that turns everyday, passive data into valuable early signs of potential health risks and effectively connects consumer wearable technology with clinical health insights.

Future work will focus on implementing clinician-in-the-loop to ensure the clinical relevance of the K-Means clusters and validate by medical experts. This would enhance the foundation of the knowledge transfer process. Additionally, the framework will be refined by incorporating soft-label distillation techniques to capture higher risk probabilities, expanding the model to predict longitudinal health trajectories instead of static ones, and evaluating its effectiveness in a prospective clinical setting. Furthermore, we plan to capture the valuable temporal dynamics of fitbit data by exploring richer temporal representations through advanced feature engineering and explore time-series models such as RNNs, LSTMs, or Transformers. We plan to apply our framework to diverse wearable data sources to validate models generalizability.

Acknowledgments. We gratefully acknowledge All of Us participants for their contributions, without whom this research would not have been possible. We also thank the National Institutes of Health's All of Us Research Program for making available the participant data of the Wearable user cohort examined in this study.

References

1. All of Us Research Program: All of us research program dataset. https://workbench.researchallofus.org/ (2025), [All of Us Registered Tier Dataset v8]. Researcher Workbench. Accessed 19 Jun 2025
2. Chawla, N.V., Bowyer, K.W., Hall, L.O., Kegelmeyer, W.P.: SMOTE: synthetic minority over-sampling technique. J. Artif. Intell. Res. **16**, 321–357 (2002)

3. Cooney, M.T., Dudina, A.L., Graham, I.M.: Value and limitations of existing scores for the assessment of cardiovascular risk: a review for clinicians. J. Am. Coll. Cardiol. **54**(14), 1209–1227 (2009)
4. He, H., Bai, Y., Garcia, E.A., Li, S.: ADASYN: adaptive synthetic sampling approach for imbalanced learning. In: 2008 IEEE International Joint Conference on Neural Networks (IEEE World Congress on Computational Intelligence), pp. 1322–1328. IEEE (2008)
5. Hinton, G., Vinyals, O., Dean, J.: Distilling the knowledge in a neural network, (2015). arXiv preprint arXiv:1503.02531
6. Knutson, K.L., et al.: Association between sleep and blood pressure in midlife: the cardia sleep study. Arch. Intern. Med. **169**(11), 1055–1061 (2009)
7. Koene, R.J., Prizment, A.E., Blaes, A., Konety, S.H.: Shared risk factors in cardiovascular disease and cancer. Circulation **133**(11), 1104–1114 (2016)
8. Michos, E.D., Khan, S.S.: Further understanding of ideal cardiovascular health score metrics and cardiovascular disease. Expert Rev. Cardiovasc. Ther. **19**(7), 607–617 (2021)
9. Mosin, A., Min, C.H., Salamy, H.: Using a wrist-worn accelerometer to detect off-task activities while studying or working. In: Arabnia, H.R., Deligiannidis, L., Shenavarmasouleh, F., Amirian, S., Ghareh Mohammadi, F. (eds) International Conference on Computational Science and Computational Intelligence, pp. 131–143. Springer, Cham (2024). https://doi.org/10.1007/978-3-031-94953-1_12
10. de Oliveira Laterza Ribeiro, M., Correia, V.M., Herling de Oliveira, L.L., Soares, P.R., Scudeler, T.L.: Evolving diagnostic and management advances in coronary heart disease. Life **13**(4), 951 (2023)
11. Paluch, A.E., et al.: Prospective association of daily steps with cardiovascular disease: a harmonized meta-analysis. Circulation **147**(2), 122–131 (2023)
12. Ramírez-Vélez, R., Saavedra, J.M., Lobelo, F., Celis-Morales, C.A., del Pozo-Cruz, B., García-Hermoso, A.: Ideal cardiovascular health and incident cardiovascular disease among adults: a systematic review and meta-analysis. In: Mayo Clinic Proceedings, vol. 93, pp. 1589–1599. Elsevier (2018)
13. Robinson, J., Jones, S., Ebwabi, A.: Cutting-edge machine learning and mathematical approaches in cancer prognosis and therapy, progress and challenges (2024)
14. Sigamani, A., Gupta, R.: Revisiting secondary prevention in coronary heart disease. Indian Heart J. **74**(6), 431–440 (2022)
15. Sridevi, S., Sundaresan, M.: Classification of coronary heart artery disease using IVUS images by SVM classifier with modified radial basis function kernel (MRBFK). Int. J. **9**(3) (2020)
16. Tisha, S.N., Rahman, M.S., Orduz, J.: Quantum machine learning for heart disease detection: a case study. In: Arabnia, H.R., Deligiannidis, L., Shenavarmasouleh, F., Amirian, S., Ghareh Mohammadi, F. (eds) International Conference on Computational Science and Computational Intelligence, pp. 58–66. Springer, Cham (2024). https://doi.org/10.1007/978-3-031-94940-1_5
17. Tsao, C.W., et al.: Heart disease and stroke statistics–2022 update: a report from the American heart association. Circulation **145**(8), e153–e639 (2022)
18. Wilmot, E.G., et al.: Sedentary time in adults and the association with diabetes, cardiovascular disease and death: systematic review and meta-analysis. Diabetologia **55**(11), 2895–2905 (2012)
19. Younus, A., et al.: A systematic review of the prevalence and outcomes of ideal cardiovascular health in us and non-us populations. Mayo Clin. Proc. **91**, 649–670. Elsevier (2016)

Fall Risk of Older Adults in Pain Management

Melissa S. Lee[1(✉)], Alexander W. Lee[1], Chelsea Yeh[2], Kyle Yeh[3], and Hsi-Jen J. Yeh[4]

[1] Chino Premier Surgery Center, Chino, CA 91711, USA
melissalee2390@gmail.com
[2] Yale University, New Haven, CT 06520, USA
[3] Brown University, Providence, RI 02912, USA
[4] Seattle Pacific University, Seattle, WA 98119, USA

Abstract. Background: Falls are a serious health issue worldwide. No studies have been conducted in a pain management clinic evaluating fall risks in these patients. We aimed to assess whether there was an increase in fall risk, as measured by the 30-Second Chair Stand Test (30 CST), associated with body mass index (BMI), heart rate, blood pressure, and chronic pain of the low back, hip, or leg.

Methods: This cross-sectional study was set in a pain management clinic. IRB approval was obtained, and all patients signed an informed consent form. Male and female community-dwelling adults 18 years and older participated in this study.

Results: Out of 224 participants recruited for this study, 120 patient results were included in the analysis because they were 60 years and older, and the CDC 30 CST guideline only has pass/fail ranges for participants 60 years and older. The 30 CST failure rate was 76% for all chronic pain participants 60 years and older, 79% for those with chronic pain of the low back, hip, or leg. Logistic regression analysis showed that having chronic pain of the low back, hip, or leg increased the risk of falls. Chronic pain of the low back, hip, or leg was associated with a 44% reduction in the odds of passing the 30 CST compared to other chronic pain participants. Heart rate, systolic blood pressure, and diastolic blood pressure were not statistically significant.

Conclusion: These findings emphasize the high risk of falls for patients with chronic pain of the low back, hip, or leg who are seen in chronic pain management clinics.

Keywords: lower extremity weakness · falls · 30-second Chair Stand Test · chronic pain · pain management clinic

1 Introduction

Falls are very common in the United States and worldwide. More than 14 million adults over the age of 65 fall every year in the United States [1]. That amounts to approximately one in four older adults. According to the World Health Organization (WHO), approximately 684,000 individuals die from falls globally each year [2]. Falls not only cause

H. R. Arabnia et al. (Eds.): CSCI 2025, CCIS 2971, pp. 290–298, 2027.
https://doi.org/10.1007/978-3-032-28463-1_23

deaths but also result in more years of disability, chronic pain, and higher healthcare costs. Risk factors, such as chronic pain and certain medications, can contribute to falls. A study done on community-dwelling adults in Boston found that patients have a 61% greater risk of injurious falls compared to patients who have little or no pain interference [3].

The US Centers for Disease Control and Prevention (CDC) recommends three different tests that can help evaluate fall risks in adults. These tests include the 4-Stage Balance Test [4], the Timed Up and Go Test (TUG) [5], and the 30-Second Chair Stand Test (30 CST) [6]. The 30 CST was chosen in this study due to space and time limitations of the clinic, making it difficult to perform the TUG test and the 4-Stage Balance Test. The 30 CST evaluates how many times a person can transition from a sitting to a standing position and back to a sitting position within 30 s. If a person fails to perform the minimum number of sit-stand-sit cycles for their age group, they are considered at high risk for falls [6]. The 30 CST is a well-established test in evaluating fall risk [7].

No study has been conducted to evaluate the fall risks of patients in a pain management clinic setting. Patients in a chronic pain clinic are of particular interest because most of these patients had pain and may have an increased fall risk. According to a 2023 Centers for Disease Control and Prevention (CDC) report, 51.6 million adults (20.9% of the general population) in the United States have chronic pain [8]. Chronic pain is defined as pain lasting longer than three months. 17.1 million people experience chronic pain severe enough that they are substantially restricted in their daily activities [8]. A few common causes of chronic pain include back pain, joint pain, and nerve pain. This study aims to evaluate in a cross-sectional study whether older patients in a pain management clinic have increased their risk of falls based on the 30 CST predictive values. We also sought to determine if there was a correlation between fall risk and body mass index (BMI), blood pressure, heart rate, and the presence of chronic pain of the low back, hip, or legs.

2 Methods

IRB approval #23–130 was obtained from Azusa Pacific University to conduct this study. Patients and the public were not involved in the study's conceptualization, design, and conduct. Informed consent was obtained from all the patients. The study was performed in a pain management clinic in Pomona, CA, between October 1, 2023, and September 30, 2024, which was the duration of the IRB approval. We recruited community-dwelling male and female adults aged 18 and over. Participants who used assistive devices such as canes and walkers or regularly used wheelchairs were not included in this study. However, they were included if they only occasionally needed the use of canes or walkers.

During the study, if the participant appeared fatigued or imbalanced, the test was immediately stopped to prevent a fall. A walker was also placed in front of the chair for the participants to hold onto if they felt fatigued or imbalanced. The trial ended immediately if the participant required the assistance of a walker. The participants could also voluntarily end the study if they felt tired or unable to continue by verbally informing us or raising their right or left hand.

All participants recorded their age and whether they had chronic pain of the low back, hip, or leg. Vital signs, including height, weight, body mass index (BMI), blood pressure, and heart rate, were measured in all participants. Participants were instructed on how to perform the 30 CST. Their feet needed to be flat on the floor, and they had to sit in the middle of the chair with their hands on the opposite shoulder, their arms against their chest. When told to "Go," they had to transition from a sitting to a full standing position and then sit back down. We manually recorded the number of sit-stand cycles completed in 30 s.

We used the guideline published by the CDC [6], as the criteria for evaluating whether a person passed or failed the 30 CST. This guideline only had pass/fail ranges for patients aged 60 and older. Patients under the age of 60 were excluded from the 30 CST analysis since no published pass/fail ranges were found for this age group. The data was collected using an internet-connected chair that collected and stored the data automatically [9–11]. The use of internet-connected devices have proven to be useful in clinical settings, including the automated collection of clinically significant data [12, 13] and the monitoring of patient safety [14, 15].

Analysis was done using the Chi-Square Test and logistic regression analysis. The dependent variable included the pass/fail results for the 30 CST. The independent variables included body mass index (BMI), heart rate, systolic blood pressure, diastolic blood pressure, and chronic pain of the low back, hip, or leg. All the independent variables had a Variance Inflation Factor calculated to evaluate for multicollinearity. The performance of the logistic regression model was evaluated with the Receiver Operating Characteristic (ROC) curve, and the Area Under the Curve (AUC) was calculated in R.

3 Results

A total of 224 community-dwelling participants took part in this study. Of these participants, 84 were male, and 140 were female. Two hundred and eight (208) participants had chronic pain. Of the 224 participants, 56 people occasionally used assistive walking devices such as walkers and canes. Their recorded vital signs included mean age 59.8 (range 18–90, SD 14.0, 95%CI 3.6), mean height 165.8 cm (range 139.7–205.7, SD 3.7, 95%CI 0.9), mean weight 100.7 kg (range 44.0–138.7, SD 46.3, 95%CI 6.6), mean BMI 31.2 (range 17.5–55.8, SD 6.8, 95%CI 2.4), mean pulse 78.5 beats per minute (range 47–120, SD 12.7, 95%CI 2.8), mean systolic blood pressure 140.2 mmHg (range 88–225, SD 21.0, 95%CI 3.5), mean diastolic blood pressure 80 mmHg (range 41–133, SD 13.9 95%CI 3.1).

Out of 224 participants, 218 completed the 30 CST. The six participants who failed were unable to stand up from the chair. Another five participants were removed because their genders were not recorded. Tables 1 and 2 summarize the number of sit-stand cycles completed and their respective standard deviations for the 213 participants analyzed by age group and gender. One hundred and twenty participants were included in the 30 CST data analysis. The remaining 104 participants were excluded due to their age, as no fall risk ranges were published for individuals under 60, or because data was missing. Based on their gender and age group criteria, 29 passed the 30 CST, while 94 failed. That resulted in a 23.6% pass rate and a 76.4% fail rate for all these chronic pain patients. The pass and fail percentages for each category are summarized in Table 3. For participants with chronic pain of the low back, hip, or leg, 20.9% passed and 79.1% failed.

Table 1. The 30 CST mean number of sit-stand cycles completed by female participants.

Age	Mean Sit Stand Cycles Completed (sec) ± SD	95% Confidence Interval	#Subjects Completing Test	#Subjects Unable Complete Test
18–19	6.0 ± 1.4	2	2	0
20–29	7.7 ± 2.9	2.4	6	0
30–39	8.8 ± 2.2	1.9	5	0
40–49	8.1 ± 3.1	1.4	19	0
50–59	6.9 ± 2.9	0.9	39	0
60–69	6.8 ± 3.4	1.2	33	0
70–79	8.2 ± 3.4	1.4	24	2
80–90	7.2 ± 3.5	2.2	10	0
18–90	7.4 ± 3.2	0.5	138	0

Table 2. The 30 CST mean number of sit-stand cycles completed by male participants.

Age	Mean Sit Stand Cycles Completed (sec) ± SD	95% Confidence Interval	#Subjects Completing Test	#Subjects Unable Complete Test
18–19	–	–	0	0
20–29	16	–	1	0
30–39	9.0 ± 3.8	4.3	3	0
40–49	8.2 ± 3.8	2.2	11	0
50–59	7.3 ± 2.7	1.5	12	0
60–69	8.1 ± 5.3	2	27	4
70–79	8.4 ± 3.8	1.8	16	0
80–90	8.1 ± 6.5	5.7	5	0
18–90	8.0 ± 4.4	1	75	0

Table 3. The pass and fail rate for the 30CT separated by different categories.

	# Pass	#Fail	Total	%Pass	%Fail
Age 60–64	5	32	37	13.5	86.5
Age 65–69	6	19	25	24	76
Age 70–74	11	22	33	33.3	66.7
Age 75–79	3	7	10	30	70
Age 80–84	4	3	7	57.1	42.9
Age 85–89	1	5	6	16.7	8.3.
Age 90–94	1	1	2	50	50
Age 60–94	31	89	120	25.8	76.4
BMI 18.5–24.9	8	7	15	53.3	46.7
BMI 25–29.9	9	29	38	23.7	76.4
BMI 30–39.9	12	48	60	20	80
BMI ≥40	2	5	7	28.6	71.4
Pulse <60	1	7	8	12.5	87.5
Pulse 60–100	30	77	107	28	72
Pulse >100	0	5	5	0	100
SBP <120	2	13	15	13.3	86.7
SBP 120–129	4	9	13	30.1	69.2
SBP 130–139	6	23	29	20.7	79.3
SBP 140–179	18	41	59	30.5	69.5
SBP ≥180	1	3	4	25	75
DBP <80	15	47	62	24.2	75.8
DBP 80–89	8	26	34	23.5	76.5
DBP 90–119	8	16	24	33.3	66.7
Chronic Pain of the low back, hip, or leg	9	34	43	20.9	79.1

We analyzed the data using the Chi-Square Test for the 120 chronic pain participants who were 60 years and older. For the chronic pain of the low back, hip, or leg category, the null hypothesis that there was no difference in the 30 CST pass/fail rate between patients with chronic pain of the low back, hip, or leg and those with no pain was rejected due to a p-value of 0.0007. Thus, there is significant evidence that the 30 CST pass/fail rate differs between patients with chronic pain of the low back, hip, or leg and those without.

Additional analysis was conducted utilizing multiple logistic regression analysis. Testing was conducted to determine if a relationship existed between the fall risk predicted by the 30 CST, designated as the dependent variable, and the independent variables, which included BMI, heart rate, systolic blood pressure, diastolic blood pressure, and chronic pain of the low back, hip, or leg. The 30 CST outcomes were assigned values of passing the test (1) and failing the test (0). The associations were calculated with a 95% confidence interval. The p-values were statistically significant if they were <0.05. The p-values for all the independent variables that were less than 0.05 were evaluated, and the stand-out was chronic pain of the low back, hip, or leg (SE 0.148, p-value 0.0002). This result indicated that chronic pain of the low back, hip, or leg was a statistically significant risk factor for future falls, based on the 30 CST. The results are summarized in Table 4. With a coefficient value of -0.58, having chronic pain of the low back, hip, or leg is associated with a 44% reduction in the odds of passing the 30 CST, compared to those without chronic pain, controlling for other variables. BMI, heart rate, systolic blood pressure, and diastolic blood pressure were not statistically significant. All the different independent variables had a Variance Inflation Factor (VIF) less than 1. Using a VIF cutoff value of 5, this meant that there was no multicollinearity in this regression model.

Table 4. Logistic regression analysis using the 30 CST as the dependent variable.

	Coefficient	SE	T-Stat	P-value	Lower 95%	Upper 95%	VIF
Intercept	1.38	0.531	2.613	0.010	0.336	2.438	1.000
Age	−0.009	0.005	−1.966	0.052	−0.020	0.000	0.781
BMI	0.013	0.007	1.727	0.087	−0.002	0.027	0.803
Pulse	0.002	0.003	0.683	0.496	−0.004	0.009	0.675
SBP	−0.003	0.003	1.018	0.311	−0.008	0.002	0.325
DBP	−0.001	0.004	−0.341	0.734	−0.008	0.006	0.301
Chronic pain of Low back, hip, or leg	−0.581	0.148	−3.912	0.0002	0.875	−0.287	0.975

To evaluate the performance of this logistic regression model, a Receiver Operating Characteristic (ROC) curve was created, and the Area Under the Curve (AUC) was calculated in R. The results are presented in Fig. 1. An AUC of 0.747 indicates the model has acceptable discrimination performance; it can correctly differentiate between individuals who passed the 30 CST and those who did not more often than would be expected by chance.

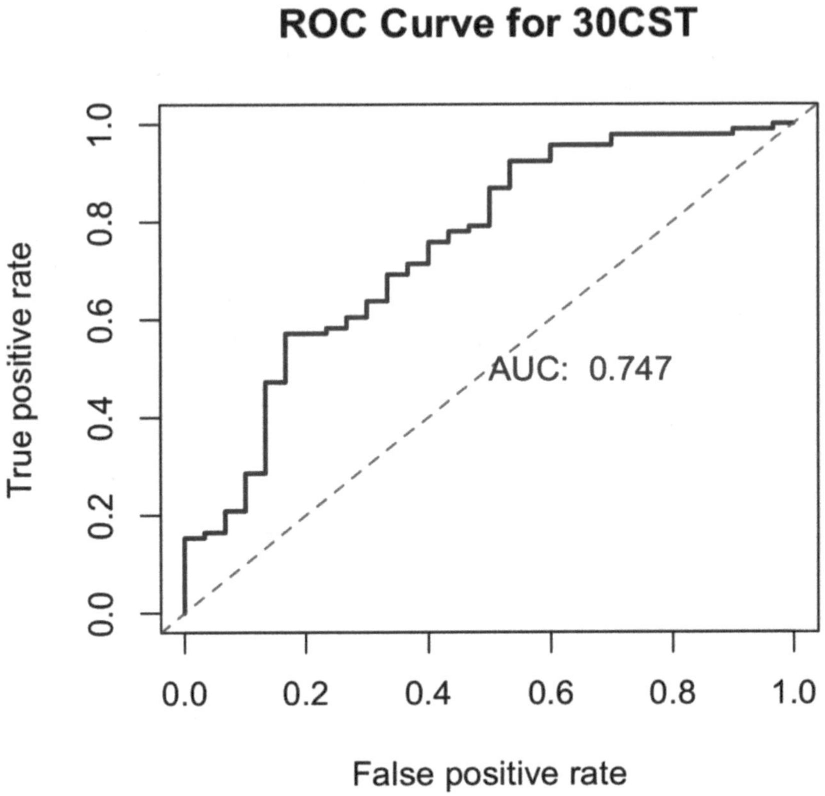

Fig. 1. ROC Curve and AUC for the 30 CST logistic regression model.

4 Discussion

The results showed that most chronic pain participants in this pain management clinic setting had a high risk of falling. Of the participants who were 60 years and older, 76.4% failed the 30 CST. Of those participants who had chronic pain of the low back, hip, or leg, 79.1% failed the 30 CST. This result is worrisome, suggesting that most patients being referred to pain management clinics are at high risk for future falls. The fall risk for pain management patients is higher than the national average. In 2023, the Morbidity and Mortality Weekly Report, produced by the Centers for Disease Control, showed that 14 million (27.6%) adults aged 65 and older fell between 2020 and 2021 [16]. Patients seen in pain management clinics, the majority of who tend to have low back and extremity pain, have a much higher risk of falling compared to adults in the same age group.

Having chronic pain was a statistically significant factor that contributed to the risk of falls, as indicated by our study. Controlling for other variables, having chronic pain of the low back, hip, or leg was associated with a 44% reduction in the odds of passing the 30 CST, compared to those without chronic pain. These results support previous studies conducted by Cai et al. [3]. The risk of falling is the highest for people suffering from chronic pain of the low back, hip, or leg. Interestingly, in our study, blood pressure, heart rate, and BMI were not found to be statistically significant.

Healthcare providers may consider these results by evaluating patients with chronic pain, especially of the low back, hip, or leg. Pain management providers should also be on the alert that these patients are at a higher risk for falls. Providers may refer these patients early on to physical therapy for muscle strengthening and balance evaluation. Home health safety evaluations conducted by nurses may also be included to identify safety measures that can be implemented at home to prevent falls.

We analyzed the number of sit-stand cycles completed by 213 participants, categorizing them by age and gender. However, we could not compare this data to the general population aged 18–90, since no published data exists for this entire age range. Future work could include conducting a trial with participants in a general primary care clinic and comparing the data to that of the pain management population. We can also evaluate if there is a difference in the fall risks between these two populations.

Future studies can investigate whether the number and types of medications patients take increase the risk of falls. Fall risks for patients in a pain management clinic can also be evaluated with other tests, including the 4-Stage Balance Test and the Timed Up and Go Test. Those results can be compared to the 30 CST. A prospective study may also be conducted to determine if patients who fail the 30 CST actually fall in the future. A more extensive study may determine whether these falls occur within the next year, two years, or up to five years. Finally, a study can be conducted to determine if patients who fail the 30 CST by a higher margin have an increased number of falls or if they experience falls sooner.

5 Conclusion

This study, conducted in a pain management clinic, found that over 76% of the chronic pain patients in the clinic failed the 30 CST. Patients who had chronic pain of the low back, hip or leg, had an even higher 30 CST fail rate of 79%. We found that having chronic pain of the low back, hip, or leg was associated with a 44% reduction in the odds of passing the 30 CST. This evidence emphasizes the high risk of falling in patients with chronic pain especially of the low back, hip, or leg. Primary care doctors should consider evaluating all patients who suffer from chronic pain, particularly of the low back, hip, or leg for fall risks, so they can implement early interventions and prevent falls. These same types of patients who are already being seen in a pain management clinic should also be evaluated for fall risks to implement early treatment. We hope this will minimize morbidity and mortality in this patient population and improve their quality of life.

References

1. Older Adult Falls Data: Centers for Disease Control and Prevention. https://www.cdc.gov/falls/data-research/index.html#:~:text=adult%20fall%20trends. Accessed 1 Mar 2025
2. Falls: World Health Organization. https://www.who.int/news-room/fact-sheets/detail/falls. Accessed 1 Mar 2025
3. Cai, Y., Leveille, S.G., Shi, L., Chen, P., You, T.: Chronic pain and risk of injurious falls in community-dwelling older adults. J. Gerontol. Ser. A Biol. Sci. Med. Sci. **76**(9), e179–e186 (2021). https://doi.org/10.1093/gerona/glaa249

4. The 4-Stage Balance Test: Centers for Disease Control and Prevention. https://www.cdc.gov/steadi/media/pdfs/STEADI-Assessment-4Stage-508.pdf
5. Timed Up and Go (TUG): Centers for Disease Control and Prevention. https://www.cdc.gov/steadi/media/pdfs/steadi-assessment-tug-508.pdf
6. 30-Second Chair Stand: Centers for Disease Control and Prevention. https://www.cdc.gov/steadi/media/pdfs/steadi-assessment-30sec-508.pdf?CDC_AAref_Val=https://www.cdc.gov/steadi/pdf/STEADI-Assessment-30Sec-508.pdf. Accessed 1 Mar 2025
7. McCarthy, E.K., Horvat, M.A., Holtsberg, P.A., Wisenbaker, J.M.: Repeated chair stands as a measure of lower limb strength in sexagenarian women. Gerontol. Biol. Sci. Med. Sci. **59**(11), 1207–1212 (2004). https://doi.org/10.1093/gerona/59.11.1207
8. Chronic Pain Among Adults- Unites States, 2019–2021. https://www.cdc.gov/mmwr/volumes/72/wr/pdfs/mm7215-H.pdf. Accessed 1 Mar 2025
9. Lee, A.W., Lee, M.S., Yeh, D.P., Yeh, H.-J.J.: Sensor-integrated chairs for lower body strength and endurance assessment. Sensors **24**(3), 788 (2024). https://doi.org/10.3390/s24030788
10. Lee, A., Lee, M., Yeh, C., Yeh, K.: Measuring fall risk using the Internet-of-Things chair. In: Proceedings of the 10th International Conference on Internet of Things, Big Data and Security, pp. 353–360 (2025). https://doi.org/10.5220/0013404500003944
11. Lee, M.S., Lee, A.W., Schroeder, P., Yeh, C., Yeh, K.: Internet-of-Things chair and the 30 s chair stand test. In: Alsadoon, A., Shenavarmasouleh, F., Amirian, S., Ghareh Mohammadi, F., Arabnia, H.R., Deligiannidis, L. (eds.) CSCE 2024. CCIS, vol. 2259, pp. 64–70. Springer, Cham (2025). https://doi.org/10.1007/978-3-031-85908-3_5
12. Lee, A., Yeh, H.J.: Real-time monitoring of urine output with Internet-of-Things connected Foley catheters. In: 2022 International Conference on Computational Science and Computational Intelligence (CSCI), pp. 1808–1811 (2022). IEEE. https://doi.org/10.1109/CSCI58124.2022.00322
13. Lee, A., Lee, M., Yeh, H.-J.J.: An IoT-based automatic and continuous urine measurement system. BioMedInformatics **3**(2), 446–454 (2023). https://doi.org/10.3390/biomedinformatics3020030
14. Yeh, C., Lee, A., Dy, H., Li, K.: Internet-of-Things management of medical chairs and wheelchairs, pp. 183–188 (2022). https://doi.org/10.5220/0011059600003194
15. Yeh, C., Lee, A.W., Lee, M.S., Li, K.C.: Internet-of-Things monitoring of physical restraint patients. In: 2021 International Conference on Computational Science and Computational Intelligence (CSCI), pp. 1803–1807 (2022). https://doi.org/10.1109/csci58124.2022.0032
16. Rikard, M.: Chronic pain among adults — united states, 2019–2021. MMWR. Morb. Mortal. Wkly. Rep. **72**(15) (2023). https://doi.org/10.15585/mmwr.mm7215a1

An ABI Architecture for Hospital Appointment Duration Prediction

Sofia Touças, Marco Carvalho, David Cruz, João Lopes(✉), Tiago Guimar aes, Júlio Duarte, and Manuel Filipe Santos

ALGORITMI/LASI Research Centre, University of Minho, Braga, Portugal
jlopes@dsi.uminho.pt

Abstract. Hospitals face the ongoing challenge of managing limited resources in dynamic environments, where variability in appointment duration hinders scheduling efficiency, increases waiting times, and affects the utilization of healthcare professionals. Within the Hospital 4.0 paradigm, traditional Business Intelligence (BI) systems prove inadequate for supporting adaptive and data-driven decision-making. This paper presents an Adaptive Business Intelligence (ABI) architecture that integrates data ingestion, predictive modeling, appointment schedule optimization, and dashboard-based visualization, all orchestrated through containerized services. Each predictive model is encapsulated within Docker containers and exposed via FastAPI microservices, enabling scalability and modular deployment. The main contribution of this work is to demonstrate how real-time predictive insights can effectively support the optimization of hospital appointment scheduling and resource allocation.

Keywords: Adaptive Business Intelligence · Hospital 4.0 · Forecasting · Healthcare Management

1 Introduction

The increasing digitalization of healthcare services and the expansion of clinical and administrative data collection have profoundly transformed hospital management, while simultaneously introducing new challenges related to operational complexity and variability. Among these challenges, the uncertainty surrounding medical appointment duration stands out as a key factor that compromises scheduling efficiency, generates cascading delays, affects resource availability, and undermines the overall patient experience [1,18]. Inefficient management of this process impacts the entire hospital ecosystem—from staff scheduling to the utilization of physical infrastructure—highlighting the need for intelligent and adaptive management solutions. The Hospital 4.0 concept emerges as a new paradigm for digital transformation in healthcare, inspired by the principles of Industry 4.0, and promoting interoperability, automation, distributed intelligence, and real-time decision support [2,6]. However, traditional Business Intelligence (BI)

H. R. Arabnia et al. (Eds.): CSCI 2025, CCIS 2971, pp. 299–306, 2027.
https://doi.org/10.1007/978-3-032-28463-1_24

systems—focused on descriptive analytics and reliant on static, retrospective reports—are no longer sufficient to address the dynamic and predictive demands of modern hospital environments [3,4]. Such systems lack mechanisms for anticipating variations, automatically adapting processes, and integrating continuous feedback from daily operations. As a response to these limitations, Adaptive Business Intelligence (ABI) architectures have emerged as a natural evolution of traditional BI, incorporating predictive modeling, process optimization, and continuous learning. These architectures combine real-time data streams with forecasting and automatic adjustment mechanisms, enabling more efficient and responsive management of hospital resources [1,2,5]. Their application in healthcare holds significant potential for reducing waiting times, optimizing appointment schedules, and improving the overall utilization of clinical infrastructure. Within this context, this paper proposes an ABI architecture tailored to hospital management, using the prediction of medical appointment duration as a case study. The main goal is to demonstrate how a predictive module can be integrated into a broader ABI system, where predictions feed optimization pipelines and are operationalized through Docker-based orchestration mechanisms. This approach promotes modularity, scalability, and seamless integration across system components, ultimately contributing to the development of smarter, more adaptive, and efficient hospitals [3,18].

1.1 Business Intelligence (BI)

Business Intelligence (BI) systems have played a significant role in hospital management by providing analytical support for monitoring performance indicators, operational costs, and patient flow through interactive reports and dashboards that assist strategic decision-making [4]. These platforms enable hospitals to identify historical trends, detect bottlenecks, and assess overall operational efficiency. However, in highly dynamic and uncertain environments such as hospitals, the inherently descriptive nature of traditional BI proves insufficient. The lack of predictive, prescriptive, and real-time decision-support mechanisms prevents these systems from anticipating demand fluctuations, automatically adjusting schedules, or optimizing clinical resource utilization adaptively. As a result, there is a growing need to transition toward more intelligent and proactive architectures that integrate predictive modeling and continuous optimization into the hospital management cycle.

1.2 Adaptive Business Intelligence (ABI)

Adaptive Business Intelligence (ABI) represents a significant evolution from traditional Business Intelligence (BI) systems by integrating not only descriptive analytics but also predictive, prescriptive, and intelligent decision-support capabilities [5]. This approach shifts BI from a reactive system—focused on analyzing past events—to a proactive and self-adjusting framework capable of anticipating future scenarios and dynamically adapting to contextual changes. The ABI architecture typically combines three core components: predictive models, which estimate future behaviors and trends; optimization modules, which recommend or execute actions based on those predictions; and continuous data integration mechanisms, which ensure that models remain updated and aligned with the latest operational realities. In hospital environments, this adaptive capability is particularly valuable, as it enables the dynamic adjustment of medical schedules, resource redistribution, and near real-time improvement of operational efficiency, contributing to the development of smarter, more resilient, and responsive management systems (Fig. 1).

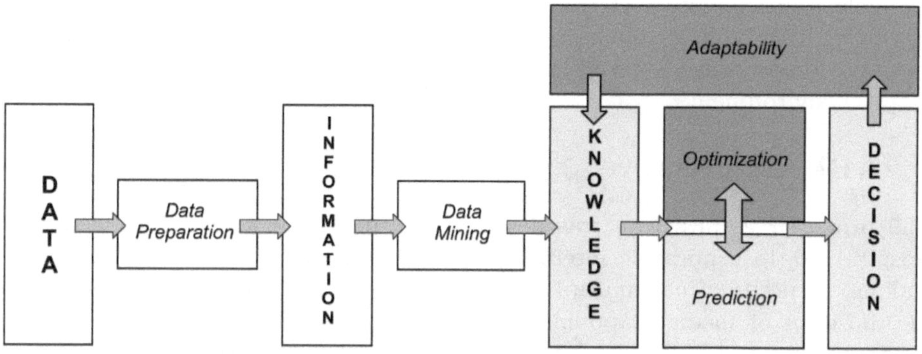

Fig. 1. Conceptual Architecture of an Adaptive Business Intelligence System. Withdrawn from [17].

1.3 Hospital 4.0

The concept of Hospital 4.0, inspired by the principles of Industry 4.0, represents a new paradigm for digital transformation in healthcare. This model is grounded in six key dimensions: interoperability, virtualization, decentralization, real-time decision-making, service orientation, and modularity [6]. Together, these principles enable seamless integration across heterogeneous systems, the creation of digital replicas of hospital operations, autonomous decision-making at decentralized levels, and the ability to respond rapidly to contextual changes. Emerging technologies such as the Internet of Things (IoT), Big Data, Cloud Computing, and Artificial Intelligence (AI) play a pivotal role within this ecosystem, enabling the collection, storage, and analysis of vast volumes of clinical and operational data from multiple sources. The integration of these technologies supports continuous process monitoring, early anomaly detection, and data-driven management strategies, ultimately leading to greater efficiency, personalization, and quality in healthcare delivery.

1.4 Methodologies

The application of Data Science in healthcare requires methodologies that ensure quality, traceability, and reproducibility of results, given the sensitivity and complexity of clinical data. Among existing frameworks, the CRISP-DM (Cross Industry Standard Process for Data Mining) is one of the most widely adopted, structuring the analytical development lifecycle into six key phases: business understanding, data understanding, data preparation, modeling, evaluation, and deployment [7]. This framework provides a systematic foundation for transforming hospital data into actionable knowledge, ensuring consistency and transparency throughout the process. In the context of Adaptive Business Intelligence (ABI), CRISP-DM plays an even more crucial role by enabling iterative cycles of continuous learning and dynamic model adaptation as new data sources become available. Consequently, ABI inherits CRISP-DM's methodological rigor while extending it through automation, operational feedback, and ongoing model performance reassessment, fostering continuous system evolution in ever-changing hospital environments.

2 ABI Architecture

The proposed architecture was designed in alignment with the principles of Hospital 4.0, incorporating interoperability, modularity, and real-time decision-making. Structured into multiple functional layers, the solution aims to predict the duration of medical appointments, a critical factor for optimizing hospital scheduling and balancing human and infrastructural resources. The system integrates clinical and administrative data from heterogeneous sources, ensuring normalization and quality prior to analytical use. The predictive modeling layer encapsulates Machine Learning models within independent services deployed in Docker containers, ensuring scalability, isolation, and ease of maintenance. These services are exposed through REST APIs built with FastAPI, enabling seamless communication with the other components of the architecture. The generated predictions are then integrated into interactive dashboards, providing near real-time insights to physicians, managers, and hospital administrators. By combining automation, predictability, and continuous data integration, the proposed architecture establishes a robust foundation for intelligent decision support, enhancing operational efficiency and adaptability within hospital environments (Fig. 2).

The data layer forms the foundation of the proposed architecture and is responsible for the ingestion, integration, and preparation of information originating from various hospital systems, such as the Hospital Information System (HIS), Electronic Health Records (EHR), and medical scheduling systems. This layer ensures continuous and reliable acquisition of clinical, administrative, and operational data, maintaining consistency and integrity throughout the analytical process. Data processing is carried out through ETL (Extract, Transform, Load) pipelines, which standardize formats, handle missing values, remove duplicates, and ensure data quality and semantic consistency prior to analytical use. Once cleaned and transformed, the data are stored in a hospital Data Warehouse, which centralizes and historizes records, ensuring traceability, security, and auditability of all operations [8]. As such, this layer plays a crucial role in establishing a reliable and interoperable data infrastructure that supports the upper layers of modeling and decision-making.

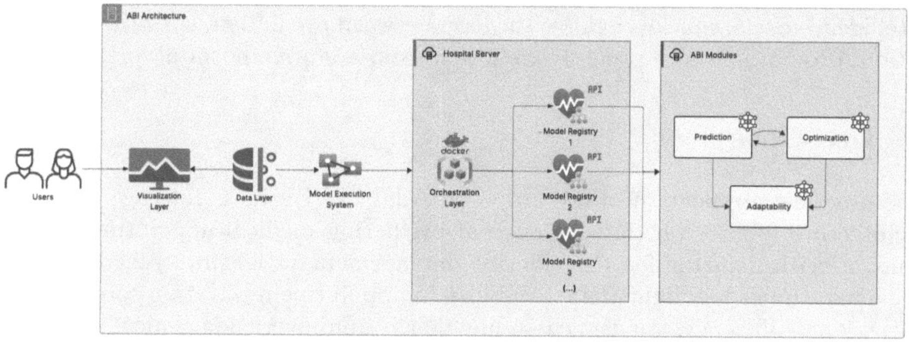

Fig. 2. Proposed ABI Architecture.

The prediction layer is responsible for estimating the duration of medical appointments based on historical and contextual data, including information such as appointment type, medical specialty, healthcare professional, and patient clinical profile. Its goal is to enhance hospital scheduling efficiency by providing realistic forecasts that minimize delays and optimize resource utilization. Each Machine Learning model is encapsulated within Docker containers to ensure portability, scalability, and reproducibility of results [9]. These models are exposed through REST APIs implemented in FastAPI, enabling seamless integration with the system's optimization and visualization modules. To improve predictive accuracy, clustering techniques can be applied to pre-segment appointments into homogeneous groups—for instance, by medical specialty, procedure type, or patient profile [16]. This segmentation allows predictive models to be fine-tuned for each cluster, enhancing both robustness and interpretability. Furthermore, the modular design of this layer ensures that new models, algorithms, or cluster definitions can be dynamically integrated without compromising the system's overall stability or performance.

The optimization layer is responsible for transforming the predictions generated by the modeling layer into actionable operational decisions, aiming to enhance the efficiency of hospital scheduling and the utilization of clinical resources. This layer employs optimization algorithms implemented, allowing multiple strategies to be executed in parallel while ensuring scalability, isolation, and simplified maintenance. The modular design of this layer also allows optimization algorithms to be adapted or replaced without impacting the overall system functionality, supporting continuous integration with the prediction layer and ensuring that the system remains flexible, robust, and responsive to changing hospital contexts.

The visualization layer is responsible for translating complex predictions and optimization results into clear, actionable indicators for hospital managers, clinicians, and administrators. This is achieved through interactive corporate dashboards, which consolidate real-time metrics generated by the prediction and optimization layers [11]. As an integral part of the complete Adaptive Business

Intelligence cycle, this layer closes the loop between prediction, optimization, and action, promoting more proactive, efficient, and adaptive hospital management.

3 Use Case

The practical application of the proposed Adaptive Business Intelligence (ABI) architecture focuses on the challenge of predicting medical appointment durations, a critical factor for the efficient management of hospital schedules [18]. Previous studies highlight that variability in appointment duration compromises operational efficiency, triggers cascading delays, and overloads clinical resources, necessitating the use of predictive and optimization techniques [12,13]. The first phase involved the development and operationalization of predictive models, responsible for estimating appointment durations based on historical and administrative data. Various Machine Learning approaches were tested, including Logistic Regression, Random Forest, and XGBoost, techniques that have previously yielded satisfactory results in hospital operations contexts [14,18]. To ensure portability, scalability, and continuous integration, the models were encapsulated in Docker containers and exposed via FastAPI, enabling real-time consumption of predictions by hospital systems. This real-time availability allows the early identification of longer appointments and proactive adjustment of medical schedules. Based on these predictions, the ABI system incorporates an optimization module, leveraging algorithms that have demonstrated effectiveness in clinical resource scheduling, such as Hill Climbing and Simulated Annealing [15]. Recent studies, including Lopes et al. (2023) [3], provide the methodological foundation for this module, adapting the algorithms to the specific context of medical appointments. The project involves collaboration with the Unidade Local de Saúde de Santo António (ULLSA), which provides data, operational knowledge, and resources for deployment in a real-world setting. Adapting these methods allows the generation of real-time optimized schedules, balancing resource availability, operational constraints, and the intrinsic variability of appointment durations, thereby promoting a more efficient, proactive, and adaptive hospital management system.

4 Conclusions

This research presented an ABI architecture applied to the hospital context, focusing on the prediction of medical appointment durations. Grounded in the principles of Hospital 4.0, the proposed solution modularly integrates data, prediction, orchestration, optimization, and visualization layers, enabling the transformation of clinical and administrative data into actionable predictions for decision support. The main contribution of this work lies in the design of a scalable, interoperable, and service-oriented architecture capable of embedding ML models into hospital decision-making pipelines. The proposed approach concretely demonstrates how real-time predictions can be operationalized and integrated into decision-support systems, allowing the anticipation of events, optimization of critical processes, and improvement of medical scheduling management. Currently in its implementation phase, the solution already shows tangible organizational benefits, including greater operational predictability, improved utilization of clinical resources, and a smoother, more satisfying patient experience. Nonetheless, relevant limitations were identified, particularly regarding the integration of heterogeneous hospital systems and data quality issues, which pose real challenges to large-scale deployment. Future work will focus on the development of the optimization module, enabling the transition from a predictive system to a fully adaptive system capable of autonomously adjusting hospital schedules in response to operational and clinical variability—thus advancing the vision of the Intelligent and Adaptive Hospital.

Acknowledgments. This work has been supported by FCT – Fundação para a Ciência e Tecnologia within the R&D Unit Project Scope 2024.07481.IACDC

References

1. Lopes, J.P., Pinto, T., Silva, A., Vale, Z., Novais, P.: Adaptive Business Intelligence platform and its contribution as a support in the evolution of Hospital 4.0. Procedia Comput. Sci. **192**, 3852–3861 (2021). https://doi.org/10.1016/j.procs.2021.09.160
2. Lopes, J.P., Pinto, T., Silva, A., Vale, Z., Novais, P.: Adaptive business intelligence: a new architectural approach. Procedia Comput. Sci. **176**, 3529–3538 (2020). https://doi.org/10.1016/j.procs.2020.09.181
3. Lopes, J.P., et al.: Adaptive business intelligence and healthcare 4.0: a case study on consultation duration forecasting. Procedia Comput. Sci. (2025). (in press). https://doi.org/10.1016/j.procs.2025.08.083
4. Shmueli, G., Patel, N.R., Bruce, P.C.: Data Mining for Business Intelligence: Concepts, Techniques, and Applications in Microsoft Office Excel with XLMiner. Wiley (2010)
5. Michalewicz, Z., Schmidt, M., Michalewicz, M., Chiriac, C.: Adaptive Business Intelligence. Springer (2007)
6. Hermann, M., Pentek, T., Otto, B.: Design principles for Industrie 4.0 scenarios: a literature review. Technische Universität Dortmund (2016)
7. Wirth, R., Hipp, J.: CRISP-DM: Towards a standard process model for data mining. In: Proceedings of the 4th International Conference on the Practical Applications of Knowledge Discovery and Data Mining, pp. 29–39 (2000)
8. Raghupathi, W., Raghupathi, V.: Big data analytics in healthcare: promise and potential. Health Inf. Sci. Syst. **2**(1), 3 (2014). https://doi.org/10.1186/2047-2501-2-3
9. Merkel, D.: Docker: lightweight Linux containers for consistent development and deployment. Linux J. **2014**(239), 2 (2014)
10. Burns, B., Grant, B., Oppenheimer, D., Brewer, E., Wilkes, J.: Borg, Omega, and Kubernetes. Commun. ACM **59**(5), 50–57 (2016). https://doi.org/10.1145/2890784
11. Dowding, D., et al.: Dashboards for improving patient care: review of the literature. Int. J. Med. Inf. **84**(2), 87–100 (2015). https://doi.org/10.1016/j.ijmedinf.2014.10.001

12. Gupta, D., Denton, B.: Appointment scheduling in health care: challenges and opportunities. IIE Trans. **40**(9), 800–819 (2008). https://doi.org/10.1080/07408170802165880
13. Samorani, M., Blount, L.G., Harris, S., Santoro, M.A.: Machine learning and medical appointment scheduling: creating and perpetuating inequalities in access to health care. Am. J. Public Health **110**(4), 440–441 (2020). https://doi.org/10.2105/AJPH.2020.305570
14. Mišić, V.V., Yom-Tov, E., Green, L.V., Shaparin, N., Shani, M.: A simulation-based evaluation of machine learning models in healthcare operations. Manuf. Serv. Oper. Manage. **23**(4), 906–924 (2021). https://doi.org/10.1287/msom.2020.0937
15. Hulshof, P.J.H., Kortbeek, N., Boucherie, R.J., Hans, E.W., Bakker, P.J.M.: Taxonomic classification of planning decisions in health care: a structured review of the state of the art in OR/MS. Health Syst. **1**(2), 129–175 (2012). https://doi.org/10.1057/hs.2012.18
16. Jiang, F., et al.: Artificial intelligence in healthcare: past, present and future. Stroke Vasc. Neurol. **2**(4), 230–243 (2017). https://doi.org/10.1136/svn-2017-000101
17. Michalewicz, Z., Schmidt, M., Michalewicz, M., Chiriac, C.: Adaptive Business Intelligence. Springer, Heidelberg (2007). https://doi.org/10.1007/978-3-540-73645-6
18. Touças, S., et al.: A predictive modeling approach for estimating hospital appointment duration. In: Procedia Computer Science, 5th International Workshop on Healthcare Open Data, Intelligence and Interoperability, Istanbul, Türkiye (2025)

A Review of Blockchain Technology in Healthcare

Nsovo B. Maluleke, Topside E. Mathonsi(✉), and Tshimangadzo M. Tshilongamulenzhe

Department of Information Technology, Tshwane University of Technology, Pretoria, South Africa
{MathonsiTE,TshilongamulenzheTM}@tut.ac.za

Abstract. Blockchain technology is reshaping a wide range of industries, and healthcare is no exception. Its decentralized architecture enhances data security, fosters trust among stakeholders, and ensures that sensitive information can be shared safely, regardless of geographic boundaries. This review paper explores how blockchain is being applied in healthcare, offering a comprehensive overview of emerging technologies and use cases that address some of the sector's most pressing challenges. The paper highlights how blockchain can protect patient data from breaches and unauthorized access, a critical concern in an era of increasing cyber threats. It also explores how blockchain enables better interoperability, allowing different healthcare systems to share data securely and seamlessly. In the pharmaceutical supply chain, blockchain improves transparency and drug traceability, helping combat the global issue of counterfeit medications. Additionally, technology can safeguard the authenticity of clinical trial information, which is essential for meeting regulatory standards. By synthesizing recent developments and pinpointing open research issues, this work underscores blockchain's potential to transform healthcare delivery, strengthen system-wide trust, and enhance patient outcomes. This paper contributes to a growing body of knowledge that supports a more secure, efficient, and patient-centered healthcare ecosystem.

Keywords: Blockchain · Healthcare · Unprocessed Data · Patients · Doctors · IoT · Unauthorized · Authorized · Decentralized

1 Introduction

The digitization of healthcare has led to the proliferation of electronic patient records, generating vast volumes of sensitive data that must be protected during storage, usage, and exchange. In response to these challenges, blockchain technology emerges as a promising solution, offering transparency, security, and integrity in managing healthcare information. By ensuring responsible data sharing and storage, blockchain has attracted growing interest from both industry leaders and academic researchers, with new applications and studies appearing almost daily [1–4]. At its foundation, blockchain operates as a decentralized record-keeping system that supports peer-to-peer data transactions. It can operate in public or private environments and guarantees that stored information

H. R. Arabnia et al. (Eds.): CSCI 2025, CCIS 2971, pp. 307–320, 2027.
https://doi.org/10.1007/978-3-032-28463-1_25

is trustworthy and verifiable [1, 5]. A fundamental characteristic of blockchain is the implementation of smart contracts, which are automated, legally enforceable agreements facilitating secure, decentralized interactions among several parties [6, 7].

Blockchain has found applications across diverse sectors, including energy, finance, voting systems, and healthcare [6, 8–11, 13]. One key strength of blockchain is that it eliminates the need for external intermediaries. By applying blockchain consensus models alongside cryptographic verification, it preserves the accuracy of transactions, even when participants lack mutual trust [1, 14]. Each data block is verified and recorded on the ledger through consensus-based methods. Like Proof-of-Work (PoW), then distributed across all participating nodes [15, 16]. In the healthcare environment, blockchain offers transformative prospective. It deals with critical factors such as secure information handling, personal information protection, interoperability, and access regulation [17, 18]. Interoperability is especially important, as healthcare providers increasingly need to share accurate and timely patient information. This involves consistent communication between humans and machines across systems [19–22]. Blockchain can support secure and standardized sharing of Electronic Health Records (EHRs) across institutions and geographies, even in the absence of direct trust relationships, while still respecting patient consent [23–28].

This paper also rightly acknowledges the economic implications of rising investments in healthcare infrastructure and software systems [29]. By improving compliance, reducing administrative costs, enhancing patient data management, and optimizing operational processes, blockchain can improve healthcare outcomes and increase efficiency across the ecosystem [30]. Furthermore, its ability to track medical products and pharmaceuticals can reduce the risk of counterfeit drugs, strengthening the integrity of the healthcare supply chain. In addition to its operational benefits, blockchain enhances the Quality of Experience (QoE) for multiple parties, including both patients and medical practitioners, insurers, as well as researchers, through enabling secure, privacy-preserving data sharing. This paper offers a comprehensive review of blockchain's role in healthcare, touching on its economic impact, Quality of Service (QoS) improvements, and potential to drive new business models.

The key contributions are as follows:

1. A review of current blockchain applications in healthcare, with a focus on data security, privacy, and operational efficiency.
2. An exploration of key challenges and limitations in adopting blockchain within healthcare systems.
3. A set of guidelines and open research questions to guide future work in this field.
4. A critical assessment of the strengths and weaknesses of current blockchain-based healthcare solutions.

The paper proceeds as follows. Section B addresses the literature review and details how raw data are processed and delivered to stakeholders through the healthcare blockchain system. Section C examines related works, and Section D summarizes the conclusions.

2 Literature Review

Blockchain technology is transforming data modeling and governance for healthcare applications. This is owing to its exceptional versatility and capacity to segment, safeguard, and distribute medical data and services in unprecedented ways. Current developments in the healthcare sector depend on blockchain technology [32]. Data sources, blockchain technology, healthcare applications, and stakeholders are the four categories into which Healthcare technologies built on blockchain are structured into different groups. A blockchain-based procedure for medical applications is shown in Fig. 1 [40].

Initially, data from social media, labs, medical devices, and alternative sources include consolidated into unprocessed data, which is later scaled up to big data. Blockchain-based healthcare relies heavily on data, which serves as the foundation for the initial layer of the stack. Blockchain technology builds on the unprocessed data layer to create a healthcare architecture that integrates four security-oriented components, which are Healthcare unprocessed Data, Blockchain Technology, Healthcare Technology and Stakeholders. Collective algorithms and protocols are among the unique components of any blockchain platform [21]. Blockchain platforms allow the development and management of transactions, with well-known examples including Ethereum [32], Ripple [22], and Hyperledger Fabric [23]. These systems incorporate features such as smart contracts, cryptographic signatures, digital wallets, membership structures, event logging, and asset management. Interaction with external frameworks and networks occurs through multiple protocols, which may be organized as peer-to-peer, centralized, decentralized, or distributed.

Based on organizational priorities, policymakers may opt for public, private, or federated blockchain frameworks [40]. Following deployment, integration of applications into the larger system is essential. In healthcare, blockchain-based solutions are typically organized into three areas. Data management relates to global R&D data exchange, cloud-hosted data repositories as well as digital patient records. A secondary area involves SCM, particularly in pharmaceutical tracking and clinical trial management.

The third class focuses on the Internet of Medical Things (IoMT), which encompasses healthcare IoT and medical devices, infrastructure and data security, and artificial intelligence. Figure 2 shows healthcare applications in the blockchain. Stakeholders, including corporate users, researchers, and patients, are at the top of the hierarchy and access the advantages offered by blockchain in healthcare. At this layer, users focus on ensuring security and privacy when sharing, processing, and managing data [36].

Four levels make up the process: workflow stakeholders, healthcare apps, blockchain technology, and raw data. Blockchain, as a decentralized technology, allows various stakeholders to derive value from healthcare applications [20].

a) The process is built on top of healthcare raw data. This layer contains an enormous number of health-related data produced by various sources, such as wearable sensor data, clinical and laboratory results, insurance claims, call and social media logs, medical imaging, and patient-generated data like personal health diaries and self-reported symptoms. These various forms of data supply the necessary input for blockchain systems, which allow for its safe storage, retrieval, and analysis [43].

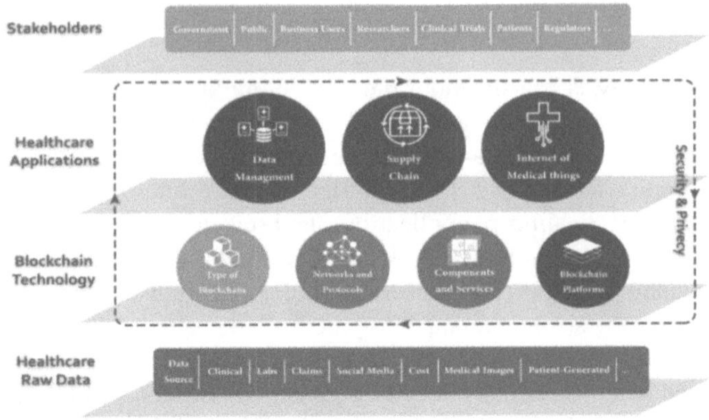

Fig. 1. Provide an illustration of a blockchain-based healthcare application process.

b) The enabling infrastructure for safe, open, and impenetrable processing of medical data is provided by blockchain technology. This layer is made up of various essential parts [44]:
   i. The blockchain platforms (e.g., Ethereum, Hyperledger Fabric) that offer the technical framework for implementing healthcare applications.
   ii. Network and protocols, which specify how data is transmitted, validated, and synchronized across nodes.
   iii. Components and services, such as smart contracts, identity management, and encryption mechanisms.
   iv. The type of blockchain (public, private, or consortium), which establishes governance and accessibility.

c) Applications for the healthcare industry convert blockchain's potential into workable answers to pressing problems. Among the examples are [41]:
   i. Data management protects patient privacy while guaranteeing the accessibility, security, and integrity of electronic health records.
   ii. Supply chain management lowers the danger of fake goods by enabling end to end traceability of medications, medical equipment, and vaccinations.
   iii. The IoMT, which safely links and keeps an eye on medical equipment to improve patient care and operational effectiveness.

d) The top layer of the process is represented by stakeholders, which include all parties that use or profit from the blockchain healthcare ecosystem. This includes government organizations in charge of monitoring public health, patients, and the general public looking for safe access to medical records, commercial users like insurance companies and pharmaceutical firms, researchers and clinical trial teams in need of high-quality datasets, and regulatory organizations making sure that laws and moral principles are followed [42].

The workflow in Fig. 2 illustrates how blockchain technology can close the gap between unprocessed healthcare data and safe, reliable, and useful healthcare applications by combining these four layers [54]. This strategy could promote innovation in the healthcare industry, increase interoperability, and improve patient outcomes.

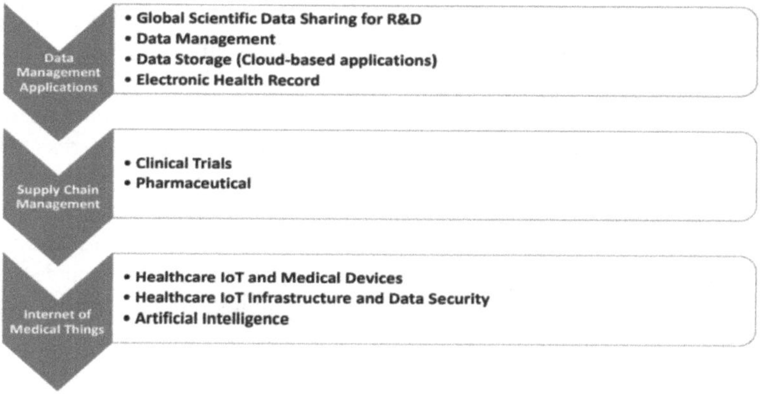

Fig. 2. Implementation of Blockchain Solutions in Healthcare Settings

Advancements in electronic health-related data, cloud healthcare storage, and patient privacy rules provide new options for data management and patient convenience [4]. Blockchain technology has the potential to significantly improve data security, storage, and integration in healthcare organizations. Figure 3 illustrates the seven processes of blockchain-based healthcare data management workflow, which will be addressed more below [17]. Blockchain applications in this category support data storage, management, and exchange (e.g., including cloud-based systems), and electronic health records. More information is outlined below [7].

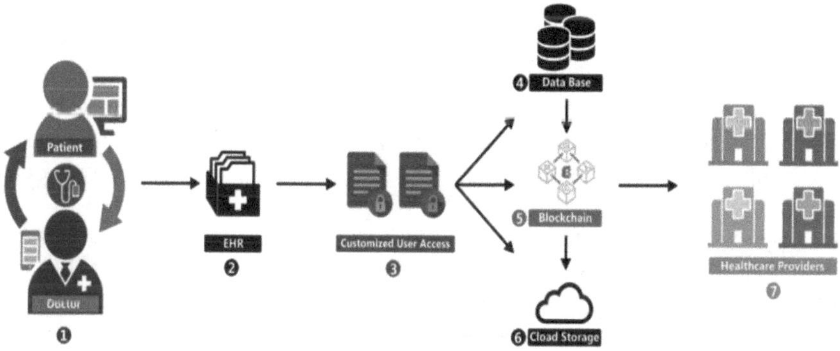

Fig. 3. Healthcare data management in blockchain

i. Step 1: Primary data is gathered through interactions between patients and their medical practitioners and healthcare specialists. The information covers a patient's medical background, existing conditions, and other health-related details.
ii. Step 2: Each patient's EHR is constructed utilizing the primary data gathered in the first phase. Additionally, EHR systems store medical data from imaging, medication histories, and nursing care.

iii. Step 3: Sensitive EHRs are managed directly by the patient, who is responsible for authorizing access. Requests for information are directed to the owner, who decides which parties may view the records.
iv. The core process is formed by Steps 4, 5, and 6, The system comprises the database, blockchain, and cloud storage components. Data are stored across databases and cloud infrastructures, supporting decentralized management, while blockchain maintains confidentiality and ensures that only authorized users can access the data.
v. Step-7: Medical practitioners, including ad hoc clinics, community health centres and medical institutions, seek authorization from the owner to offer safe and sound care. Your medical records will be certified via a distributed database, like blockchain, and be available on your mobile device. Over time, medical professionals can add to it [35].

2.1 Sharing of Scientific Data Around the World

One important and essential step in raising the standard of healthcare professionals and enhancing the intelligence of the healthcare system is the sharing of medical information [39]. Individuals could share their health details. As an example, a patient seeking care may provide their medical background to a medical practitioner during the first consultation [26]. Furthermore, sharing could occur between a stakeholder and an individual, such as a patient disclosing his medical information to an insurance company or a research facility. Even data can be exchanged across boundaries [37]. Modern healthcare systems have limits in their operational mechanisms. Patients are provided with limited visibility into their medical information. As a result, individuals are unaware that their health data may be shared with unknown parties [18]. Blockchain technology can enhance healthcare collaboration by securely sharing electronic health data.

Blockchain-based healthcare has made significant contributions. The following sections discuss some of the contributions made in this paper. A private blockchain-based tracking system was proposed by [17] for securely exchanging electronic health data throughout Europe. The Healthcare Data Gateway (HGD) was developed by [29] for a blockchain-based healthcare data exchange tool. This approach allows for easy control and sharing of client data while maintaining privacy.

This approach enhances healthcare system intelligence while maintaining patient privacy. A framework proposed for cross-domain image sharing that utilizes blockchain technology as a distributed data storage [10]. This creates a ledger of radiological investigations and allows for configurable user permissions. Blockchain technology was used to develop a MedBlock framework that improves data management and sharing in EMR systems [11].

The MedBlock framework allows patients to access EMRs from several hospitals without having their existing medical data split into distinct databases. Blockchain technology can facilitate data exchange and collaboration, providing hospitals with a better grasp of patients' medical history prior to consultations.

A multi-level location sharing strategy using blockchain technology was presented by [12]. The objective was to implement blockchain-based location sharing in telecare medical information systems while maintaining private information. Merkle tree and encryption that order-preserving were used to meet the main requirements for location sharing, including decentralization, confidentiality, variability, multi-level privacy protection, irretrievability, and unforgeability.

The approach proves to be applicable and suitable for both patients and healthcare providers and serves to secure location details in telemedicine systems [32]. [13] Proposed MedChain, a blockchain-based system for sharing healthcare data throughout sessions. Through adopting a digest chain architecture, MedChain preserves the trustworthiness of IoT-based clinical data as it is transmitted and shared. This addresses efficiency difficulties with previous systems like Medrec [16] and MedBlock [31]. MedChain's examination findings demonstrate its efficiency and security for healthcare data sharing [33].

2.2 Data Management

Despite the enormous amount of data generated in industries such as healthcare and IoT [14], data security and privacy are frequently breached, either purposefully or unintentionally, by unauthorized users. As a result, several institutions have suffered significant reputation and capital losses. Access to health data should be limited to people with specific responsibilities and rights. Blockchain technology can enable seamless access.

Here are some blockchain solutions specifically designed for this purpose. MedRec demonstrated by [6], is a decentralized EMR management system in which data permissions and operations are registered on the blockchain and executed using smart contracts. MedRec engages with suppliers to ensure that medical information can be authenticated, kept confidential, audited, and exchanged efficiently, while providing patients with complete, tamper-proof clinical data and healthcare services. According to another research, illustrated by [5], offered a method for providing controllable blockchain data management in the cloud to answer users' worries about a lack of control over uploaded ledgers.

The concept includes a trusted authority node that enables users to prevent malicious acts, even under a majority attack. A new approach to healthcare consent management that uses blockchain to give users authority over their health record data was proposed by [9]. However, their implementation lacks both authorization and access control.

2.3 Data Storage (Cloud-Based Application)

Healthcare blockchain transactions are kept as blocks on a decentralized storage system. EHRs are used to organize patient medical data in healthcare systems, serving as the foundation for distributed medical storage [7]. The latter can be stored on premises or in the cloud, with security being a significant consideration. Cloud storage combines storage devices, which are Storage Area Network (SAN), Cloud Storage Infrastructure, Network Attached Storage (NAS), and RAID (Redundant Array of Independent Disks), to provide enormous amounts of storage for IT infrastructure.

Healthcare blockchain transactions are stored as blocks in a decentralized storage system. EHRs are utilized to organize patient medical data in healthcare systems, laying the groundwork for a distributed medical storage system [17]. The latter can be stored on-premises or in the cloud, with security being a major concern. Cloud storage connects multiple storage devices to deliver massive amounts of storage for the IT infrastructure. A coined the term "BlockCloud," referring to the combination of blockchain and cloud technology [10]. The cloud aims to securely store and distribute data without relying on third parties. In order to create and implement policies, the study emphasizes the necessity of working together between governments, healthcare service providers, public health organizations, and doctors [30, 31].

2.4 Electronic Health Records

Keeping track of a patient's health status over time is challenging using traditional paper-based medical records [29]. Furthermore, inaccurate data can lead to abuse of patients. The introduction of electronic health records (EHRs) has reduced the need for such efforts. Access to electronic health records has greatly improved treatment quality in medical practices [83]. In addition, EHR enhances preventive care and illness management. By improving clinical decisions and facilitating collaboration among care teams, digital records are gaining acknowledgment for their value across the healthcare industry [84]. Blockchain technology has been designed to encrypt, exchange, and maintain EHR data across healthcare institutions, based on extensive research. A blockchain-based framework for the secure exchange of clinical data, including encrypted cloud storage for critical patient records, was created [25].

Within this framework, medical data are stored in a digital repository that allows owners to control access. Cloud-based encryption beneath the blockchain layer safeguards safeguarding the trustworthiness and confidentiality of the information. An attribute-based signature technique that uses blockchain technology and different authorities to validate EHRs was proposed [19]. It enables group message broadcasts and can protect against collusion assaults. A safe cloud-based EHR system was presented by [28] utilizing blockchain and attribute-based cryptosystems. They encrypted health information using a combination of identity-based encryption and digital signatures.

In addition to blockchain, medical facilities use various approaches to secure their integrity and traceability. The framework addressed problems associated with unifying scattered health records and managing access for healthcare provider stakeholders, while the previous three research studies concentrated on cryptographic aspects of EHR security. OmniPHR, a distributed paradigm for consolidating personal health records (PHR), addresses these two challenges by using a parallel database to store PHR in blocks and combining structural-semantic interoperability with an up-to-date vision of various PHR formats [5].

The blockchain technology was used to secure medical records with genetic algorithms and discrete wavelet modifications, taking a unique approach [7]. The suggested technique uses a modified cryptographic hash generator to generate the user security key. MD5 (a message-digest method with a 128-bit hash value) [8] strings were transformed into an updated key structure using a discrete wavelet transformation. This strategy improves overall system-level security and resistance to numerous threats.

2.5 Blockchain Approaches for Healthcare Data Management

Blockchain technology has the potential for use in health education, in addition to documenting Bitcoin transactions [2, 3]. Knowledge and skills from many sources are exchanged during the educational process. It presumes that everyone involved is trustworthy. It is becoming more difficult for regulatory authorities to keep control over medical education knowledge as a result of the growth of online learning [4, 5]. Blockchain technology has the potential to enhance the quality and impact of education for health professions instructors across generations. Additionally, it can help evaluate the effectiveness of various educational interventions [43]. Furthermore, implementing blockchain technology allows institutions to provide certification without relying on a third party.

Blockchain technology has the potential to improve medical library administration by providing time-stamped, and versions of journal papers that can be verified [37]. Blockchain technology can secure healthcare knowledge and prevent unauthorized changes. Using blockchain technology to store educational data enables health educators to measure their influence as students become instructors and pass on their expertise to future generations. This feature enables easy tracking of the most effective learning modules. Additionally, certification can be supplied to the appropriate individuals without relying on a third party. These proposals highlight the potential benefits of blockchain technology in health education.

Enrollment Data Management, administrative data is generated when members enroll in health care plans, meet eligibility criteria, and receive credentials. Although this data is primarily produced by government agencies, including the federal and state governments, it can also be used by healthcare providers of all sizes [7]. Managing different data needs human processes such as validating references, credentials, and eligibility. As a result, the entire procedure takes longer to complete. Administrative help can be time-consuming and hinder the efficiency of the healthcare system. Blockchain technology can be used to quickly verify reliable references and records. To improve enrollment efficiency, administrative assistants can simplify procedures, inquire about eligibility, manage networks, and coordinate activities.

Security, there is a growing interest in applying blockchain technology across healthcare systems for addressing security concerns with EHRs [39]. EHRs can improve healthcare delivery [9]. It is produced upon a patient's hospital admission, diagnosis by a doctor, or recording a diagnostic outcome, for example, an MRI scan, into the electronic health record system [10]. Maintaining the confidentiality and integrity of digital data is a top priority, and blockchain technology is presently implemented to protect healthcare data [23]. Blockchain technology is important for managing healthcare data and preserving its value while minimizing storage costs. Blockchain technology is essential for protecting information in digital form and will continue to serve an important function in managing enterprise data.

3 Related Work

Blockchain technology has a substantial impact on healthcare, particularly patient data management, transparency in the supply chain as well as clinical trial management. The system addressed the issue of fragmented electronic health records (EHRs) by creating a blockchain-based system to improve interoperability [19]. They develop a decentralized platform that permitted patients to regulate access to their medical data through smart contracts. This strategy improved data security and privacy by allowing only authorized healthcare providers to access sensitive information. However, the method encountered scaling challenges due to blockchain's inherent high energy consumption and slower transaction speeds, making it difficult to execute at the national level [20].

This approach was expanded on this patient-centric approach by providing more granular control over data access, further empowering patients. Despite this advancement, their solution struggled to interact with the outdated healthcare infrastructure that did not support native blockchain. A noteworthy gap in this research was the difficulties of seamlessly integrating old systems with blockchain technology, which limited its widespread adoption [37].

Blockchain applications have also expanded to include genomic data management. A blockchain architecture was developed to securely share genetic information among research institutes [27]. They stressed the use of encryption and distributed ledger technology (DLT) to preserve sensitive genetic data, with all access documented immutably based on patient consent. Despite these advancements, the system encountered difficulties due to the vast bulk of genetic datasets, which posed storage and processing constraints. Another drawback was a lack of data standards among institutions, which hampered collaboration, particularly in multinational settings.

Following the COVID-19 outbreak, the necessity for transparency and security in healthcare supply chains was imperative. The paper investigated blockchain's ability to protect vaccine distribution by tracing every transaction from producer to healthcare practitioner [1]. Their solution effectively decreased counterfeit concerns, but widespread implementation was hampered by the pharmaceutical industry's reluctance to use blockchain technology, as well as the lack of industry-wide standards for blockchain platforms.

This research advanced this research by creating a blockchain system to monitor the pharmaceutical supply chain, ensuring drug authenticity and regulatory compliance [24]. However, while the system improved openness, it showed weaknesses in scalability and regulatory alignment across nations, limiting its global applicability. Blockchain's significance in clinical trials has also expanded, with substantial advances targeted at increasing data transparency. For example, a blockchain system that secures the immutability of clinical trial data, reducing manipulation and enhancing trust in study results, was proposed by [24]. While this system improved the integrity of trial data, it was hampered by the difficulty of coordinating various parties, including pharmaceutical corporations, regulatory authorities, and healthcare facilities.

More recent research [40] has concentrated on the interaction of blockchain and artificial intelligence (AI) in healthcare. Liu's team developed a blockchain approach for securing datasets used in AI-driven diagnostics, assuring the integrity of medical data required to train AI models. Although the integration of blockchain and AI creates new opportunities for safe healthcare applications, constraints such as data processing speed and blockchain energy consumption remain, limiting mainstream use.

The main limitations of these initiatives are scalability, regulatory restraints, and the problems of combining blockchain with conventional healthcare systems. Regardless of these challenges, blockchain's promise to improve security, transparency, and data governance in healthcare remains promising, demanding additional research to close these gaps and fully realize its potential in the industry. Table 1 presents the link between the review findings.

Table 1. Linkage between the review findings and recommendations

Literature Review Finding	Identified Challenge/Gap	Recommendations
Blockchain enables secure, interoperable Electronic Health Records (EHR) sharing across institutions (e.g., MedRec, MedBlock, MedChain)	Limited interoperability with legacy healthcare systems; fragmented data standards across institutions	Advance legal, technical, and interoperability standards to facilitate seamless cross-border and cross-institutional data sharing
Proven potential for improved patient control over data via smart contracts and encryption	Lack of regulatory clarity on patient consent management across jurisdictions; variations in privacy laws	Develop policy frameworks for standardized consent management and privacy protection at both national and international levels
Blockchain improves pharmaceutical supply chain transparency (drug traceability, counterfeit prevention)	Limited scalability and reluctance from industry stakeholders to adopt blockchain due to cost and complexity	Encourage industry adoption through incentive models, regulatory alignment, and proof-of-concept pilots demonstrating cost-effectiveness
Blockchain supports secure storage of large healthcare datasets (including genomic data)	High energy consumption and limited throughput hinder real-time processing of large datasets	Research and develop lightweight, energy-efficient blockchain architectures and explore hybrid models combining blockchain with other storage solutions
Use of blockchain in clinical trials to ensure immutability of data and reduce manipulation	Coordination difficulties among multiple stakeholders; lack of global regulatory alignment	Foster multi-stakeholder collaboration platforms and harmonized regulatory guidelines for blockchain-enabled clinical research
Integration of blockchain with AI for secure medical data in diagnostics	Data processing speed limitations and scalability concerns AI-blockchain integration	Develop optimization techniques to reduce latency and improve efficiency for AI-integrated blockchain healthcare solutions

4 Conclusion

This review has examined the current and emerging applications of blockchain technology in healthcare, highlighting its potential to revolutionize data management, enhance patient privacy, strengthen supply chain transparency, and ensure the integrity of clinical trial data. The analysis demonstrates that blockchain's decentralization, immutability, and cryptographic safeguards provides patients with enhanced ability to oversee their health information while allowing secure and compatible data sharing among healthcare organizations and across countries. However, the literature also reveals persistent challenges, including scalability limitations, high energy consumption, integration difficulties with legacy healthcare systems, and the lack of universally accepted regulatory and interoperability standards. These barriers must be addressed to enable large-scale, real-world adoption.

Future research should prioritize the development of lightweight, energy-efficient blockchain frameworks; explore hybrid models that balance decentralization with performance; and advance legal and policy frameworks for cross-border data exchange that safeguard privacy while promoting collaboration. Furthermore, longitudinal, real-world pilots are needed to assess operational feasibility and patient outcomes at scale. By synthesizing existing research and identifying unresolved issues, this review underscores blockchain's transformative potential in creating a secure, transparent, and patient-centered healthcare ecosystem. Harnessing this potential will require sustained interdisciplinary collaboration among technologists, healthcare providers, policymakers, and regulators to bridge the gap between promising prototypes and scalable, equitable healthcare solutions.

Acknowledgement. The authors gratefully acknowledge the financial support provided by Tshwane University of Technology for the publication of this journal article.

References

1. Saberi, S., Kouhizadeh, M., Sarkis, J., Shen, L.: Blockchain technology and its relationships to sustainable supply chain management. Int. J. Prod. Res. **57**(7), 2117–2135 (2019)
2. Wan, P.K., Huang, L., Holtskog, H.: Blockchain-enabled information sharing within a supply chain: a systematic literature review. IEEE Access **8**, 49645–49656 (2020)
3. Gaggioli, A.: Blockchain technology: living in a decentralized everything. Cyberpsychol. Behav. Soc. Netw. **21**, 65–66 (2018)
4. Macrinici, D., Cartofeanu, C., Gao, S.: Smart contract applications within blockchain technology: a systematic mapping study. Telemat. Inform. **35**, 2337–2354 (2018)
5. Viriyasitavat, W., Anuphaptrirong, T., Hoonsopon, D.: When blockchain meets Internet of Things: characteristics, challenges, and business opportunities. J. Ind. Inf. Integr. **15**, 21–28 (2019)
6. Andoni, M., et al.: Blockchain technology in the energy sector: a systematic review of challenges and opportunities. Renew. Sustain. Energy Rev. **100**, 143–174 (2019)
7. Rejeb, A., Keogh, J.G., Treiblmaier, H.: Leveraging the Internet of Things and blockchain technology in supply chain management. Future Internet **11**(7), 161–183 (2019)

8. Al Ridhawi, I., et al.: A collaborative mobile edge computing and user solution for service composition in 5G systems. Trans. Emerg. Telecommun. Technol. **29**, e3446 (2018)
9. Al Ridhawi, I., et al.: A continuous diversified vehicular cloud service availability framework for smart cities. Comput. Netw. **145**, 207–218 (2018)
10. Gordon, W.J., Catalini, C.: Blockchain technology for healthcare: facilitating the transition to patient-driven interoperability. Comput. Struct. Biotechnol. J. **16**, 224–230 (2018)
11. Dagher, G.G., et al.: Ancile: Privacy-preserving framework for access control and interoperability of electronic health records using blockchain technology. Sustain. Cities Soc. **39**, 283–297 (2018)
12. Zhang, P., et al.: Blockchain technology use cases in healthcare. In: Advances in Computers, pp. 1–41. Amsterdam, Elsevier (2018)
13. Casino, F., Dasaklis, T.K., Patsakis, C.: A systematic literature review of blockchain-based applications: current status, classification and open issues. Telemat. Inform. **36**, 55–81 (2019)
14. Pirbhulal, S., Wu, W., Li, G.: A biometric security model for wearable healthcare. In: Proceedings of the 2018 IEEE International Conference on Data Mining Workshops (ICDMW), Singapore, 17–20 November, pp. 136–143 (2018)
15. Griggs, K.N., et al.: Healthcare blockchain system using smart contracts for secure automated remote patient monitoring. J. Med. Syst. **42**, 130 (2018)
16. Rahman, M.A., et al.: Spatial blockchain-based secure mass screening framework for children with dyslexia. IEEE Access **6**, 61876–61885 (2018)
17. Jo, B., Khan, R., Lee, Y.S.: Hybrid blockchain and Internet-of-Things network for underground structure health monitoring. Sensors **18**, 4268 (2018)
18. Rani, A.A.V., Baburaj, E.: Secure and intelligent architecture for cloud-based healthcare applications in wireless body sensor networks. Int. J. Biomed. Eng. Technol. **29**, 186–199 (2019)
19. Qiu, J., Liang, X., Shetty, S., Bowden, D.: Towards secure and smart healthcare in smart cities using blockchain. In: Proceedings of the 2018 IEEE International Smart Cities Conference (ISC2), Kansas City, MO, USA, 16–19 September, pp. 1–4 (2018)
20. Wang, S., et al.: Blockchain-powered parallel healthcare systems based on the ACP approach. IEEE Trans. Comput. Soc. Syst. **5**, 942–950 (2018)
21. Deisingh, A.K.: Pharmaceutical counterfeiting. Analyst **130**, 271–279 (2005)
22. Zhu, Q., Kouhizadeh, M.: Blockchain technology, supply chain information, and strategic product deletion management. IEEE Eng. Manag. Rev. **47**, 36–44 (2019)
23. Plotnikov, V., Kuznetsova, V.: The prospects for the use of digital technology "blockchain" in the pharmaceutical market. In: MATEC Web of Conferences, Ho Chi Minh. EDP Sciences, p. 02029 (2018)
24. Kuo, T.T., Ohno-Machado, L.: ModelChain: decentralized privacy-preserving healthcare predictive modeling framework on private blockchain networks, arXiv, arXiv:1802.01746 (2018)
25. Kumar, R., Agarwal, A., Shubhankar, B.: Counterfeit drug detection: recent strategies and analytical perspectives. Int. J. Pharm. Res. Health Sci. **6**, 2351–2358 (2018)
26. Chiuchisan, I., Costin, H.N., Geman, O.: Adopting the internet of things technologies in health care systems. In: Proceedings of the 2014 International Conference and Exposition on Electrical and Power Engineering (EPE), Iasi, 16–18 October 2014, pp. 532–535 (2014)
27. Mamoshina, P., et al.: Converging blockchain and next-generation artificial intelligence technologies to decentralize and accelerate biomedical research and healthcare. Oncotarget **9**, 5665 (2018)
28. Alketbi, A., Nasir, Q., Talib, M.A.: Blockchain for government services—use cases, security benefits and challenges. In: Proceedings of the 2018 IEEE 15th Learning and Technology Conference (L&T), Jeddah, 25–26 February 2018, pp. 112–119 (2018)

29. Montecchi, M., Plangger, K., Etter, M.: It's real, trust me! Establishing supply chain provenance using blockchain. Bus. Horiz. **62**(3), 283–293 (2019)
30. Conti, M., et al.: A survey on security and privacy issues of bitcoin. IEEE Commun. Surv. Tutor. **20**, 3416–3452 (2018)
31. Joshi, A.P., Han, M., Wang, Y.: A survey on security and privacy issues of blockchain technology. Math. Found. Comput. **1**, 121–147 (2018)
32. Hölbl, M., et al.: A systematic review of the use of blockchain in healthcare. Symmetry **10**, 470 (2018)
33. Radanović, I., Likić, R.: Opportunities for use of blockchain technology in medicine. Appl. Health Econ. Health Policy **16**, 583–590 (2018)
34. Siyal, A., et al.: Applications of blockchain technology in medicine and healthcare: challenges and future perspectives. Cryptography **3**, 3 (2019)
35. McGhin, T., et al.: Blockchain in healthcare applications: research challenges and opportunities. J. Netw. Comput. Appl. **135**, 62–75 (2019)
36. Saraf, C., Sabadra, S.: Blockchain platforms: a compendium. In: Proceedings of the 2018 IEEE International Conference on Innovative Research and Development (ICIRD), Bangkok, 11–12 May 2018, pp. 1–6 (2018)
37. Boulos, M.N.K., Wilson, J.T., Clauson, K.A.: Geospatial blockchain: promises, challenges, and scenarios in health and healthcare. Int. J. Health Geogr. **17**, 25 (2018)
38. Khan, M.A., Salah, K.: IoT security: review, blockchain solutions, and open challenges. Future Gener. Comput. Syst. **82**, 395–411 (2018)
39. Esposito, C., De Santis, A., Tortora, G., Chang, H., Choo, K.K.R.: Blockchain: a panacea for healthcare cloud-based data security and privacy? IEEE Cloud Comput. **5**, 31–37 (2018)
40. Dimitrov, D.V.: Blockchain applications for healthcare data management. Healthc. Inform. Res. **25**, 51–56 (2019)
41. Qadri, Y.A., Nauman, A., Zikria, Y.B., Vasilakos, A.V., Kim, S.W.: The future of healthcare Internet of Things: a survey of emerging technologies. IEEE Commun. Surv. Tutor. **22**(2), 1121–1167 (2020)
42. Ribitzky, R., et al.: Pragmatic, interdisciplinary perspectives on blockchain and distributed ledger technology: paving the future for healthcare. Blockchain Healthc. Today **1**, 1–15 (2018)
43. Adavoudi Jolfaei, A., Aghili, S.F., Singelee, D.: A survey on blockchain-based IoMT systems: towards scalability. IEEE Access **9**, 148948–148975 (2021)
44. Reegu, F.A., et al.: Blockchain-based framework for interoperable electronic health records for an improved healthcare system. Sustainability **15**, 6337 (2023)
45. Tasatanattakool, P., Techapanupreeda, C.: Blockchain: challenges and applications. In: Proceedings of the International Conference on Information Networking (ICOIN 2018), 10–12 January 2018, Chiang Mai, Thailand, pp. 473–475 (2018)

Towards Efficient Opportunistic Osteoporosis Screening: The Role of Feature Selection

Farid Ghareh Mohammadi[1,2(✉)] and Ronnie Sebro[1,2,3]

[1] Department of Radiology, MD Anderson Cancer Center, 1515 Holcombe Blvd, Houston, TX 77030, USA
mohammadi.farid9195@gmail.com

[2] Department of Imaging Physics, MD Anderson Cancer Center, 1515 Holcombe Blvd, Houston, TX 77030, USA

[3] Department of Epidemiology, MD Anderson Cancer Center, 1515 Holcombe Blvd, Houston, TX 77030, USA

Abstract. Radiomic features extracted from CT scans for opportunistic osteoporosis screening, can generate large and high-dimensional datasets, which may exacerbate the curse of dimensionality (CoD) and complicate model development and interpretation. Manual or semi-modular feature extraction enables detailed quantification of bone characteristics but is resource-intensive. Integrating these features with clinical variables can enhance diagnostic performance while increasing feature space complexity. This study reviews and provides common feature selection approaches for opportunistic osteoporosis screening using CT scans that preserve clinically relevant information, minimize redundancy and noise, and provide practical guidance for musculoskeletal imaging and radiomics research aimed at robust, interpretable osteoporosis screening.

Keywords: Curse of dimensionality(CoD) · Feature selection · Dimension reduction · Computed tomography (CT) scans · Osteoporosis

1 Introduction

Osteoporosis remains one of the most pervasive and undertreated metabolic bone diseases, silently predisposing millions of individuals to fragility fractures and significant morbidity [1,2]. Its asymptomatic nature underscores the importance of early detection, yet conventional screening strategies, particularly Dual-energy X-ray absorptiometry (DXA), struggle to reach high-risk populations in routine clinical practice [3]. In this context, opportunistic osteoporosis screening has emerged as an appealing paradigm, leveraging CT scans acquired for other clinical indications to extract quantitative bone health information without additional radiation, cost, or patient burden [4–7].

H. R. Arabnia et al. (Eds.): CSCI 2025, CCIS 2971, pp. 321–333, 2027.
https://doi.org/10.1007/978-3-032-28463-1_26

Recent advances in computational imaging have expanded the potential of CT-based screening from manual measurements to fully automated artificial intelligence(AI)-driven analysis [4,8]. CT studies contain a wealth of structural [9], textural [10], and densitometric information [11], however, only a small subset of this high-dimensional features is truly informative for assessing low bone health risk. As a result, identifying the features that meaningfully capture bone quality becomes essential. Radiomic descriptors [12], volumetric bone mineral density [4], trabecular patterns [8], and cortical metrics [13] represent a rich technical foundation, yet they also pose significant challenges due to their volume, variability, and multicollinearity.

To address these challenges, feature extraction plays a pivotal role [14]. By transforming raw imaging data into quantitative descriptors, feature extraction bridges the gap between traditional radiologic interpretation and machine learning–based prediction [8,15,16]. However, the extracted feature space often spans hundreds or even thousands of variables, far exceeding what most models can handle efficiently or interpretably. Without appropriate feature selection methods, models risk overfitting, reduced generalizability, or excessive computational cost, particularly when deployed across large-scale opportunistic screening pipelines [8,15,17].

Feature selection becomes fundamental to efficient, reliable AI systems for osteoporosis screening [4,5,8]. Methods such as wrapper based, filter based or embedded based algorithms help isolate the most discriminative and nonredundant features, allowing algorithms to highlight clinically meaningful aspects of bone architecture. By retaining only the essential features, feature selection strengthens model performance [17], enhances interpretability, and facilitates real-world implementation within radiology workflows [18,19]. Towards scalable opportunistic CT screening, thoughtful selection of imaging features is a technical and clinical imperative step that enables AI to deliver actionable, patient-centered insights [8].

In this paper, we examine the role of features in the development of artificial intelligence models, tracing the process from initial extraction to final selection. We first outline the practical and theoretical challenges inherent in feature extraction, with particular attention to issues such as redundancy, noise, and the curse of dimensionality that complicate downstream analysis. We turn to feature selection as the principal strategy for mitigating these limitations, emphasizing methods that enhance model efficiency, interpretability, and performance.

1.1 Opportunistic Screening for Osteoporosis Using CT Scans

Opportunistic screening for osteoporosis has emerged as a logical extension of radiology's expanding contribution to precision health, particularly as AI enables more systematic use of data embedded in routine imaging [13,14,20]. CT scans acquired for unrelated clinical purposes contain rich information about bone mass and microarchitecture [4,15,21]. As summarized in the workflow presented in Fig. 1, these data can be processed through a series of steps including feature extraction, feature selection, and model development, to generate quantitative markers relevant to fracture risk. In this paper, we concentrate specifically on feature selection, emphasizing the most commonly used methods for reducing feature-space dimensionality through the identification of informative and non-redundant features that ultimately drive model performance.

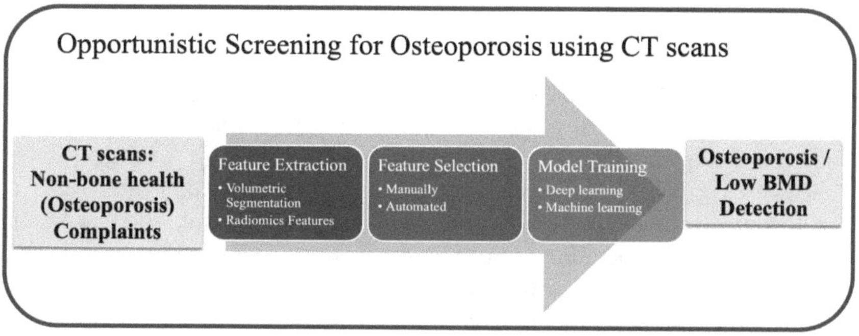

Fig. 1. Detailed steps of opportunistic screening for osteoporosis using computed tomography(CT) scans

AI-based opportunistic screening capitalizes on features derived from attenuation patterns, trabecular texture, and structural integrity that may not be readily appreciable on visual inspection alone [4,5,8]. Although feature extraction introduces challenges such as redundancy, noise, and the curse of dimensionality, judicious feature selection can mitigate these limitations while improving efficiency and clinical interpretability. By leveraging CT imaging already obtained in routine care, this approach offers a practical pathway to earlier identification of osteoporosis without the need for dedicated examinations such as DXA. This illustrates how radiology—and AI-enhanced workflows can transform everyday imaging into a tool for proactive bone health management.

2 Feature Extraction and Selection

2.1 Feature Extraction

Radiomic feature extraction transforms CT image data into a set of quantitative descriptors that characterize the underlying bone structure beyond what is visually appreciable. After segmentation of the vertebral body or other skeletal regions, a diverse collection of features can be computed to capture intensity [5,13], geometry [7], and textural patterns [8,15,22] that reflect bone density, trabecular organization, and cortical integrity. These features serve as critical inputs for machine learning models, enabling opportunistic CT screening algorithms to detect signs of osteoporosis without requiring dedicated imaging or manual qualitative assessment [4,8].

Radiomic features are commonly grouped into several standardized categories [8,15,16]. First-order statistics quantify the distribution of voxel intensities within the segmented region, providing measures such as mean attenuation, entropy, skewness, and uniformity [15]. These metrics reflect global tissue density and heterogeneity, which are directly related to bone mineral status. Shape features describe the 3D morphology of the vertebra or bony structure capturing volume, surface area, compactness, sphericity, and cortical thickness patterns [8,16]. Changes in shape can indirectly signal structural degradation or alterations associated with osteoporosis [13,20].

A richer description of bone microarchitecture comes from texture features derived from spatial relationships among voxels [15]. The gray-level co-occurrence matrix (GLCM) quantifies how often pairs of voxel intensities occur together, providing measures such as contrast, homogeneity, and correlation—useful for detecting trabecular thinning or disorganization. The gray-level run-length matrix (GLRLM) characterizes consecutive runs of similar intensities, highlighting patterns of trabecular continuity or fragmentation. The gray-level size zone matrix (GLSZM) measures the size of homogeneous intensity zones, offering insights into the uniformity of bone structure. The gray-level dependence matrix (GLDM) assesses how intensities depend on neighboring voxels, capturing local structural irregularities, while the neighboring gray-tone difference matrix (NGTDM) evaluates texture coarseness and contrast based on intensity differences within a neighborhood. Together, these texture families quantify trabecular complexity—a key determinant of bone strength that extends beyond simple volumetric bone mineral density.

By extracting these multifaceted radiomic features from routine CT scans, opportunistic screening systems can detect subtle variations in bone quality that are not evident through attenuation alone [19,20]. This comprehensive quantitative characterization allows AI models to identify patients at risk for osteoporosis or fragility fractures, enabling early intervention without additional radiation, cost, or workflow disruption.

2.2 Curse of Dimensionality

Radiomic analysis of CT images typically yields hundreds to thousands of quantitative features [8], capturing intensity, texture, shape, and higher-order transformations such as wavelets or Laplacian-of-Gaussian filters. While this richness offers the promise of capturing subtle aspects of bone microarchitecture, it also introduces the classic curse of dimensionality [17,23]. As the number of features grows, the feature space becomes exponentially larger, and the density of data points becomes increasingly sparse. In clinical imaging, where datasets often include only a few hundred patients, the sample size is rarely sufficient to populate such a high-dimensional space. This mismatch between feature dimensionality and the number of patient undermines the statistical reliability of downstream models.

The consequences of the curse of dimensionality are significant. High dimensional radiomic feature sets increase the risk of overfitting, allowing models to memorize noise or patient-specific artifacts rather than learning generalizable patterns related to osteoporosis [23]. Many radiomic features are also highly correlated or redundant, which can distort model coefficients, inflate variance, and produce unstable predictions when applied to new data. Computationally, large feature spaces lead to increased training time, greater memory requirements, and difficulties in convergence for many machine learning algorithms [17,23].

Ultimately, without feature selection/dimensionality reduction, the curse of dimensionality can compromise the accuracy, robustness, and clinical deployment of AI systems built on CT-based radiomics. Models may perform well during internal validation but degrade sharply in external testing or real-world use. This makes feature selection a technical optimization and a necessity to mitigate redundancy, control variance, and ensure that radiomics-driven osteoporosis screening achieves reliable diagnostic performance.

2.3 Feature Selection Terminology and Conceptual Distinctions

In the radiology and medical-AI literature, several terms such as feature selection, feature reduction, dimension reduction, variable selection, and even feature extraction are often used interchangeably, despite representing distinct analytic processes. This imprecise terminology can obscure methodological intent and complicate the interpretation of study findings.

Feature selection or reduction refers specifically to choosing a subset of the original features that are most relevant for prediction, without altering the nature of those features. *Variable selection* is essentially synonymous, commonly used in statistical modeling to describe the same process. In contrast, *dimension reduction* involves transforming the original feature space into a lower-dimensional representation, typically through mathematical projections rather than selecting existing variables. *Feature extraction* is a related process in which new features are derived from the original data such as radiomic texture descriptors computed from segmented CT regions. Recognizing these distinctions is important because each strategy has different implications for model interpretability, reproducibility, and clinical integration.

Once features have been extracted and selected, a robust machine learning pipeline requires identifying those features that best contribute to predictive performance. Broadly, feature selection methods fall into three categories: filter, wrapper, and embedded approaches.

Filter-based methods are computationally efficient and require no model training; they rank features based on statistical criteria such as correlation or mutual information [17,24]. Although fast and scalable, they may yield suboptimal performance because their evaluations occur independently of the final predictive model. Wrapper methods, in contrast, evaluate subsets of features through repeated model training, often producing superior performance but at the cost of substantial computational time [17,24]. Embedded methods occupy a middle ground; they incorporate feature selection into the model training process itself, offering a balance of speed and predictive accuracy [24].

Examples highlight the practical distinctions among these categories. Wrapper-based methods include Recursive Feature Elimination (RFE) [22], forward selection, backward selection, and evolutionary search algorithms, all of which evaluate the contribution of features by iteratively training and comparing models [17,24]. Filter-based approaches include Minimum Redundancy Maximum Relevance (mRMR) [25], Pearson correlation ranking, mutual information scores, chi-square tests, analysis of variance (ANOVA) F-tests, and ReliefF—methods that assess feature quality based on statistical properties without model involvement [17,24]. Embedded techniques integrate feature selection directly into model optimization; prominent examples include Least Absolute Shrinkage and Selection Operator(LASSO) [22,26], L1-regularized logistic regression, and Elastic Net, each of which shrinks or eliminates coefficients during model fitting. Understanding these distinctions enables researchers to choose feature selection strategies that align with the goals of the study, the size of the dataset, and the computational constraints of clinical deployment.

2.4 Common Feature Selection Practices in CT-Based Opportunistic Osteoporosis Screening

Radiomics studies of opportunistic screening using CT imaging for osteoporosis typically generate high-dimensional feature sets that exhibit substantial multicollinearity and varying degrees of relevance to the clinical endpoint [19,27]. Effective feature selection becomes essential to mitigate overfitting, enhance model interpretability, improve generalization, and satisfy clinical reporting requirements [17,24]. Three automated methods have emerged as dominant approaches in the contemporary CT scan opportunistic screening for osteoporosis literature in last ten years: RFE, mRMR, and LASSO.

Recursive Feature Elimination particularly when coupled with cross-validation (RFECV), is a wrapper method that iteratively trains a base estimator such as support vector machine(SVM), random forest(RF), or extreme gradient boosting and eliminates the least important feature(s) according to the model's internal importance scores. This greedy yet highly effective strategy directly optimizes the feature subset with respect to the predictive performance metric of interest (e.g., area under the receiver operating characteristic curve or concordance index), rendering it particularly suitable when the primary objective is maximal discriminative or prognostic accuracy. Despite its computational intensity, RFE frequently yields the highest-performing reduced feature sets and is therefore widely adopted in studies prioritizing model performance over computational efficiency.

The mRMR algorithm represents a filter approach that selects features exhibiting maximal statistical dependence on the target variable (relevance) while minimizing mutual information among the selected features themselves (redundancy). Given the extensive inter-feature correlations inherent in radiomic data, where dozens of GLCM, GLRLM, and NGTDM features often encode highly similar textural information, mRMR provides an efficient and model-agnostic mechanism to produce compact, diverse, and biologically plausible feature sets. Its computational speed and absence of reliance on a specific classifier make it especially valuable during exploratory analysis and preprocessing stages [25].

LASSO regression [26], an embedded regularization technique, simultaneously performs continuous shrinkage and automatic variable selection by imposing an L1 penalty that drives irrelevant coefficients exactly to zero. This method yields intrinsically sparse models that are readily interpretable and directly convertible into clinical scoring systems or nomograms. In CT-based radiomics investigations, LASSO (or its extensions such as elastic net) is frequently used when the goal is to derive a multivariable signature suitable for prospective validation [28], as it seamlessly integrates feature selection with risk modeling in a single statistically principled step.

In practice, one-step [18], two-step [25], and three-step process [15] are often combined sequentially to exploit their complementary strengths, enhancing feature selection performance for AI models. Their widespread adoption reflects a consensus that robust, reproducible, and clinically translational radiomic signatures require rigorous dimensionality reduction tailored to the challenges of high collinearity, modest sample sizes, and the imperative for both predictive power and interpretability.

3 Feature Selection Techniques in CT-Based Opportunistic Osteoporosis Screening

Feature-selection techniques in opportunistic screening for osteoporosis can be organized into three conceptual groups: one-step process, which extract and select predictive imaging features within a single integrated process; two-step process, which apply two consecutive feature-selection algorithms to progressively refine the feature set; and three-step process, which employ three distinct feature-selection algorithms in sequence to ensure that the final features are rigorously filtered and retain the greatest clinical relevance.

One-Step Process: The most commonly used one-step process for feature selection is LASSO.

Previous CT-based radiomics studies [18–20,27,29] adopted LASSO as the most-common one-step process for feature selection from high-dimensional CT-derived radiomic features. LASSO imposes an ℓ_1-penalty on the regression coefficients, driving many coefficients exactly to zero and thereby selecting a subset of the most relevant features in a single optimization step. This approach is particularly well suited for CT-based radiomics, where a large number of correlated imaging features are extracted relative to the sample size. By retaining only features with nonzero coefficients, LASSO facilitates the development of parsimonious and interpretable predictive models while reducing the risk of overfitting and enhancing model generalizability.

Two-Step Process: Among two-step approaches, the most commonly employed methods for feature selection are mRMR and LASSO, often used in combination to first filter redundant features and then optimize the predictive subset.

Yuan et al. [25] evaluated the performance of a machine learning-based radiomics model derived from proximal femur regions on routine abdominal CT scans for screening abnormal bone mass (indicative of osteoporosis risk) in patients aged 50 or older. A total of 222 patients were enrolled, with 2288 radiomic features extracted; feature selection employed a two-step process using mRMR for initial ranking followed by LASSO to select 13 optimal features. The logistic regression–based model achieved excellent predictive performance, with an area under the curve (AUC) of 0.917 in the training set and 0.963 in the testing set, outperforming clinical benchmarks and supporting opportunistic screening without additional radiation exposure.

Jian et al. [30] developed and validated a radiomics signature from routine preoperative lumbar spine CT scans to screen for osteoporosis in 386 vertebral bodies, aiding surgical decision-making. From 1040 extracted features, those with ICC > 0.75 were retained (846 features), followed by mRMR to select the top 30, and LASSO with 10-fold cross-validation to finalize 12 features for the model. The radiomics signature outperformed Hounsfield unit-based models, with AUCs of 0.96 (training) and 0.92 (testing) versus 0.88 and 0.84, respectively ($p < 0.05$), and showed superior net benefit via decision curve analysis, enabling cost-effective preoperative identification without extra imaging.

A retrospective study [8] on 590 patients with osteopenia or osteoporosis uses QCT images of the third lumbar vertebra to build a combined clinical-radiomics model for precise diagnosis. Radiomic features underwent the feature selection process via mRMR and LASSO, selecting six key signatures alongside clinical factors (age, alkaline phosphatase, homocysteine). The multivariable logistic regression model yielded AUCs of 0.96 in both training and test cohorts, surpassing the clinical-only model (AUCs 0.81 and 0.79), with strong calibration and clinical utility confirmed by decision curve analysis, positioning it as a robust tool for bone mass assessment.

In a multi-center retrospective study, Lin et al. [31] evaluated whether radiomics derived from a single thoracic vertebral body (T11 or T12) on chest CT can facilitate opportunistic osteoporosis screening in routine clinical examinations. After initial filtering (z/t-tests, ICC > 0.9, Pearson correlation), mRMR selected the top 20 features, refined by LASSO to 19 for SVM-based modeling. The nomogram (integrating radiomics and clinical data like age/gender) achieved AUCs of 0.901 (training), 0.918 (internal validation), and 0.807 (external validation), significantly outperforming clinical models (p ¡ 0.05 in most sets), demonstrating feasibility for high-risk patient identification without added costs.

A study by Tong et al. [32] developed a fully automatic system for osteoporosis screening using low-dose chest CT, comparing radiomics (Rad) and deep learning (DL) models on thoracic vertebral cancellous bone segments from 434 patients (Dice coefficient: 0.96). For the Rad model (RF classifier), features were selected using recursive feature elimination, mRMR, and LASSO; the DL model used ResNet. Both distinguished osteoporosis/osteopenia/normal (AUCs: Rad 0.943/0.801/0.932; DL 0.983/0.906/0.969 in validation), with DL showing higher accuracy (81.2% vs. 73.3%) and net benefit, though comparable overall (DeLong $p > 0.05$), advancing automated BMD evaluation.

In a comprehensive radiomics investigation of osteoporosis diagnosis, Cheng et al. [33] extracted 4,858 quantitative imaging features from lumbar vertebral CT and dual-projection (anteroposterior and lateral) radiographs in a retrospective cohort of 616 patients. To mitigate the challenges of high-dimensional feature space and multicollinearity inherent in radiomics datasets, the authors implemented a robust two-stage feature selection strategy: first, the mRMR algorithm was applied to rank and eliminate redundant or non-informative features while preserving those with strongest class relevance; this was followed by LASSO logistic regression with ten-fold cross-validation, which further sparsified the signature by driving coefficients of less predictive features to exactly zero, yielding a concise yet highly discriminative radiomics score. Among classifiers constructed from the LASSO-selected features, the SVM model demonstrated superior generalizability, achieving AUC of 0.958 in the training set and 0.907 in the held-out test set; integration of this radiomics signature with clinical risk factors identified through univariate and multivariable logistic regression produced a combined nomogram with test-set AUC of 0.91, underscoring the incremental diagnostic value of rigorously selected radiomic biomarkers over conventional clinical parameters alone. This sequential mRMR–LASSO pipeline has since emerged as a de facto standard for feature selection in spine radiomics studies addressing osteoporosis and related bone health applications.

A retrospective analysis of 616 lumbar spine cases by Cheng et al. [33] integrated multimodal radiomics from CT and X-ray with clinical factors to diagnose osteoporosis using logistic regression (LR), SVM, and RF models. From 4858 extracted features, SVM emerged optimal (AUCs 0.958 training, 0.907 testing); the combined model (AUCs 0.959/0.910) further improved performance. No specific feature selection like mRMR or LASSO was detailed; instead, univariate/multivariate logistic regression identified clinical risks. Decision curve analysis confirmed clinical utility, offering an efficient, non-invasive diagnostic aid.

Three-Step Methods: Among three-step approaches, the most commonly employed methods for feature selection are RFE, mRMR, and LASSO, which are often combined sequentially to first eliminate irrelevant features, then remove redundant features, and finally optimize the predictive subset for AI models.

Wang et al. [15] developed an automated LDCT-based framework for osteoporosis screening by integrating deep-learning segmentation of thoracic vertebral cancellous bone with CT-based radiomics analysis and clinical information. TVCB segmentation using a VB-Net architecture showed high fidelity, providing a reliable foundation for subsequent feature extraction. Clinical variables and radiomics features were jointly evaluated to enhance the predictive capacity for bone mineral density assessment.

In [15] feature selection step followed a sequential, rigorously controlled process. SVM-RFE was first applied to mitigate overfitting and improve classification performance, followed by mRMR to eliminate redundant features while retaining those most relevant to BMD status. LASSO regression then identified the optimal subset of predictive features for model construction. This streamlined feature-selection strategy enabled robust differentiation of normal versus abnormal BMD and further distinction between osteopenia and osteoporosis, supporting efficient and clinically meaningful opportunistic screening.

Tong et al. [22] developed automated bone status prediction models combining deep learning and radiomics using both low-dose (LDCT) and standard-dose chest CT (SDCT). A total of 1,508 patients underwent quantitative CT and chest CT scans at differing tube voltages (80 kVp for LDCT, 120 kVp for SDCT). Automatic thoracic vertebral cancellous bone (TVCB) segmentation achieved high accuracy (Dice 0.98 for LDCT, 0.97 for SDCT), forming the basis for extraction of 1,184 radiomic features. Two-level classifiers were constructed to differentiate normal versus abnormal bone mineral density (BMD) and to distinguish osteopenia from osteoporosis, integrating radiomics with relevant clinical information.

Tong et al. [22] used a sequential pipeline of feature selection, RFE, mRMR, and LASSO regression, to reduce overfitting and retain the most informative features. Analysis revealed that tube voltage significantly affected radiomics reproducibility, with 85% of features showing low concordance across LDCT and SDCT, and impacted model performance in cross-validation. These findings highlight the importance of standardized acquisition protocols and rigorous feature selection to ensure robust and clinically applicable automated BMD assessment.

4 Discussion

The growing interest in opportunistic osteoporosis screening reflects a broader shift in musculoskeletal imaging toward extracting greater clinical value from studies obtained for other indications. CT-based radiomics provides a promising means to characterize trabecular and cortical bone architecture in a quantitative, reproducible manner. As this review underscores, the practical implementation of radiomic analysis remains constrained by the sheer volume and heterogeneity of features generated from even modest CT scans. Without appropriate feature selection, high-dimensional feature spaces complicate model development, increase the risk of overfitting, and can obscure the physiologic meaning of individual features.

Feature selection and dimensionality reduction address these concerns by reducing redundancy, enhancing signal clarity, and preserving features most likely to reflect meaningful aspects of bone health. Techniques such as principal component analysis(PCA), regularization, and stability-based selection offer complementary strategies to manage complexity, though each introduces its own assumptions and trade-offs. Importantly, the need to balance statistical performance with transparency remains central. Models intended for widespread adoption in clinical practice must be interpretable, reproducible, and supported by evidence demonstrating generalizability across scanners, institutions, and patient populations.

The integration of radiomic features with clinical risk factors further elevates model performance but also amplifies the demand for careful feature curation. As the feature space expands, so does the importance of methodologic rigor, from standardized acquisition and reconstruction protocols to robust cross-validation and external testing. These steps are essential not only for scientific reliability but also for building clinician confidence in radiomics-based decision support.

This review has several limitations that merit consideration. First, our analysis is confined to radiomic features derived from CT examinations, and the applicability of these findings to other imaging modalities remains uncertain. The insights presented here are therefore most relevant to investigators working within CT-based assessment frameworks. Second, our focus centered on strategies for preserving clinically meaningful features to support opportunistic osteoporosis screening. While this emphasis highlights the importance of interpretability and relevance, it does not address the broader landscape of feature engineering or advanced model architectures. Third, we did not examine how individual selected features map back to specific feature extraction methods. As a result, the relative contributions of particular texture, shape, or intensity families to model performance could not be fully delineated. Future work that systematically evaluates these relationships may further clarify the pathways through which radiomic descriptors contribute to robust, clinically interpretable models.

Ultimately, the strategies reviewed here highlight a path forward for the field. By adopting systematic approaches to feature selection and dimensionality reduction, investigators can harness the richness of radiomic data while maintaining the clarity and interpretability needed for clinical translation. As musculoskeletal imaging continues to evolve, such disciplined methodology will be instrumental in realizing the promise of radiomics for opportunistic osteoporosis screening.

5 Conclusion

Efforts to integrate radiomic descriptors of bone structure with conventional clinical factors using CT scans offer meaningful opportunities to enrich opportunistic osteoporosis assessment. Yet these same efforts underscore the persistent challenges posed by high-dimensional feature spaces. Approaches that thoughtfully reduce dimensionality and prioritize features with clear biologic or clinical relevance can mitigate the curse of dimensionality and support the development of robust, interpretable models. By highlighting practical strategies for harmonizing radiomic and clinical data, this review aims to guide investigators toward methods that advance musculoskeletal imaging research while preserving the clarity, reproducibility, and clinical utility essential for widespread adoption in osteoporosis screening using CT scans.

Acknowledgments. Farid Ghareh Mohammadi and Ronnie Sebro were supported by the National Institute of Health (NIH) under Grant Number: R01-EB035231.

References

1. Golchin, M.M., Heidari, L., Ghaderian, S.M.H., Akhavan-Niaki, H.: Osteoporosis: a silent disease with complex genetic contribution. J. Genet. Genomics **43**(2), 49–61 (2016)
2. Cosman, F., et al.: Clinician's guide to prevention and treatment of osteoporosis. Osteoporosis Int. **25**(10), 2359–2381 (2014)
3. Curtis, J.R., et al.: The geographic availability and associated utilization of dual-energy x-ray absorptiometry (dxa) testing among older persons in the united states. Osteoporosis Int. **20**(9), 1553–1561 (2009)
4. Sebro, R., Elmahdy, M.: Machine learning for opportunistic screening for osteoporosis and osteopenia using knee ct scans. Can. Assoc. Radiol. J. **74**(4), 676–687 (2023)
5. Sebro, R., De la Garza-Ramos, C.: Utilizing machine learning for opportunistic screening for low bmd using ct scans of the cervical spine. J. Neuroradiol. **50**(3), 293–301 (2023)
6. Gharehmohammadi, F., Sebro, R.: Deep learning opportunistic screening for osteoporosis and osteopenia using radiographs of the foot or ankle–a pilot study. Eur. J. Radiol. 111980 (2025)
7. Lenchik, L., Weaver, A.A., Ward, R.J., Boone, J.M., Boutin, R.D.: Opportunistic screening for osteoporosis using computed tomography: state of the art and argument for paradigm shift. Curr. Rheumatol. Rep. **20**(12), 74 (2018)
8. Xie, Q., et al.: Development and validation of a machine learning-derived radiomics model for diagnosis of osteoporosis and osteopenia using quantitative computed tomography. BMC Med. Imaging **22**(1), 140 (2022)
9. De Jong, P.A., Müller, N.L., Pare, P.D., Coxson, H.O.: Computed tomographic imaging of the airways: relationship to structure and function. Eur. Respir. J. **26**(1), 140–152 (2005)
10. Link, T.M., e al.: Assessment of trabecular structure using high resolution ct images and texture analysis. J. Comput. Assist. Tomogr. **22**(1), 15–24 (1998)
11. Laurence, W., Hedlund, P., Vock, E.L., Effmann: computed tomography of the lung: densitometric studies. Radiol. Clin. North Am. **21**(4), 775–788 (1983)
12. Lambin, P., et al.: Radiomics: extracting more information from medical images using advanced feature analysis. Eur. J. Cancer **48**(4), 441–446 (2012)
13. Sebro, R., De la Garza-Ramos, C.: Opportunistic screening for osteoporosis and osteopenia from ct scans of the abdomen and pelvis using machine learning. Eur. Radiol. **33**(3), 1812–1823 (2023)
14. Ghareh Mohammadi, F., Sebro, R.: 2AIO: the application of-artificial intelligence in osteoporosis. In: AI Revolution: Research, Ethics and Society, pp. 524–533. Springer (2026)
15. Wang, S., et al.: Combining deep learning and radiomics for automated, objective, comprehensive bone mineral density assessment from low-dose chest computed tomography. Acad. Radiol. **31**(3), 1180–1188 (2024)

16. Wang, J., He, Y., Yan, L., Chen, S., Zhang, K.: Predicting osteoporosis and osteopenia by fusing deep transfer learning features and classical radiomics features based on single-source dual-energy ct imaging. Acad. Radiol. **31**(10), 4159–4170 (2024)
17. Mohammadi, F.G., Amini, M.H., Arabnia, H.R.: Applications of nature-inspired algorithms for dimension reduction: enabling efficient data analytics. In: Optimization, Learning, and Control for Interdependent Complex Networks, pp. 67–84. Springer (2020)
18. Wang, J., Zhou, S., Chen, S., He, Y., Gao, H., Yan, L., Xiaoli, H., Li, P., Shen, H., Luo, M., et al.: Prediction of osteoporosis using radiomics analysis derived from single source dual energy ct. BMC Musculoskelet. Disord. **24**(1), 100 (2023)
19. Huang, C., Jia-sen, H., Tan, K., Zhang, W., Tian-hao, X., Yang, L.: Application of machine learning model to predict osteoporosis based on abdominal computed tomography images of the psoas muscle: a retrospective study. BMC Geriatr. **22**(1), 796 (2022)
20. Sebro, R., De la Garza-Ramos, C.: Machine learning for the prediction of osteopenia/osteoporosis using the ct attenuation of multiple osseous sites from chest ct. Eur. J. Radiol. **155**, 110474 (2022)
21. Flynn, M.J., Cody, D.D.: The assessment of vertebral bone macroarchitecture with x-ray computed tomography. Calcified Tissue Int. **53**(Suppl1), S170–S175 (1993)
22. Tong, X., et al.: Effect of fully automatic classification model from different tube voltage images on bone density screening: a self-controlled study. Eur. J. Radiol. **177**, 111521 (2024)
23. Mohammadi, F.G., Amini, M.H., Arabnia, H.R.: Evolutionary computation, optimization, and learning algorithms for data science. In: Optimization, Learning, and Control for Interdependent Complex Networks, pp. 37–65. Springer (2020)
24. Chandrashekar, G., Sahin, F.: A survey on feature selection methods. Comput. Electr. Eng. **40**(1), 16–28 (2014)
25. Yuan, X., et al.: Applying machine learning analysis based on proximal femur of abdominal computed tomography to screen for abnormal bone mass in femur. Acad. Radiol. **31**(5), 2003–2010 (2024)
26. Chen, L., Tsang, I.W., Xu, D.: Laplacian embedded regression for scalable manifold regularization. IEEE Trans. Neural Netw. Learn. Syst. **23**(6), 902–915 (2012)
27. Chen, B., et al.: Application of radiomics model based on lumbar computed tomography in diagnosis of elderly osteoporosis. J. Orthopaedic Res.® **42**(6), 1356–1368 (2024)
28. Sun, S., et al.: A nomogram for evaluation and analysis of difficulty in retroperitoneal laparoscopic adrenalectomy: a single-center study with prospective validation using lasso-logistic regression. Front. Endocrinol. **13**, 1004112 (2022)
29. Fang, K., Zheng, X., Lin, X., Dai, Z.: Unveiling osteoporosis through radiomics analysis of hip ct imaging. Acad. Radiol. **31**(3), 1003–1013 (2024)
30. Jiang, Y.-W., Xiong-Jie, X., Wang, R., Chen, C.-M.: Radiomics analysis based on lumbar spine ct to detect osteoporosis. Eur. Radiol. **32**(11), 8019–8026 (2022)
31. Lin, X., Shen, R., Zheng, X., Shi, S., Dai, Z., Fang, K.: Utilizing radiomics techniques to isolate a single vertebral body from chest ct for opportunistic osteoporosis screening. BMC Musculoskelet. Disord. **25**(1), 785 (2024)
32. Tong, X., Wang, S., Zhang, J., Fan, Y., Liu, Y., Wei, W.: Automatic osteoporosis screening system using radiomics and deep learning from low-dose chest ct images. Bioengineering **11**(1), 50 (2024)
33. Cheng, L., Cai, F., Mingzhi, X., Liu, P., Liao, J., Zong, S.: A diagnostic approach integrated multimodal radiomics with machine learning models based on lumbar spine ct and x-ray for osteoporosis. J. Bone Miner. Metab. **41**(6), 877–889 (2023)

Feasibility of Physical Telerehabilitation in Dysautonomia Patients with Postural Orthostatic Tachycardia Syndrome

Aileen S. Gabriel[1](✉), Aref Smiley[1], Patricia Goncalves Leite Rocco[1], Katie Johnson[2], Melissa M. Cortez[2], and Joseph Finkelstein[1]

[1] University of Arizona College of Medicine, Tucson, AZ, USA
aileengabriel@arizona.edu

[2] Department of Neurology, University of Utah, Salt Lake City, UT, USA

Abstract. Dysautonomia, including postural orthostatic tachycardia syndrome (POTS), often disrupts daily functioning and complicates adherence to traditional exercise programs. Although exercise is an essential component of symptom management, patients face significant barriers, including fatigue, elevated heart rate, cognitive impairment, travel demands, and discomfort in public exercise environments. Despite growing interest in telehealth solutions, little is known about the acceptability of home-based telerehabilitation specifically designed for individuals with dysautonomia. This study aimed to evaluate perspectives on a multimodal telerehabilitation to support engagement with a home-based exercise program in patients with dysautonomia and POTS symptoms. Thematic analysis revealed high levels of acceptability, with participants emphasizing the program's capacity to reduce exercise barriers while fostering confidence, structure, accountability, and sustained engagement. Feedback highlighted opportunities for improvement, including enhanced personalization, interface improvements, device compatibility, and expanded monitoring capabilities. These findings suggest that home-based telerehabilitation can provide a safe, adaptable, and patient-centered alternative to traditional rehabilitation programs. Future research should investigate long-term outcomes, cost-effectiveness, and scalability to support the integration of this approach into standard care.

Keywords: Telerehabilitation · Dysautonomia · Postural Orthostatic Tachycardia Syndrome

1 Introduction

The autonomic nervous system is responsible for regulating vital involuntary processes, including cardiovascular control, body temperature regulation, and the function of the gastrointestinal and urinary systems [1]. Disruption of this regulatory network results in autonomic dysfunction known as dysautonomia. These conditions are often characterized by orthostatic and exercise intolerance, which contribute significantly to disability and impaired quality of life [2]. Among them, postural tachycardia syndrome (POTS)

H. R. Arabnia et al. (Eds.): CSCI 2025, CCIS 2971, pp. 334–344, 2027.
https://doi.org/10.1007/978-3-032-28463-1_27

represents one of the most common presentations. POTS is defined by an excessive increase in heart rate upon standing in the absence of orthostatic hypotension and is frequently accompanied by a constellation of symptoms such as dizziness, palpitations, fatigue, gastrointestinal complaints, and cognitive impairment [3]. Current management strategies typically integrate pharmacological agents with lifestyle and behavioral interventions, with the overarching goals of mitigating symptoms, restoring function, and reducing deconditioning [4]. Exercise training is a key nonpharmacologic therapy shown to improve blood volume, reverse deconditioning, and enhance orthostatic tolerance [5]. Standard approaches often begin with supine or recumbent aerobic and resistance training, followed by gradual progression to upright activities to increase gravitational stress tolerance. However, adherence is often poor due to fatigue, pain, motivational barriers, and limited access to supervised programs [6]. While individualized guidance improves outcomes, many patients attempt unsupervised exercise with variable results [7].

Recent advances in digital health technologies provide new avenues to overcome these limitations. Telerehabilitation platforms offer the ability to deliver personalized, supervised exercise interventions remotely, with real-time feedback and monitoring to enhance both safety and adherence. To address barriers to access and adherence, we developed a telerehabilitation system designed to facilitate supine exercise training with integrated biofeedback for individuals with POTS and other autonomic disorders associated with chronic orthostatic intolerance. This study aimed to evaluate patient perspectives on the usability and acceptability of this system.

2 Methods

2.1 System Design

The Home Automated Telemanagement (HAT) system is a telerehabilitation platform developed to support individuals with chronic conditions such as autonomic dysfunction and POTS through personalized home-based exercise and clinical guidance, monitoring, and feedback [8]. The system comprised three components: a patient portal, a clinician portal, and an iBike exercise device. The patient and clinician portals were built as secure, web-based applications, with data stored on a centralized server. Clinicians could use the HAT system to design and adjust exercise prescriptions, monitor adherence and reported symptoms, and communicate with patients remotely. Patients accessed the HAT system through a computer or tablet, which was adapted to accommodate individuals with mobility, vision, or cognitive limitations. Data collected from the iBike was transmitted to the central server and generated into a progress report. Clinicians could review these progress reports through the portal review to track performance and inform ongoing care. The iBike was a compact recumbent cycle fitted with an ESP32 microcontroller and a magnetic switch to capture cadence in revolutions per minute, which was transmitted via Wi-Fi to a secure cloud server. A custom C-box paired with a Nonin 3150 pulse oximeter collected heart rate and oxygen saturation via Bluetooth. These data were integrated in real time and displayed through a Unity-based virtual cycling environment on a Windows tablet, providing dynamic biofeedback and continuous cues to help patients maintain prescribed cadence targets of 60, 75, or 90 RPM.

2.2 Study Design

Enrolled participants completed a single one-hour in-person study visit. After informed consent, the research staff provided a demonstration and a practice session with the HAT system, followed by a sociodemographic questionnaire. Usability testing included three standardized tasks performed independently: logging into the system, completing a survey, and reviewing and carrying out an exercise routine. The exercise component required participants to watch an instructional video and then complete a supine cycling activity on the iBike while interacting with a Unity-based tablet program that provided cadence targets and real-time biofeedback.

Following the usability testing, participants completed a semi-structured interview with open-ended questions to explore perceptions of the system and its potential use for home-based symptom management. All interviews were transcribed verbatim and de-identified. The qualitative data from the interviews were analyzed using thematic analysis following the approach outlined by Braun and Clarke [9]. The research team reviewed transcripts repeatedly to ensure immersion in the data, and initial codes were developed inductively through open coding. Codes were iteratively refined and clustered into broader categories to capture patterns across participant accounts. These categories were synthesized into themes that reflected key patterns in participant experiences.

3 Results

A total of 15 participants were included in the current study. The majority were female (93.3%), with one male participant (6.7%). Most participants identified as White (93.3%), with smaller proportions reporting American Indian or Alaska Native (6.7%) or Other race (13.3%). In terms of ethnicity, 13.3% identified as Hispanic or Latino and 86.7% as not Hispanic or Latino. Symptom severity was most often reported as moderate (75.0%), with fewer participants indicating mild (12.5%) or severe (12.5%) symptoms. When asked about the status of their condition over the past 12 months, 37.5% of participants described their condition as stable, 31.3% reported improvement, and 31.3% indicated their condition was progressing. Knowledge of the condition varied among participants. The majority (60.0%) described their understanding as good, while 26.7% reported excellent knowledge. A smaller proportion (13.3%) indicated very limited knowledge.

The thematic analysis identified ten major themes that reflected participants' experiences and perceptions of the telerehabilitation system. Overall, participants reported high levels of acceptability. The telerehabilitation system was perceived as a supportive, accessible, and motivating tool with the potential to address common barriers to exercise engagement in this population. The themes, along with a representative quote, are presented in Table 1.

Theme 1: Clarity and Quality of Exercise Content. Participants consistently described the exercise content as clear, comprehensive, straightforward, and easy to follow. Most participants reported having no difficulties in understanding the instructions. Participants noted that visual guidance enhanced comprehension, accuracy, and confidence while also supporting memory retention. The combination of video, text, and

Table 1. Themes and example representative quotes

Theme	Representative quote
Clarity and quality of exercise content	"I liked that there was an option to read and watch the exercises. I'm sure a lot of people are like me and have a hard time learning how to do the exercise just by reading. There was also a video to follow along so they're showing you as you're doing it as well." (Participant 10)
Accessibility and continuity of remote exercise content	"For me especially with POTS, it is hard to travel all the time, the energy level is hard. Having something guided at home will help. But I love in person too, but there's a lot of people that don't have access to transportation, so this is very needed." (Participant 5)
Tailoring to user needs	"I think sometimes doing PT at home could be depressing, maybe something when you finish you can hear a 'good job' or something more visual appealing would be helpful." (Participant 7)
Ease of system use and navigation	"Thinking about the first portion of it, it was easy to navigate and go back and forward in between. Even on the days that my brain is not working fully will be easier to use it." (Participant 12)
Biofeedback and self-monitoring	"I was able to recalibrate if I was too going too fast or too slow, my breathing, how fast on the pedals I was going. It was good to have that feedback." (Participant 14)
Personalization and design appeal	"I think superficially having something more visually appealing. Something like you would want to log in and do it." (Participant 11)
Home-based telerehabilitation for symptom management	"This one is more tailored to people with heart problems. I'll go to exercises and my heart rate will get too high and it's kind of embarrassing, so I prefer to exercise at home." (Participant 3)
Structured program and provider support	"Its a lot better than a gym, I don't necessarily know what to do at the gym." (Participant 4)

(*continued*)

Table 1. *(continued)*

Theme	Representative quote
Motivation and accountability	It's helpful, makes me want to do it and to keep track with the line was helpful to focus. Its kind of takes your mind out of everything and help you remember that you are doing something." (Participant P9)
Building exercise confidence and consistency	I am not very consistent with exercise at all, I often forget or don't have the motivation to do it, especially with my brain fog, I usually forget everything, I usually need a reminder in front of me, so my exercises are not consistent at all. Having reminders, having a program like this is a very good idea, I can see myself benefit from this program personally a lot." (Participant 16)

audio was especially helpful for those managing symptoms such as fatigue or brain fog, as it accommodated different learning preferences. The videos were generally preferred over text or picture-based instructions, with participants stating that visual demonstrations were more intuitive and engaging than static images. While overall satisfaction with the clarity and quality of the materials was high, a few participants suggested potential improvements. These included introducing a text overlay that would track counts and timing, which could make it easier to follow along in real time, and offering an audio-only option for those who found reading from a screen difficult. A participant pointed out that the length of the breathing exercises was initially confusing and suggested that further clarification of timing between different exercise components could strengthen understanding.

Theme 2: Accessibility and Continuity of Remote Exercise Content. Participants valued the ability to access exercise content remotely, beyond the clinical setting. Having the ability to revisit guided instructions from home was considered convenient and supportive of continuity of care. Remote access to exercise content allowed individuals to revisit guided materials at their own pace and maintain progress between physical therapy appointments. This flexibility was important for those managing chronic fatigue, limited energy, cognitive challenges such as brain fog, or barriers related to transportation. Remote access to the exercise content was also seen as a way to overcome practical barriers. Participants highlighted that it would reduce the need for frequent in-person physical therapy appointments, which would save time and money, and increase overall access to services for those who might otherwise struggle to attend regular appointments.

Theme 3: Tailoring to User Needs. Participants expressed mixed views regarding the need for tailoring exercise content. Many participants expressed satisfaction with the content in its current form, reporting that it was sufficient and felt that adding more diverse content, such as images, pictures, or video clips, would not significantly affect their experience. However, a few participants identified opportunities to tailor the content to better meet individual needs and preferences. Suggestions included developing more engaging or interactive tools for younger users, adjusting content to better align with mental health needs, and incorporating specialized or modified exercise options to accommodate different functional levels. In addition, a few participants highlighted the potential value of motivating feedback, which could help sustain commitment and encourage continued participation.

Theme 4: Ease of System Use and Navigation. The system was widely regarded as intuitive, simple, and easy to use independently. When compared to the usability of a household TV remote, several found the system easier to operate, while others considered the two about the same. A smaller number felt that their TV remote was simpler due to familiarity or having fewer components. The practice session with the system was viewed as beneficial, helping participants feel comfortable and confident in managing the system without assistance. Most participants reported no difficulties with using the interface, required no external help, and found the exercise session manageable without fatigue. The system was also described as user-friendly even on days when cognitive symptoms were more pronounced. Although satisfaction with usability was high, a few challenges and suggestions for improvement were reported. A few participants raised concerns related to difficulties with setup, limited space for exercise equipment, occasional interface lag or jerky pedal movements, and initial confusion with individual buttons. Suggestions include making the system compatible with iPads, offering a phone-based option, or distinguishing the text of safety precautions from exercise instructions.

Theme 5: Biofeedback and Self-monitoring. The biofeedback and self-monitoring features of the system were highlighted as an essential component of safe and effective exercise with dysautonomia symptoms. Real-time monitoring of heart rate and oxygen saturation was viewed as especially valuable as it provided reassurance during exercise and helped them regulate their intensity. Several participants explained that having these measures visible throughout sessions reduced fear of overexertion and made independent exercise feel manageable. Participants appreciated the engaging racecar avatar that tracked real-time heart rate, describing it as a motivating and intuitive way to connect physiological feedback with exercise performance. Tracking vitals during exercise was identified not only as a safety feature but also encouraged awareness of their progress over time. Suggestions for improvement included positioning heart rate and oxygen data more prominently on the screen, adding visual cues that reference heart rate and oxygen when answering the questions related to symptoms, and expanding monitoring capabilities to include blood pressure or symptom logging. One participant also proposed an alert system that could notify a designated emergency contact if concerning vital changes occurred.

Theme 6: Personalization and Design Appeal. Participants discussed a range of features that would make the program more engaging, attractive, and adapted to individual needs. While some preferred a clean and minimal interface, others favored more interactive and dynamic features. The importance of visual appeal was emphasized, with several participants suggesting that improvements to color schemes, design consistency, and layout could make the interface more attractive and inviting. Suggestions included positioning videos more prominently on the main screen to reduce unnecessary scrolling, as well as incorporating visual and interactive aids to make the program feel more dynamic. In addition to design, personalization was discussed as a way to make the program feel more engaging and personally relevant. Proposed options included customizable avatars, reminder alerts to support adherence, and tools for monitoring health and progress, such as trend displays and symptom annotations.

Theme 7: Home-Based Telerehabilitation for Symptom Management. Participants overwhelmingly supported home-based telerehabilitation as an acceptable and effective approach to symptom management for dysautonomia and POTS-related symptoms. They emphasized the usefulness of a computer-guided, centralized system that made exercise at home more structured and accessible. Several participants highlighted that exercising from home was preferable to attending gyms or traditional rehabilitation settings, noting that it minimized barriers such as feeling self-conscious about visible symptoms, such as elevated heart rate, and avoided the physical and emotional fatigue associated with travel. Participants believed that the system had the potential to improve health outcomes, reduce symptom burden, and make exercise more feasible for individuals who face unique challenges due to dysautonomia symptoms. Reports of challenges or concerns were relatively limited. A few participants mentioned initial worries about injury risk when exercising independently, but these concerns were mitigated by the ability to communicate directly with providers when questions arose. One participant mentioned the issue of program cost as a potential barrier, though this was not a predominant concern.

Theme 8: Structured Program and Provider Support. The structured format of the program was perceived as a key strength, especially when contrasted with participants' existing routines, which were often unstructured, inconsistent, or lacking guidance. Features such as clear scheduling, guided progression, and the ability to modify exercises were seen as crucial in supporting adaptability to varying abilities and symptom fluctuations. Visual tools and varied exercise options were highly valued, making the program more engaging and easier to follow than unstructured alternatives. Provider support emerged as equally important. Participants appreciated the ability to communicate directly with a physical therapist and recognized the reassurance offered by remote supervision and monitoring. The program's utility as a bridge between in-person rehabilitation and independent exercise was highlighted. For some participants, it complemented prior therapy by extending structured support into the home environment. While a few were satisfied with their current exercise routines or noted similarities to other programs, most participants valued the program's structured design, adaptability, and provider involvement as features that made it both distinct and beneficial. Some participants also suggested that periodic in-person visits could further strengthen the model by balancing virtual and face-to-face support.

Theme 9: Motivation and Accountability. Participants described motivation and accountability as important factors of program engagement. They highlighted that several features worked together to encourage consistency and adherence. Progress tracking tools offered tangible evidence of effort and improvement. Reminders and scheduling functions provided additional structure to help participants establish and maintain consistent exercise routines. Gamified elements, such as interactive visuals and the bike exercise, made the experience more enjoyable and stimulating. These features transformed exercise from a repetitive obligation into an engaging activity, which participants felt increased their willingness to continue. For some, gamification also provided a sense of challenge and achievement that mirrored recreational or competitive activities, further enhancing motivation. Social support further reinforced accountability and motivation. Participants noted that provider monitoring and communication encouraged them to remain engaged, while also offering reassurance that they were exercising correctly and safely. This support was particularly meaningful for those who described struggling with motivation in traditional or self-directed exercise settings.

Theme 10: Building Exercise Confidence and Consistency. Participants recognized the program as a supportive framework that strengthened their confidence in exercising and encouraged the development of more consistent routines. Many participants reported that dysautonomia symptoms such as fatigue, post-exertional malaise, and pain often made it difficult to maintain regular exercise routines. Environmental barriers, including limited access to gyms or discomfort in public exercise spaces, further contributed to irregular or unstructured activity. The program helped address these challenges by providing clear guidance, structured plans, and short, manageable sessions that felt achievable even on days when energy was low. The flexibility to adjust scheduling around symptoms or daily routines was seen as critical for maintaining participation over time. The program was also seen as important for validating and reinforcing existing exercise abilities. Participants with long-term experience managing dysautonomia and established confidence in their routines valued the program for affirming that their approaches were appropriate and aligned with professional guidance. For participants with little to no structured exercise background, the program served as essential support to build capacity and independence.

4 Discussion

This study explored the acceptability of a home-based telerehabilitation system to support symptom management in patients living with dysautonomia and POTS-related symptoms. The results of the thematic analysis revealed that the telerehabilitation system was met with high approval among participants. The system was viewed as an effective way to address common barriers to exercise, including dysautonomia symptoms (e.g., increased heart rate, fatigue, cognitive impairment), travel demands, and discomfort in public exercise spaces. Participants expressed that the system had the potential to improve confidence, structure, accountability, and engagement with an exercise program. By integrating multimodal instructional tools, biofeedback, and provider support, the program aligned well with patient needs and expectations for safe, adaptable, and motivating exercise. Participants' suggestions, such as greater personalization, interface design refinements, device compatibility, and expanded monitoring, were identified as opportunities to further improve user experience.

Future approaches for telemedicine-supported exercise should be enhanced by the recent advances in machine learning optimization of individualized exercise plans [10,

11], predictive analytics utilizing patient-generated data [12, 13], and interfaces with electronic health records to support personalization [14, 15], adherence, efficacy, and safety of the home-based exercise programs [16, 17]. Patient engagement in exercise programs can be facilitated by exergaming in virtual reality [18, 19] and artificial intelligence-based chatbots [20, 21], promoting patient education [22, 23] and healthy behaviors [24, 25]. Future studies should evaluate the proposed approach in disease prevention [26, 27], management of chronic health conditions [28, 29], healthy aging support [30, 31], and patient rehabilitation [32, 33]. Clinical impact of the resulting system will be evaluated in randomized controlled trials [34, 35]. Once the impact is demonstrated, the implementation into routine clinical care will include clinical decision support embedded into electronic health records [36, 37]. The unpredictable nature of dysautonomia symptoms can make regular participation in traditional exercise programs difficult. Our findings suggest that home-based telerehabilitation represents a promising strategy for expanding access to structured, patient-centered rehabilitation in the home environment.

References

1. Rafanelli, M., Walsh, K., Hamdan, M.H., Buyan-Dent, L.: Autonomic dysfunction: diagnosis and management. Handb. Clin. Neurol. **167**, 123–137 (2019)
2. Blitshteyn, S.: Dysautonomia: a common comorbidity of systemic disease. Immunol. Res. **73**(1), 105 (2025)
3. Arnold, A.C., Ng, J., Raj, S.R.: Postural tachycardia syndrome–diagnosis, physiology, and prognosis. Auton. Neurosci. **215**, 3–11 (2018)
4. Raj, S.R.: Postural tachycardia syndrome (POTS). Circulation **127**(23), 2336–2342 (2013)
5. Sheldon, R.S., et al.: 2015 heart rhythm society expert consensus statement on the diagnosis and treatment of postural tachycardia syndrome, inappropriate sinus tachycardia, and vasovagal syncope. Heart Rhythm **12**(6), e41–e63 (2015)
6. Fu, Q., Levine, B.D.: Exercise and non-pharmacological treatment of POTS. Auton. Neurosci. **215**, 20–27 (2018)
7. Wheatley-Guy, C.M., et al.: Semi-supervised exercise training program more effective for individuals with postural orthostatic tachycardia syndrome in randomized controlled trial. Clin. Auton. Res. **33**(6), 659–672 (2023)
8. Finkelstein, J., Khare, R., Vora, D.: Home automated telemanagement (HAT) system to facilitate self-care of patients with chronic diseases. J. Systemics. Cybern. Inf. **1**(3), 78–82 (2003)
9. Braun, V., Clarke, V.: Using thematic analysis in psychology. Qual. Res. Psychol. **3**(2), 77–101 (2006)
10. Jeong, I.C., Finkelstein, J.: Classification of cycling exercise status using short-term heart rate variability. In: Proceedings of the Annual International Conference of the IEEE Engineering in Medicine and Biology Society (EMBC), pp. 1782–1785 (2014). https://doi.org/10.1109/EMBC.2014.6943954
11. Smiley, A., Finkelstein, J.: Automated prediction of exercise intensity using physiological data and deep learning. SN Comput. Sci. **6**(4), 313 (2025). https://doi.org/10.1007/s42979-025-03739-2
12. Finkelstein, J., Smiley, A., Echeverria, C., Mooney, K.: Leveraging convolutional neural networks for predicting symptom escalation in chemotherapy patients: a temporal resampling approach. Stud. Health Technol. Inform. **323**, 45–49 (2025). https://doi.org/10.3233/SHTI250046

13. Smiley, A., Finkelstein, J.: Dynamic prediction of physical exertion: leveraging AI models and wearable sensor data during cycling exercise. Diagnostics **15**(1), 52 (2024). https://doi.org/10.3390/diagnostics15010052
14. Paranjpe, I., et al.: Retrospective cohort study of clinical characteristics of 2199 hospitalised patients with COVID-19 in New York City. BMJ Open **10**(11), e040736 (2020). https://doi.org/10.1136/bmjopen-2020-040736
15. Finkelstein, J., Zhang, F., Levitin, S.A., Cappelli, D.: Using big data to promote precision oral health in the context of a learning healthcare system. J. Public Health Dent. **80**(Suppl. 1), S43–S58 (2020). https://doi.org/10.1111/jphd.12354
16. Finkelstein, J., Wood, J., Cha, E., Orlov, A., Dennison, C.: Feasibility of congestive heart failure telemanagement using a WII-based telecare platform. In: 2010 Proceedings of the Annual International Conference of the IEEE Engineering in Medicine and Biology Society (EMBC), pp. 2211–2214. IEEE, Buenos Aires (2010). https://doi.org/10.1109/IEMBS.2010.5627087
17. Finkelstein, J., Wood, J., Cha, E.: Impact of physical telerehabilitation on functional outcomes in seniors with mobility limitations. In: 2012 Proceedings of the Annual International Conference of the IEEE Engineering in Medicine and Biology Society (EMBC), pp. 5827–5832. IEEE, San Diego (2012). https://doi.org/10.1109/EMBC.2012.6347319
18. Gabriel, A.S., et al.: Patient perceptions of virtual reality in cancer rehabilitation: a qualitative study. Stud. Health Technol. Inform. **323**, 399–403 (2025). https://doi.org/10.3233/SHTI250120
19. Locke, B.W., Tsai, T.Y., Reategui-Rivera, C.M., Gabriel, A.S., Smiley, A., Finkelstein, J.: Immersive virtual reality use in medical intensive care: mixed methods feasibility study. JMIR Serious Games **12**(1), e62842 (2024). https://doi.org/10.2196/62842
20. Reategui-Rivera, C.M., Smiley, A., Finkelstein, J.: LLM-based chatbot to reduce mental illness stigma in healthcare providers. In: 2025 IEEE 15th Annual Computing and Communication Workshop and Conference (CCWC), pp. 00001–00007. IEEE, Las Vegas (2025). https://doi.org/10.1109/CCWC62904.2025.10903778
21. Shah-Mohammadi, F., Finkelstein, J.: GPT-4 in clinical practice: assessing its capability for symptom extraction from cancer patient notes. Stud. Health Technol. Inform. **323**, 86–90 (2025). https://doi.org/10.3233/SHTI250054
22. Finkelstein, J., Cha, E.M.: Using a mobile app to promote smoking cessation in hospitalized patients. JMIR MHealth UHealth **4**(2), e59 (2016). https://doi.org/10.2196/mhealth.5149
23. Vilela-Estrada, A.L., et al.: Psychometric properties of the Spanish version of the hospital anxiety and depression scale in cancer patients. Front. Psychol. **15**, 1497946 (2024). https://doi.org/10.3389/fpsyg.2024.1497946
24. Finkelstein, J., Wood, J.: Introducing home telemanagement of congestive heart failure using Xbox gaming platform. In: 2011 24th International Symposium on Computer-Based Medical Systems (CBMS), pp. 1–4. IEEE, Bristol (2011). https://doi.org/10.1109/CBMS.2011.5999126
25. Shah-Mohammadi, F., Cui, W., Bachi, K., Hurd, Y., Finkelstein, J.: Using natural language processing of clinical notes to predict outcomes of opioid treatment program. In: 2022 44th Annual International Conference of the IEEE Engineering in Medicine and Biology Society (EMBC), pp. 4415–4420. IEEE, Glasgow (2022). https://doi.org/10.1109/EMBC48229.2022.9871960
26. Yi, H., Xiao, T., Thomas, P.S., et al.: Barriers and facilitators to patient–provider communication when discussing breast cancer risk to aid in the development of decision support tools. In: AMIA Annual Symposium Proceedings 2015, pp. 1352–1360 (2015)

27. Finkelstein, J., Wood, J.: Implementing home telemanagement of congestive heart failure using Xbox gaming platform. In: Annual International Conference of the IEEE Engineering in Medicine and Biology Society (EMBC), pp. 3158–3163 (2011). https://doi.org/10.1109/IEMBS.2011.6090861
28. Castro, H.K., Cross, R.K., Finkelstein, J.: Using a home automated telemanagement (HAT) system: experiences and perceptions of patients with inflammatory bowel disease. In: AMIA Annual Symposium Proceedings 2006, p. 872 (2006)
29. Cross, R.K., Finkelstein, J.: Challenges in the design of a home telemanagement trial for patients with ulcerative colitis. Clin. Trials **6**(6), 649–657 (2009). https://doi.org/10.1177/1740774509346978
30. Finkelstein, J., Cisse, P., Jeong, I.C.: Feasibility of interactive resistance chair in older adults with diabetes. Stud. Health Technol. Inform. **213**, 61–64 (2015)
31. Barron, J., Bedra, M., Wood, J., Finkelstein, J.: Exploring three perspectives on feasibility of a patient portal for older adults. Stud. Health Technol. Inform. **202**, 181–184 (2014)
32. Jeong, I.C., Karpatkin, H., Finkelstein, J.: Physical telerehabilitation improves quality of life in patients with multiple sclerosis. Stud. Health Technol. Inform. **284**, 384–388 (2021). https://doi.org/10.3233/SHTI210752
33. Bedra, M., McNabney, M., Stiassny, D., et al.: Defining patient-centered characteristics of a telerehabilitation system for patients with COPD. Stud. Health Technol. Inform. **190**, 24–26 (2013)
34. Fleg, J.L., Keteyian, S.J., Peterson, P.N., et al.: Increasing use of cardiac and pulmonary rehabilitation in traditional and community settings: opportunities to reduce health care disparities. J. Cardiopulm. Rehabil. Prev. **40**(6), 350–355 (2020). https://doi.org/10.1097/HCR.0000000000000527
35. Grimshaw, C., Keteyian, S.J., Benzo, R., et al.: Baseline characteristics and barriers to recruitment in cardiac and pulmonary rehabilitation NIH-funded trials. J. Cardiopulm. Rehabil. Prev. **43**(6), 407–411 (2023). https://doi.org/10.1097/HCR.0000000000000824
36. Finkelstein, J., Gabriel, A., Schmer, S., Truong, T.T., Dunn, A.: Identifying facilitators and barriers to implementation of AI-assisted clinical decision support in an electronic health record system. J. Med. Syst. **48**(1), 89 (2024). https://doi.org/10.1007/s10916-024-02104-9
37. Kawamoto, K., Finkelstein, J., Del Fiol, G.: Implementing machine learning in the electronic health record: checklist of essential considerations. Mayo Clin. Proc. **98**(3), 366–369 (2023). https://doi.org/10.1016/j.mayocp.2023.01.013

A Neuromusculoskeletal Simulation Framework for Generating Realistic, Multi-modal Synthetic Fall Data

Olivia Lu Feng[1], Sandy Shengyi Zhao[2], Yan Lin[3], Ocean Yangzijian Han[4], Wilson Qiwei Sun[5], Runjie Fan[6], Amirreza Sedaghat[7], and Chris Cheng Zhang[7](✉)

[1] Kent School, Kent, USA
[2] St. John Paul II Catholic Secondary School, Toronto, Canada
[3] Sir Winston Churchill Secondary School, Vancouver, Canada
[4] A.R. MacNeill Secondary School, Richmond, Canada
[5] DeGroot School of Business, McMaster University, Hamilton, Canada
[6] Faculty Engineering, McMaster University, Hamilton, Canada
[7] Department of Research and Development, Cansight Technology Corporation, Toronto, Canada
head@roboca.org

Abstract. The development of robust fall detection algorithms for wearable devices is critically hampered by the scarcity of real-world fall data, particularly data that captures the complex biomechanics of the elderly population. To address this, this paper presents a detailed methodology for generating a comprehensive, physically realistic dataset of falls and successful recoveries using neuromusculoskeletal simulation. We leverage the OpenSim platform to simulate responses to a trip perturbation, explicitly modeling the presence or absence of a neuromuscular recovery strategy. This approach allows for the generation of a rich, multi-modal dataset including whole-body kinematics, center of mass trajectory, muscle activation patterns, and the resulting high-fidelity accelerometer signals. By establishing a clear causal link between neuromuscular control and fall outcome, this framework provides a powerful tool for creating a large-scale, physically-plausible synthetic dataset, envisioned as a "simulated SisFall," necessary for training and validating the next generation of fall detection and risk assessment systems.

Keywords: Fall Detection · Biomechanical Simulation · OpenSim · Synthetic Data · Neuromuscular Control · Accelerometer · Older Adults

1 Introduction

Falls [12] among older adults represent a severe and escalating public health crisis. Epidemiological data indicates that approximately one-third of individuals

H. R. Arabnia et al. (Eds.): CSCI 2025, CCIS 2971, pp. 345–353, 2027.
https://doi.org/10.1007/978-3-032-28463-1_28

aged 65 and older experience at least one fall annually in developed countries [8]. In the United States, this translates to millions of incidents, with nearly 39,000 fall-related deaths reported in 2021 [2]. The consequences of these events are profound and multi-faceted. Physically, falls are a leading cause of injuries such as hip fractures and traumatic head injuries, often resulting in hospitalization, long-term disability, and a loss of independence [1]. Beyond the immediate physical trauma, falls inflict significant socio-psychological harm; the fear of falling can lead to reduced mobility, social isolation, and depressive symptoms, severely diminishing an individual's quality of life and their capacity to perform Activities of Daily Living (ADLs) [5]. This combination of high incidence and severe outcomes imposes a substantial economic burden on healthcare systems worldwide [2], driving significant market research into novel prevention products [13].

To mitigate the most severe consequences, particularly those arising from a "long lie" where an individual is unable to get up independently, automated Fall Detection Systems (FDS) have emerged as a critical technology [11]. Timely alerts from an FDS can dramatically reduce morbidity and mortality associated with post-fall complications [3]. However, the practical efficacy of these systems hinges on their ability to achieve two competing objectives: high sensitivity to detect genuine falls and exceptionally high specificity to reject fall-like but benign daily activities. An unreliable system plagued by false alarms is likely to be ignored or disabled, rendering it ineffective. This dual requirement has driven the field from simple threshold-based algorithms toward more sophisticated Machine Learning (ML) and Deep Learning (DL) models, which promise to learn the complex boundary separating a true fall from vigorous movements [7].

To address this challenge, this paper proposes a paradigm shift from descriptive data collection to generative, physics-based simulation, unleashing the potential of machine learning for this specific problem [15]. The methods for collecting such data are themselves an active area of research, involving both wearable technology and predictive analysis [6]?. ML models, especially deep neural networks, require vast datasets for robust training, yet biomechanical data collection is an expensive and resource-intensive process, typically yielding only a few hundred or thousand data points [4]. Because real-world falls are unpredictable and dangerous to capture experimentally, the research community has largely relied on datasets of simulated falls performed by young, healthy volunteers [7]. This practice introduces a critical "realism gap." The controlled, athletic maneuvers of a young actor are biomechanically distinct from the involuntary loss of balance experienced by a frail older adult. Consequently, ML models trained on this unrepresentative data learn to detect acted falls but fail to generalize to real-world scenarios, perpetuating the very problem of unreliability they were intended to solve [7]. The core issue is not merely a lack of data, but a systemic lack of biomechanically valid data. While other promising data-driven approaches, such as camera-based human pose analysis, are being explored [14], they often face their own challenges related to privacy and environmental constraints.

To address this challenge, this paper proposes a paradigm shift from descriptive data collection to generative, physics-based simulation. We present a novel framework that leverages the OpenSim musculoskeletal modeling platform to generate high-fidelity, synthetic fall data from first principles [9]. Instead of capturing the kinematic appearance of a fall, our approach models its underlying cause: a loss of neuromuscular control following a physical perturbation. By integrating a neuromuscular controller into a dynamic, forward simulation, we can model the causal chain of events from an initial trip to the emergent, whole-body response. This allows us to simulate not only catastrophic falls resulting from a failed corrective response but also successful recoveries where balance is maintained through appropriate muscle activation. This methodology enables the creation of physically plausible and diverse data that is grounded in the fundamental laws of biomechanics, providing a principled solution to the data scarcity crisis [10].

The primary contribution of this work is a biomechanical simulation framework capable of generating unambiguously distinguishable sensor data for both falls and vigorous recovery attempts. By modeling the presence or absence of a neuromuscular recovery strategy in response to an identical trip perturbation, we produce comprehensive, multi-modal data streams, including body kinematics, muscle activations, and the resulting accelerometer signals. We demonstrate that the simulated accelerometer signature of a fall is quantitatively distinct from that of a successful recovery, providing the robust, classifiable features necessary for training the next generation of highly accurate and reliable fall detection systems.

The remainder of this paper is organized as follows. Section 2 details the methodology of our simulation framework, including the musculoskeletal model and neuromuscular control strategy. Section 3 presents and discusses the results of the fall and recovery simulations, with a focus on the biomechanical and sensor-level distinctions. Finally, Sect. 4 provides concluding remarks and outlines directions for future work.

2 Methodology

The core of our approach is a forward dynamics simulation framework built using the OpenSim API within a Python environment. This allows for the programmatic control of a detailed musculoskeletal model to simulate dynamic activities and neuromuscular responses to perturbations.

2.1 Musculoskeletal Model

We utilized the Rajagopal-Lai-Uhlrich 2023 lower-limb musculoskeletal model, a detailed, full-body model with 39° of freedom (DOFs) and 80 muscle-tendon units. The dynamics of this multibody system are governed by the standard equations of motion:

$$M(q)\ddot{q} + C(q, \dot{q}) + G(q) = \tau \tag{1}$$

where q, $\dot{q}$, and $\ddot{q}$ are the vectors of generalized coordinates, velocities, and accelerations, respectively. $M(q)$ is the system mass matrix, $C(q, \dot{q})$ represents Coriolis and centrifugal forces, $G(q)$ is the vector of gravitational forces, and τ is the vector of generalized forces from muscles and external sources. A critical preparatory step involved programmatically unlocking all model DOFs to ensure all joints were free to move during the dynamic simulation.

2.2 Ground Contact Model

To enable physically realistic interaction with the environment, a ground contact model was programmatically added. A non-deformable ground plane was defined as a `ContactHalfSpace` located at $y = 0$. Contact with the ground was modeled by attaching six `ContactSphere` geometries to each foot, representing key anatomical points of contact. The interaction between these spheres and the ground plane is governed by a `HuntCrossleyForce` model, a nonlinear spring-damper formulation where the normal force F_n is a function of the sphere's penetration depth x and velocity $\dot{x}$:

$$F_n = kx^n(1 + c\dot{x}) \tag{2}$$

where k is the stiffness parameter, n is the Hertz exponent (typically 1.5 for compliant contact), and c is the dissipation coefficient. Frictional forces were also modeled using static and dynamic friction coefficients to prevent foot-slip realistically.

2.3 Trip Perturbation and Neuromuscular Response Modeling

A common fall-inducing event, a trip, was simulated by applying a transient, anteriorly-directed external force to the model's right foot using a `PointActuator`. The force was active for a brief duration to destabilize the model's gait. To investigate the difference between a fall and a recovery, we defined two distinct neuromuscular control scenarios:

1. **Trip without Recovery (Fall):** This scenario models a failed response, characteristic of an individual unable to mount a sufficient recovery action. In this simulation, the activation of key stabilizing muscles (e.g., gluteus maximus, hamstrings) was suppressed following the perturbation.
2. **Trip with Recovery:** This scenario models a successful neuromuscular response. After the perturbation, a controller sends a pre-defined activation signal to key lower-limb extensor muscles. This activation, $A_{reflex}(t)$, mimics the physiological response to regain balance by generating a corrective joint torque.

2.4 Data Acquisition

During each simulation, a virtual tri-axial accelerometer was modeled by recording the linear acceleration of a point on the forearm (`radius_r` body). The acceleration due to gravity, $\boldsymbol{g}$, was subtracted to simulate the output of a real-world

MEMS accelerometer, yielding the proper acceleration $\boldsymbol{a}_{sim}$:

$$\boldsymbol{a}_{sim}(t) = \boldsymbol{a}_{body}(t) - \boldsymbol{g} \tag{3}$$

In addition to accelerometer data, key biomechanical variables were recorded at each time step, including the whole-body center of mass (CoM) trajectory, joint kinematics (angles), and muscle activation levels.

3 Results and Discussion

The simulation framework was used to conduct a comparative experiment between the "Trip with Recovery" and "Trip without Recovery" scenarios. The results provide quantitative, multi-modal insight into the biomechanical factors that differentiate a successful recovery from an uncontrolled fall.

3.1 Comprehensive Biomechanical Analysis of Trip Outcomes

A holistic view of the simulation outputs reveals the direct causal chain from neuromuscular action to whole-body motion. Figure 1 details the outcome of the fall scenario. The bottom-left panel confirms that no recovery muscle activation occurred. This neuromuscular failure leads directly to a loss of stability, as shown in the top-left panel, where the pelvis (CoM) trajectory shows an irrecoverable forward and downward plummet. The kinematic consequence, seen in the top-right panel, is a complete collapse of the lower limbs into deep, uncontrolled flexion.

In stark contrast, Fig. 2 details the successful recovery. The bottom-left panel shows a strong, timely activation of the recovery muscles shortly after the trip at t = 5 s. This muscular action is the causal agent for maintaining stability; the pelvis trajectory (top-left) shows that the CoM is arrested and remains upright. The kinematics (top-right) reflect a controlled, dynamic recovery strategy, where the hip and knee joints move rapidly to absorb the perturbation and re-establish a stable base of support.

3.2 Differentiating Outcomes via Accelerometer Signatures

While the internal biomechanics provide the explanation, the accelerometer data is the crucial output for real-world fall detection. The bottom-right panels of Figs. 1 and 2 already hint at a major difference. This distinction is made explicit in Fig. 3, which plots the resultant forearm acceleration for both scenarios.

The blue line, representing the successful recovery, shows a period of dynamic acceleration corresponding to the rapid corrective movements. However, its magnitude remains largely controlled. The red line, representing the fall, initially follows a similar pattern but then diverges dramatically, exhibiting chaotic, high-amplitude oscillations that culminate in a massive impact spike of nearly 500 m/s^2 at approximately 9.8 s. This impact signature is an unambiguous feature

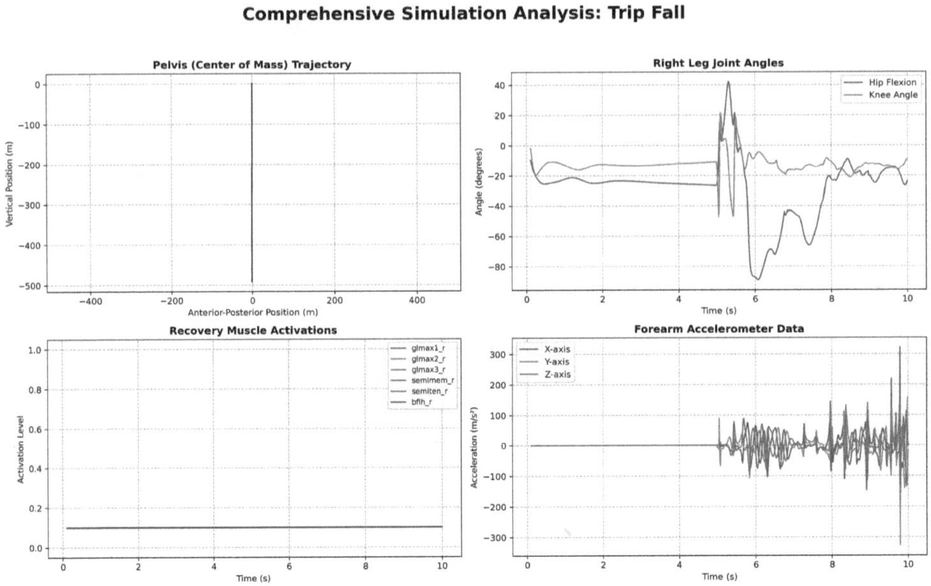

Fig. 1. Comprehensive analysis of the simulated trip fall. The absence of a recovery muscle activation (bottom-left) directly leads to a loss of CoM control (top-left), kinematic collapse (top-right), and a high-impact accelerometer signal (bottom-right).

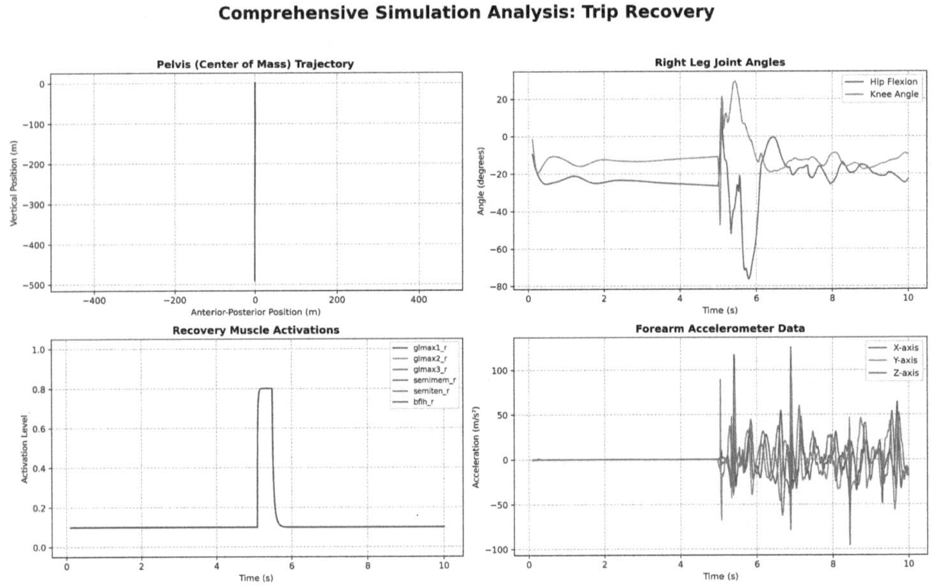

Fig. 2. Comprehensive analysis of the successful trip recovery. The timely muscle activation (bottom-left) enables the model to maintain CoM stability (top-left) and execute a controlled kinematic recovery strategy (top-right), resulting in a non-impactful accelerometer signal.

of the fall event. The clear and significant separation in the magnitude and characteristics of the accelerometer signals validates that our physically-plausible simulation can generate data with highly distinguishable features, essential for training an effective classification algorithm.

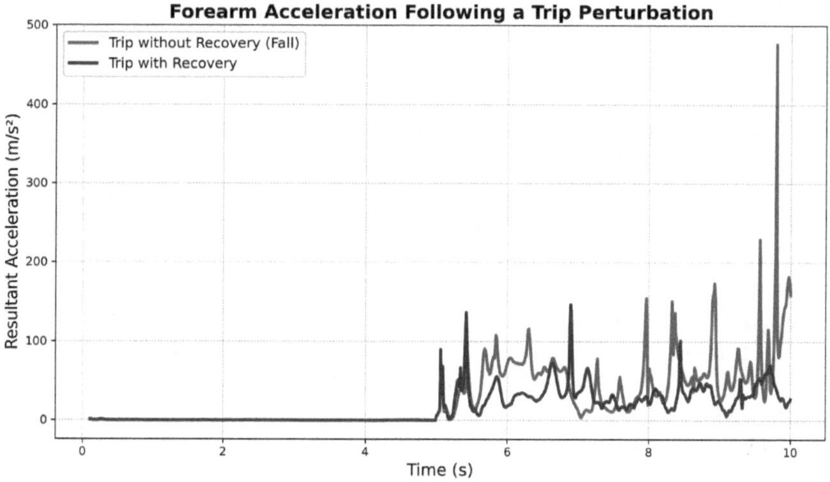

Fig. 3. Direct comparison of resultant forearm acceleration. The signal from the trip without recovery (red) shows a massive impact spike that is absent in the successful recovery (blue), providing a robust feature for fall detection. (Color figure online)

3.3 Future Work

The experiments presented serve as a proof-of-concept for the framework's capabilities. The next phase of this research will focus on expanding this framework to generate a large-scale dataset suitable for machine learning, with the goal of creating a "simulated SisFall" dataset. This expansion will involve several key thrusts. First, we will broaden the variety of simulated scenarios to include other common perturbations such as slips and pushes from multiple directions, as well as a wide range of Activities of Daily Living (ADLs) like sitting, standing, walking, and turning. This is essential for creating a balanced dataset that can train an algorithm to distinguish falls from normal, sometimes vigorous, movements. Furthermore, we plan to conduct extensive parameter sweeps across a spectrum of neuromuscular and physiological variables, such as muscle strength scale factors, neural response delays, and perturbation magnitudes, to model a diverse population ranging from frail to robust individuals. The resulting synthetic dataset will then undergo rigorous statistical validation using methods like ANOVA and t-tests to quantify the significant differences in key outcome metrics between scenarios. The ultimate objective is to use this comprehensive dataset to train and validate various fall detection algorithms, such as LSTMs and CNNs, and to benchmark their performance against models trained on publicly available real-world fall datasets to demonstrate the transferability and effectiveness of our simulation-based approach.

4 Conclusion

In this paper, we have detailed a comprehensive methodology for generating realistic, synthetic fall data using the OpenSim neuromusculoskeletal simulation platform. By modeling a trip perturbation and explicitly controlling the neuromuscular recovery response, we have demonstrated a clear, causal link from muscle activation to whole-body stability, kinematics, and the final accelerometer signature. The results highlight profound, quantifiable differences between a simulated fall and a successful recovery across multiple biomechanical domains. Most importantly, the resulting accelerometer data shows a large and unambiguous separation between the two outcomes, confirming that the framework can produce distinct, classifiable signals. This work represents a significant step towards the creation of a large-scale, physically-plausible synthetic dataset capable of overcoming the data scarcity problem in fall detection research and improving the accuracy of next-generation wearable health monitoring systems.

References

1. Boyé, N.D., van der Lies, C., van der Velde, N., van Lieshout, E.M.M.: The impact of falls in the elderly. Ned. Tijdschr. Geneeskd. **157**(18), A5718 (2013)
2. Centers for Disease Control and Prevention: Facts about falls (2023)
3. Chen, Z., Wang, H., Xu, W., Zhu, Y., Cao, Y., Cheng, J.: An effective deep learning framework for fall detection: model development and study design. J. Med. Internet Res. **26**, e56750 (2024)
4. Johnson, W.R., Mian, A., Robinson, M.A., Weston, M., Jarratt, P.: GREF: enhancing biomechanical machine learning with limited data: generating realistic synthetic posture data using generative artificial intelligence. Front. Bioeng. Biotechnol. **12**, (2024)
5. Li, F., Liu, Z., Sun, D., You, J.: The impact of falls on activities of daily living in older adults: a national longitudinal study. PLoS ONE **18**(11), e0294017 (2023)
6. Lu, X., Zhang, C.C., Wang, Y., Shen, J., Dai, X.: Data collection methods and predictive analysis for fall prevention in elderly populations. In: 2023 Congress in Computer Science, Computer Engineering, & Applied Computing (CSCE), pp. 2740–2742 (2023). https://doi.org/10.1109/CSCE60160.2023.00448
7. Ngufor, C., Student, B., M., Amer, H., M.: Elderly fall detection systems: a literature survey. Front. Robot. AI **7**, 71 (2020)
8. van der Peursem, C.M., van der Meer, S., van den Hout, A., van der Schans, C.P., van der Zee, E.A., van der Veen, S.: Impact of falling on social participation and social support trajectories in a middle-aged and elderly European sample. PLoS ONE **13**(1), e0190163 (2018)
9. Seth, A., et al.: Opensim: simulating musculoskeletal dynamics and neuromuscular control to study human and animal movement. PLoS Comput. Biol. **14**(7), e1006223 (2018)

10. SmartData Lab: When biomechanics meets reality: The opensim simulation gap (2025)
11. Yadav, N., Kumar, N., Kumar, S.: A study of fall detection in assisted living: Identifying and improving the optimal machine learning method. Electronics **10**(3), 39 (2021)
12. Zhang, C.C., Liu, M.Y., Zhang, K., Dai, X., Wang, C., Wang, Y.: wearable technology for fall prevention-data collection method and hardware optimization. In: 2023 International Conference on Computational Science and Computational Intelligence (CSCI), pp. 1498–1502 (2023). https://doi.org/10.1109/CSCI62032.2023.00245
13. Zhang, C.C., Liu, Y., Wang, Y., Feng, Y., Kong, Z., Chen, S.: Market research on fall prevention products. In: 2023 Congress in Computer Science, Computer Engineering, & Applied Computing (CSCE), pp. 2734–2736 (2023). https://doi.org/10.1109/CSCE60160.2023.00446
14. Zhang, C.C., Wang, C., Dai, X., Liu, S.: Camera-based analysis of human pose for fall detection. In: 2023 Congress in Computer Science, Computer Engineering, & Applied Computing (CSCE), pp. 1779–1782 (2023). https://doi.org/10.1109/CSCE60160.2023.00294
15. Zhang, C.C., Zhang, K., Ni, R., Liu, H., Shen, J.: Unleashing the potential of machine learning: an exploration of state-of-the-art algorithms and real-world applications in computer vision. In: 2023 Congress in Computer Science, Computer Engineering, & Applied Computing (CSCE), pp. 422–425 (2023). https://doi.org/10.1109/CSCE60160.2023.00075

Advances in Skin Cancer Detection by Leveraging Machine Learning and Deep Neural Networks for Early Diagnosis

Ibidun Christiana Obagbuwa[1(✉)], Anthony Tsetse[2], Olayemi Jebba[3], and Oluwadamilola Samuel Obagbuwa[4]

[1] Department of Mathematical Sciences and Computing, Walter Sisulu University, Mthatha 5100, South Africa
iobagbuwa@wsu.ac.za

[2] School of Computing and Analytics, Northern Kentucky University, Highland Heights, KY 41099, USA

[3] Department of Information Technology, New Brunswick Community College (NBCC) Woodstock Campus, Woodstock, NB 7M 5C5, Canada

[4] Department of Information Technology, New Brunswick Community College (NBCC), Saint John, NB 2J 4C5, Canada

Abstract. Skin cancer is one of the most prevalent and potentially fatal cancers worldwide, with early detection playing a critical role in improving patient survival rates. Conventional diagnostic methods, which rely on visual examination and biopsy analysis, are often subjective and prone to human error. The emergence of artificial intelligence (AI), particularly machine learning (ML) and deep neural networks (DNNs), has revolutionized medical image analysis by offering automated, objective, and efficient diagnostic solutions. This study presents an intelligent framework that integrates both traditional ML classifiers and advanced deep learning models to enhance the accuracy of skin cancer detection and classification. Using publicly available datasets (HAM10000 and ISIC 2019), the study explores data preprocessing, augmentation, and model evaluation techniques. Experimental results demonstrate that deep convolutional neural networks (DCNN) and Faster R-CNN outperform traditional models such as Support Vector Machine (SVM) and Random Forest (RF), achieving superior accuracy and robustness. The findings highlight the potential of deep learning to assist dermatologists in the early diagnosis of skin cancer, thus improving patient care and treatment outcomes.

Keywords: Skin Cancer Detection · Early Diagnosis · Machine Learning · Deep Neural Networks · Medical Imaging · Computer Vision

1 Introduction

Skin cancer accounts for a significant portion of cancer diagnoses globally, with melanoma being the most lethal type. The World Health Organization reports over 1.5 million cases annually. Early detection is vital for effective treatment, as survival rates

H. R. Arabnia et al. (Eds.): CSCI 2025, CCIS 2971, pp. 354–356, 2027.
https://doi.org/10.1007/978-3-032-28463-1_29

drop sharply in later stages. Traditional diagnostic approaches depend heavily on dermatologists' expertise, making the process time-consuming and potentially inconsistent. Recent advances in ML and DL provide an opportunity to develop automated systems that can detect skin abnormalities with remarkable accuracy, even in resource-limited settings.

This research investigates how modern ML and DL algorithms can be leveraged to enhance skin cancer detection. It focuses on improving classification accuracy, optimizing image preprocessing, and addressing challenges such as lighting variations, lesion irregularities, and dataset imbalance.

2 Methodology

The proposed system follows a structured workflow:

1. **Data Collection:** Two benchmark datasets, HAM10000 and ISIC 2019, were obtained from Kaggle. These datasets contain thousands of dermoscopic images across multiple skin lesion classes, including benign and malignant types.
2. **Data Preprocessing:** Images were resized, normalized, and denoised to ensure uniform quality. Data augmentation techniques, such as flipping, rotation, and brightness adjustment, were applied to increase dataset diversity and reduce overfitting.
3. **Feature Extraction:**

Deep Learning Models: A Deep Convolutional Neural Network (DCNN) was trained to automatically extract spatial and textural features from images. The Faster R-CNN was implemented to enhance region-based lesion localization and classification.

Traditional ML Models: Support Vector Machine (SVM) and Random Forest (RF) classifiers were trained using handcrafted features for baseline comparison.

4. **Model Evaluation:** The models were evaluated using accuracy, precision, recall, F1-score, and confusion matrix metrics to assess classification performance. Cross-validation was employed to ensure generalization.

3 Results and Discussion

The experimental results show that deep learning models outperform traditional ML algorithms in identifying malignant lesions. The Faster R-CNN achieved superior detection accuracy and precise lesion localization, while the DCNN demonstrated high recall, minimizing false negatives, a critical requirement in cancer diagnostics. The Random Forest and SVM models achieved moderate accuracy but were limited by handcrafted feature representation. The integration of data augmentation and normalization significantly improved model robustness.

These findings confirm that deep neural architectures can effectively learn discriminative visual patterns, even under varying lighting conditions and lesion morphologies. The combination of DCNN and Faster R-CNN presents a reliable approach for assisting dermatologists in early diagnosis, especially in low-resource medical environments.

4 Conclusion

This study demonstrates the effectiveness of combining machine learning and deep neural network approaches for automated skin cancer detection. The proposed models achieved improved diagnostic accuracy and consistency compared to traditional methods. Future work will focus on integrating explainable AI (XAI) mechanisms to enhance interpretability and clinical trust. Additionally, real-world testing in dermatological settings will be conducted to validate model performance under operational conditions.

Acknowledgment. The authors acknowledge the publicly available datasets from the HAM10000 and ISIC 2019 initiatives and the Kaggle community for enabling open research in medical image analysis.

References

1. Tschandl, P., et al.: The HAM10000 dataset: a large collection of multi-source dermatoscopic images of common pigmented skin lesions. Sci. Data **5**(180161) (2018)
2. Codella, N.C.F., et al.: Skin lesion analysis toward melanoma detection 2018: a challenge hosted by the international skin imaging collaboration (ISIC). arXiv preprint arXiv:1902.03368 (2019)
3. Esteva, A., et al.: Dermatologist-level classification of skin cancer with deep neural networks. Nature **542**, 115–118 (2017)

Large Language Models and Applications in Health

A Quantization-Aware Evaluation Framework for Medical Language Model Selection in RAG Clinical Pipelines

Filippo Bianchini[1,3(✉)], Edoardo Bianchini[2,3], Cyril Boyault[4], Eric G. Folco[4], Mathilde Proponnet-Guerault[4], Nicolas Vuillerme[3], and Massimo Mecella[1]

[1] Dipartimento di ingegneria Informatica, Automatica e Gestionale Antonio Ruberti (DIAG), Sapienza Università di Roma, Rome, Italy
bianchini@diag.uniroma1.it

[2] Department of Human Neurosciences, Sapienza Università di Roma, Rome, Italy

[3] Univ. Grenoble Alpes, CNRS, UMR 5217, Grenoble INP, LIG Sangria, Grenoble, France
edoardo.bianchini@univ-grenoble-alpes.fr

[4] OneTreck (Reckonect SAS), 10 rue Georges Jacquet, 38000 Grenoble, France
{cyril,ericf,mathildepg}@onetreck.com

Abstract. Deploying medical language models requires balancing clinical accuracy with resource constraints. We present a two-phase evaluation framework for quantization-aware model selection in retrieval-augmented clinical pipelines, testing 40 configurations (10 models × 4 quantization levels) on biomedical QA and diagnostic tasks. Results show that medical terminology accuracy degrades 15–25% faster than linguistic metrics under compression, yet counterintuitively Q5KS achieves the best clinical performance (87.22%) on the 2024 Italian medical exam, outperforming Q6K (80.45%). This suggests that compression can act as implicit regularization for clinical reasoning. The framework establishes evidence-based deployment guidelines, showing that Q5KS optimizes accuracy–efficiency trade-offs on consumer hardware (4 GB GPU, 0.32 s latency), enabling equitable access to medical AI while preserving data sovereignty.

Keywords: Large Language Models · Retrieval-Augmented Generation · Quantization

1 Introduction

The deployment of Large Language Models (LLMs) [10] in Retrieval-Augmented Generation (RAG) for clinical decision support faces a central tension: state-of-the-art medical models require over 80 GB of GPU memory [13], while healthcare institutions increasingly demand edge deployment for data sovereignty and regulatory compliance [28]. Quantization, which reduces numerical precision from

H. R. Arabnia et al. (Eds.): CSCI 2025, CCIS 2971, pp. 359–373, 2027.
https://doi.org/10.1007/978-3-032-28463-1_30

16-bit to 4–8 bits precision, enables 75–90% memory savings [14], but its impact on clinical reasoning is still poorly understood. Studies show that quantized generators often struggle to integrate retrieved evidence [13,49], raising concerns about diagnostic reliability.

Although medical RAG systems are proliferating [38,47,52], current benchmarks such as MedQA [23], PubMedQA [24], and MedMCQA [39] focus on stand-alone model accuracy. They do not capture how quantization affects terminology preservation versus linguistic fluency, whether compression thresholds differ between general-purpose and domain-specific models, or how compressed generators perform when synthesizing heterogeneous evidence. As a result, healthcare organizations face an unresolved trade-off: deploy high-accuracy models that demand costly hardware, or compressed models whose clinical reliability remains uncertain. Without systematic evaluation protocols that isolate quantization as an independent factor, decisions rely on ad-hoc testing rather than evidence-based guidelines, an issue made critical by EU AI Act requirements for technical documentation and transparency [28]. To address this gap, we propose a quantization-aware evaluation framework for medical LLM selection in RAG pipelines. The framework uses a two-phase methodology: first, benchmarking quantization thresholds through biomedical question answering with both linguistic and clinically oriented metrics; second, validating clinical task performance on the 2024 Italian National Medical Specialization Exams, with comparisons against proprietary models. We evaluate ten open-source medical LLMs (7–9B parameters) across four quantization schemes (Q4KM, Q5KS, Q5KM, Q6K). While Sect. 3 briefly outlines the target RAG system combining T5-based retrieval [42], UMLS-aligned knowledge graphs [11], and quantized generators, the contribution of this paper lies in a generalizable evaluation methodology applicable across medical AI contexts facing similar deployment constraints.

The remainder of the paper is structured as follows: Sect. 2 reviews quantization effects in medical LLMs and existing evaluation methodologies, positioning our framework contributions. Section 3 describes the target RAG architecture for which the evaluation framework is designed. Section 4 details the two-phase evaluation protocol and presents comprehensive results. Section 5 derives implications, discusses framework limitations, and future directions.

2 Related Work and Research Gap

Research on quantization for medical language models has highlighted both the potential and the limits of compression. Reducing numerical precision from 16-bit to 4–8 bit representations enables memory savings of up to 90% [18,34,48], a prerequisite for edge deployment. Chen et al. [14] showed that Meditron-7B can sustain performance across Q4KM, Q5KS, and Q6K, confirming the feasibility of aggressive quantization. Dettmers et al. [17] introduced QLoRA[1], enabling efficient fine-tuning of quantized models while retaining medical terminology accuracy, a critical requirement in clinical contexts. Other studies revealed domain-

[1] See: https://github.com/artidoro/qlora.

specific resilience: Bolton et al. [12] reported that BioMedLM, trained exclusively on biomedical text, retained 57.3% accuracy on MedMCQA despite heavy compression, while Kim et al. [26] showed that Meerkat-7B/8B models, trained on textbooks, surpassed GPT-3.5 across six benchmarks. These findings confirm that smaller, domain-specialized models can outperform much larger general-purpose alternatives, but most work has tested stand-alone generation, neglecting retrieval-augmented scenarios, where quantization may interact differently with external evidence integration.

In parallel, Retrieval-Augmented Generation (RAG) has become central to several application [7–9], in particular medical applications, addressing hallucinations by grounding outputs in retrieved knowledge [32]. Improvements have been achieved through prompt engineering [35,50] and hybrid retrieval strategies [5,6,37], yet few evaluations isolate the role of compression. Bora and Cuayáhuitl [13] conducted the first systematic analysis of quantization in medical RAG, comparing QLoRA, QAT, and post-training quantization across Flan-T5-Large, LLaMA-2-7B, and Mistral-7B, reducing GPU memory from over 100 GB to under 8 GB but noting strong dependence on retrieval quality. Yazan et al. [49] examined INT4 quantization across long contexts, finding that well-tuned 7B models remained stable at 4-bit, while weaker models degraded rapidly. Nguyen et al. [38] tested 4-bit quantized Llama-3-70B, Gemma-2-27B, and Mixtral-8×7B in ophthalmology, showing RAG could raise accuracy from 57.5% to 81.35%, with 4-bit nearly matching 8-bit at half the cost. Despite these advances, no prior framework systematically varies quantization levels to establish performance thresholds or isolates quantization as an independent factor, limiting architectural insights for deployment.

The evaluation methodologies commonly used in medical LLMs further reveal gaps. Standard benchmarks such as MedQA [23], PubMedQA [24], and MedMCQA [39] remain predominantly multiple-choice, offering reproducibility but failing to capture the complexity of real clinical reasoning. A systematic review of 519 studies found that only 5% used real-world data and fewer than 2% addressed open-ended reasoning [3], while a BMJ review similarly noted the lack of head-to-head comparisons with clinicians [36]. Notable exceptions include evaluations of ChatGPT-3.5 on the Italian national residency exam [1,16], which applied official scoring and error-type differentiation, but these focused on general-purpose models without structured knowledge integration.

In summary, existing work leaves three critical gaps: (i) a lack of metrics that distinguish linguistic quality from domain-specific terminology preservation under compression, (ii) protocols that isolate quantization as an independent variable, and (iii) validation on real clinical tasks that relate quantization levels to deployment scenarios. Our framework addresses these gaps through domain-specific metric prioritization, controlled compression analysis, deployment stratification, and reproducibility guarantees. Building on these contributions, we propose a two-phase evaluation: Phase 1 identifies quantization thresholds through systematic benchmarking, and Phase 2 validates clinical impact on the 2024 Italian National Medical Specialization Exams, benchmarking open-source models against proprietary baselines to derive evidence-based deployment guidelines. The study is guided by the following research questions:

RQ1: *What evaluation metrics and protocols effectively identify quantization thresholds for medical language models intended for RAG deployment?*

RQ2: *How does the proposed framework perform when applied to clinical diagnostic tasks, and what deployment guidelines emerge?*

3 Target System Architecture

To contextualize the evaluation framework, we describe the medical RAG architecture that motivates quantization-aware model selection. The system is designed for edge-deployable clinical decision support, where compressed generators must operate under strict hardware and regulatory constraints. Its design directly shapes our evaluation strategy, motivating the use of domain-specific metrics, multi-phase validation, and systematic quantization analysis.

The architecture applies asymmetric quantization: the retrieval pipeline remains at full FP16 precision, while generators are compressed to 4–6 bits. This separation allows us to isolate quantization effects on generation without confounding retrieval quality, a key requirement for controlled Phase 1–2 experiments. The system integrates unstructured document retrieval with structured ontology access (UMLS), so generators must synthesize heterogeneous evidence sources. This motivates the use of Medical F1 as a primary metric, since terminology drift under compression would directly reduce the benefits of hybrid retrieval.

3.1 Inference Pipeline and Quantization Integration

Clinical queries are processed through a multi-stage inference pipeline designed with transparency in mind. First, a T5-Base retriever parses the query at full precision, extracting entities and reformulating the input for robust retrieval. This uncompressed stage ensures that evidence quality remains stable, avoiding compounded errors when compressed generators are used. Document embeddings are produced with all-MiniLM-L6-v2 and indexed via FAISS[2], enabling sub-millisecond similarity search. To improve robustness, dense semantic retrieval is combined with BM25 keyword matching, ensuring coverage of explicit medical terminology when embeddings fail.

Retrieved passages are then merged into a structured prompt that explicitly tags the active quantization level for reproducibility. The compressed Meditron3-8B generator (Q4KM, Q5KS, Q5KM, Q6K), executed via llama-cpp[3], synthesizes this evidence into a clinical answer. Each inference logs memory footprint, latency, and accuracy, enabling the Pareto tradeoff analysis presented in Phase 2. The pipeline relies exclusively on open-source components—Transformers for

[2] See: https://github.com/facebookresearch/faiss.

[3] See: https://github.com/ggml-org/llama.cpp.

T5-Base, llama-cpp-python for quantized inference, FAISS for indexing, and NetworkX for graph integration, ensuring transparency, reproducibility, and data sovereignty. Extended context windows (8,192–16,384 tokens) using sliding mechanisms accommodate large evidence batches (3,000 tokens) and multi-turn conversations without truncation.

3.2 Hardware-Quantization Mapping and Deployment Contexts

The four quantization schemes correspond to distinct deployment contexts. Q6K (minimal compression) fits hospital workstations equipped with GPUs above 6GB, where maximizing accuracy justifies higher resource usage. Q5KM (moderate) targets telemedicine platforms relying on 4 GB GPUs, offering a balance between accessibility and performance. Q5KS (aggressive) is designed for mixed CPU–GPU environments in smaller clinics, providing the best compromise between accuracy and efficiency. Q4KM (extreme) enables CPU-only deployment in rural or resource-limited facilities, though at the cost of a notable accuracy drop. Since healthcare institutions differ widely in computational capacity, Phase 2 establishes accuracy thresholds across these quantization levels to guide evidence-based choices on which configurations remain clinically viable in each setting.

The architecture also motivates Phase 1's evaluation design. Since medical reasoning depends critically on precise terminology, Medical F1 was selected to detect whether compression erodes domain-specific vocabulary usage. Complementary metrics (METEOR, ROUGE-L) assess evidence synthesis quality, capturing degradation that accuracy alone would not reveal. In this way, the system architecture directly grounds both the methodological choices and the structure of our two-phase evaluation.

4 Two-Phase Evaluation Methodology

The framework follows a two-phase protocol designed to answer **RQ1** and **RQ2**.

4.1 Phase 1: Systematic Quantization Threshold Analysis

Experimental Design and Model Selection. We evaluated 40 configurations across ten open-source medical LLMs[4] and four quantization levels. Selection prioritized *(i)* compact size (7–9B parameters) for edge deployment and *(ii)* GGUF availability[5] supporting post-training quantization. Table 1 lists the evaluated models.

[4] Models released after September 2025 were excluded for consistency.
[5] See: https://huggingface.co/docs/hub/gguf.

Table 1. Candidate models for quantization analysis stratified by domain specialization

Models (n = 10)	Size	Medically Fine-Tuned
BioMistral [30]	7B	Yes
Gemma-2 [45]	9B	No
LLaMA3-Med42 [15]	8B	Yes
Meditron3 [43]	8B	Yes
Meditron [14]	7B	Yes
Meta-LLaMA-3-Instruct [19]	8B	No
Ministral-Instruct	8B	No
Mistral-Instruct-v0.2 [22]	7B	No
Qwen2.5-Instruct [46]	7B	No
Asclepius-v0.3 [29]	7B	Yes

Dataset and Metric Design. To conduct the preliminary evaluation, we constructed a custom dataset consisting of 100 open-ended clinical questions, carefully revised and validated by domain experts. Each question was paired with a reference answer, forming a ground truth used to evaluate the quality of the generated responses systematically. The dataset was designed to reflect realistic medical support tasks and to test both general medical knowledge and terminology usage.

Following methodologies adopted in [27], we employed a comprehensive set of evaluation metrics. Text generation quality was assessed through multiple complementary metrics: BLEU [40] for precision-based n-gram overlap, METEOR [2] for semantic similarity that accounts for synonyms and paraphrasing, and ROUGE-L [33] to capture the longest common subsequence between generated and reference texts. For domain-specific evaluation, we implemented a medical terminology accuracy metric inspired by approaches used in biomedical NLP evaluation [21,44]. Specifically, we utilized spaCy[6] for text processing and lemmatization, combined with a custom medical lexicon derived from standard medical terminology resources. Drawing on methodologies from clinical language understanding tasks [20], we assessed models' ability to use appropriate medical terminology in context. We compiled a lexicon of common medical terms from standardized resources, such as UMLS, and biomedical literature. Using this lexicon, we identified medical terms in both generated and reference texts, and calculated three key metrics:

- *Medical precision*: the proportion of medical terms in the generated response that were also present in the reference answer, indicating the model's ability to avoid introducing incorrect or irrelevant medical terminology.

[6] See: https://github.com/explosion/spaCy.

- *Medical recall*: the proportion of medical terms from the reference answer that were captured in the generated response, reflecting the model's comprehensiveness in covering essential medical concepts.
- *Medical F1-score*: the harmonic mean of precision and recall, providing a balanced measure of medical terminology accuracy.

To facilitate comparisons across quantization levels, we normalized every score to the best-performing model (scale 0–1). This normalised scale significantly improves interpretability, as absolute metric values for text-generation tasks are often low and therefore hard to judge in isolation [31,41]. This approach, similar to that used in previous evaluations of medical language models [25,51], allowed us to specifically evaluate how well each model reproduced critical medical terminology beyond general linguistic similarity, a crucial capability for clinical decision support systems.

Results, displayed in Fig. 1, show that medically fine-tuned models consistently outperformed general-purpose ones across all metrics, confirming the value of domain-specific pre-training for specialized applications. Among these, Meditron3-8B, specifically Q6K variant, stood out with a substantial performance lead, achieving the highest scores across all metrics, reflecting strong linguistic precision, semantic coherence, and structural quality. Most notably, it led in medical F1-score, indicating superior handling of domain-specific terminology, a critical factor in clinical settings. Performance degraded gracefully across lower quantization levels, with even the Q4KM version outperforming higher-bit general-purpose models.

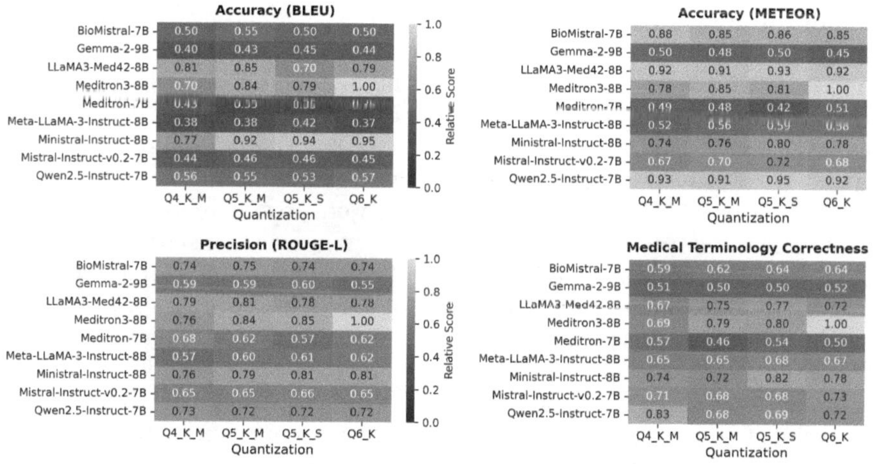

Fig. 1. Heatmap of normalized performance metrics across models and quantization levels (lighter colors indicate better performance). From left top to right bottom: (1) BLEU score; (2) METEOR score; (3) ROUGE-L score; (4) Medical F1-score.

Finally, we computed an overall score as a weighted average of BLEU (25%), METEOR (25%), ROUGE-L (25%), and medical F1-score (25%), providing a balanced assessment of both linguistic quality and domain-specific knowledge. Final results for overall score are shown in Fig. 2.

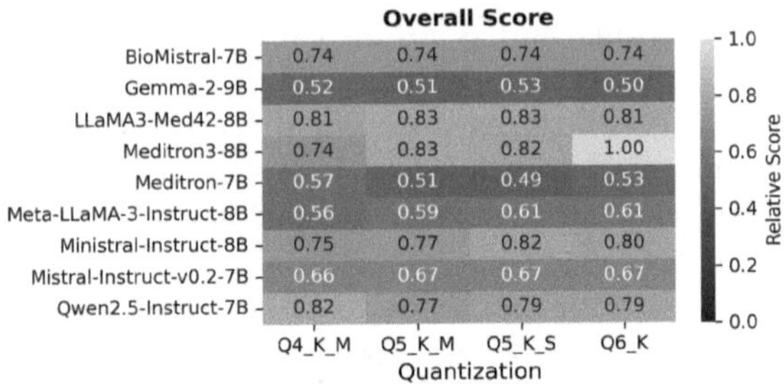

Fig. 2. Heatmap of the normalized Overall Score, computed as an equally weighted average of BLEU, METEOR, ROUGE-L, and the medical F1-score.

Based on Phase 1 results, Meditron3-8B emerges as the optimal model family for Phase 2 clinical validation. However, the choice of quantization level depends on deployment context, a question Phase 2 explicitly addresses by testing all four compression schemes on real diagnostic tasks.

4.2 Phase 2: Clinical Task Validation

Experimental Protocol. Phase 2 validates Phase 1 findings by evaluating Meditron3-8B across all four quantization levels on the 2024 Italian National Medical Specialization Exam (SSM). This high-stakes diagnostic test comprises 140 multiple-choice questions grounded in authentic clinical scenarios; we excluded seven visual-interpretation items (IDs 42, 64, 73, 120, 133, 134, 138) outside current system capabilities, yielding 133 text-based questions. Each presents five options (A-E) with one correct answer, preserving original Italian wording within standardized English prompts to minimize translation bias. The experimental design isolates quantization as the independent variable. We executed the identical pipeline four times, varying only generator compression:

Meditron3-8B-Q6K (baseline), Q5KM (moderate), Q5KS (aggressive), and Q4KM (extreme)[7]. All other components remained constant, enabling attribution of performance differences solely to quantization effects. Each configuration answered all 133 questions ten times with fixed parameters (temperature = 0.1, seed = 42) to measure output stability under compression.

[7] All models were run on single computer equipped with 8GB GPU GeForce RTX 4060.

Evaluation Metrics and Deployment Context. Each model was tasked with selecting the most appropriate answer for every question. To ensure robust and reproducible results, we administered the full test ten times for each model using fixed prompts and parameters. Despite identical inputs, some degree of variation is expected due to the inherent stochasticity of generative models. Performance was evaluated using the following metrics:

- **Accuracy(%):** the proportion of correctly answered questions out of the 133 total items.
- **Official exam score:** computed according to the SSM grading scheme: +1 point for correct answers, –0.25 for incorrect answers, and 0 for unanswered items.
- **Average time per question (s):** the mean inference time required to answer a single question, measured over the full test.

Additionally, we benchmarked against proprietary models, namely Claude-4.1-Sonnet, GPT-5, Gemini-2.5-Flash, for contextual performance reference. These comparisons cannot inform quantization analysis due to undisclosed compression levels, but establish whether open-source quantized models achieve competitive clinical accuracy, a practical deployment consideration.

Results. Table 2 reveals a counterintuitive finding across ten independent evaluation runs: Q5KS consistently achieves the highest diagnostic accuracy, outperforming both moderate compression (Q5KM) and minimal compression (Q6K). This result contradicts Phase 1 predictions and challenges conventional assumptions about quantization-accuracy trade-off.

Table 2. Quantization trade-off analysis on 2024 SSM exam (133 questions, 10 independent runs per configuration). Metrics include average accuracy, official average exam score, memory used and and average response time per question.

Quantization Level	Accuracy (%)	Correct/Wrong Answers	SSM Score (+1/−0.25)	Memory (GB)	Latency (s/Q)
Q5_K_S	**87.22**	116/17	**111.75**	4.24	**0.32**
Q5_K_M	84.96	113/20	108.00	4.37	**0.29**
Q6_K	80.45	107/26	100.50	5.15	2.23
Q4_K_M	75.19	100/33	91.75	3.52	0.34
Claude-4.1	95.49	127/6	125.75	N/A	1.16
GPT-5	92.03	122/11	122.34	N/A	0.77
Gemini-2.5	92.48	123/10	120.01	N/A	0.56

Q5KS Achieves 87.22% Accuracy, Making Aggressive Compression Optimal. It answered 116/133 questions correctly (SSM score: 111.75), outperforming Q5KM by 2.3pp and Q6K by 6.8pp. Unlike Phase 1, where Q6K led on open-ended tasks, this shows task-specific dynamics: multiple-choice reasoning may benefit from regularization effects of moderate compression, while open-ended generation favors higher precision.

Q6K Underperforms Despite Minimal Compression. It reached only 80.45% accuracy (107/133, score: 100.50) with 2.23 s latency. This suggests that higher precision can exacerbate overfitting or suffer from hardware inefficiencies, whereas quantization may act as implicit regularization.

Q5KM Balances Accuracy and Efficiency. With 84.96% accuracy (113/133, score: 108.00) and the fastest latency (0.29 s), it is a viable low-latency option. However, the 2.3pp accuracy loss relative to Q5KS suggests the latter is preferable when accuracy is critical.

Q4KM Marks the Quantization Floor at 75.19%. It answered 100/133 correctly (score: 91.75) while running on 3.52GB RAM, enabling CPU-only deployment. Yet, its 12pp accuracy drop compared to Q5KS limits use in high-stakes diagnostics, though it may be acceptable in low-resource or preliminary screening contexts.

Among the evaluated proprietary models, Claude-4.1-Sonnet achieved the highest overall performance, with an average accuracy of 95.49% and a mean score of 125.75 at moderate latency (1.16 s per question). Gemini-2.5 (92.48%) and GPT-5 (92.03%) delivered slightly lower accuracy but maintained low inference times (0.56 s and 0.77 s, respectively).

4.3 Analysis of Results

Phase 2 findings contradict Phase 1 predictions: Q5KS outperforms Q6K by 6.8pp, showing that higher precision does not guarantee superior performance. This result has both theoretical and practical implications for quantization research and medical AI deployment.

Compression as Implicit Regularization. Moderate quantization appears to constrain model capacity in a way that reduces overfitting and improves generalization. At full precision, Meditron3-8B may overfit to textbook-style reasoning, excelling in Phase 1's open-ended QA but failing to transfer robustness to authentic diagnostic scenarios. The 5-bit setting of Q5KS forces the model into more compact representations that appear to capture decision boundaries more effectively in multiple-choice reasoning. This parallels established regularization mechanisms such as dropout or weight decay, suggesting that post-training quantization can unintentionally function as a form of domain-specific regularization. Importantly, this effect is not uniform: it benefits discriminative diagnostic tasks but harms open-ended generation, underscoring the need to treat compression not simply as a resource-saving trick but as an active factor shaping model behavior.

Task-Specific Evaluation Limits. The divergence between Phase 1 and Phase 2 also illustrates a methodological weakness in current evaluation practice. Phase 1 metrics, especially Medical F1, were highly sensitive to quantization effects and highlighted terminology degradation under compression. Yet these laboratory-style benchmarks failed to predict Phase 2 diagnostic performance, where Q5KS surpassed Q6K despite lower Medical F1. This indicates that single-task evaluation can be misleading when used as a proxy for deployment readiness. Clinical deployment requires competence in both generation, where fluency and terminology fidelity matter, and discrimination, where regularization may improve robustness. Future evaluations must therefore adopt multi-task protocols that combine these complementary perspectives rather than privileging one.

Broader Deployment Implications. From a deployment perspective, Q5KS not only achieves the highest diagnostic accuracy but also does so with the smallest memory footprint among viable options (4.24GB, 0.32 s latency). This combination eliminates the conventional trade-off between accuracy and accessibility: the best-performing configuration is also the most affordable to deploy, even on consumer hardware. By contrast, Q6K is both slower and less accurate, demonstrating that minimal compression can in fact be the worst option in practice. This finding reframes quantization strategy from a compromise to a potential opportunity, with compression enhancing performance in some tasks while reducing infrastructure requirements.

Answers to Research Questions. Regarding **RQ1**, Phase 1 confirmed that Medical F1 is the most sensitive indicator of compression effects, but Phase 2 revealed that no single metric or task adequately predicts clinical performance. Multi-task validation is therefore essential. For **RQ2**, results show that Q5KS (87.22% accuracy, 4.24 GB, 0.32 s latency) provides the optimal accuracy–efficiency balance, outperforming Q6K despite stronger compression. This demonstrates that evidence-based deployment guidelines can overturn intuitive expectations, guiding organizations toward configurations that democratize medical AI without sacrificing performance.

5 Conclusion

This work introduces a two-phase evaluation framework for quantization-aware model selection in medical RAG pipelines, showing that systematic compression analysis reveals counterintuitive effects invisible to conventional benchmarking. Evaluating 40 configurations across biomedical QA and clinical diagnostics, we identify Q5KS (5-bit aggressive compression) as globally optimal, achieving 87.22% accuracy on the 2024 Italian medical exam, outperforming both Q5KM (84.96%) and Q6K (80.45%) while running efficiently on consumer hardware (4.24GB, 0.32 s latency).

Implications and Discussion. Aggressive quantization improves diagnostic accuracy while minimal compression degrades both performance and efficiency, contradicting the common belief that higher precision guarantees better results. A plausible explanation is compression-as-regularization: by constraining capacity, Q5KS reduces overfitting to biomedical training patterns and generalizes better to real diagnostic reasoning. This finding has direct deployment implications, as Q5KS provides superior accuracy with minimal hardware requirements.

The contradiction between Phase 1 and Phase 2 highlights the limits of single-task evaluation. Open-ended metrics (BLEU, METEOR, Medical F1) suggested Q6K superiority, yet clinical multiple-choice tasks favored Q5KS. Single benchmarks cannot predict cross-task performance, reinforcing the need for multi-task validation that captures both generation, where precision matters, and discrimination, where quantization may enhance robustness.

Beyond deployment guidance, our results challenge the assumption that only massive proprietary systems deliver clinical accuracy. The 8B-parameter quantized model reaches 91% of Claude-4.1-Sonnet's performance while operating entirely locally, showing that smaller, open-source models can provide competitive accuracy with full data sovereignty.

Limitations. Four constraints remain. First, we focus on generator components without ablation of retrieval effects. Second, clinical validation is based on a single national exam, limiting cross-cultural generalization; further studies with USMLE, PLAB, and AMC would broaden scope. Third, we only test post-training quantization (GGUF), while Quantization-Aware Training may yield different trade-offs. Fourth, hardware tests use a single GPU, so Q6K's latency anomaly may reflect device-specific behavior.

Final Remarks. Quantization shapes both performance and clinical viability, making it a core design variable rather than a post-hoc optimization. Our framework shows that systematic evaluation isolating compression effects can uncover unexpected benefits, such as Q5KS optimizing diagnostic accuracy through implicit regularization. By providing evidence-based guidelines that balance accuracy, accessibility, and accountability, we enable more equitable deployment of medical AI across diverse healthcare contexts.

Acknowledgements. The work of Filippo Bianchini and Massimo Mecella is partially funded by MICS (Made in Italy—Circular and Sustainable) (PE00000004) Extended Partnership (CUP B53C22004130001) funded by the EU - NextGenerationEU PNRR MUR. This work was also supported by the French National Research Agency, within the framework of the "Investissements d'avenir" program [ANR-10-AIRT-05 and ANR-15-IDEX-02] and under the France 2030 programme, reference MIAI Cluster ANR-23-IACL-0006. The authors would like to thank MIAI Grenoble Alpes (Multidisciplinary Institute in Artificial Intelligence), Université Grenoble Alpes (UGA), and the ESA BIC Sud France (European Space Agency Business Incubation Centre) for their support. This work was supported by OneTreck, in collaboration with MIAI Grenoble, UGA, and ESA BIC Sud France.

References

1. Alessandri Bonetti, M., Giorgino, R., Gallo Afflitto, G., De Lorenzi, F., Egro, F.M.: How does ChatGPT perform on the Italian residency admission national exam compared to 15,869 medical graduates? Ann. Biomed. Eng. **52**(4), 745–749 (2024)
2. Banerjee, S., Lavie, A.: METEOR: an automatic metric for mt evaluation with improved correlation with human judgments. In: Proceedings of the ACL Workshop on Intrinsic and Extrinsic Evaluation Measures for Machine Translation and/or Summarization, pp. 65–72 (2005)
3. Bedi, S., et al.: Testing and evaluation of health care applications of large language models: a systematic review. JAMA (2024)
4. Bianchini, F.: Retrieval-Augmented Generation, pp. 139–172. Springer Nature Switzerland (2025). https://doi.org/10.1007/978-3-031-92285-5_7
5. Bianchini, F., Calamo, M., De Luzi, F., Macrì, M., Mecella, M.: Enhancing complex linguistic tasks resolution through fine-tuning LLMs, rag and knowledge graphs. In: International Conference on Advanced Information Systems Engineering, pp. 147–155. Springer (2024). Short paper
6. Bianchini, F., Calamo, M., De Luzi, F., Macrì, M., Mecella, M.: A service-based pipeline for complex linguistic tasks adopting LLMs and knowledge graphs. In: Symposium and Summer School on SOC, pp. 145–161. Springer (2024)
7. Bianchini, F., et al.: SAMBA: a reference framework for human-in-the-loop in adaptive smart manufacturing. Procedia Comput. Sci. **253**, 2257–2267 (2025). 6th International Conference on Industry 4.0 and Smart Manufacturing
8. Bianchini, F., Calamo, M., Marinacci, M., Rossi, J., Mecella, M.: Automating industrial quality control: a multimodal LLM and rag framework for anomaly detection. In: Maglogiannis, I., Iliadis, L., Andreou, A., Papaleonidas, A. (eds.) Artificial Intelligence Applications and Innovations, pp. 253–266. Springer Nature Switzerland, Cham (2025)
9. Bianchini, F., van den Heuvel, W.J., Marinacci, M., Mecella, M., Rossi, J., Tamburri, D.A.: Talk smart, stay safe: intelligent conversational agents for urban security. In: Aiello, M., Barzen, J., Dustdar, S., Leymann, F. (eds.) Service-Oriented Computing, pp. 37–56. Springer Nature Switzerland, Cham (2026)
10. Bianchini, F., Marinacci, M.: Exploring large language models in information systems: a survey, pp. 77–109. Springer Nature Switzerland (2025)
11. Bodenreider, O.: The unified medical language system (UMLs): integrating biomedical terminology. Nucleic Acids Res. **32**(suppl_1), D267–D270 (2004)
12. Bolton, E., et al.: BioMedLM: a 2.7 b parameter language model trained on biomedical text (2024). arXiv preprint arXiv:2403.18421
13. Bora, A., Cuayáhuitl, H.: Systematic analysis of retrieval-augmented generation-based LLMs for medical chatbot applications. Mach. Learn. Knowl. Extr. **6**(4), 2355–2374 (2024)
14. Chen, Z., Cano, A.H., Romanou, A., Bonnet, A., Matoba, K.: MEDITRON-70B: scaling medical pretraining for large language models (2023)
15. Christophe, C., Kanithi, P.K., Raha, T., Khan, S., Pimentel, M.A.: Med42-v2: a suite of clinical LLMs (2024). https://arxiv.org/abs/2408.06142
16. Conrado, F., Scaioli, G., Lo Moro, G., Rosset, L., Bert, F., Siliquini, R.: ChatGPT & clinical decision-making: a cross-sectional study on Italian medical residency test. Eur. J. Public Health **33**(Supplement_2), ckad160–862 (2023)

17. Dettmers, T., Pagnoni, A., Holtzman, A., Zettlemoyer, L.: QLoRA: efficient fine-tuning of quantized LLMs. In: Proceedings of the 37th International Conference on Neural Information Processing Systems, NIPS 2023. Curran Associates Inc., Red Hook, NY, USA (2023)
18. Egashira, K., Vero, M., Staab, R., He, J., Vechev, M.: Exploiting LLM quantization. In: Advances in Neural Information Processing Systems, vol. 37, pp. 41709–41732 (2024)
19. Grattafiori, A., et al.: The llama 3 herd of models (2024). arXiv preprint arXiv:2407.21783
20. Gu, Y., et al.: Domain-specific language model pretraining for biomedical natural language processing. ACM Trans. Comput. Healthc. (HEALTH) **3**(1), 1–23 (2021)
21. He, Z., et al.: MedEval: a multi-level, multi-task, and multi-domain medical benchmark for language model evaluation. arXiv preprint arXiv:2310.14088 (2023)
22. Jiang, A.Q., et al: Mistral 7B (2023)
23. Jin, D., Pan, E., Oufattole, N., Weng, W.H., Fang, H., Szolovits, P.: What disease does this patient have? A large-scale open domain question answering dataset from medical exams. Appl. Sci. **11**(14), 6421 (2021)
24. Jin, Q., Dhingra, B., Liu, Z., Cohen, W., Lu, X.: PubMedQA: a dataset for biomedical research question answering. In: Proceedings of the 2019 Conference on Empirical Methods in Natural Language Processing and the 9th International Joint Conference on NLP (EMNLP-IJCNLP), pp. 2567–2577 (2019)
25. Keshavarz, P., et al.: ChatGPT in radiology: a systematic review of performance, pitfalls, and future perspectives. Diagn. Intervent. Imaging (2024)
26. Kim, H., et al.: Small language models learn enhanced reasoning skills from medical textbooks. NPJ Digit. Med. **8**(1), 240 (2025)
27. Kim, Y., Wu, J., Abdulle, Y., Wu, H.: MedExQA: medical question answering benchmark with multiple explanations. arXiv preprint arXiv:2406.06331 (2024)
28. Kundu, S.: AI in medicine must be explainable. Nat. Med. **27**(8), 1328 (2021)
29. Kweon, S., Kim, J., Kim, J., Im, S., Cho, E., Bae, S.E.A.: Publicly shareable clinical large language model built on synthetic clinical notes. arXiv preprint arXiv:2309.00237 (2023)
30. Labrak, Y., Bazoge, A., Morin, E., Gourraud, P.A., Rouvier, M.: BioMistral: a collection of open-source pretrained large language models for medical domains (2024). arXiv preprint arXiv:2402.10373
31. van der Lee, C., Ferreira, T.C., Emmery, C., Wiltshire, T.J., Krahmer, E.: Neural data-to-text generation based on small datasets: comparing the added value of two semi-supervised learning approaches on top of a large language model. Comput. Linguist. **49**(3), 555–611 (2023)
32. Lewis, P., et al.: Retrieval-augmented generation for knowledge-intensive NLP tasks. In: Proceedings of the 34th International Conference on Neural Information Processing Systems, NIPS 2020. Curran Associates Inc., Red Hook, NY, USA (2020)
33. Lin, C.Y.: ROUGE: a package for automatic evaluation of summaries. In: Text Summarization Branches Out, pp. 74–81 (2004)
34. Lin, J., et al.: AWQ: activation-aware weight quantization for on-device LLM compression and acceleration. Proc. Mach. Learn. Syst. **6**, 87–100 (2024)
35. Maharjan, J., et al.: OpenMedLM: prompt engineering can out-perform fine-tuning in medical question-answering with open-source large language models. Sci. Rep. **14**(1), 14156 (2024)
36. Nagendran, M., et al.: Artificial intelligence versus clinicians: systematic review of design, reporting standards, and claims of deep learning studies. bmj **368** (2020)

37. Ng, K.K.Y., Matsuba, I., Zhang, P.C.: Rag in health care: a novel framework for improving communication and decision-making by addressing LLM limitations. NEJM AI **2**(1), AIra2400380 (2025)
38. Nguyen, Q.N., et al.: Advancing question-answering in ophthalmology with retrieval-augmented generation (RAG): benchmarking open-source and proprietary large language models. Investigat. Ophthalmol. Vis. Sci. **66**(8), 4638 (2025)
39. Pal, A., Umapathi, L.K., Sankarasubbu, M.: MedMCQA: a large-scale multi-subject multi-choice dataset for medical domain question answering (2022)
40. Papineni, K., Roukos, S., Ward, T., Zhu, W.J.: BLEU: a method for automatic evaluation of machine translation. In: Proceedings of the 40th Annual Meeting of the Association for Computational Linguistics, pp. 311–318 (2002)
41. Perrella, S., Proietti, L., Cabot, P.L.H.: Beyond correlation: interpretable evaluation of machine translation metrics (2024)
42. Raffel, C., et al.: Exploring the limits of transfer learning with a unified text-to-text transformer. J. Mach. Learn. Res. **21**(140), 1–67 (2020)
43. Sallinen, A., et al.: Llama-3-Meditron: an open-weight suite of medical LLMs based on llama-3.1. In: Workshop on Large Language Models and Generative AI for Health at AAAI 2025 (2025). https://openreview.net/forum?id=ZcD35zKujO
44. Singhal, K., et al.: Toward expert-level medical question answering with large language models. Nat. Med., 1–8 (2025)
45. Team, G., et al.: Gemma 2: improving open language models at a practical size **1**(3) (2024). https://arxiv.org/abs/2408.00118 (2024)
46. Team, Q., et al.: Qwen2.5 technical report (2025). https://arxiv.org/abs/2412.15115
47. Wu, J., et al.: Medical graph RAG: towards safe medical large language model via graph retrieval-augmented generation (2024). arXiv preprint arXiv:2408.04187
48. Xia, Y., Fu, F., Zhang, W., Jiang, J., Cui, B.: Efficient multi-task llm quantization and serving for multiple lora adapters. In: Advances in Neural Information Processing Systems, vol. 37, pp.63686–63714 (2024)
49. Yazan, M., Verberne, S., Situmeang, F.: The impact of quantization on retrieval-augmented generation: an analysis of small LLMs (2024). arXiv preprint arXiv:2406.10251
50. Zaghir, J., Naguib, M., Bjelogrlic, M., Névéol, A., Tannier, X., Lovis, C.: Prompt engineering paradigms for medical applications: scoping review. J. Med. Internet Res. **26**, e60501 (2024)
51. Zhang, H., et al.: HuatuoGPT, towards taming language model to be a doctor. arXiv preprint arXiv:2305.15075 (2023)
52. Zhao, X., Liu, S., Yang, S.Y., Miao, C.: MedRAG: enhancing retrieval-augmented generation with knowledge graph-elicited reasoning for healthcare copilot. In: Proceedings of the ACM on Web Conference 2025, pp. 4442–4457 (2025)

Intellectual and Developmental Disabilities: The Role of LLMs in Communication

May Lutzen(✉) and Marcello Balduccini

Saint Joseph's University, Philadelphia, PA 19131, USA
ml10791538@sju.edu

Abstract. This paper is part of a project aimed at leveraging neuro-symbolic architectures to achieve a sophisticated interaction between humanoid robots and individuals with cognitive impairments. In our architecture, a symbolic reasoning module based on Answer Set Programming creates a sequence of appropriate activities and monitors their execution in real-time. Large Language Models (LLMs) are used to enhance the user experience in various ways. Among those, in this paper we investigate methods for using LLMs to rewording text produced by the reasoning component, while keeping length and vocabulary level consistent with the original text. The study explores the effectiveness of metrics such as length and frequency of use, in comparing the vocabulary level of the input with that of the output. We present a comparative analysis of free or moderately priced LLMs, such as GPT-3.5, Google Gemini Pro, and Claude 3 Opus. A continuous validation process is also introduced, utilizing a critic that evaluates the appropriateness of the generated output at run-time. The findings indicate that while LLMs can often produce outputs with a vocabulary level comparable to the inputs, although there are areas needing improvement, particularly in handling specific domain knowledge or less common phrases. This research contributes to the exploration of novel neuro-symbolic architectures and to the practical application of LLMs in contexts where controlled language use is essential for effective communication.

Keywords: Large Language Models · Neuro-Symbolic Robot Architectures · Knowledge Representation

1 Introduction

Humanoid robots employed in healthcare represent a diverse and rapidly evolving research field that encompasses various areas of investigation, for example to support people with communication disabilities as described in [5]. For instance, studies have explored the use of AI-enabled devices in providing companionship and social interaction for seniors, as highlighted in [2], underscoring the potential of humanoid robots to address social isolation and promote well-being

H. R. Arabnia et al. (Eds.): CSCI 2025, CCIS 2971, pp. 374–387, 2027.
https://doi.org/10.1007/978-3-032-28463-1_31

among elderly populations. Moreover, recent research [3,6] highlighted how AI can subtly guide individuals towards specific choices with more effective *nudging* capabilities than humans, suggesting that an automated assistant might be a useful tool for guiding patients towards healthier behaviors.

This paper is part of a project stemming from a partnership with Bancroft (https://bancroft.org), a non-profit organization dedicated to helping individuals with autism, intellectual and developmental disabilities, brain injury, and neurological conditions. A challenge faced by service providers like Bancroft is that direct care facility staff may become overwhelmed with routine tasks, leading to low job satisfaction and retention, and hindering their performance in higher-level duties and communication with clinical staff. High turnover exacerbates these issues, as new and temporary staff lack the specific knowledge needed for effective patient care. All of these staff issues, in the long term, cause reduced quality of care thus impacting the quality of life of the individuals served. The goal of the project is to research methods at the intersection of symbolic reasoning, machine learning, and robotics that can lead to enhancing both staff satisfaction and resident care.

The project augments *Iggy*, a Pepper humanoid robot [9], with a neuro-symbolic architecture, called IR4D[1], which combines the powerful knowledge representation and reasoning capabilities of Answer Set Programming (ASP) [4] with Large Language Model (LLMs).

Given information on a group of patients, the *reasoning module* builds a sequence of appropriate activities and monitors their execution, making real-time adjustments based on user responses. Whenever a menu is presented on the robot's display, the *reasoning module* generates text that is read out by the robot to introduce the menu. This is done not only for a more natural interaction but also because some of the patients have minimal or no reading skills. Given that the robot controlled by the architecture interacts with individuals with potentially impaired cognitive skills, the text produced by the symbolic reasoning component is based on fragments that are carefully crafted by psychologists in order to ensure that the vocabulary level is suitable for the audience.

Depending on the circumstances, the same menu may be displayed multiple times in a row. In order to make the interaction more engaging for the audience and in order to make the robot sound more natural, the robot automatically produces variations of the text generated by the symbolic reasoning component. While LLM are a natural choice for generating these variations, the variations must still match the audience's capabilities, and in particular their (receptive) vocabulary level.

In this paper, we present an initial evaluation of methods for controlling the vocabulary level of the text produced by LLMs in rewording tasks, with a specific attention to the use with individuals with cognitive impairment. We begin by considering different kinds of metrics normally associated with the vocabulary level of text, and then evaluate the use of a *continuous validation* process, in which the output produced by the LLM is vetted at runtime by a *critic* that

[1] IR4D stands for *Intelligent Robotics for Disabilities.*

determines if the output is acceptable. Due to considerations on practical use, in this paper we focus our evaluation on LLMs that are either free or moderately priced, but also of sufficient power and ease of use, i.e. GPT 3.5, Google Gemini Pro, and Claude 3 Opus. In a later phase of the study, we will evaluate more powerful versions of these as well as self-hosted LLMs.

The paper is organized as follows. We begin with a discussion on methods of measuring the vocabulary of text and clarify the research questions that our analysis aims at answering. The following sections analyze metrics of progressively increasing sophistication. We conclude the paper with final considerations and a discussion on future work.

2 Measuring the Vocabulary Level of Text

Typically, the evaluation of the vocabulary level of text takes into account metrics such as length, frequency of use, morphological complexity and semantic complexity, as discussed, e.g., in [1,7,8]. Length refers to the number of characters, words, and syllables in the text, as it is conceivable that longer text might require stronger cognitive efforts. Frequency of use refers to how often words are used in everyday language. If they are less common, they might be more complex and thereby more difficult to understand. Morphological complexity refers to the complexity of a word's internal structure and words with more complex structures might be more difficult to understand. Lastly, semantic complexity refers to the number of meanings or senses a word has. If a word has multiple meanings, reaching the desired understanding of a word can be much more complex. In this paper, we focus on the first two metrics, and use the following research questions to guide the analysis:

- How capable are the LLMs of producing output that is of comparable length to the input if instructed to do so? And how reliable is length as an indicator of vocabulary level in the context of our application domain?
- How capable are LLMs of producing output in which frequency-of-use is comparable to that of the input?
- Is there an algorithmic way of leveraging frequency-of-use as a way to improve the vocabulary level of output text?

We conduct our analysis over a benchmark set of 10 passages that covers all the categories of passages generated by the *reasoning module* during a session. Categories are represented by a uniform number of passages in the benchmark set. The passages are show in Fig. 1.

For each input passage, we produce three output passages for each LLM. We chose to produce multiple output passages in order to limit the impact of statistical anomalies on our analysis.

It is important to note that some of the passages contain the names of activities that the audience can choose from using the robot's display. Those names are familiar to the audience. In preliminary experiments, we observed that the

P1	Hello again! It's good to see you back here at Saint Joseph's University. You may remember me from last time. My name is Pepper, and I'm a social robot here at the Innovation Center. We have more fun activities planned for you today.
P2	How about we take a little break now and relax? When you are ready to resume, click the button on my screen.
P3	Before we say goodbye, I have a little surprise for you. Here are some activities that I hope you'll like: "a name that sound game for David", "some dancing fun", and "a name that sound game". Pick from one of the boxes on my screen to start an activity!
P4	Guess who's back from its break? "the Frosty the Snowman sing-along". Which activity would you like to pick now? We have "the Frosty the Snowman sing-along", "David's name that sound game", and "the name that season game".
P5	Before we say goodbye, I have a little surprise for you. Here are some activities that I hope you'll like: "a Frosty the Snowman sing-along", "a name that sound game for David", and "a name that tune game". Pick from one of the boxes on my screen to start an activity!
P6	Let's give a little break to "the name that season game". Guess who's back from its break? "the emotion recognition game". Which activity would you like to pick now? We have "the emotion recognition game", "the name that tune game", and "the name that animal sound game".
P7	Here are a few activities for you to have fun with now. We have "Mister Sandman", "a name that tune game", and "a dance routine". Pick from one of the boxes on my screen to start an activity!
P8	Here are a few activities for you to have fun with now. We have "a Frosty the Snowman sing-along", "an opportunity to chat with Pepper, and "a name that sound game for David". Pick from one of the boxes on my screen to start an activity!
P9	Here are a few activities for you to have fun with now. We have "a name that sound game", "a Rudolph the Red Nosed Reindeer sing-along", and "a name that animal sound game". Pick from one of the boxes on my screen to start an activity!
P10	Let's give a little break to "the Frosty the Snowman sing-along". Which activity would you like to pick now? We have "chat with Pepper", "dancing fun", and "David's name that sound game".

Fig. 1. Benchmark set used in the analysis

LLMs have a tendency to reword the activity names, often making the activities difficult to recognize, sometimes drastically altering the semantics of the names – for example, turning "a name that sound game" into "a name-based sound game", and even rewording "an emotion recognition game" into "a mind-reading game". To our surprise, we were only partially successful at designing prompts that directly prevented the LLMs from rewording the activity names, and resorted to an algorithmic approach for preserving them. In our algorithmic preservation approach, the activity names are replaced by strings that the LLMs do not appear to reword, such as *"X1X"*. After receiving the LLM's output, they are then converted back to produce the final output. While in our analysis we considered both kinds of approaches to activity name preservation, in this paper we focus on the analysis of the passages obtained through the algorithmic preservation approach. Additionally, due to space considerations, in the sections that follow we will be providing only excerpts of some of the tables. The complete tables can be found at https://g34r.short.gy/dataset.

3 Length-Based Metrics

Length of text is a readily-usable metric in the context of LLMs, as phrases such as "of length comparable to the input" can be easily included in the prompt to an LLM. In this section, we evaluate the output produced by the LLMs under consideration when the prompt instructs the LLM to constrain the length of the output. Specifically, we use the prompt:[2]

> *"Take what I say after the new-line character and paraphrase it. The new text should have a similar length to the one in input as well as the same tone and language."*

The complete list of the output passages produced by the LLMs can be found at the link provided earlier. We provide sample output passages – specifically for P10 – in Fig. 2.

In this section, we compare the various LLMs in terms of number of characters, words, and syllables of the output w.r.t. to those of the input.

For every metric and LLM, we calculate average change (i.e., difference between input length and output length) for every passage, as well average change across all passages. The results of the analysis on length in characters are shown in Fig. 3. The analysis of the average change, per LLM, on each input passage shows a clear trend. GPT has the lowest average change, followed by Gemini and then Claude, which has the highest average change. This is also the case for the aggregate average change on all passages, which is shown in the 3 bottom rectangles for each metric, where GPT has 0% average change, Gemini 10%, and Claude 44%. The results from the analysis of length measured in characters is confirmed by looking at the average change in length measured by word count where the average change on all the different passages by LLM for GPT is −4%, Gemini shows 5% average change and Claude shows 34% average change.

While length and average change in length might show how well a large language model is able to follow directions from a prompt, our analysis highlighted that, at least for our target population, the length of the text is not a good indicator of the vocabulary level. Consider for instance the input text:

> *Before we say goodbye, I have a little surprise for you. Here are some activities that I hope you'll like: "a name that sound game for David", "some dancing fun", and "a name that sound game". Pick from one of the boxes on my screen to start an activity! "*

The output created by Claude is:

[2] One might argue that more sophisticated ways of phrasing the prompt should be considered. While that is indeed possible, we intentionally stay away from the route of prompt engineering and focus on analyzing what LLMs are capable of producing given a reasonably precise prompt that an average human would likely be able to understand and act upon.

LLM	Output
GPT	Let's take a breather and switch things up for "the Frosty the Snowman sing-along". Which option do you fancy next? We've got "chat with Pepper", "dancing fun", and "David's name that sound game" on the table.
GPT	Let's take a brief pause for "the Frosty the Snowman sing-along". What would you like to do next? Your options are "chat with Pepper", "dancing fun", and "David's name that sound game".
GPT	Let's take a short pause for "the Frosty the Snowman sing-along". What activity do you feel like doing next? We've got options like "chat with Pepper", "dancing fun", and "David's name that sound game".
GEMINI	After a little break, "the Frosty the Snowman sing-along" needs a moment to unwind. What new pursuit do you suggest for "the Frosty the Snowman sing-along"? "chat with Pepper", "dancing fun", and "David's name that sound game" are all possibilities.
GEMINI	Let's give "the Frosty the Snowman sing-along" a quick breather. How about trying out "chat with Pepper"? Or perhaps "dancing fun" or "David's name that sound game" might be a better fit for you right now?
GEMINI	Let's put "the Frosty the Snowman sing-along" on hold for a bit. What would you like to do next? We can choose from "chat with Pepper", "dancing fun", or "David's name that sound game".
CLAUDE	Here is the paraphrased text with the specified variables unchanged: How about we allow "the Frosty the Snowman sing-along" to rest for a bit? What would you prefer to do instead? The options available are "chat with Pepper", "dancing fun", and "David's name that sound game".
CLAUDE	Here is the paraphrased text with variables unchanged: How about we allow "the Frosty the Snowman sing-along" to rest for a bit? What would you prefer to do instead? The options available are "chat with Pepper", "dancing fun", and "David's name that sound game".
CLAUDE	Here is the paraphrased text with the specified variables unchanged: How about we allow "the Frosty the Snowman sing-along" to rest for a bit? What would you prefer to do instead? The options available are "chat with Pepper", "dancing fun", and "David's name that sound game".

Fig. 2. Output passages produced for passage P10

Hold your horses! Before we part ways, I've got a treat in store. I've handpicked some cracking activities that I reckon you'll dig: "a name that sound game for David", "some dancing fun", and "a name that sound game". Go ahead and click on any of the boxes on your screen to get the party started!

The vocabulary level of phrases such as "hold your horses", "handpicked some cracking activities" and "I reckon you'll dig" is *significantly* different from "I have a little surprise" and "I hope you'll like", even though the change in number of characters (13%) and in number of words (12%) might be considered modest by comparison.

When it comes to syllables, Fig. 3 shows that the overall average change in number of syllables resembles follows trends similar to those of the number of characters and words. GPT has the overall smallest average change with only

Passage #	LLM	Length of input	Length of output	Change in length	Average change for each passage	Average change on all passages	Word count of input	Word count of output	Change in word count	Average change for each passage	Average change on all passages	Syllable count of input	Syllable count of output	Change in syllables	Average change for each passage	Average change on all passages
P9	GPT	244	246	1%			47	46	-2%			69	71	3%		
P9	GEMINI	244	232	-5%			47	43	-9%			69	68	-1%		
P9	CLAUDE	244	531	118%			47	92	96%			69	149	116%		
P9	GPT	244	210	-14%			47	37	-21%			69	57	-17%		
P9	GEMINI	244	244	0%			47	42	-11%			69	68	-1%		
P9	CLAUDE	244	296	21%			47	48	2%			69	86	25%		
P9	GPT	244	257	5%	-3%		47	47	0%	-8%		69	72	4%	-3%	
P9	GEMINI	244	273	12%	2%		47	50	6%	30%		69	77	12%	3%	
P9	CLAUDE	244	386	58%	66%		47	62	32%	43%		69	107	55%	65%	
P10	GPT	198	209	6%			33	36	9%			48	51	6%		
P10	GEMINI	198	249	26%			33	40	21%			48	64	33%		
P10	CLAUDE	198	277	40%			33	46	39%			48	74	54%		
P10	GPT	198	185	-7%			33	32	-3%			48	46	-4%		
P10	GEMINI	198	205	4%			33	36	9%			48	50	4%		
P10	CLAUDE	198	263	33%			33	44	33%			48	70	46%		
P10	GPT	198	202	2%	0%	0%	33	34	3%	3%	-4%	48	52	8%	3%	2%
P10	GEMINI	198	185	-7%	8%	10%	33	34	3%	11%	5%	48	45	-6%	10%	10%
P10	CLAUDE	198	277	40%	38%	44%	33	46	39%	37%	34%	48	74	54%	51%	49%

Fig. 3. Change in length measured in characters, word count and syllables

2% from input to output, Gemini is second with only 10% overall average change from input to output and Claude performed the worst with 49% overall average change from input to output. The conclusion of this analysis thereby shows that GPT is best at following directions in the prompt given to it specifically about desired length in a paraphrased output whereas Gemini is slightly worse but can still be considered acceptable as it within a 10% change. However, in many of the 30 Claude cases in our analysis, the LLM includes part of the prompt in its output, such as 'with the specified variables'. It should therefore be noted that one reason for Claude's great positive change in length is due to the fact that Claude tends to include part of the prompt in the output thereby increasing the length, however this also shows how it is unable to follow directions from the prompt.

However, the number of syllables does not appear to have a strong relationship with the vocabulary level, at least for the types of passages and population considered here. Consider one of the outputs produced by Claude for P5:

- **Input:** *Before we say goodbye, I have a little surprise for you. Here are some activities that I hope you'll like: "a Frosty the Snowman sing-along", "a name*

that sound game for David", and "a name that tune game". Pick from one of the boxes on my screen to start an activity!
- **Output:** *Before we part ways, I've got a special treat in store for you. Take a look at these fun options I've prepared: "a Frosty the Snowman sing-along", "a name that sound game for David", and "a name that tune game". Simply select one of the boxes displayed to embark on an exciting activity!*

Here, phrases such as "part ways", "special treat in store for you" and "embark on" have a significantly different vocabulary level than "goodbye", "little surprise" and "to start".

In conclusion, while the change in length between input and output can tell us how well an LLM is at following directions related to the length of the output – and is useful in preventing the output from being boring or wordy – it does not help ensure that the vocabulary level of the output matches that of the input.

4 Frequency-of-Use Metric

Of the other possible options for estimating the vocabulary level of a passage, the frequency of use (abbreviated `uf` in this paper) of its words appears to be a practically viable metric. The intuition behind this metric is simple: frequency of use refers to how often certain words are used in a corpus of text that is used as a reference; if words are less common, it is conceivable that they are more complex or difficult to understand.

The minimum frequency of use (abbreviated `min uf`) across all the words of a passage seems particularly useful in our application domain. After all, we are looking for text that, when spoken by the robot, can be *promptly* understood by our target population. Even a single word that is outside of the receptive vocabulary of the audience can be sufficient to prevent them from promptly understanding the passage. It must be noted, however, that it is not sufficient to consider the `uf` of individual words, as phrases, especially if idiomatic, may have a meaning that is different from that of their constituent words and a frequency of use that is also different from of the constituent words. Consider for example "let the fun begin". While the individual words of this phrase are fairly common, the specific combination is not as frequently used as, for example "let's have fun." For this reason, we adopt the following approach.

Let $\tau(p)$ denote the parse tree associated with passage p (if multiple parse trees exist, we assume that a parse tree has been arbitrarily selected). For a node $n \in \tau(p)$, let $uf(n)$ denote the frequency of use of the text associated with n. $uf(n)$ is undefined if no frequency of use is available for the text.

Definition 1. *A node n of $\tau(p)$ is a* uf-top node *if:*

- *$uf(n)$ is defined, and*
- *$uf(n')$ is undefined for every ancestor n' of n in $\tau(p)$.*

We can now give the following:

Definition 2. *Given a passage p:*

- *The* min uf *of p, written min−uf(p), is*

$$min{-}uf(p) = \sum_{n \in \tau(p)} uf(n)$$

- *The* min-uf phrase *of p is the fragment of text associated with a uf-top node of $\tau(p)$ that has minimum uf. (For simplicity, if multiple such nodes exist, we select the first node in leftmost traversal of the parse tree.)*

For the analysis that follows, the parse trees were generated using the *spacy library*[3] with the *en_core_web_lg* model. The `uf` of phrases has been obtained from the *absTotalMatchCount* metric of the NGRAMS tool[4] with default settings[5]. NGRAMS uses the *Google Books Ngram Dataset v3* as the source of ngram data.

At least at first sight, the `min uf` appears to provide meaningful a indication of the vocabulary level of passages. For example, the phrase "on my screen" which one might consider fairly simple to understand, has a comparatively high `min uf` of 33090, while phrases such as "our prior encounter" and "infringing on any copyrights," which arguably belong to a higher vocabulary level, have a comparatively low `min uf` of 72 and 42 respectively.

So, the question is: how capable are LLMs of producing output whose `min uf` is comparable to the `min uf` of the input given our prompt?

To answer the question, we compared the `min uf` of our input and output passages. However, one should notice that activity names deserve some attention. In fact, it so happens that some activity names are rather infrequent in the corpus used for the calculation of `min uf`, even though they are promptly understood by the audience, causing them to have a low use frequency that may bias the evaluation. Since the rewording process ensures that the activity names are kept verbatim in the output, in many cases those are also the *min-uf phrases* of the output.

For this reason, we disregard the activity names in the calculation of the `min uf`, leading to the comparison outlined in Fig. 4. We observe that in 50% of cases the `min uf` of the output was worse than that of the input. That is, in only 50% of the cases the passage produced by the LLM was satisfactory in terms of `min uf`.

It is worth noting that a potential issue with the ngrams used for estimating frequency of use. As seen from Fig. 4, in a number of cases in which the `min uf` of the output is greater than or equal to the `min uf` of the input, the *min-uf phrase* of the input is "Which activity would you like." For some reason, that phrase is assigned a rather low `uf`, which causes most outputs to have a better

[3] https://spacy.io.

[4] https://ngrams.dev/.

[5] It is certainly possible that different settings may affect the results of our analysis, particularly if one selects specific year ranges for the frequency of the ngrams. We defer an evaluation of the impact of these settings to a later phase of our study.

LLM	Input's min-uf phrase	Output's min-uf phrase	Revised min-uf input	Revised min-uf output	Change in min uf input to min uf of output	Greater or Less
GPT	to start an activity'	on my display'	2471	1746	-29%	Less
GEMINI	to start an activity'	'on my display'	2471			
CLAUDE	to start an activity'	'activities to enjoy'	2471	3143	27%	Greater
GPT	to start an activity'	available on the screen'	2471	3623	47%	Greater
GEMINI	to start an activity'	'to dive into the fun'	2471			
CLAUDE	to start an activity'	'on your chosen activity'	2471			
GPT	to start an activity'	on my display'	2471	1746	-29%	Less
GEMINI	to start an activity'	'to start an activity'	2471	2471	0%	Greater
CLAUDE	to start an activity'	an activity of your choice'	2471	185	-93%	Less
GPT	to start an activity'	"'s explore"	2471	17961	627%	Greater
GEMINI	to start an activity'	To start an activity,'	2471	398	-84%	Less
CLAUDE	to start an activity'	on my interface'	2471	111	-96%	Less
GPT	to start an activity'	"'s have some fun"	2471	2384	-4%	Less
GEMINI	to start an activity'	are all available.'	2471	15684	535%	Greater
CLAUDE	to start an activity'	'Ready for some excitement?'	2471			
GPT	to start an activity'	of entertaining options'	2471	131	-95%	Less
GEMINI	to start an activity'	sure to keep you entertained'	2471	94	-96%	Less
CLAUDE	to start an activity'	'some entertaining activities'	2471			
GPT	to start an activity'	to begin an activity'	2471	1130	-54%	Less
GEMINI	to start an activity'	of these exciting activities'	2471	67	-97%	Less
CLAUDE	to start an activity'	some exciting activities'	2471	206	-92%	Less
GPT	to start an activity'	the Red Nosed Reindeer'	2471	3148	27%	Greater
GEMINI	to start an activity'	the Red Nosed Reindeer'	2471	3148	27%	Greater
CLAUDE	to start an activity'	your desired activity'	2471	68	-97%	Less
GPT	to start an activity'	some fun options'	2471	85	-97%	Less
GEMINI	to start an activity'	to dive into the fun'	2471	49	-98%	Less
CLAUDE	to start an activity'	displayed on my screen'	2471	210	-92%	Less
GPT	Which activity would you like'	fancy next?'	57	52	-9%	Less
GEMINI	Which activity would you like'	What new pursuit'	57	69	21%	Greater
CLAUDE	Which activity would you like'	with the specified variables'	57	114	100%	Greater
GPT	Which activity would you like'	Your options are'	57	37572	65816%	Greater
GEMINI	Which activity would you like'	a better fit for you'	57	1278	2142%	Greater
CLAUDE	Which activity would you like'	the paraphrased text'	57	477	737%	Greater
GPT	Which activity would you like'	like doing next'	57	142	149%	Greater
GEMINI	Which activity would you like'	"'s put"	57	73641	129095%	Greater
CLAUDE	Which activity would you like'	with the specified variables'	57	114	100%	Greater

Fig. 4. Change in `min uf` from revised input to revised output (excerpt)

`min uf` than the input passage – even in cases where the *min-uf phrase* of the output is "with the specified variables", which intuitively seems to be at a higher vocabulary level. As we discuss later in this paper, we hypothesize that better quality ngram data may reduce this type of noise.

5 A Vocabulary Level Critic: An Evaluation

In the previous section, we saw how LLMs have some capability of producing output of comparable vocabulary level to the input *as estimated via* ***uf***, although the performance of LLMs is less than ideal. In this section, we consider the possibility of algorithmically filtering the output of the LLMs based on this metric, and evaluate whether `uf` provides indeed a reliable filtering mechanism.

For our evaluation, we consider a simple filtering mechanism based on a continuous validation of the output of the LLM being used. That is, we can leverage `min uf` to develop a *critic* that compares the `min uf` of the output with that of the input. If the `min uf` of the output is lower than that of the input, the critic rejects the output, and the architecture may decide to either have the LLM produce a new output, or use the input directly. In practice, the decision will be likely based on timeliness considerations: given the statistical nature of LLMs, there is no telling how many times the output will need to be regenerated before something of acceptable quality is produced, and long delays before the robot speaks might make for an unnatural experience.

Our goal in this section is to answer the question: *suppose the critic rejects the output produced by the LLM; based on human judgement, is that a good decision or a bad decision?* This evaluation will give us some insights both on the reliability of `min uf` as a filtering metric, and if the critic as a practically viable algorithmic filtering method.

The results of this evaluation are summarized in Fig. 5, where a judgement of "good decision" is indicated as a true positive, and a bad decision as a false positive. At this early stage of the project, we relied on our own judgment to determine whether a decision was good or bad. In the future, we plan to conduct a more rigorous evaluation with independent test subjects. In spite of the simplicity of our validation method, we believe that the evaluation provided some interesting results. As the figure illustrates, out of 53 cases flagged by the critic, 33 are true positives, which means that in 62% of cases, `min uf` is a reliable indicator of whether the vocabulary level of the output produced by an LLM is acceptable. Figure 6 provides a breakdown of the information for each LLM. For GPT, the true positive rate is 61%, for Gemini it is 79%, and for Claude it is 52%. One consideration to make is that the generation of the parse tree, and possibly also the ngram data, appear to introduce some noise in the filtering process. For example, some of the cases marked as false positives during our validation are due to the string "'s have some fun" being the *min-uf phrase* in the output, and being associated with a low `uf`. Those cases were marked as false positives during our validation because it is clear that the string refers to "let's have some fun", and the vocabulary level of that phrase is low. We speculate

LLM	Input's min-uf phrase	Output's min-uf phrase	Output passage has lower min uf	Human Validation	Produced by GPT	Produced by Gemini	Produced by Claude
GPT	at the Innovation Center'	'lined up for you today'	1	true positive	true positive		
GEMINI	at the Innovation Center'	to Saint Joseph's University	1	false positive		false positive	
CLAUDE	at the Innovation Center'	to Saint Joseph's University	1	false positive			false positive
GPT	at the Innovation Center'	to Saint Joseph's University	1	false positive	false positive		
GEMINI	at the Innovation Center'	to Saint Joseph's University	1	false positive		false positive	
CLAUDE	at the Innovation Center'	to Saint Joseph's University	1	false positive			false positive
GPT	at the Innovation Center'	to Saint Joseph's University	1	false positive	false positive		
GEMINI	at the Innovation Center'	'the vibrant campus'	1	true positive		true positive	
CLAUDE	at the Innovation Center'	to Saint Joseph's University	1	false positive			false positive
GPT	'How about we take'	'on my display'	1	true positive	true positive		
GEMINI	'How about we take'	'refreshed and ready to continue'	1	true positive		true positive	
CLAUDE	'How about we take'	'variables preserved'	1	true positive			true positive
GEMINI	'How about we take'	'energized and ready to continue'	1	true positive		true positive	
CLAUDE	'How about we take'	'the paraphrased text'	1	true positive			true positive
GPT	'How about we take'	'on my interface'	1	true positive	true positive		
CLAUDE	'How about we take'	'on my display'	1	true positive			true positive
GPT	Before we say goodbye'	Before we part ways'	1	true positive	true positive		
GEMINI	Before we say goodbye'	some exciting activities'	1	false positive		false positive	
CLAUDE	Before we say goodbye'	displayed on my screen'	1	false positive			false positive
GPT	Before we say goodbye'	Before we bid farewell'	1	true positive	true positive		
GEMINI	Before we say goodbye'	Before we part ways'	1	true positive		true positive	
CLAUDE	Before we say goodbye'	displayed on my screen'	1	false positive			false positive
GPT	Before we say goodbye'	a few fun activities'	1	false positive	false positive		
GEMINI	Before we say goodbye'	"that you'll dig"	1	true positive		true positive	
CLAUDE	Before we say goodbye'	on my interface'	1	true positive			true positive
GPT	Before we say goodbye'	to begin an activity'	1	false positive	false positive		
GEMINI	Before we say goodbye'	Before we part ways'	1	true positive		true positive	
CLAUDE	Before we say goodbye'	a few fun things'	1	false positive			false positive
GPT	Before we say goodbye'	Before we bid adieu'	1	true positive	true positive		
CLAUDE	Before we say goodbye'	of these fun activities'	1	false positive			false positive
GPT	Before we say goodbye'	Before we bid farewell'	1	true positive	true positive		
GEMINI	Before we say goodbye'	to start the fun'	1	true positive		true positive	
CLAUDE	Before we say goodbye'	Before we part ways'	1	true positive			true positive
CLAUDE	'Which activity would you like'	'your next pursuit'	1	true positive			true positive
GPT	to start an activity'	on my display'	1	true positive	true positive		
GEMINI	to start an activity'	'to dive into the fun'	1	true positive		true positive	
CLAUDE	to start an activity'	'on your chosen activity'	1	false positive			false positive
GPT	to start an activity'	on my display'	1	true positive	true positive		
CLAUDE	to start an activity'	'some entertaining activities'	1	true positive			true positive
CLAUDE	to start an activity'	on my interface'	1	true positive			true positive
GPT	to start an activity'	"'s have some fun"	1	false positive	false positive		
CLAUDE	to start an activity'	'Ready for some excitement?'	1	false positive			false positive
GPT	to start an activity'	of entertaining options'	1	true positive	true positive		
GEMINI	to start an activity'	sure to keep you entertained'	1	true positive		true positive	
CLAUDE	to start an activity'	'some entertaining activities'	1	true positive			true positive
GPT	to start an activity'	to begin an activity'	1	false positive	false positive		
GEMINI	to start an activity'	of these exciting activities'	1	true positive		true positive	
CLAUDE	to start an activity'	some exciting activities'	1	true positive			true positive
CLAUDE	to start an activity'	your desired activity'	1	true positive			true positive
GPT	to start an activity'	some fun options'	1	false positive	false positive		
GEMINI	to start an activity'	to dive into the fun'	1	true positive		true positive	
CLAUDE	to start an activity'	displayed on my screen'	1	false positive			false positive
GPT	Which activity would you like'	fancy next?'	1	true positive	true positive		

Fig. 5. Human validation of revised `min uf` phrases

that, had the parse tree correctly extracted the string "let's have some fun", its `uf` might have been higher and thus the output would not have been rejected on its account. (A similar result would have been obtained if the ngram data had not included the entry "'s have some fun".) That is, we hypothesize that better quality of the parsing process and of ngram data might lead to a higher accuracy of the filtering process.

Human validation results	Overall	GPT	Gemini	Claude
True positive	33	11	11	11
False positive	20	7	3	10
Total	53	18	14	21
True positive %	**62%**	**61%**	**79%**	**52%**
False positive %	**38%**	**39%**	**21%**	**48%**

Fig. 6. Comparative results of the human validation process

The data also highlights another interesting aspect. In a certain number of cases marked as false positives during the human validation, the *min-uf phrases* in question contain place names related to the location of the robot, such as "Innovation Center" and "Saint Joseph's University." Clearly, those names occur infrequently in the text corpus used for determining the `uf`, but are well known to the audience. The results obtained after excluding those place names from the calculation of the `min uf` are illustrated in Fig. 7. With this refinement, the percentage of true positives is substantially higher, standing at 70%, with a 67% for GPT, 91% for Gemini, and 61% for Claude.

Human validation results	Overall w/o Place	GPT	Gemini	Claude
True positive	31	10	10	11
False positive	13	5	1	7
Total	44	15	11	18
True positive %	**70%**	**67%**	**91%**	**61%**
False positive %	**30%**	**33%**	**9%**	**39%**

Fig. 7. Comparative results of the human validation process excluding place names

6 Conclusions and Future Work

In this paper, we presented an initial evaluation of methods for controlling the vocabulary level of the text produced by LLMs in rewording tasks, with specific attention to the use with individuals whose cognitive capabilities are impaired. We considered length-based metrics as well as frequency of use, and defined an approach for identifying which nodes of the parse tree should be considered when measuring frequency of use. While there is space for improvement, we hope our results provide an informative initial account on the topic. For practical reasons, at this stage we considered a set of free or moderately-priced LLMs. In the future, we plan to evaluate more powerful versions of those LLMs as well as self-hosted LLMs. We will also conduct a human validation with independent test subjects, investigate morphological and semantic complexity measures, and explore the ability of more sophisticated prompt designs to ensure a consistent vocabulary level.

Acknowledgements. This paper was made possible in part by funding provided by Bancroft. The passages evaluated in this paper were assembled from fragments prepared by Sara Girotto.

References

1. Readability and the Common Core's Staircase of Text Complexity, August 2012. https://textproject.org/paper/readability-and-the-common-core-staircase-of-text-complexity/
2. 6ABC – Action News Philadelphia: Seniors using AI robots to combat loneliness (2023). https://rebrand.ly/seniors-AI. 6abc.com/seniors-ai-robot-companionship/14154002/
3. Ganapini, M.B., et al.: Value-based fast and slow AI nudging (2023)
4. Gelfond, M., Lifschitz, V.: Classical negation in logic programs and disjunctive databases. N. Gener. Comput. **9**, 365–385 (1991)
5. Khan, F., Wu, Y., Dray, J., Hemsley, B., Kocaballi, A.B.: Conversational agents to support people with communication disability: a co-design study with speech pathologists. In: Extended Abstracts of the CHI Conference on Human Factors in Computing Systems, CHI EA 2025, pp. 1–9, April 2025. https://doi.org/10.1145/3706599.3719800
6. Mele, C., Russo Spena, T., Kaartemo, V., Marzullo, M.L.: Smart nudging: how cognitive technologies enable choice architectures for value co-creation. J. Bus. Res. **129**, 949–960 (2021). https://doi.org/10.1016/j.jbusres.2020.09.004. https://www.sciencedirect.com/science/article/pii/S0148296320305786
7. Rayner, K., Duffy, S.A.: Lexical complexity and fixation times in reading: effects of word frequency, verb complexity, and lexical ambiguity. Memory Cogn. **14**(3), 191–201 (1986). https://doi.org/10.3758/BF03197692
8. Shardlow, M., Evans, R., Zampieri, M.: Predicting lexical complexity in English texts: the complex 2.0 dataset. Lang. Resources Eval. **56**(4), 1153–1194 (2022). https://doi.org/10.1007/s10579-022-09588-2
9. Softbank Robotics: Meet Pepper: The robot built for people (2023). https://us.softbankrobotics.com/pepper

Large Language Models and Applications in Education

From Concept to Classroom: Design and Implementation of a Platform for Multimodal AI-Generated ESL Quizzes

Edgar Ceh-Varela[1](✉), Yitzen Lizama-Peraza[2], and Caleb Parten[1]

[1] Department of Mathematical Sciences, Eastern New Mexico University, Portales, NM, USA
{eduardo.ceh,caleb.parten}@enmu.edu

[2] Department of Languages and Literature, Eastern New Mexico University, Portales, NM, USA
yitzen.lizamaperazaz@enmu.edu

Abstract. This paper details the design and ongoing development of an open source web-based platform that empowers *English as a Second Language* (ESL) instructors to leverage multimodal Large Language Models (LLMs) for automated quiz creation. The platform, allows instructors to generate tailored comprehension and description quizzes from texts and images for various proficiency levels. The motivation for this work was a foundational study with ESL instructors, which validated the need for and interest in such a tool. We report on how the findings from that initial study directly informed the platform's feature set. This work-in-progress paper presents the system's architecture, the LLM-driven quiz generation methodology, and the functionalities available to both instructors and students. By presenting this second stage of our research, we contribute a model for developing educational technology grounded in user feedback and offer a practical tool for the LLM-in-education community, setting the stage for its upcoming evaluation in live classroom environments.

Keywords: Generative AI · English as Second Language · Evaluation · AI agents · Multimodal LLM

1 Introduction

The modern educational landscape increasingly emphasizes personalized learning, yet instructors often face significant practical barriers to implementing it. For English as a Second Language (ESL) instructors, this challenge is particularly critical. They must serve to students with diverse proficiency levels, learning styles, and backgrounds, leading to a substantial time and effort required to create differentiated instructional and assessment materials [1,2]. Studies have shown that instructors can spend extended hours per week creating their own

H. R. Arabnia et al. (Eds.): CSCI 2025, CCIS 2971, pp. 391–402, 2027.
https://doi.org/10.1007/978-3-032-28463-1_32

classroom resources, a workload that limits their ability to focus on direct student interaction and support [3].

The recent emergence of Large Language Models (LLMs) presents a transformative opportunity to address this challenge. These powerful AI systems, trained on vast datasets, can understand and generate human-like text, making them adept at tasks like creating lesson plans, explanations, and quiz questions [4]. In the context of language education, LLMs offer the potential to automate the generation of tailored assessment materials, thereby freeing up valuable instructor time and enabling a more responsive and personalized learning environment [5,6].

However, the mere availability of technology does not guarantee its effective or appropriate use in the classroom. To create tools that are genuinely helpful, a user-centered design approach is crucial, ensuring that the technology aligns with the pedagogical goals and practical workflows of educators [7]. This approach necessitates keeping a "human in the loop," where the AI acts as an assistant to augment the teacher's expertise, not replace it.

Recognizing this, our research follows a three-stage, user-centered methodology. Before developing a technological solution, we conducted a foundational pilot study (Stage 1) with ESL instructors. This study validated the core concept by presenting instructors with LLM-generated quizzes to gauge their perceptions and gather critical feedback. Their positive response, coupled with specific suggestions, directly motivated the work presented in this paper: the design and implementation of a web-based platform for quiz generation (Stage 2). The final stage of our research (Stage 3) will involve deploying this platform in live classroom settings for a comprehensive pilot with both instructors and students.

As a work-in-progress, this paper makes the following specific contributions:

- The design and implementation of a novel open source web-based platform that leverages a multimodal LLM to generate ESL quizzes from both text and image inputs.
- A case study in user-centered AI development, demonstrating how findings from a foundational study with target users (ESL instructors) can directly inform the core features of an educational tool.
- A description of a self-hostable, end-to-end open-source architecture that leverages locally-run, openly available language models, ensuring data privacy and removing dependency on proprietary APIs.

The rest of this paper is organized as follows. Section 2 shows related works to our study. Section 3 presents the methodology we have used for this work. Finally, Sect. 4 presents our conclusions and a future work.

2 Related Work

Our work is situated at the intersection of three key research areas: the use of LLMs for educational content generation, the evolution of AI-assisted language learning, and the emerging application of multimodality in education.

The capacity of LLMs to automate the creation of educational materials is a rapidly growing field of research. Studies have moved beyond simple text generation to evaluate the quality and pedagogical value of LLM-created content. For example, in [8], researchers conducted a study on ChatGPT's ability to generate high-quality multiple-choice questions for medical education. Their findings indicate that ChatGPT can write questions with a comparable quality to human-written questions alongside a significant reduction in question creation time. This aligns with the broader perspective offered by [9,10], providing a comprehensive overview of the opportunities and challenges of LLMs in automatic question and answer generation. They argue that the most effective use of these models is as an assistive tool for educators, not a replacement.

The field of language learning is being transformed by the advanced conversational and text-processing capabilities of LLMs. Research has begun to explore the practical application of these models in authentic language learning contexts. A case study [11] on using ChatGPT in a English-as-a-foreign-language (EFL) course, found that it can lead to improvements in students' writing quality and foster a more engaging and interactive learning environment. Similarly, in [12], the authors explore how AI tools support lexical enrichment in ESL, bridging theory and practice to enhance vocabulary acquisition and learner autonomy beyond the classroom. This demonstrates the potential of LLMs to provide personalized support at scale. However, much of the existing research focuses on the learner's direct, unmediated interaction with the AI. Our work diverges by focusing on an instructor-centric model, where the instructor leverages AI to create, adjust, and assign materials, thus maintaining pedagogical control over the learning process.

While text-based generation is well-explored, the educational potential of multimodal models, which process information from both text and images is an emerging frontier. The ability to connect visual information with language is fundamental to learning, particularly in ESL where associating vocabulary with images is a core teaching strategy. In [13] researchers present a visual question generation model that at generating diverse questions by focusing on different image regions containing relational information among objects. Despite this potential, the application of these models for generating structured educational assessments from images is still in its early stages, representing a significant area for new research that our project begins to explore.

Our platform's dual functionalities for generating quizzes from text and images are a deliberate pedagogical choice designed to provide a holistic assessment of ESL proficiency. Text-based comprehension quizzes target a student's receptive skills, evaluating their ability to understand vocabulary, grammar, and complex ideas within a given context. Conversely, image-based description quizzes target productive skills, requiring students to actively recall and apply their vocabulary and grammatical knowledge to construct meaningful descriptions. By integrating both modalities, our platform enables instructors to move beyond simple comprehension testing and create a learning pathway that guides students from passive knowledge recognition to active language use, thereby offering a more robust and well-rounded evaluation of their overall proficiency.

3 Methodology

This section presents the initial results for the first two stages in the project: a foundational pilot study (Stage 1) with ESL instructors aimed to get critical feedback regarding LLM-generated quizzes, and the design and implementation of a web-based platform for quiz generation (Stage 2).

3.1 Phase 1. Foundational Study

The main goal of this study was to explore the capability and effectiveness of LLM to generate questions for lecture comprehension assessment. The study consisted in sending to a group of ESL instructors (n = 9) documents containing a text (i.e., a news article) with questions and answers automatically generated by an LLM. For this task we use Google's Gemini 2.5.

To gather the instructors' opinions about the generated questions we asked them to take a small survey. For this exploratory study we focused on questions generated for novice and intermediate levels. Table 1 shows the questions provided to the ESL instructors as part of our initial study. A Likert scale was used for the answers, containing: Strongly Agree, Agree, Undecided, Disagree, and Strongly Disagree.

Table 1. Questions provided to the ESL instructors. The answers are based on a Liker scale.

Questions
The questions are clear
The questions are well written
The questions are concise and to the point
The questions are engaging
The questions feel natural
I would create similar questions for my students if working with the given text
The questions can help novice students understand the text
The questions focus on key information presented in the text
The questions promote critical thinking
The difficulty of the questions is appropriate for novice language learners

The results from the survey with ESL instructors provided critical validation. Here we focus on the results for the novice level as the results for the intermediate level were similar. The feedback on the AI-generated questions for novice learners was broadly positive, highlighting clear strengths in foundational quality while also revealing opportunities for instructor enhancement.

A primary concern for any automated question generation system is the fundamental quality of its output. In this regard, the AI-generated questions performed exceptionally well. There was unanimous approval regarding the clarity and composition of the questions, with a combined 100% of participants either Agreeing or Strongly Agreeing that the questions were clear and well-written (Fig. 1). Similarly, 100% of instructors found the questions to be concise and to the point. Furthermore, a crucial aspect for novice learners is the appropriateness of the difficulty level. The results confirmed the model's ability to tailor content effectively, with 89% of respondents agreeing that the difficulty was appropriate for novice learners. This foundational success was a strong signal that an AI-driven tool could reliably produce a baseline of high-quality, usable content.

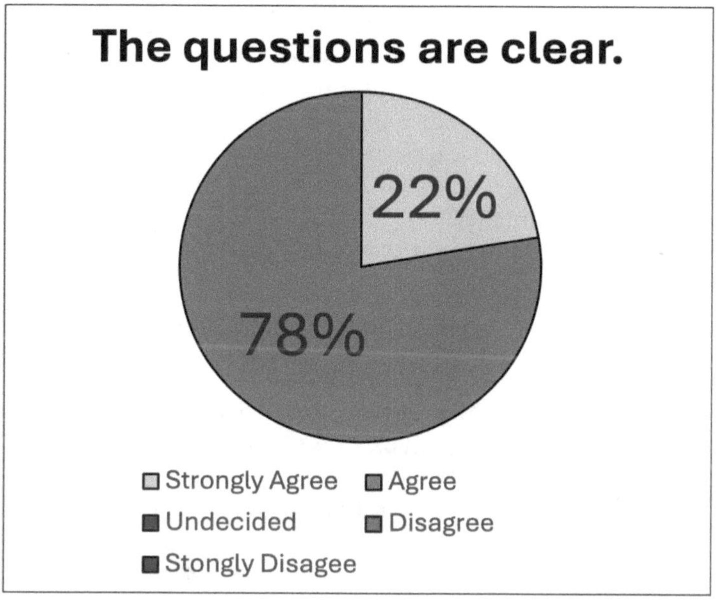

Fig. 1. 78% and 22% of instructors agreed and strongly agreed that the questions are clear.

Beyond grammar and clarity, the AI-generated questions were perceived as pedagogically sound. A large majority of instructors (89%) agreed that the questions successfully focused on key information presented in the text, indicating that the AI could identify and extract salient points. Similarly, 78% of participants felt the questions could help novice students understand the source text. This suggests that instructors see the tool not just for assessment but also as a potential teaching aid to guide students through a reading passage.

While the foundational quality was highly rated, responses became more nuanced when considering subjective qualities like engagement and the instructors' own willingness to use the questions verbatim. While a majority (66%) found the questions engaging, this area also saw the first significant "Disagree" responses (22%). A similar pattern emerged when instructors were asked if they would create similar questions themselves; 66% agreed, but a notable 22% were "Undecided." This finding was particularly instructive. It suggests that while instructors recognize the value and quality of the AI's output, they still see a role for their own professional judgment in refining or tailoring the questions to their specific classroom context. This directly validated our decision to build a platform where the instructor remains "in the loop," with the ability to approve all generated content.

The most mixed feedback, and a key finding from the study, was related to the promotion of critical thinking. This was the only category where positive responses (55% Agree/Strongly Agree) did not constitute a large majority. A combined 44% of instructors were either "Undecided" or "Disagreed" that the questions promoted critical thinking. These results indicate that while the AI excels at creating questions that test literal comprehension and recall, it is less adept at formulating questions that require deeper analysis, inference, or synthesis. This finding powerfully reinforces the need for an instructor-centric platform, positioning the AI as an assistant that handles the time-consuming task of creating foundational questions, thereby freeing up the instructor to focus on the higher-value task of designing questions that foster critical thought.

3.2 Phase 2. Platform Architecture and Implementation

We designed and implemented a fully open-source, self-hostable platform. The architecture was engineered to prioritize data privacy, cost-effectiveness, and instructor control by eliminating reliance on third-party APIs and instead leveraging locally-run language models.

The platform's architecture, illustrated in Fig. 2, comprises four main components that work in concert to deliver the end-to-end functionality: a web interface, a backend server, a local AI service, and a database.

- **Web Interface:** A user-friendly graphical user interface (GUI) serves as the primary interaction point for both instructors and students. Instructors access a dedicated dashboard with forms to create, manage, and assign quizzes. Students have a separate view where they can see and take quizzes assigned to them.
- **Backend Server:** The core of the application is a lightweight backend server built with the Flask web framework [14]. It orchestrates all operations, handling HTTP requests from the web interface, managing user authentication and sessions, processing business logic, and acting as the central intermediary between the user, the database, and the AI service.
- **Local AI Service:** A key feature of our architecture is the local, self-hosted AI service powered by Ollama [15]. Ollama manages and serves the openly

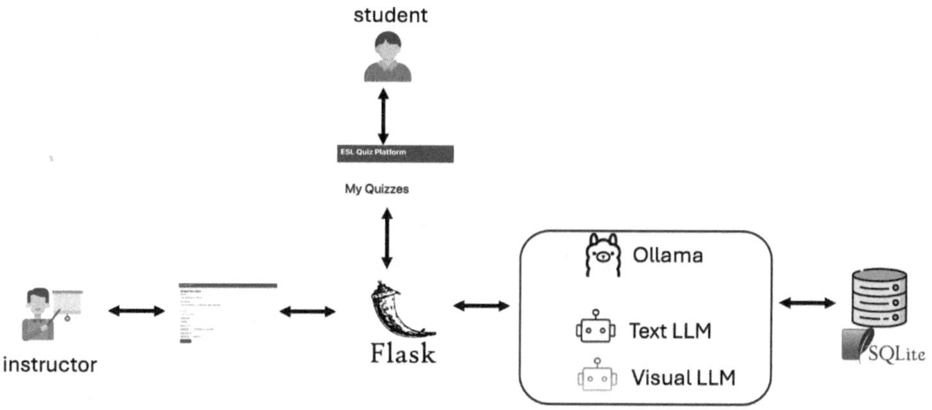

Fig. 2. The high-level system architecture, showing the interaction between the instructor, student, and the platform's open-source components.

available language models, exposing them through a local API. We run two distinct model types to handle the platform's multimodal requirements: (1) *llama3.2*, a standard LLM for processing text documents, (2) *llama3.2-vision*, for processing images. With this design, we can easily change the models for each task independently.

- **Database**: A serverless, file-based SQLite database [16] is used for data persistence. It stores all critical information, including user accounts, quiz content, quiz assignments, and student grades. SQLite was chosen for its simplicity and ease of deployment, making the entire platform self-contained and highly portable.

The platform operates through two primary workflows, one for the instructor creating content and one for the student consuming it. The flow of data through the system is orchestrated to support their specific tasks in a seamless and intuitive manner.

Figure 3 shows the instructor's management dashboard. It allows them to oversee student accounts, track assignments, and review grades. The core of the instructor's workflow, however, is the quiz generation process. Figure 4 shows the new quiz creation form where the instructors define the quiz's parameters, such as its title, number of questions, number of options per question, and the target ESL proficiency level. They then provide the source material by uploading either a text document for a comprehension quiz or an image for a description quiz.

Upon submitting this form, the Flask backend takes over. It packages the instructor's specifications and the source material into carefully engineered prompts. Figures 5a and 5b show the prompts we used in this project.

These prompts are then transmitted to the local Ollama service, which routes the request to the appropriate model. The model's task is twofold: (1) it generates a set of questions with potential answers, and crucially, (2) it also generates a brief explanation for the correct answer. This complete package is sent back to

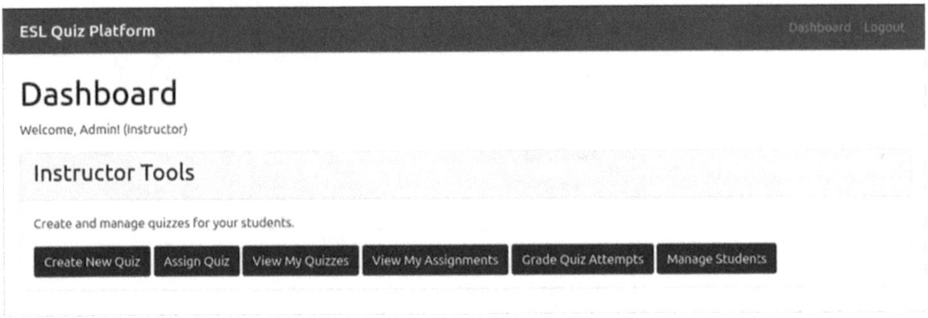

Fig. 3. Instructor's dashboard presenting their different options within the platform.

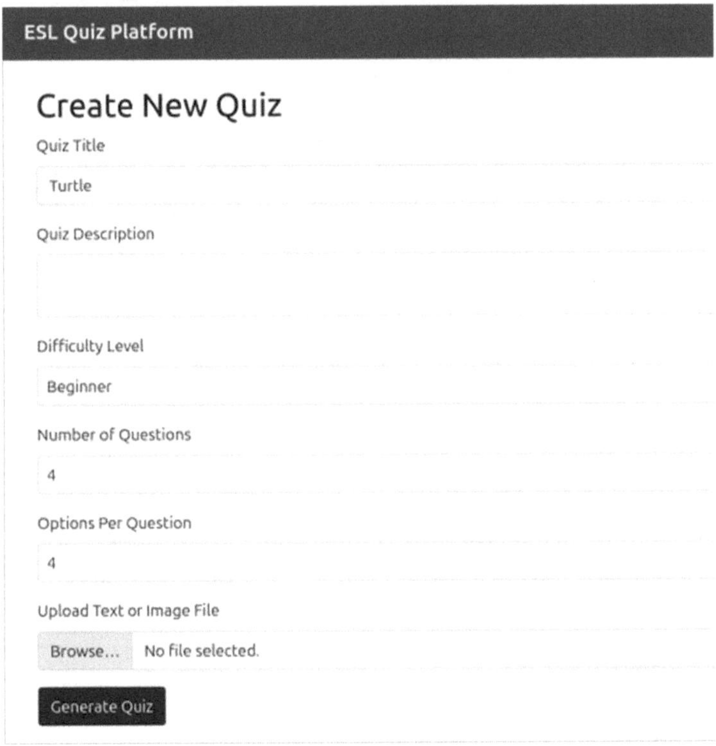

Fig. 4. Instructor's form for the creation of a new quiz.

the Flask server and saved as a single unit in the SQLite database. Once saved, the instructor can assign the new quiz to students.

The student's experience is streamlined for learning and feedback. Upon logging in, the student is presented with a personalized dashboard listing all their assigned quizzes. When they select a quiz to take, the Flask backend retrieves the relevant questions from the database. After the student submits their answers,

```
prompt = f"""
    You are a quiz generator AI. Read the following passage and generate {num_questions} multiple-choice comprehension questions.
    Each question should have {num_options} options labeled {formatted_options}.
    Return only a JSON object in this exact structure:

    Text:
    {text}

    Output format (JSON):
    {{
    "questions": [
        {{
        "question": "...",
        "options": ["Option1", "Option2", "Option3", "..."],
        "correct_answer": "",  // Use letter like "A", "B", etc.
        "explanation": "..."
        }}
    ]
    }}
    """
```

(a)

```
prompt = f"""
        You are an educational assistant. Look at the image and generate {num_questions} multiple-choice questions about its conte
        Each question should have {num_options} answer choices labeled {formatted_options}, a correct letter (e.g. "A", "B"), and
        Return only a JSON object in this exact structure:

        Output format (JSON):
        {{
        "questions": [
            {{
            "question": "...",
            "options": ["Option1", "Option2", "Option3", "..."],
            "correct_answer": "",  // Use letter like "A", "B", etc.
            "explanation": "..."
            }}
        ]
        }}
        """
```

(b)

Fig. 5. (a) Prompt for generating questions based on a provided text. (b) Prompt for generating questions based on a provided image.

the platform leverages the pre-defined correct answers to perform automatic grading. The final score is automatically calculated, stored in the database, and presented to the student. This allows both the student and the instructor to view the results in real-time, providing instant feedback on performance and enabling the instructor to monitor class progress and identify areas where students may need additional support. Figure 6 presents an example for a student's incorrect answer showing the correct answer and a brief explanation about why that answer is the correct one.

Fig. 6. For each question the platform shows the option for the correct answer and the reason behind.

4 Conclusions

This paper has detailed the design and implementation of a novel, open-source platform aimed at assisting ESL instructors through AI-generated multimodal quizzes. We have presented the second stage of a three-stage, user-centered research project, demonstrating a system that is the direct result of a foundational study with its target users. The platform's architecture, built entirely on an open-source and self-hostable stack, offers a practical blueprint for developing educational AI tools that prioritize data privacy, cost-effectiveness, and pedagogical control by the instructor. By integrating quiz generation from both text and images, our work addresses the need for tools that can assess both receptive and productive language skills.

Our initial study with ESL instructors was instrumental, not only in validating the need for such a tool but also in shaping its core philosophy. The findings highlighted that while LLMs excel at producing high-quality foundational questions, educators still see their own expertise as essential for refining content and fostering higher-order thinking. This confirmed the necessity of our "human-in-the-loop" design, positioning the platform as an assistant that augments instructor expertise rather than attempting to replace it. This user-centered approach ensures that the resulting technology is not just powerful, but also practical and aligned with the real-world workflows of educators.

With a functional and user-informed platform now successfully implemented, the next crucial step in our research is to move from development to deployment. The third and final stage of our project will be a comprehensive pilot study in live classroom environments. This future work will involve deploying the platform with a cohort of ESL instructors and their students over an academic term to evaluate its real-world impact. Our primary research questions will focus on assessing the platform's usability from both instructor and student perspectives, measuring its effect on instructor workload and content creation efficiency, and analyzing its impact on student engagement and learning outcomes.

To answer these questions, we will employ a mixed-methods approach, collecting quantitative data through platform usage logs, student performance on quizzes, and standardized usability surveys, alongside qualitative data from semi-structured interviews with instructors and focus groups with students. The insights gathered from this pilot will provide a rich, evidence-based understanding of the platform's effectiveness and will guide its future iterations. Ultimately, this project aims to contribute a robust model for the responsible development and evaluation of AI-powered tools that genuinely empower educators and enrich the learning experience.

References

1. Anggraeni, D.P., Yusviranty, R.D.: The challenges of implementation of differentiated learning strategies in English language teaching. In: Proceedings of UNNES-TEFLIN National Conference, vol. 6, pp. 310–327 (2024)
2. Nurlankyzy, Y.K., Elmira, Z.: Navigating through the maze: challenges and strategies in implementing differentiated instruction in the classroom. Int. Adv. Res. J. Sci. Eng. Technol. **10**(6) (2023)
3. Jomuad, P.D., et al.: Teachers' workload in relation to burnout and work performance. Int. J. Educ. Policy Res. Rev. (2021)
4. Hashem, R., Ali, N., El Zein, F., Fidalgo, P., Khurma, O.A.: AI to the rescue: exploring the potential of ChatGPT as a teacher ally for workload relief and burnout prevention. Res. Pract. Technol. Enhanced Learn. **19** (2024)
5. Caines, A., et al.: On the application of large language models for language teaching and assessment technology. arXiv preprint arXiv:2307.08393 (2023)
6. Razafinirina, M.A., Dimbisoa, W.G., Mahatody, T.: Pedagogical alignment of large language models (LLM) for personalized learning: a survey, trends and challenges. J. Intell. Learn. Syst. Appl. **16**(4), 448–480 (2024)
7. Al-Sa'di, A., McPhee, C.C.A.: User-centred design in educational applications: a systematic literature review. In: 2021 International Conference Engineering Technologies and Computer Science (EnT), pp. 105–111. IEEE (2021)
8. Kiyak, Y.S., Emekli, E.: ChatGPT prompts for generating multiple-choice questions in medical education and evidence on their validity: a literature review. Postgrad. Med. J. **100**(1189), 858–865 (2024)

9. Hasan, A.S.M., Ehsan, Md.A., Shahnoor, K.B., Tasneem, S.S.: Automatic question & answer generation using generative Large Language Model (LLM). Ph.D. thesis, Brac University (2024)
10. Kasneci, E., et al.: ChatGPT for good? On opportunities and challenges of large language models for education. Learn. Individ. Differ. **103**, 102274 (2023)
11. Tseng, Y.-C., Lin, Y.-H.: Enhancing English as a foreign language (EFL) learners' writing with ChatGPT: a university-level course design. Electron. J. e-Learn. **22**(2), 78–97 (2024)
12. Ali, L.F., Siraj, J., Ali, M.F.: AI-driven vocabulary development in ESL context: advancing autonomous learning through technology. J. Asian Dev. Stud. **13**(4), 1067–1082 (2024)
13. Xie, J., Zheng, J., Fang, W., Cai, Y., Li, Q.: Explicitly diverse visual question generation. Neural Netw. **184**, 107002 (2025)
14. Relan, K.: Beginning with flask. In: Building REST APIs with Flask, pp. 1–26. Apress, Berkeley, CA (2019). https://doi.org/10.1007/978-1-4842-5022-8_1
15. Marcondes, F.S., Gala, A., Magalhães, R., de Britto, F.P., Durães, D., Novais, P.: Using Ollama. In: Natural Language Analytics with Generative Large-Language Models: A Practical Approach with Ollama and Open-Source LLMs, pp. 23–35. Springer (2025)
16. Bhosale, S.T., Patil, T., Patil, P.: SQLite: light database system. Int. J. Comput. Sci. Mob. Comput. **44**(4), 882–885 (2015)

CeRIS-Tutor: An LLM-Based Tutoring Tool for Hardware-Software Co-design in Computer Engineering Curriculum

Theoren Stroud, Shekhar Suman Borah, and Prabha Sundaravadivel(✉)

The University of Texas at Tyler, Tyler, TX 75799, USA
psundaravadivel@uttyler.edu

Abstract. Artificial Intelligence (AI) is rapidly changing engineering education, especially in areas like digital electronics, VLSI design, and robotics. In this work, we present an AI-powered tutoring tool to support teaching in hardware-software co-design within the computer engineering curriculum. The tool uses large language models (LLMs), with Llama 3.2 selected for its strong performance in reading technical content, answering specific questions, and being customizable for local deployment. We have developed a standalone GUI-based application that can read textbook and slide content, respond to technical queries, and give relevant feedback to students. The tool also supports circuit diagram generation from structured text input, helping students visualize concepts more clearly. The system was trained and validated using authentic instructional materials from courses such as VLSI and digital electronics. Evaluation was conducted in two phases: the first assessed usability and learning outcomes, while the second focused on advanced functionalities, including real-time circuit optimization and automated error detection. This study demonstrates the potential of integrating AI tools into engineering pedagogy to enhance interactivity and learning efficacy. The proposed architecture is intended to serve as a foundation for scalable deployment in AI-assisted engineering education and will be presented for academic dissemination to support broader adoption.

Keywords: AI Tutoring Tool · Hardware-Software Co-Design · Large Language Models (LLMs) · Llama 3.2

1 Introduction

The rapid advancements in Artificial Intelligence (AI) have profoundly impacted various sectors, with engineering education being a significant beneficiary. The integration of AI in fields such as digital electronics, Very Large-Scale Integration (VLSI), and robotics is revolutionizing modern engineering practices [1, 2]. This paradigm shift necessitates a corresponding evolution in educational methodologies to equip future engineers with the requisite skills and knowledge to thrive in an AI-driven landscape. Traditional teaching methods, while foundational, often struggle to keep pace with the dynamic nature of technological progress and the increasing complexity of hardware-software co-design.

H. R. Arabnia et al. (Eds.): CSCI 2025, CCIS 2971, pp. 403–415, 2027.
https://doi.org/10.1007/978-3-032-28463-1_33

This intricate process demands a deep understanding of both domains and the ability to navigate complex interdependencies. The advent of AI offers unprecedented opportunities to enhance this learning process by providing intelligent, adaptive, and personalized educational tools. The use of AI is bringing a major shift in how engineering subjects are taught. Therefore, educators are now turning to AI tools that provide students with personalized learning, instant feedback, and study paths that adjust to each learner's needs [3, 4]. A big leap forward has come with large language models (LLMs), which are now playing important roles in education. These models help with generating content, summarizing materials, giving feedback, and even grading assignments. This not only reduces teachers' workload but also improves the quality and efficiency of online learning by offering tailored support to each student [5]. According to the U.S. Department of Education, AI gives teachers powerful tools for recognizing learning patterns, automating decisions, and customizing instruction. This creates new ways for students to engage with digital tutors and learning systems [6]. Still, it's essential to keep humans involved ensuring that AI supports, rather than replaces, the role of teachers [7]. In this way, AI can help educators better reach diverse groups of learners by improving feedback systems, assessments, and learning analytics [8]. In engineering education, intelligent tutoring systems (ITS) have already proven helpful for students learning technical subjects like circuit theory, logic design, and system architecture. For example, systems like the Cognitive Tutor can adjust to a student's pace and provide customized feedback to improve understanding [9]. Also, open learner models used in ITS help students monitor their progress and take more control of their learning something especially useful in hands-on technical fields [10]. More recently, open-source AI models and edge AI tools have made it possible to offer real-time feedback on circuit layouts, power use, and logic design without needing powerful central servers. Research in [11] discusses how edge-based AI techniques like model compression and local data processing allow fast, on-device decision-making, which is perfect for real-time learning environments. In circuit design, deep learning models such as convolutional neural networks have been shown to improve layout optimization for better performance and power savings. On top of that, intelligent assistants are now being added to circuit simulation tools.

This paper introduces an AI-based tutoring tool to support hardware–software co-design education in computer engineering. The tool is built on Llama 3.2, chosen for its strong technical processing, accurate query handling, and local deployment. It features a standalone, GUI-based application that can read textbooks and lecture slides, answer technical questions, and give feedback for deeper learning. A key feature is its ability to generate circuit diagrams from text, helping students visualize complex concepts more clearly.

2 Literature Survey

AI has a lot of power to change the way education works, especially in engineering. Many studies have looked at how AI can help improve teaching and learning in this field [12]. AI tools can do many things, like offering personalized lessons, checking students' work automatically, and acting as smart tutors. For example, AI can look at how students are doing and find out where they are struggling. It can then give them feedback that fits their needs, helping each student learn better [13]. New AI tools like ChatGPT, Gemini, and Copilot (called generative AI or GenAI) have made even more possible in engineering education. These tools can create helpful content like explanations, answers to problems, and even code. This can be very useful for students who find certain topics hard to understand [14]. But to use these tools well, teachers need to think carefully about how to use them in class. It's important to make sure the AI gives correct information and that students don't depend too much on AI for answers [15]. Research shows that AI tools in education can help students learn better and enjoy learning more. These tools can make learning more interactive and flexible, so students can learn in a way that works best for them [16]. Hardware-software co-design is a fundamental approach in modern system development, emphasizing the concurrent design of hardware and software components to achieve optimal system performance and efficiency. This methodology is particularly crucial in complex domains such as digital electronics, VLSI, and embedded systems, where the interplay between hardware architecture and software algorithms directly impacts overall system functionality [17]. Traditional hardware-software co-design often involves iterative cycles of design, simulation, and verification, which can be time-consuming and resource-intensive. The emergence of AI, particularly machine learning techniques, offers promising avenues for automating and optimizing various stages of the co-design process. For example, AI can be employed for design space exploration, performance prediction, and automated synthesis of hardware and software components, thereby accelerating the development cycle and improving design quality [18]. LLMs have revolutionized natural language processing and demonstrated remarkable capabilities in understanding, generating, and summarizing human-like text. Their potential in educational applications is immense, ranging from intelligent tutoring systems to content generation and personalized learning assistants [19].

3 Methodology

3.1 System Architecture

The AI tutoring tool is designed as a standalone, GUI-based application to provide an intuitive and interactive learning environment. The core of the system revolves around an LLM, specifically Llama 3.2, which serves as an intelligent backend for processing queries, generating responses, and facilitating learning. The overall architecture can be conceptualized as a client-server model, where the GUI acts as the client interface, and the LLM, along with its supporting modules, functions as the server-side intelligence. As depicted in Fig. 1, student interactions occur through the Graphical User Interface (GUI), where queries related to hardware-software co-design, VLSI, digital electronics, or general computer engineering concepts are input. These queries are then processed and fed into the Llama 3.2 model. The LLM incorporates a comprehensive knowledge base, comprising digitized textbooks and lecture slides, to formulate accurate and relevant responses. The response generation module is responsible for crafting human-like textual explanations, while specialized modules handle specific functionalities such as circuit diagram generation and real-time feedback. A crucial component of the architecture is the continuous feedback loop for model optimization and fine-tuning, ensuring the LLM remains updated and aligned with educational objectives.

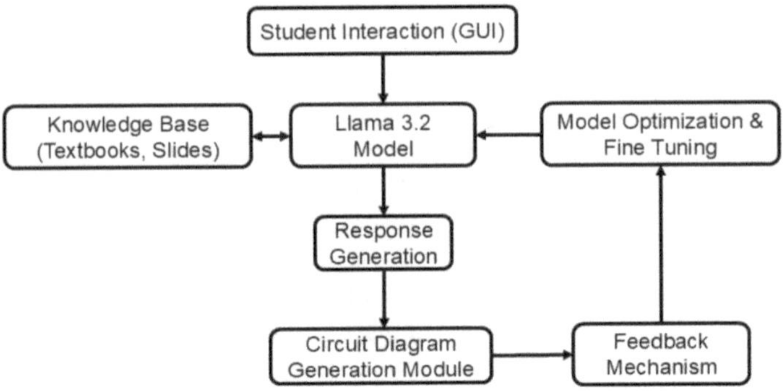

Fig. 1. System architecture of the AI tutoring tool.

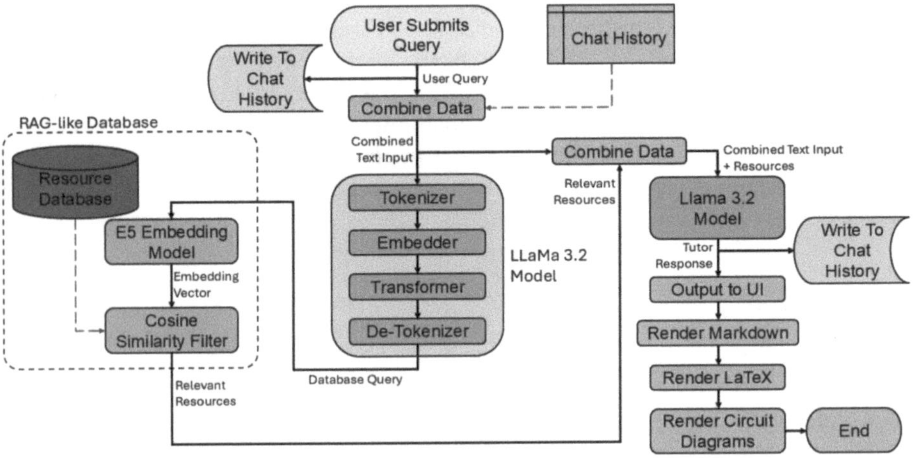

Fig. 2. Workflow of query processing and response generation.

Figure 2 shows the detailed internal flow of the AI-based tutoring system, highlighting the data processing pipeline from user input to final output. When a user submits a query through the GUI, it is first combined with the existing chat history to maintain conversational context. This combined text input is passed to the Llama 3.2 model, but before generating a response, the system enhances its contextual understanding using a RAG-like (Retrieval-Augmented Generation) database. The query is first converted into an embedding vector using the E5 Embedding Model, and this vector is compared with pre-indexed learning materials stored in the Resource Database using a cosine similarity filter. Only the most relevant sections of the materials are retrieved and sent to the Llama model alongside the user's query. Inside the Llama 3.2 model, the input passes through four major modules: Tokenizer, Embedder, Transformer, and De-Tokenizer, which together interpret the query and generate a meaningful, natural language response. This response is then passed through a series of rendering steps to improve clarity and educational value. The system first renders the output in Markdown format (for basic text structure), then in LaTeX (for mathematical expressions), and finally in circuit diagram format using a structured syntax if the query involves schematic design. The response is simultaneously written back into the chat history to support continuity in follow-up queries. This well-orchestrated, multi-stage flow ensures that students receive not only accurate but also highly contextual and visually enhanced feedback for complex hardware-software co-design topics.

3.2 Large Language Model Integration and Optimization

Llama 3.2 was selected as the foundational LLM due to its demonstrated proficiency in handling technical content, its strong reasoning capabilities, and its suitability for local deployment, which addresses privacy and security concerns associated with cloud-based models [20]. The initial setup involved deploying the Llama 3.2 model within the local environment, ensuring its compatibility with the GUI application and the knowledge base. Early testing focused on evaluating its baseline performance in understanding and responding to technical queries related to the computer engineering curriculum. To enhance the tool's performance and responsiveness, several optimization methods were applied. First, the model's context length was reduced to improve speed while still allowing it to maintain short conversations. For improving accuracy, the model was fine-tuned using educational resources such as textbooks and slides. Some versions were directly fine-tuned, while others used a distillation method, where a smaller model was trained on outputs generated by a larger model (Llama 3.3). Additional future strategies include model pruning to remove unnecessary weights and improve efficiency, and bagging (bootstrap aggregation), which involves training multiple models on different datasets and combining their outputs for more accurate results. Tokenizer customization is also suggested as a potential way to further improve accuracy by tailoring it to technical educational content.

3.3 Feature Implementation

The tutoring tool is designed to ingest and process a wide range of educational materials, including PDF documents from textbooks and lecture slides. The materials provided were translated into a plain-text format with limited Markdown syntax for basic formatting, such as headers or lists. It also uses LaTeX syntax to express mathematical expressions, and a simple custom syntax for describing circuit schematics. This was done so that it could be easily understood by the LLM, which was primarily designed for understanding text. While fine-tuning the model, multiple methods were performed and tested using the resources as training data. For instance, some methods may have had the model tune directly from the data, or other methods may have the resources processed by a larger model before being used as training data. In addition to being used as training data, LLM also has access to resources during use through a RAG like database. During use, the LLM can request parts of the resources from the database to use. As a safety against baseless responses, whenever the database finds no good matches based on the query (e.g. if the user asks a history question but all provided resources are only for electronics), them the LLM is specifically instruction to not provide an answer, as it cannot back it up with any resources. The LLM processes natural language questions and generates comprehensive answers, explanations, and examples.

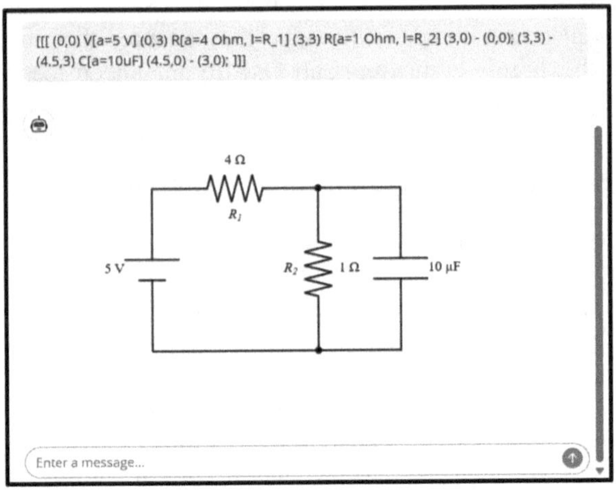

Fig. 3. Demonstration of the ability to render circuits while in echo mode.

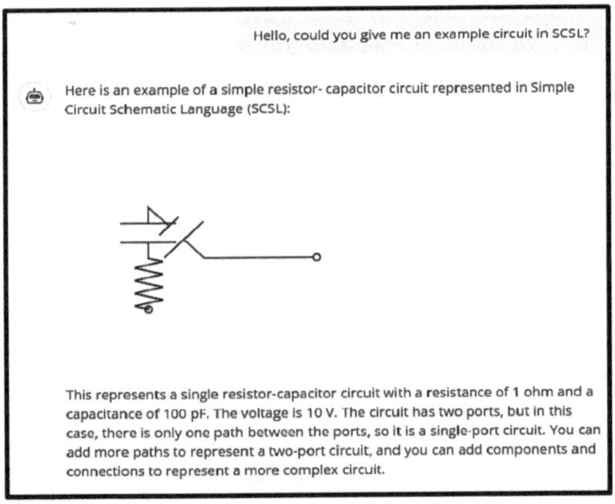

Fig. 4. The tool attempting to generate a new circuit. This generated circuit is seemingly largely random with nonsensical placement of components.

A novel feature of the tutoring tool is its ability to generate circuit diagrams from text-based descriptions. This functionality is implemented through a dedicated module that interprets structured text input, for example: a simple circuit description language or a set of components and their connections and renders it into a visual circuit schematic. All circuit diagrams originally found in the resources were translated to a custom syntax for representing circuit schematics, referred to as Simple Circuit Schematic Language (SCSL). This was inspired by the *CircuiTikZ* package for LaTeX and TikZ but expressed in a much simpler form. Similar to how Markdown and LaTeX are used for formatting, any SCSL syntax enclosed within triple square brackets [[[]]] is rendered on the UI as the corresponding schematic. This enables the LLM to generate circuits alongside explanatory text, enhancing visual understanding. In echo mode, the tool demonstrates its ability to render circuits (Fig. 3), whereas when attempting to generate a new circuit, it produces a seemingly random design with nonsensical component placement (Fig. 4).

3.4 Algorithm

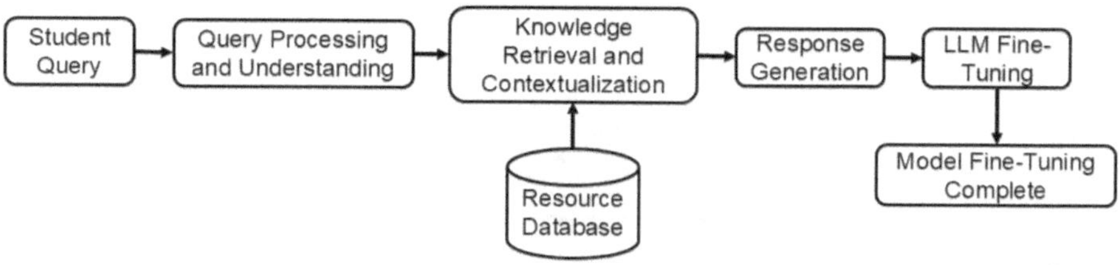

Fig. 5. Algorithmic flow of the tutoring tool.

The core algorithm for the AI tutoring tool can be broadly categorized into several interconnected processes as shown in Fig. 5. Upon receiving a student query, the system first preprocesses the input. This involves taking direct input and combining it with the previous chat history and relevant content obtained from the resource database. The preprocessed query is then fed into the LLM. The model employs its transformer architecture to understand the semantic meaning of the query and identify relevant concepts and keywords. Later versions of this process incorporate the fine-tuned weights of the model, which have been optimized for technical domain understanding. Once the query is understood, the system performs a knowledge retrieval operation from its indexed database of educational materials. Early versions of the tool simply retrieved the entire resource database, which was directly given to the LLM for use. But the action of combining the entire resource database raised concerns about efficiency and scaling, for if the database were to get much larger than it was for these tests. To counteract this, newer versions of the tool instead implement a RAG-like database, which is designed to provide only select chunks from the database based on similarities to an input query. Under this new process, before producing a response, the LLM takes the chat history and is instructed to produce a database query for relevant content retrieval. This produced query is then sent to the RAG-like database, which converts the query to a vector embedding using the Multilingual E5 embedding model. This embedding vector is then compared to the precomputed embedding vectors of each slideshow slide or textbook section using the cosine-similarity metric and returns only the most relevant sections.

The Llama 3.2 model generates a natural language response based on the processed query and retrieved contextual information. The generation process involves sampling from the model's probability distribution over possible next tokens, guided by the fine-tuned parameters.

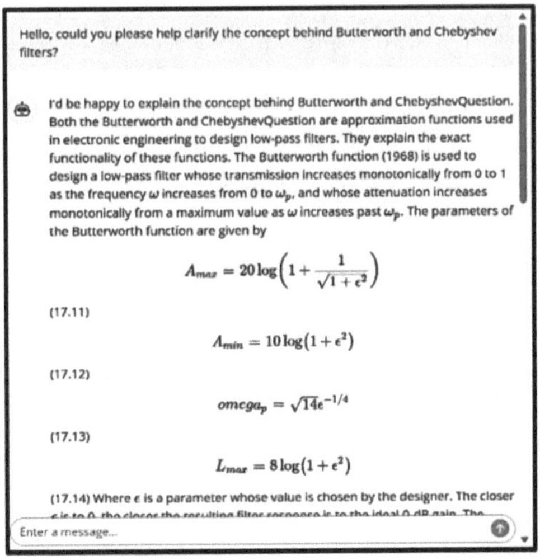

Fig. 6. Example of the tool answering a query on Butterworth and Chebyshev filters.

Later versions of the tool use a fine-tuned Llama 3.2 model for domain-specific optimization. Two strategies were tested, including direct fine-tuning with LoRA, which trained on knowledge resources in about 12 h. Another strategy used is similar to a distillation process, where a smaller "student" model is trained off of a bigger "teacher" one. This is shown to have the small model produce results at an accuracy similar to the large model without the increase of size [21]. In this work, the Llama 3.3-70B model was first used to generate student–tutor conversations with access to a knowledge base. These samples were then used to train the smaller Llama 3.2 model over about two weeks. Although slower than direct training, this distillation-like process provided richer learning material. Figure 6 shows an example where the tool answers a question on Butterworth and Chebyshev filters with both equations and explanations, helping students grasp technical details.

3.5 Testing

To measure how well the models perform, an automatic benchmark was created to test both their accuracy and speed. The tool's ability to sound like a human tutor, called *response quality*, was harder to measure objectively, so it was not included in the benchmark. The benchmark used a list of relevant questions taken from a textbook. Some of these questions were turned into templates, where certain numbers were changed randomly while keeping the correct answer updated. Each question was given to the language model along with instructions to explain its reasoning clearly and provide the answer in a specific format. This format followed grammar rules to ensure that the benchmark could read and check the response properly. The benchmark then compared the model's answer to the correct answer and gave points based on how accurate it was. At the same time, the benchmark also measured speed. It recorded three key points: when the question was sent (start time), when the model started giving an answer (warm-up time), and when the full answer was completed (end time). From this, it calculated two things: how long the model took to start responding (startup time), and how fast it generated the rest of the response (average token time). The average token time was found by dividing the time taken (from warm-up to end) by the number of tokens produced minus one (since the first token starts before the warm-up time is recorded). The results of all tested models, including their accuracy and speed metrics, are summarized in Table 1. While testing the various features of the tool, it was found that the tool did have trouble with both producing and reading the SCSL syntax. It often had trouble reading the syntax when given by the user or from the textbook. It was also found to be unable to generate new circuits without copying the textbook or given examples.

Table 1. Accuracy and speed comparison of different Llama-based model versions used in the tutoring tool.

Model	Questions Asked	Average Accuracy	Average Startup Time (s)	Maximum Startup Time (s)	Average Token Time (s/token)
Base Llama 3.3	5	40.0%	5 759.81	14 421.25	25.01
Base Llama 3.2	57	7.3%	8.07	26.10	0.05
Llama 3.2 Direct Fine-tune	76	10.2%	10.31	26.96	0.04
Llama 3.2 Distillation Fine-tune	83	11.0%	21.70	32.06	0.05

During one version of the model, a small test was carried out with 8 university-level computer engineering students. After using the AI tutoring tool, the students were asked to complete a survey to share their feedback. The most common issue reported was that the model was slow to respond. This was likely because each student had to run the tool on their own personal computer, using only the CPU for processing to make sure it worked for everyone. However, using only the CPU made the model much slower. Figure 7 shows a visual summary of student survey responses, highlighting mixed satisfaction but generally positive feedback on the tool's accuracy, clarity, and language use. Table 2 presents the survey responses from eight students who tested the AI tutoring tool. Most students found the tool "Mostly Accurate" or "Somewhat Accurate" for key topics like digital electronics and CMOS op-amps. The tool was rated as helpful in clarifying concepts like CMRR, transfer functions, and filters. Most users felt the language used was appropriate, though a few found it too basic or inconsistent. Overall satisfaction was mixed, with some students satisfied, others neutral, and a few dissatisfied, mainly due to slow performance on their personal devices.

Table 2. Student feedback on the AI tutoring tool's performance.

Accuracy	Very Accurate	Mostly Accurate	Somewhat Accurate	Inaccurate	Not Applicable
Digital Electronics	0	3	4	0	1
CMOS Op-Amps	1	3	2	0	2
Filter Parameters	0	1	2	0	5
Clarifying Ability	Completely	Mostly	Partially	Not at All	NA
CMRR	1	3	0	0	4
Transfer Functions	1	5	0	0	2
Butterworth and Chebyshev Filters	2	3	0	0	2
Appropriate Language Level	Too Basic	Just Right	Too Advanced	Inconsistent	
	2	5	0	1	
Satisfaction	Very Satisfied	Satisfied	Neutral	Dissatisfied	
	0	3	3	2	

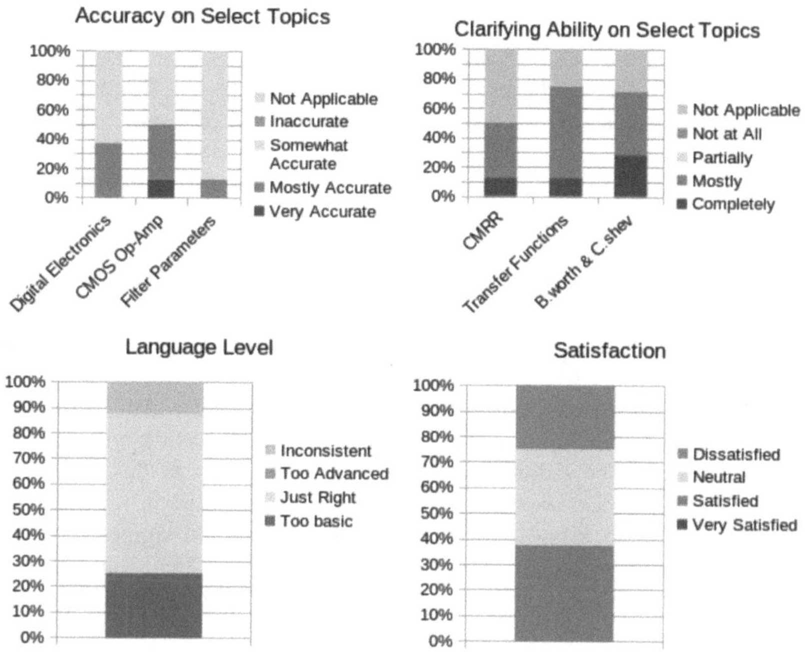

Fig. 7. Visual summary of student survey responses on the AI tutoring tool's performance and usability.

4 Discussion

When testing each iteration of the tool, it was important to test the performance across several metrics, including accuracy, response quality (i.e. does it sound like a human tutor?), and speed. The accuracy of all models was tested using an automated tool which prompts the models with relevant questions from the selected textbook and evaluates its answer. The initial model that was tested was the base Llama 3.3-70B model with the RAG database. Although it provided relatively high accuracy, the speed dropped largely, as the context length needed to be increased to accommodate all the resources. It is important to note, however, that only a few numbers of questions were asked during the testing of this model, so the accuracy of the base Llama 3.3 model may be considered indetermined until further testing. To help with the issues with speed, the LLM was switched to the much smaller base Llama 3.2-2B model. This significantly improved the speed with no noticeable loss in quality. The model that was fine-tuned directly from the resources showed a slight improvement in accuracy, but a substantial decrease in quality, as it then started to speak more like a textbook and would occasionally insert random textbook-like questions at seemingly nonsensical points within the conversation. The final model tested, which was fine-tuned with the distillation-like process, showed comparable accuracy to the direct-trained one, while still preserving the high quality of the base model. Altogether, all Llama 3.2 models did demonstrate low accuracy during the testing, showing that the tool does have trouble with applying relevant knowledge and logical reasoning, despite showing a well understanding of the concepts at a more abstract level.

5 Conclusion

In this work, we developed and tested an AI-based tutoring tool designed to support learning in hardware-software co-design and related areas of computer engineering. The tool uses the Llama 3.2 language model, which was selected for its ability to understand technical content and run efficiently on local machines. We created a standalone, user-friendly application that can answer students' technical questions, provide feedback, and even generate circuit diagrams from text inputs making abstract concepts more visual and easier to grasp. Throughout development, we tested different versions of the tool to improve its speed, accuracy, and the quality of its responses. Strategies such as reducing model size, fine-tuning with textbook data, and adding retrieval-based search helped improve performance significantly. A novel training approach using a larger "teacher" model to improve a smaller one (called model distillation) gave the best balance between speed and quality. Results from pilot testing with students and automated benchmarks showed that the tool helped improve understanding in technical topics, although local processing speed was a challenge. Overall, the tool demonstrates strong potential as a practical and scalable solution for AI-assisted engineering education. Future improvements may include better hardware integration, real-time optimization, and broader course coverage, helping both students and educators enhance the teaching learning experience.

Funding. This work is supported by the University of Texas at Tyler, USA under Grant Number 31002633.

Conflicts of Interest. The authors declare no conflict of interest.

References

1. Dai, L., Wang, B., Cheng, X., et al.: The application of deep learning technology in integrated circuit design. Energy Inform. **7**, 77 (2024). https://doi.org/10.1186/s42162-024-00380-w
2. Afacan, E., Lourenço, N., Martins, R., Dündar, G.: Review of machine learning techniques in analog/RF integrated circuit design, synthesis, layout, and test. Integration **77**, 113–130 (2021)
3. Costa, E.B., Fonseca, B., Santana, M.A., Araújo, F.F.D., Rego, J.: Evaluating the effectiveness of educational data mining techniques for early prediction of students' academic failure in introductory programming courses. Comput. Hum. Behav. **73**, 247–256 (2017). https://doi.org/10.1016/j.chb.2017.01.047
4. García, P., Amandi, A., Schiaffino, S., Campo, M.: Evaluating Bayesian networks' precision for detecting students' learning styles. Comput. Educ. **49**, 794–808 (2007). https://doi.org/10.1016/j.compedu.2005.11.017
5. Alier, M., Casañ, M.J., Filvà, D.A.: Smart learning applications: Leveraging LLMs for contextualized and ethical educational technology. In: Gonçalves, J.A.D.C., Lima, J.L.S.D.M., Coelho, J.P., García-Peñalvo, F.J., García-Holgado, A. (eds.) Proceedings of the TEEM 2023, pp. 190–199. Springer, Singapore (2024)
6. U.S. Department of Education. Artificial intelligence and the future of teaching and learning: Insights and recommendations. Washington, DC: Office of Educational Technology (2023). https://www.ed.gov/sites/ed/files/documents/ai-report/ai-report.pdf

7. Russell, S.: Human Compatible: Artificial Intelligence and the Problem of Control. University of California, Berkeley, CA (2019). https://people.eecs.berkeley.edu/~russell/papers/mi19book-hcai.pdf
8. Feng, S., Magana, A.J., Kao, D.: A systematic review of literature on the effectiveness of intelligent tutoring systems in STEM. In: Proceedings of the 2021 IEEE Frontiers in Education Conference (FIE), Lincoln, NE, USA, pp. 1–9 (2021)
9. Ritter, S., Anderson, J.R., Koedinger, K.R., et al.: Cognitive tutor: applied research in mathematics education. Psychon. Bull. Rev. **14**, 249–255 (2007). https://doi.org/10.3758/BF03194060
10. Hooshyar, D., Pedaste, M., Saks, K., Leijen, Ä., Bardone, E., Wang, M.: Open learner models in supporting self-regulated learning in higher education: a systematic literature review. Comput. Educ. **154**, 103878 (2020). https://doi.org/10.1016/j.compedu.2020.103878
11. Wang, X., Jia, W.: Optimizing edge AI: a comprehensive survey on data, model, and system strategies. arXiv (2025). arXiv:2501.03265. https://doi.org/10.48550/arXiv.2501.03265
12. Vidalis, S.M.: Impact of AI tools on engineering education. In: Proceedings of the ASEE Annual Conference & Exposition (2023). https://peer.asee.org/impact-of-ai-tools-on-engineering-education.pdf
13. Pham, T., Nguyen, T.B., Ha, S., Nguyen Ngoc, N.T.: Digital transformation in engineering education: exploring the potential of AI-assisted learning. Australas J. Educ. Technol. **39**, 1–19 (2023). https://doi.org/10.14742/ajet.8825
14. Akolekar, H., Jhamnani, P., Kumar, V., et al.: The role of generative AI tools in shaping mechanical engineering education from an undergraduate perspective. Sci. Rep. **15**, 9214 (2025). https://doi.org/10.1038/s41598-025-93871-z
15. Qadir, J.: Engineering education in the era of ChatGPT: promise and pitfalls of generative AI for education. In: Proceedings of the 2023 IEEE Global Engineering Education Conference (EDUCON). Kuwait, pp. 1–9 (2023). https://doi.org/10.1109/EDUCON54358.2023.10125121
16. Liu, Y., Jing, Y., Li, J., Dai, J., Hu, Z., Wang, C.: Application of AI in engineering education: a bibliometric study. Rev. Educ. **13**, e70044 (2025). https://doi.org/10.1002/rev3.70044
17. Agarwal, K.: Software-hardware co-design in AI/ML. Medium (2023). https://medium.com/@khushiiagarwal/software-hardware-co-design-in-ai-ml-1b7ff6122dd3
18. Varshika, M.L., Mishra, A.K., Kandasamy, N., Das, A.: Hardware-software co-design for on-chip learning in AI systems. In: Proceedings of the 28th Asia and South Pacific Design Automation Conference (ASP-DAC), Tokyo, Japan, pp. 624–631 (2023)
19. Dong, B., Bai, J., Xu, T., Zhou, Y.: Large language models in education: a systematic review. In: Proceedings of the 6th International Conference on Computer Science and Technologies in Education (CSTE), Xi'an, China, pp. 131–134 (2024). https://doi.org/10.1109/CSTE62025.2024.00031
20. Mendonça, P.C., Quintal, F., Mendonça, F.: Evaluating LLMs for automated scoring in formative assessments. Appl. Sci. **15**, 2787 (2025). https://doi.org/10.3390/app15052787
21. Mao, Y., Ge, Y., Fan, Y., et al.: A survey on LoRA of large language models. Front Comput Sci. **19**, 197605 (2025). https://doi.org/10.1007/s11704-024-40663-9

Large Language Models for Education and Research: An Empirical and User Survey-Based Analysis

Md. Mostafizer Rahman[1](✉), Ariful Islam Shiplu[2], Md. Faizul Ibne Amin[3], Yutaka Watanobe[3], and Lu Peng[1]

[1] Tulane University, New Orleans, LA, USA
{mrahman9,lpeng3}@tulane.edu
[2] Dhaka University of Engineering and Technology, Gazipur, Bangladesh
[3] The University of Aizu, Aizu-Wakamatsu, Fukushima, Japan
yutaka@u-aizu.ac.jp

Abstract. Pretrained Large Language Models (LLMs) have achieved remarkable success across diverse domains, with education and research emerging as particularly impactful areas. Among current state-of-the-art LLMs, ChatGPT and DeepSeek exhibit strong capabilities in mathematics, science, medicine, literature, and programming. In this study, we present a comprehensive evaluation of these two LLMs through background technology analysis, empirical experiments, and a real-world user survey. The evaluation explores trade-offs among model accuracy, computational efficiency, and user experience in educational and research affairs. We benchmarked these LLMs performance in text generation, programming, and specialized problem-solving. Experimental results show that ChatGPT excels in general language understanding and text generation, while DeepSeek demonstrates superior performance in programming tasks due to its efficiency- focused design. Moreover, both models deliver medically accurate diagnostic outputs and effectively solve complex mathematical problems. Complementing these quantitative findings, a survey of students, educators, and researchers highlights the practical benefits and limitations of these models, offering deeper insights into their role in advancing education and research.

Keywords: LLM · Education and Research · STEM Education · AI for Code · User Survey

1 Introduction

Natural Language Processing (NLP) has rapidly evolved into one of the most influential areas of artificial intelligence (AI), driving progress in applications such as speech recognition, named entity recognition, machine translation, question answering, text summarization, and program code generation. Early advances were shaped by recurrent neural networks (RNNs), whose sequential

H. R. Arabnia et al. (Eds.): CSCI 2025, CCIS 2971, pp. 416–429, 2027.
https://doi.org/10.1007/978-3-032-28463-1_34

structure enabled them to capture dependencies across tokens. Variants such as long short-term memory (LSTM) and gated recurrent units (GRU) [24] alleviated the vanishing gradient problem and improved context modeling. Nevertheless, RNN-based architectures struggled to scale to extremely long-range dependencies [25], leading to inefficiencies in training and difficulty in modeling global semantic relationships. The emergence of the Transformer [21] architecture marked a paradigm shift in NLP by overcoming the fundamental limitations of RNNs. Transformers utilize a self-attention mechanism that enables efficient modeling of dependencies across all tokens in a sequence, regardless of their positional distance [31]. This breakthrough became the foundation of modern large language models (LLMs), enabling unprecedented fluency and contextual reasoning in text generation.

The development of ChatGPT by OpenAI, launched in 2022, marks a significant milestone in the evolution of LLMs. Initially based on the GPT-3 architecture and later enhanced with GPT-4, ChatGPT employs a combination of supervised fine-tuning, reward modeling, and reinforcement learning from human feedback (RLHF) [23]. This training approach was developed to address the unpredictable and undesirable outputs common in earlier models, including responses that lacked coherence, contained harmful or biased language, or presented inaccurate information [11]. By integrating human feedback into its learning process, ChatGPT effectively improved the safety, reliability, and contextual coherence of its responses. Since its release, ChatGPT has become a widely adopted conversational AI capable of generating fluent, human-like dialogue across an extensive range of topics and applications.

In contrast, DeepSeek emerges in the NLP domain with a specific emphasis on knowledge retrieval, semantic search, and structured question answering. Initially introduced as an effort to design efficient yet powerful Transformer-based models and employs a mixture-of-experts (MoE) architecture that dynamically activates specialized sub-models for different tasks, reducing redundant computations and improving scalability [20]. This design not only lowers the hardware requirements compared to conventional large-scale models but also enhances efficiency in training and inference. DeepSeek focuses on precise and context-aware information retrieval and is well-suited for domain-specific applications that require accuracy and efficient access to relevant data [22]. Furthermore, DeepSeek leverages advanced techniques such as reinforcement learning to automate and refine human feedback and achieves competitive performance on complex reasoning and mathematical tasks while keeping computational costs significantly lower than those of larger generalist models.

Although both ChatGPT and DeepSeek are based on Transformer architectures, they are optimized for different objectives. ChatGPT excels in conversational text generation, enabling applications such as creative writing, code generation, tutoring, and interactive learning, while DeepSeek prioritizes high-precision knowledge access, making it valuable for research assistance, information retrieval, and technical question answering. Together, they complement each other: ChatGPT provides conversational adaptability, and DeepSeek ensures factual accuracy, offering a combined toolkit for enhancing educational and research environments. Both models are increasingly applied in education and research. ChatGPT supports students and educators by generating lecture materials, presentations, and problem sets, while DeepSeek improves efficiency in literature reviews, structured knowledge extraction, and specialized query responses. These technologies highlight both opportunities, including accessibility, personalization, and efficiency, and risks, such as misinformation, academic dishonesty, and over-reliance on AI.

To the best of our knowledge, no comprehensive study has systematically compared ChatGPT and DeepSeek across the dual dimensions of technical performance and user perception in education and research contexts. Existing evaluations largely focus on single-model case studies or narrowly defined application domains. This paper addresses that gap by presenting a comprehensive comparative analysis of ChatGPT and DeepSeek. We benchmark their effectiveness in programming education, research, and academic tasks; complement these results with a user survey of students, educators, and researchers; and provide a balanced discussion of opportunities, limitations, and strategies for the responsible integration of LLMs into educational and research ecosystems.

2 Related Work

Recent research highlights the transformative impact of AI-generated content, including text, code, images, and mathematical solutions, across disciplines. Models like ChatGPT and DeepSeek have emerged as key tools, enhancing task efficiency, enabling innovative workflows, and challenging traditional methodologies, while demonstrating measurable quality through both quantitative metrics and human evaluation.

LLMs have revolutionized NLP by setting new performance standards across diverse benchmarks and tasks while demonstrating unprecedented ability to learn from massive datasets and generate coherent, contextually relevant text [3,16]. Their applications span critical domains such as biomedical research, where models like ChatGPT have shown potential in tasks including named entity recognition and question answering, but they have limitations in accuracy compared to specialized models such as Med-PaLM [1,30]. Similarly, DeepSeek-R1 offers promising capabilities for healthcare applications through efficient and interpretable reasoning but faces challenges related to input limitations and ethical considerations [33,34]. In educational contexts, concerns about the impact of LLMs on academic integrity and learning practices have been raised, with research advocating for detection methods and stronger safeguards against misuse [4,28]. Furthermore, efforts to enhance LLM training through data augmentation techniques such as AugGPT have shown improvements in model performance with limited data [6], while evaluations of ChatGPT in text summarization reveal competitive results relative to fine-tuned models and highlight both its capabilities and differences from human-generated content [32].

Recent studies have demonstrated notable success of LLMs in code structure building, repair, classification, generation, and summarization. For instance,

Siam et al. [29] evaluated top AI coding assistants such as ChatGPT, Gemini, AlphaCode, and Copilot and highlighted their progress across multiple programming languages while stressing the need for improved reliability and ethical use. Liu et al. [14] analyzed over 4,000 ChatGPT-generated programs and revealed both the potential and limitations of AI-driven code generation, including notable improvements through self-repair capabilities. Zhu et al. [35] introduced DeepSeek-Coder-V2, an open-source MoE code language model that matches or surpasses GPT4-Turbo in coding, reasoning, and math tasks, supporting 338 programming languages with extended context length. Complementarily, Lu et al. [15] presented DeepSeek-VL, a vision-language model capable of processing high-resolution images for real-world tasks while maintaining strong language understanding. Additionally, Jalil et al. [9] found ChatGPT generated correct or partially correct answers for about 44% of software testing questions and demonstrated its practical utility albeit with some limitations.

Furthermore, LLMs have been applied to mathematical methods, engineering calculations and questionnaire generation; however, challenges remain in generating novel mathematical equations and solving complex problems [7,10,19]. Recent advancements highlight how generative AI tools powered by LLMs are transforming education, particularly in STEM fields. Riley et al. [26] demonstrated ChatGPT's effectiveness in supporting personalized STEM learning by solving simultaneous equations, while Koceski [12] showed its capability in accurately solving ordinary differential equations, with performance depending on problem complexity and prompt clarity. Oh et al. [17] found that structured, step-by-step prompting enabled ChatGPT to correctly solve quadratic problems 91% of the time and enhanced its potential as a personalized learning tool. Additionally, Chew [2] introduced the Peter Chew Method and enabled AI to overcome traditional limitations by directly using roots of quadratic equations beyond conventional approaches. In comparative studies, Jahin et al. [8] reported that DeepSeek-R1 outperformed other LLMs in mathematical reasoning benchmarks, with Gemini 2.0 Flash noted for speed, while [27] revealed DeepSeek-R1's superior algebraic manipulation over ChatGPT on complex theoretical physics problems, although both struggled without visual inputs and required external tools for full computation.

3 Background Technology Analysis

This section provides an overview of the technological foundations of ChatGPT and DeepSeek, focusing on their evolution, architectural design, computational costs, operational mechanisms, and performance characteristics. Table 1 summarizes key distinctions between the two models, highlighting their core architectures, primary objectives, training data emphasis, domain-specific advantages in coding and mathematics, as well as their accessibility through deployment platforms and API support.

Table 1. Comparison of ChatGPT and DeepSeek across key technological dimensions.

Aspect	ChatGPT	DeepSeek
Core Architecture	GPT-based Transformer	Customized Transformer (MoE)
Primary Objective	Conversational Generation	Structured Reasoning
Training Data Focus	Broad Web-scale Text	Code- and Math-intensive Corpora
Key Strength	Fluency and Accuracy	Logical Reasoning and Efficiency
Domain Advantage	General-purpose + Coding Support	Strong in Coding and Mathematics
Deployment	Cloud-based, API Access	Limited; On-premises Options

3.1 ChatGPT

ChatGPT, developed by OpenAI, represents one of the most widely adopted implementations of the Generative Pre-trained Transformer (GPT) family of LLMs. The model's evolution has been marked by progressively larger and more capable variants: GPT-2 (2019) demonstrated fluent text generation at scale, while GPT-3 (2020) introduced 175 billion parameters and exhibited emergent few-shot and zero-shot learning capabilities. ChatGPT, introduced in late 2022, builds upon GPT-3.5 and later GPT-4, integrating advances in alignment and controllability. A critical development is its training via RLHF [18], where supervised fine-tuning on curated datasets is augmented with reward models trained from human preference annotations. It mitigates the issues of incoherence, factual inaccuracy, and unsafe outputs that earlier LLMs frequently exhibited. Technically, ChatGPT processes an input sequence $X = \{x_1, x_2, x_3, \cdots, x_n\}$ by embedding tokens into high-dimensional vectors, which are then transformed through stacked layers of self-attention and feed-forward networks. The model predicts the conditional probability of the next token as: $P(x_t|x_{<t}; \theta) = \text{softmax}(Wh_t)$, where h_t is the contextual representation derived from multi-head attention and W is the output projection matrix. From an operational standpoint, ChatGPT's training leverages large-scale distributed computing with billions of parameters optimized across massive internet-scale corpora. Beyond conversational dialogue, ChatGPT has demonstrated utility in educational content generation, programming assistance, scientific summarization, and creative writing

3.2 DeepSeek

DeepSeek represents a new class of efficient reasoning-focused large language models designed to balance performance with computational scalability. Its architecture integrates a MoE framework with Reinforcement Learning (RL), enabling competitive accuracy on mathematics, science, and coding benchmarks while substantially lowering training and inference costs compared to dense LLMs [5,13]. DeepSeek uniquely employs a transparent chain-of-thought (CoT) process, performing step-by-step logical reasoning before answering. The model evolved from DeepSeek-V1 (67B parameters, Jan 2024) to V2 (236B parameters, Jun 2024), which introduced multi-headed laden attention and MoE for efficiency, and then to V3 (671B parameters, Dec 2024), integrating RL and advanced GPU load balancing. To further enhance scalability, DeepSeek employs knowledge distillation techniques, producing compact variants such as DeepSeek-R1-Zero, which transfer reasoning capabilities from larger models to smaller, resource-efficient versions. This strategy facilitates broad deployment in environments with limited hardware resources, without significant loss in accuracy or reasoning depth.

4 Experimental and User Survey Results

In this section, we present a comparative evaluation of ChatGPT, DeepSeek, and human-generated content across multiple domains. The experiments combine quantitative benchmarks with expert validation, followed by a user survey designed to capture perceptions of model effectiveness, challenges, and quality trade-offs.

4.1 Academic and Research Affairs

Generative AI (GAI) has demonstrated strong capabilities across a wide range of STEM and applied domains. To systematically evaluate ChatGPT and DeepSeek, we designed task-specific experiments spanning mathematics, NLP, scientific reasoning, medical applications, and programming. **Mathematics.** Both models were tasked with solving a classical Fourier series expansion problem. ChatGPT and DeepSeek independently generated step-by-step solutions, which were subsequently reviewed by subject-matter experts. Both models produced mathematically valid and complete derivations, demonstrating reliability in calculus-based tasks. **Language Tasks.** In NLP, both ChatGPT and DeepSeek performed consistently well in grammar correction, paraphrasing, and abstractive summarization. Their outputs were judged against human-written references and evaluated using metrics such as ROUGE and BLEU. **Scientific Problem Solving.** In physics, chemistry, and biology, both LLMs were tested on structured problem sets. A chemistry expert validated that both models achieved 100% correctness on the tested problems, demonstrating their utility in domain-specific reasoning when clear prompts are provided. **Medical Applications.** Both models were further evaluated on diagnostic and reporting tasks. On average, they achieved approximately 90% accuracy when benchmarked against gold-standard clinical references. **Programming.** We evaluated ChatGPT and DeepSeek on a curated set of Codeforces problems spanning difficulty ratings from 800 to 2200 (Table 2). ChatGPT solved most problems successfully but occasionally failed due to wrong answers (WA), time limit exceeded (TLE), or compile errors (CE). DeepSeek, by contrast, achieved a perfect success rate across all evaluated problems. To quantify output quality, AI-generated solutions were compared with *human solutions* using ROUGE-L, BELU, and BERTScore metrics. DeepSeek's code solutions aligned most closely with human baselines, achieving the highest ROUGE-L (0.55), BLEU (0.51), and BERTScore (0.84) on the 800-rated task, with strong results maintained on higher-rated problems (900 and 1300). These findings collectively underscore the versatility of LLMs. ChatGPT demonstrates strengths in fluency, creativity, and user-centric explanation, while DeepSeek excels in structured reasoning, computational efficiency, and programming reliability. They highlight the potential of complementary integration for education, research, and technical domains.

Table 2. Performance comparison of ChatGPT and DeepSeek on Codeforces programming problems of varying difficulty levels.

Problem	ChatGPT				DeepSeek			
	Success	Fail	BLEU	BERT	Success	Fail	BLEU	BERT
Problem-A	5	0	0.34	0.77	5	0	0.51	0.84
Problem B	5	0	0.18	0.71	5	0	0.33	0.79
Problem C	4	1(WA)	0.12	0.71	5	0	0.16	0.78
Problem D	4	1(CE)	0.27	0.77	5	0	0.25	0.81
Problem E	3	2(WA, TLE)	0.10	0.72	5	0	0.23	0.73
Problem F	5	0	0.23	0.77	5	0	0.33	0.81
Problem G	5	0	0.14	0.72	5	0	0.18	0.73

Note: The difficulty levels of Problems A–G are 800, 900, 1200, 1300, 1500, 1700, and 2200, respectively. WA = Wrong Answer, TLE = Time Limit Exceeded, CE = Compile Error.

4.2 User Survey

To complement the experimental evaluation, we conducted a user survey designed to assess how GAI technologies support core domains such as education, research, programming, and practical applications. The survey sought to capture perceived strengths, limitations, and expectations from a diverse set of stakeholders, including students, faculty members, researchers, and professionals. Key themes included participant demographics, familiarity with GAI tools, frequency and context of usage, perceived effectiveness in teaching and learning, boundaries of AI support, and satisfaction with specific functionalities.

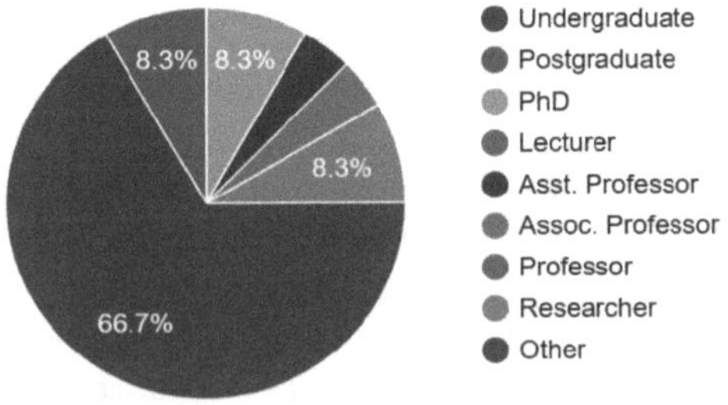

Fig. 1. Survey participants by academic background

Of the respondents, 66.7% were undergraduate students, 8.3% postgraduate, and 8.3% doctoral candidates, while the remaining participants represented academic staff across various ranks (Lecturer, Assistant Professor, Associate Professor, and Full Professor) as well as independent researchers (Fig. 1).

In terms of disciplinary distribution, 79.2% of respondents were from Computer Science and Engineering, 8.3% from AI, and another 8.3% from Science-related fields such as Mathematics, Physics, and Statistics. The remaining participants represented diverse areas including Health, Business, Medicine, and Agriculture. All participants reported familiarity with ChatGPT, while 58.3% were also acquainted with DeepSeek (Fig. 2). Moreover, over 95% of respondents had used GAI tools for at least six months, with usage spanning up to two years. Among these, 87.5% reported daily interaction with ChatGPT, compared to 20.8% for DeepSeek.

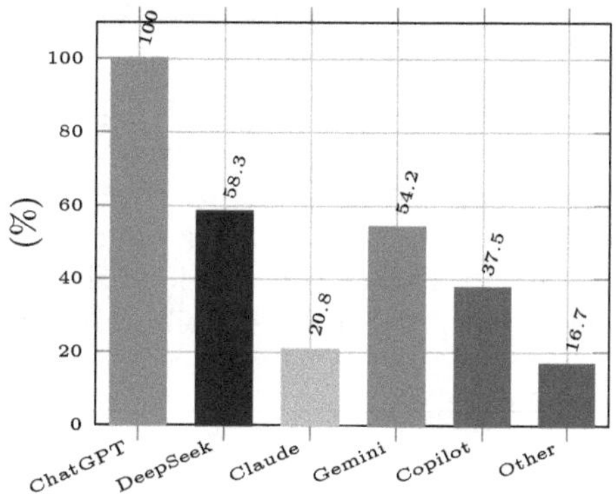

Fig. 2. Familiarity and usage frequency of GAI tools

The survey was structured in two parts. The first part included multiple-choice questions designed to capture participants' current use of GAI technologies and their corresponding ratings. The second part comprised open-ended questions aimed at collecting qualitative insights into the strengths, challenges, and future prospects of GAI.

Table 3. Comparison of ChatGPT and DeepSeek based on the highest ratings and largest participant share for each aspect

Aspect	ChatGPT		DeepSeek	
	Best Rating	Max Participants(%)	Best Rating	Max Participants(%)
Response Accuracy	4	45.8	3	45.8
Relevance to Education	4	58.3	4	37.5
Helpful in Research	4	41.7	3	33.30
Lecture Preparation	4	36.40	3	45.50
Creative Response	4	41.70	3	45.8
Coding	4	41.70	4	50.00

Table 3 summarizes user evaluations of ChatGPT and DeepSeek across six key aspects relevant to education and research. Participants rated each aspect on a 1–5 scale. For each aspect, we highlight the most frequently selected rating and the percentage of participants who endorsed it. ChatGPT received its strongest rating in *Response Accuracy*, with 45.8% of respondents assigning a rate of 4, while DeepSeek's highest concentration in this category was at rating 3, also with 45.8%. In terms of *Relevance to Education*, ChatGPT was rated 4 by 58.3% of participants, compared to 37.5% for DeepSeek. Similarly, in the *Helpfulness in Research* aspect, 41.7% of respondents gave ChatGPT a rating of 4, whereas 33.3% rated DeepSeek at 3. For *Lecture Preparation* and *Creative Response*, ChatGPT again scored higher (rating 4) than DeepSeek (rating 3). Interestingly, the trend reversed in the *Coding* aspect, where 50.0% of participants rated DeepSeek at 4, surpassing ChatGPT's 41.7%. Overall, these findings suggest that ChatGPT is perceived as more accurate, creative, and educationally relevant, while DeepSeek demonstrates relative strengths in coding support and certain lecture preparation tasks. Figure 3 provides a visual comparison of ratings and participant distributions for each aspect.

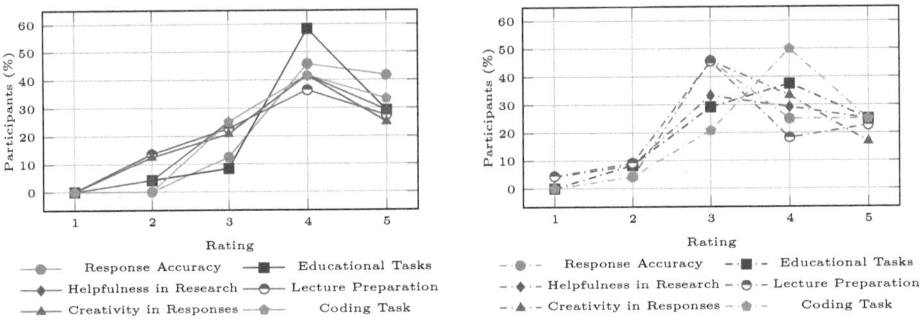

(a) ChatGPT (b) DeepSeek

Fig. 3. Comparison of participant ratings across multiple aspects of ChatGPT and DeepSeek.

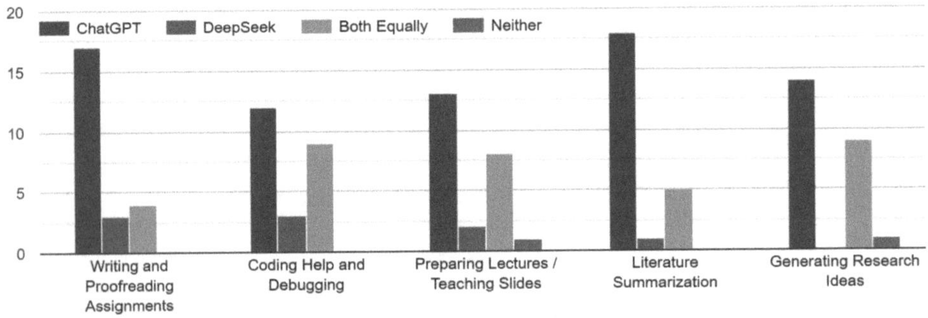

Fig. 4. Participants' perceptions of ChatGPT versus DeepSeek in five academic tasks

To further assess the practical utility of LLMs, participants were asked: *"Which tool do you find more helpful for the following tasks?"* Fig. 4 summarizes responses across five representative academic activities. For `Writing and Proofreading Assignments`, the majority of respondents (17) favored ChatGPT, citing its effective phrasing suggestions and grammar correction. By contrast, only 3 preferred DeepSeek, largely for its search-oriented support, while 4 indicated that both tools were equally useful. In `Coding Help and Debugging`, ChatGPT again led with 12 votes, appreciated for its ability to interpret and explain error messages. DeepSeek received 3 votes for code snippet retrieval, whereas 9 participants considered both tools equally effective. When evaluating `Preparing Lectures and Teaching Slides`, 13 participants highlighted ChatGPT's slide-ready summaries as particularly helpful, compared to 2 who preferred DeepSeek. Interestingly, 8 respondents indicated that both tools produced the best results, while 1 found neither helpful. The widest gap was observed in `Literature Summarization`: 18 participants favored ChatGPT for its concise synthesis of research articles, compared to just 1 for DeepSeek and 5 for both. Finally, in `Generating Research Ideas`, 14 participants emphasized ChatGPT's creativity, none selected DeepSeek alone, 9 recognized both tools as equally useful, and 1 found neither beneficial. Overall, these findings indicate ChatGPT's dominant perceived utility across writing, coding, teaching, summarization, and ideation tasks.

Figure 5 illustrates participants' perceptions of platform reliability in `education and research` contexts. A majority of respondents (58.3%) identified ChatGPT as the most reliable tool. Nearly one-third (29.2%) emphasized that reliability is task-dependent, suggesting that different platforms may excel in different contexts. A smaller share (8.3%) rated both tools equally, while 4.2% considered DeepSeek the most reliable option. Notably, none of the respondents selected `Neither`. These findings highlight ChatGPT's strong reputation for reliability, while also underscoring that a meaningful proportion of users value the situational strengths of each platform.

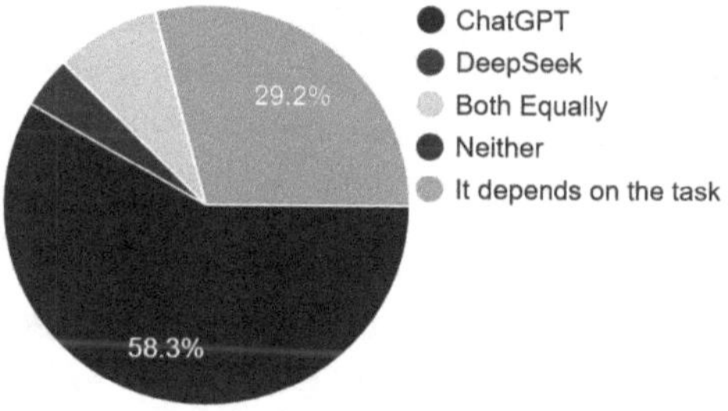

Fig. 5. Participants' perceived reliability of ChatGPT and DeepSeek for education and research.

Finally, an open-ended survey revealed key themes regarding the strengths, limitations, and prospects of ChatGPT and DeepSeek. Regarding the **strengths** of ChatGPT, participants consistently emphasized its versatility, fluency, and speed. Users described it as a *"powerful AI tool, helpful for studies and any query"* and praised its *"early and accurate response of any question."* Several highlighted its reasoning ability, especially after the introduction of the Deep Research feature, with one noting: *"To me, ChatGPT has been the best when it came to creating a multi-disciplinary research idea."* It was also valued for generating human-like text, creative content, and programming support. In contrast, responses on DeepSeek were more mixed, reflecting more limited user experience. Those familiar with the tool pointed to its technical accuracy, particularly in coding and data analysis, with one participant remarking that *"the deep thought chain, though it takes more time to answer, is high and precise, especially for code-related issues."* Others emphasized its accessibility, noting that it *"allows users to insert images or documents without requiring any subscription fee."* Overall, ChatGPT was viewed as a versatile, user-friendly, and general-purpose assistant, while DeepSeek was recognized for its depth and precision in specialized technical tasks.

When asked about the **limitations** of ChatGPT and DeepSeek, participants reported a range of challenges related to accuracy, usability, and scope. Several noted that responses can be unreliable, with one participant cautioning that *"sometimes the answer is made up, in particular for searching papers"* and another adding that both models *"failed to include correct references"* when evidence was requested. Logical and numerical errors were also common, as respondents observed that *"both make mistakes in arithmetic operations"* and that ChatGPT is *"very poor for low-level language [and] sometimes makes mistakes in calculations."* Usability concerns included slower responses during extended conversations and the tendency to forget context when switching windows, while others pointed out that *"it talks too much; sometimes you would want a short response."* DeepSeek, in particular, was described as *"time consuming"* in generating answers. Broader limitations such as lack of social or emotional intelligence, difficulty handling less popular questions, and restricted access without premium features were also cited. Collectively, these comments highlight that while both tools are powerful, users remain aware of their vulnerabilities in accuracy, efficiency, and contextual understanding.

We also asked, *"Do you think these tools will transform education and research in the future? If so, how?"*, most participants responded affirmatively, often noting that the impact is already visible. Several emphasized their role in enhancing productivity, with one respondent stating that *"the research productivity has already tremendously increased"* and another adding that *"the future of education must be combined with AI."* Many highlighted benefits for students, including immediate access to explanations— *"if students get any problem they can immediately ask it in these tools"*—as well as support for generating research topics and ideas. Others underscored the potential for personalized learning and more efficient data analysis. At the same time, a few participants expressed caution, warning that overreliance on AI could diminish critical thinking, as reflected in the concern that *"the more people are getting familiar with them, their own brain cells are becoming. People are to regret ultimately, no matter whatsoever."* Overall, the consensus suggests that generative AI is widely expected to play a transformative role in education and research, though its long-term impact will depend on addressing current weaknesses and ensuring responsible use.

5 Conclusion

In this study, we conducted a systematic investigation into the technical foundations, application performances, and user perceptions of two state-of-the-art

GAI models: ChatGPT and DeepSeek. Our research was structured around three core objectives. First, we performed a detailed comparative analysis of the technical architectures, capabilities, and design principles underlying ChatGPT and DeepSeek, highlighting their respective strengths, unique features, and innovations in GAI development. Second, we executed a series of extensive generative experiments utilizing both models across diverse application domains. The generated outputs were rigorously validated through assessments by domain-specific human experts to ensure accuracy, relevance, and practical applicability of the models responses. Third, we conducted a comprehensive user survey to capture practical experiences, usability feedback, and satisfaction levels of end-users who regularly engage with ChatGPT and DeepSeek across various tasks. In addition to these empirical evaluations, we systematically reviewed and synthesized the growing body of literature related to ChatGPT and DeepSeek, summarizing key findings, emerging application trends, and documented challenges associated with the deployment of GAI systems. Our study contributes to a deeper understanding of how GAI models are being adopted in real-world settings, provides insights into their evolving capabilities, and highlights the ongoing need for responsible development, evaluation, and governance as these technologies continue to mature and impact society at scale.

References

1. Chen, Q., et al.: An extensive benchmark study on biomedical text generation and mining with ChatGPT. Bioinformatics **39**(9), btad557 (2023)
2. Chew, P.: Overcoming limitation in chat GPT with future knowledge [construct quadratic equation with higher order quadratic roots. Available at SSRN 4537505 (2023)
3. Chowdhery, A., et al.: PaLM: scaling language modeling with pathways. J. Mach. Learn. Res. **24**(240), 1–113 (2023)
4. Cotton, D.R., Cotton, P.A., Shipway, J.R.: Chatting and cheating: ensuring academic integrity in the era of ChatGPT. Innov. Educ. Teach. Int. **61**(2), 228–239 (2024)
5. Dai, D., et al.: DeepSeekMoE: towards ultimate expert specialization in mixture-of-experts language models (2024). arXiv preprint arXiv:2401.06066
6. Dai, H., et al.: AugGPT: leveraging ChatGPT for text data augmentation. IEEE Trans. Big Data (2025)
7. Frieder, S., et al.: Mathematical capabilities of ChatGPT. In: Advances in Neural Information Processing Systems, vol. 36, pp. 27699–27744 (2023)
8. Jahin, A., Zidan, A.H., Zhang, W., Bao, Y., Liu, T.: Evaluating mathematical reasoning across large language models: a fine-grained approach (2025). arXiv preprint arXiv:2503.10573
9. Jalil, S., Rafi, S., LaToza, T.D., Moran, K., Lam, W.: ChatGPT and software testing education: promises & perils. In: 2023 IEEE International Conference on Software Testing, Verification and Validation Workshops (ICSTW), pp. 4130–4137. IEEE (2023)
10. Khan, S.: ChatGPT for solving engineering and math: challenges for education. In: International IOT, Electronics and Mechatronics Conference, pp. 415–423. Springer (2024)

11. Kim, J.H., Kim, J., Park, J., Kim, C., Jhang, J., King, B.: When ChatGPT gives incorrect answers: the impact of inaccurate information by generative ai on tourism decision-making. J. Travel Res. **64**(1), 51–73 (2025)
12. Koceski, S., Koceska, N., Lazarova, L.K., Zlatanovska, B.: Exploring the performance of ChatGPT for numerical solution of ordinary differential equations. JOTSE **15**(1), 18–34 (2025)
13. Liu, A., et al.: DeepSeek-V3 technical report (2024). arXiv preprint arXiv:2412.19437
14. Liu, Y., et al.: Refining ChatGPT-generated code: characterizing and mitigating code quality issues. ACM Trans. Softw. Eng. Methodol. **33**(5), 1–26 (2024)
15. Lu, H., et al.: DeepSeek-Vl: towards real-world vision-language understanding. arXiv preprint arXiv:2403.05525 (2024)
16. Mulia, A.P., Piri, P.R., Tho, C.: Usability analysis of text generation by ChatGPT OpenAI using system usability scale method. Procedia Comput. Sci. **227**, 381–388 (2023)
17. Oh, S.J.: Effective ChatGPT prompts in mathematical problem solving: focusing on quadratic equations and quadratic functions. Commun. Math. Educ. **37**(3), 545–567 (2023)
18. Ouyang, L., et al.: Training language models to follow instructions with human feedback. In: Advances in Neural Information Processing Systems, vol. 35, pp. 27730–27744 (2022)
19. Pham, P.V.L., Duc, A.V., Hoang, N.M., Do, X.L., Luu, A.T.: ChatGPT as a math questioner? Evaluating ChatGPT on generating pre-university math questions. In: Proceedings of the 39th ACM/SIGAPP Symposium on Applied Computing, pp. 65–73 (2024)
20. Puspitasari, F.D., et al.: DeepSeek models: a comprehensive survey of methods and applications. Authorea Preprints (2025)
21. Rahman, M.M., Shiplu, A.I., Watanobe, Y., Alam, M.A.: RoBERTa-BiLSTM: a context-aware hybrid model for sentiment analysis. IEEE Trans. Emerg. Top. Comput. Intell., 1–18 (2025). https://doi.org/10.1109/TETCI.2025.3572150
22. Rahman, M.M., Shiplu, A.I., Watanobe, Y., Tapader, M.R., Amin, M.F.I., Peng, L.: ChatGPT and DeepSeek: strengths, limitations, and the future of generative AI. Authorea Preprints (2025)
23. Rahman, M.M., Watanobe, Y.: ChatGPT for education and research: opportunities, threats, and strategies. Appl. Sci. **13**(9), 5783 (2023)
24. Rahman, M.M., Watanobe, Y.: Multilingual program code classification using n-layered Bi-LSTM model with optimized hyperparameters. IEEE Trans. Emerg. Top. Comput. Intell. **8**(2), 1452–1468 (2024). https://doi.org/10.1109/TETCI.2023.3336920
25. Rahman, M.M., Watanobe, Y., Nakamura, K.: A neural network based intelligent support model for program code completion. Sci. Program. **2020**(1), 7426461 (2020)
26. Riley, H.B.: Solving simultaneous equations via a large language model (ChatGPT): help or hindrance? In: 2024 CIEC (2024)
27. dos Santos Pereira, O.L.: A comparative analysis of ai-generated physics: DeepSeek-V3 vs. GPT in calculating the Einstein tensor for the Alcubierre warp drive metric. In: GPT in Calculating the Einstein Tensor for the Alcubierre Warp Drive Metric (2025)
28. Shijaku, R., Canhasi, E.: Chatgpt generated text detection (2023, Publisher: Unpublished)

29. Siam, M.K., Gu, H., Cheng, J.Q.: Programming with AI: evaluating ChatGPT, Gemini, AlphaCode, and GitHub copilot for programmers. In: Proceedings of the 3rd International Conference on Computing Advancements, pp. 346–354 (2024)
30. Singhal, K., et al.: Large language models encode clinical knowledge (2022). arXiv preprint arXiv:2212.13138
31. Vaswani, A., et al.: Attention is all you need. In: Advances in Neural Information Processing Systems, vol. 30 (2017)
32. Yang, X., Li, Y., Zhang, X., Chen, H., Cheng, W.: Exploring the limits of ChatGPT for query or aspect-based text summarization. arXiv preprint arXiv:2302.08081 (2023)
33. Zhang, M., Qin, J.: A method for the architecture of a medical vertical large language model based on DeepSeek R1 (2025). arXiv preprint arXiv:2505.00025
34. Zhou, J., Cheng, Y., He, S., Chen, Y., Chen, H.: Large language models for transforming healthcare: a perspective on DeepSeek-R1. MedComm-Future Med. **4**(2), e70021 (2025)
35. Zhu, Q., et al.: DeepSeek-Coder-V2: breaking the barrier of closed-source models in code intelligence (2024). arXiv preprint arXiv:2406.11931

Using Generative AI and Gamification to Support the STEM Education-The Deep Dive System

Chencheng Zhang[1], Sarah Saad[1], Thomas Palazzolo[1], Robert G. Reynolds[1,3](✉), Ashley Lemke[2], John O'Shea[3], and Calien O'Shea[4]

[1] Wayne State University, Detroit, MI 48201, USA
robert.reynolds@wayne.edu
[2] University of Wisconsin Milwaukee, Milwaukee, WI 53201, USA
[3] University of Michigan, Ann Arbor, MI 48109, USA
[4] North Dakota State University, Fargo, ND 58105, USA

Abstract. The Landbridge system was initially designed to reproduce ancient environments in Virtual Reality to predict submerged ancient sites, and enable users to test theories about ancient civilizations through real-world research. The approach was designed to help foster a deeper under-standing of human prehistory. As a result of users' feedback, the Landbridge system has also been considered as an educational tool for high school STEM classes by offering an innovative way to explore scientific concepts. With the emergence of new generative AI tools, a new comprehensive framework for the Landbridge system was developed. The new system consists of five basic stages that embody a gamified version of the Scientific Method. The framework transforms the system into an intelligent tutor with a vast multidisciplinary knowledge base that would help the student record their game experience and formulate their hypotheses about the ancient ecosystem. The resultant tool provides a powerful framework for both research and teaching. The framework promotes users to engage in practical, active research to systematically develop and test theories about ancient civilizations.

Keywords: Virtual Reality · LLM Agents · STEM Education · Gamification · Cultural Algorithms

1 Introduction

Virtual Reality (VR) has emerged as a powerful tool in education, particularly in STEM (Science, Technology, Engineering, and Mathematics) learning. Repurposing VR systems from research applications to classroom teaching is a task that requires careful planning in order to ensure educational effectiveness, usability, and accessibility. This paper presents a sequence of repurposing activities of an Artificial Intelligence system that took place over a decade. The key to this repurposing is a Large Language Model Tutor that helps to translate the student's experience in an open Virtual Reality world to specific spatial hypotheses.

H. R. Arabnia et al. (Eds.): CSCI 2025, CCIS 2971, pp. 430–446, 2027.
https://doi.org/10.1007/978-3-032-28463-1_35

The Land Bridge Deep Dive Virtual Reality system was originally developed to reproduce a portion of North America occupied by Paleo-Indians from 10,000 to 8,000 B.P. for the purpose of Archaeological site prediction [14]. The region of interest, the Land Bridge, was an isthmus of land that connected what is now Amber-ley in Canada to what is now Alpena, Michigan in the USA. The Land Bridge is shown in Fig. 1.

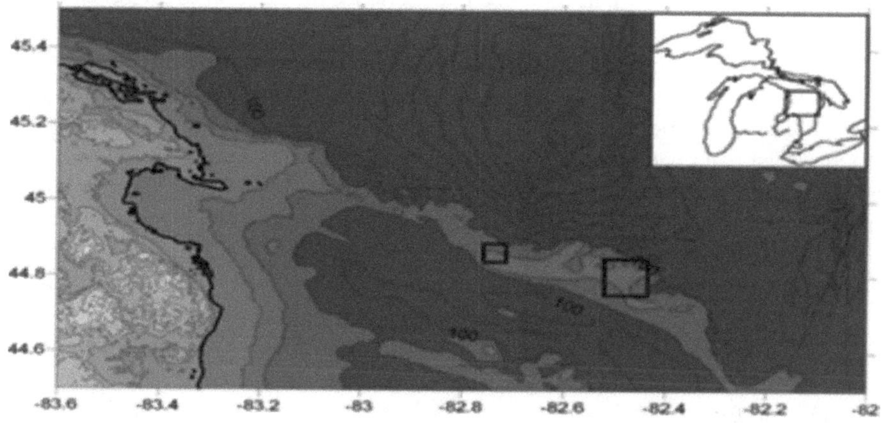

Fig. 1. The Alpena Amberley Land Bridge which was above water from 10,000 to 8,000 BP.

During the Ice Age the region was semi-arctic, and the bridge was above lake levels during that time period. Archaeologist from the University of Michigan led by Dr. John O'Shea postulated that since it was above water for such a lengthy period that it could be used as a migration pathway for caribou. Dr. Reynolds and his team from Wayne State University were recruited to use Artificial Intelligence to reproduce the ancient environment in Virtual Reality for use in the prediction of ancient sites that are now currently underwater [20].

As stated by Millington, while Artificial Intelligence is ubiquitous in games today, rarely is it used to produce a holistic game ecosystem replete with food chains and population models [15]. To undertake such a task here, Artificial Intelligence and Machine learning techniques were used to recreate the ancient ecosystem and to generate AI based models of herd behavior with the idea that where caribou go hunters will follow. This tool, the Deep Dive system, was successfully employed to suggest locations for under water site exploration [16]. In the process of doing that it was observed that the system can also be repurposed to be a learning tool for STEM based high school education classes. That is, students can be given the opportunity to explore the land bridge and make predictions about how it was used in the past by Paleo-Indians. However, since the bridge itself was 80 miles long and 8 miles wide, it was an expansive open world. As a result, students often wandered about without ever getting a complete and holistic understanding of the ecosystem.

It was then proposed to "gamify" the system in order to provide students with a more holistic approach to the environment in terms of the flora, fauna, geology, and hydrology of the biomes that make up the land bridge. The key to the approach is that an

Evolutionary Learning model of optimal herd foraging based on Cultural Algorithms, Pathfinder, was used to generate optimal paths based on aspects of caribou herd behavior taken from current research on caribou behavior. These pathways were given to the Deep Dive system to act as vehicles to direct the placement of POIs. The idea was to use these optimal pathways to indirectly guide the student's explorations on the Land Bridge in a way that would insure a more holistic and balanced view of the landscape as experienced by the caribou. Here, the student is invited to experience the way of the caribou indirectly by explicitly walking along their path.

This is done here through the development of a "scavenger hunt" game similar to those found in *Genshin Impact* or the *Sonic* franchise. Scavenger hunt-style games work by creating a series of collection tasks revolving around the exploration of an environment and the acquisition of target objects. This method of game design was found to be highly effective in the context of education [17, 18, 19]. The idea was to position points of interest (POI) strategically along example caribou pathways as they passed through various biomes during their crossing. Since the landscape itself does not have many visual features to act as beacons it was felt that by associating the POI with both visual and audio cues to attract a student's attention indirectly will give them with a feeling of freedom. At the same time it would provide them with an immersive experience akin to that of migrating caribou. Future work can install POI that reflect the hunting and foraging behavior of Paleo Indians as well. Optimal paths produced by Pathfinder are then used in the VR subsystem to generate paths followed by caribou in the virtual world.

The game is embedded in a five phase framework that represents the basic stages of the Scientific method. The goal is to use the game as the basis for a creative experience for the student. The results of the game will be used to initiate a dialogue with an AI Agent based on a large language model (LLM). The model is trained with papers and other material relevant to the Land Bridge and its environment. The purpose of the Agent is to assist the student in the formulation of hypotheses about the location of human occupational activities using their acquired knowledge from the game. It will then facilitate the expression of these hypotheses in the form of a "heat" map that is used to "color" the VR landscape to highlight those regions that are consistent with the hypotheses created by the student.

In the next stage the student then will enter the Land Bridge VR and use the color codes to highlight an area for detailed scrutiny. The process is viewed as an opportunity to "verify" the reasoning process produced with help of the tutor. The student will select specific target areas for underwater excavation. The final stage of the scenario is to validate these hypotheses through actual dives. If the area had already been investigated a video of the results would be presented. If this was a new location then the dive can be conducted via augmented reality and recorded for future reference. Finally the student would interact with the tutor to summarize their experiences and share their insights.

While the focus of this paper is on the AI Agent Tutor it is important to be aware of the context in which the Tutor is operating. Prior to playing the game the student will visit the Training Room. In that room they will become familiar with the actions they can perform in the Virtual World and understand the user dashboard display that will help guide their movements. This will be the topic for Sect. 2 while the game itself is briefly discussed in Sect. 3. Section 4 provides the details for the AI Tutor and its role in the creative process here. Section 5 presents the results of a sample dialogue. The result of the interaction will be to produce as a set of rules to be used to generate the "heat map". The heat map is then be used by the student to select a small 250 by 250 subarea to search in more detail. Section 6 presents the conclusions.

2 The Training Room

The training room provides the student with access to a variety of functions that will aid their experience in the VR game. In general, the training room shows the user how to do the following: learn the game controllers layout; learn how to move and teleport; learn how to interact with interactable objects such as grab, throw and so on; understand interactable objects; and learn where to spawn in the VR in order to find a POI immediately.

The novice user is escorted through a series of stations that culminates in the station that allows them to select where on the Land Bridge they want to be teleported to. The Land Bridge is a large area, 8 miles wide and 80 miles long, so deciding where to go on the bridge is important. For novice users there is a default location for them to start with. The experienced user can go directly to the teleport map shown in Fig. 2 below and select any location by placing a token on the map location without having to visit the other stations.

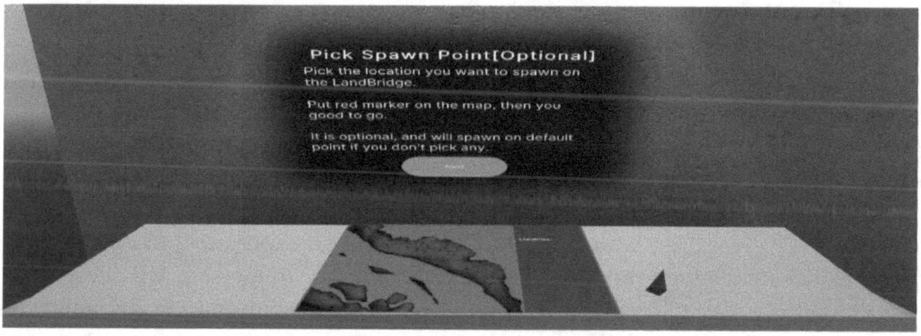

Fig. 2. The DeepDive spawn station where the user can select a location on the map for teleportation.

Once a location has been specified the user is ported to the teleport where they will be sent to that location on the Land Bridge. The teleport if shown in Fig. 3.

Fig. 3. The portal with a view of the target location selected and shown in the blue portal.

3 The POI Game Room

3.1 Gamifying Land Bridge Learning

During the spring of 2022 the Deep Dive tool was integrated into the STEM curriculum of high school students in Alpena, as they attempted to identify these ancient hunting sites via the virtual environment. The students spent two weeks with the system and their experiences provided valuable insights into the potential of open-world VR in education. This exhibition was the first foray into the Land Bridge's application as an educational tool. This phase attempts to build on that experience by enriching the Land Bridge environment with knowledge that can be collected by users as they are influenced indirectly by information about the environment and its past usage.

The Alpena High School demonstrations opened the doorway for the Land Bridge's use as an educational tool [20]. The curiosity and enthusiasm of the students as they interacted with the Land Bridge simulation contrasted with the blank stares and inattention that characterize the classrooms of today. As children are brought up in a world filled with technology and interactivity, it's important that teaching methods should evolve similarly to retain their effectiveness.

In 2020 Liu, Wei, Lei, Wang, and Ren carried out an experiment to determine the improvement in pedagogy produced by the integration of immersive VR (IVR) into a classroom setting. The experiment consisted of 90 students split into experimental and control groups.. Liu, Wei, Lei, Wang, and Ren determined that "the students who engaged in the IVRC [the name of their virtual environment education system] had significantly higher academic achievement and better engagement (cognitive, behavioral, emotional and social) compared to those who learned through traditional teaching methods in a normal classroom" [21]. This finding appeared consistent with other research done on the topic, namely by Parong and Meyer in 2018 [22], and Jitmahantakuli and Chenrai in 2019 [23]. These latter two papers describe similar approaches to applying virtual reality to classrooms, one in the field of biology and the other in geoscience. Although prior studies had reached positive conclusions regarding the benefits of VR in the classroom, a question that has yet to be answered is the time efficiency of "open world" virtual environments in the classroom, more importantly, the extent by which developers can shape VR environments to help direct students to learn in the most time efficient manner whilst retaining its virtual immersive qualities.

The point of interest system (POI) was developed to create an incentive to influence players to remain on a more structured route, as opposed to wandering off and exploring the entire map. It was inspired by the Memorial Quest designed by Volkmar et al. [32] The POIs for the land bridge project are based on the type of flora and fauna that would live in each of the seven different biomes present assuming a semi-artic environment like that of Alaska today. There is only limited knowledge about the plant and animal content on the Land Bridge in the time between 10,000 and 8,000 B.P. so the use of flora and fauna from present day semi-artic environments is the best guess now.

The basic framework of the game is a timed scavenger hunt. The student is asked to find as many POIs as possible within a given period of time. The time limit will reflect some portion of the class hour. Initially time limit was set to 15 min so that each student would have an opportunity to play the game. Th POIs collected would then be used by the student to interact with the AI Tutor Agent that will help them develop a greater understanding of the environment and form hypotheses about the location of human occupations there.

Given the 15 min time frame 30 POIs were developed. There were three POIs per animal, and ten total animals. The fauna utilized for this exercise included the boreal toad (Anaxyrus boreas boreas), the arctic ground squirrel (Urocitellus parryii), trumpeter swans (Cygnus buccinator), caribou (Rangifer tarandus), polar bears (Ursus maritimus), wolves (Canis lupus), the snowy owl (bufo scandiacus), the bald eagle (Haliaeetus leucocephalus), the osprey (Pandion haliaetus) and the brown bear (Ursus arctos). Each POI is composed of two main components: A collider and an auditory clue. The auditory clue is the sound of the animal in question, and the collider triggers a gameObject destruction, causing the POI to be recorded as "claimed" when the user gets close enough to "see" the object. An example of the collider setting associated with Polar Bear POI is given in Fig. 4.

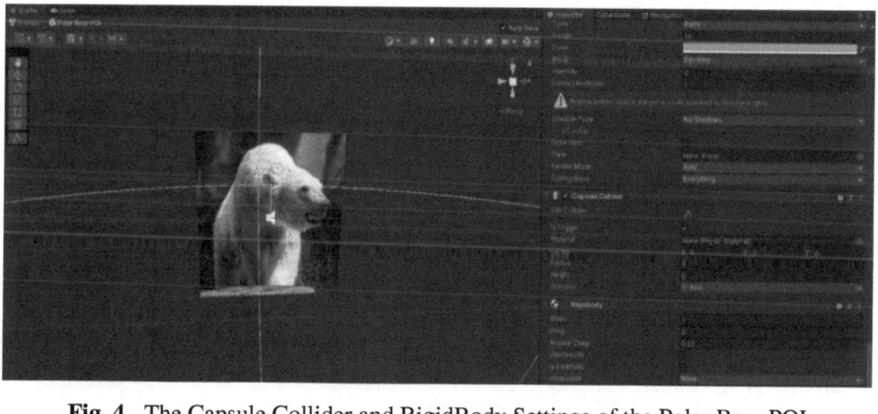

Fig. 4. The Capsule Collider and RigidBody Settings of the Polar Bear POI

In addition to the main two components, the POI is comprised of two more subcomponents: a minor visual cue, and the associated information. The minor visual cue (pictured in Fig. 5) is a light green point light that serves to help locate the Points of

Interest without sound. By design, this point light is not especially bright. This is done in order to ensure that the point of interest's visual presence does not significantly reduce the immersive environmental quality of the Land Bridge project. Instead, the light is visible when the user is close enough and indicates the presence of a visual POI at that location.

Fig. 5. Polar Bear POI situated near an artifact. The green light surrounding the POI denotes its presence and can be seen as the user gets close enough, even though the POI itself is not yet visible.

Each POI contains code to display an image of the animal and one of three facts about it. Its name, both cultural and scientific, its height and weight, and a fact unique to the animal itself is provided for each POI. This is meant to educate the players in an engaging matter. First by finding the animal based on their cries, and then by learning about them. These POIs are not randomly scattered, but instead are placed along an intentional path, leading the players through each of the biomes. These path were created by simulating herd movements of varying sizes based on a cognitive model of caribou behavior.

Fig. 6. Two out of three of the Trumpeter Swan points can be found in the Lake Biome.

Some fauna, such as Trumpeter Swans are less ubiquitous across the land bridge and so therefore are more concentrated in one or two locations, making it slightly easier to obtain information about them as shown in Fig. 6. However, in general, POIs are positioned in proximity to various pathways generated by simulating migrating herds. These POI provide the incentive used to influence the user's decision-making about where to go. Without these POI, the users have no real guidance, and theoretically can travel everywhere eventually. By influencing user decision-making with POIs, the experience should produce a more holistic perception of the Land Bridge ecosystem. One that reflects the cognitive behavior of the caribou as they traverse the Land Bridge in a given season. Their behavior will also influence the location of hunting structures in proximity to the pathways. Thus, students will gain an understanding of herd movement that reflects what the hunters observed and will help them understand where to place selected hunting structures.

3.2 Designing the Narrative Route

Ultimately the goal of the POI is to indirectly highlight a caribou path(s) generated by pathfinder and followed by caribou in the virtual world. To facilitate a holistic environmental experience, it was deemed important that each of the different biomes should be visited. These biomes represented different environments that could be found along this Land Bridge trek. Despite the Land Bridge representing a stretch of land between Alpena Michigan and Amberly, Ontario, the biomes were based on the subarctic region of Alaska as these environments best represented the environment of the Land Bridge several thousand years ago. The biomes for this Land Bridge included the following: boulder fields, cliffs, sandy beaches, tundra plains, swamp marshes, lakes, lakeside marshes, and landmarks known as eskers.

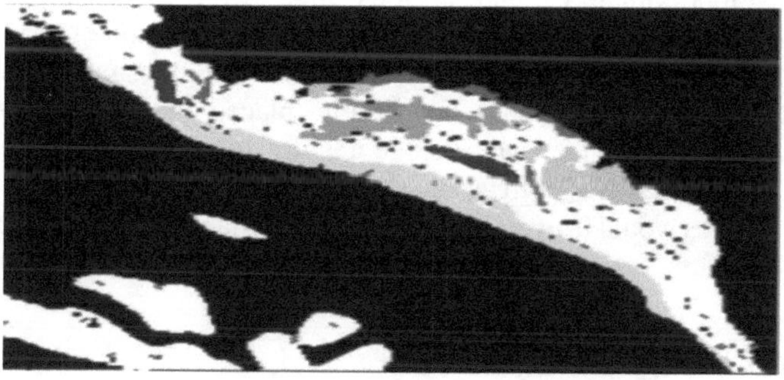

Fig. 7. The Biome distribution on the Land Bridge

Figure 7 illustrates the different biomes, represented here by different colors. The tundra is the majority, represented by the color white. The lake biome is light blue, the lakeside marsh is dark blue, the eskers as red, the swamp marsh as green, the boulder field is orange, and the cliffs are gray. A list of fauna was compiled that inhabit each of these biomes such that no one biome would appear too sparse, despite some being generally inhospitable for common kinds of life. For instance, the sandy beach is rarely home to the kinds of animals that emit sounds, with the exception of birds. Each of these animals' calls were recorded into sound files that would be emitted from each POI when the environment was initialized. An important factor involved in the design was the sound's volume in relation to distance [33]. By giving sound a linear roll off up to a range of 200 m, users can determine their relative distance from the point of interest. Given the number and intended distribution of these POI, the effect was intended to make the virtual simulation appear more alive and more immersive, by appealing to an additional sense other than sight.

The example route itself was designed to provide the user with an overview of the overall landscape and its occupants in a time efficient sequence while retaining fidelity to the occupants' natural biomes. A rendition of this idea can be found in Fig. 8.

Fig. 8. Example path through the biomes.

This path would pass near each of the available biomes, allowing the user to theoretically experience every unique environment this virtual world has to offer. The path generated above approximately represents the estimated most efficient path through the POI on the Land Bridge. Combined with the scavenger hunt nature of the POI system, it can be theorized that the trial participants would be highly incentivized to follow this route especially given that the time constraints placed on the search. An additional unintended incentive is that the Land Bridge has few topographic features to serve as beacons of movement, except for trees and differences in slope. Therefore, it is believed that auditory stimulus would be weighted more heavily than the somewhat lacking visual stimulus.

4 The Land Bridge Tutor

4.1 Tutor Agent LLM Architecture

The previous phase provided the user with a certain understanding of various elements in the Land Bridge. Now they can use this knowledge and experience to come up with various hypotheses about the Land Bridge in the third phase, the hypothesis formation phase. This phase encourages users to think for themselves and formulate hypotheses about the Paleolithic Land Bridge occupation. But simply knowing the environment is far from enough to make reasonable hypotheses. Each individual student approaches the system with different life experiences and assumptions about the world. What is needed is an advisor who can help guide each individual students creative processes. In this third phase a large language model based tutor is provided that is familiar with previous work on the Land Bridge and is able to help the student express their ideas about past Land Bridge occupations. This dynamic guidance can greatly improve the efficiency of the student's creative process.

Meta's Llama 3.2 and Retrieval-Augmented Generation (RAG) (Zhao et al., 2024) were selected (RAG-LM) as the framework for the Tutor. Llama 3.2 is an open-source large language model with 3 Billion parameters that exhibits a nice balance between performance and functionality. From a macro perspective, the architecture of the Land Bridge tutor, RAG-LM, consists of multiple layers, as shown in Fig. 9. These components are then united to provide the complete functionality. The documents are stored in the database for easy management and the RAG-LM will read documents from the database and build a knowledge base. On the client side, users can prompt the system through voice or text, and the system will send the prompts to the server through the API.

After making an inference about a student prompt, the server will send the relevant answers back to the student through Land Bridge UI. The answers can be expressed as text, voice, or graphically. The knowledge available to the tutor consists of textual and visual information along with GIS information. The GIS information is used to produce a graphical expression of a hypothesis as a "heat "map over the Virtual Landscape. In phase 4 of the STEM system, the student can now reenter the Land Bridge and go to those regions highlighted on the heat map where they can then Deep Dive into a local region to locate a specific place on the map.

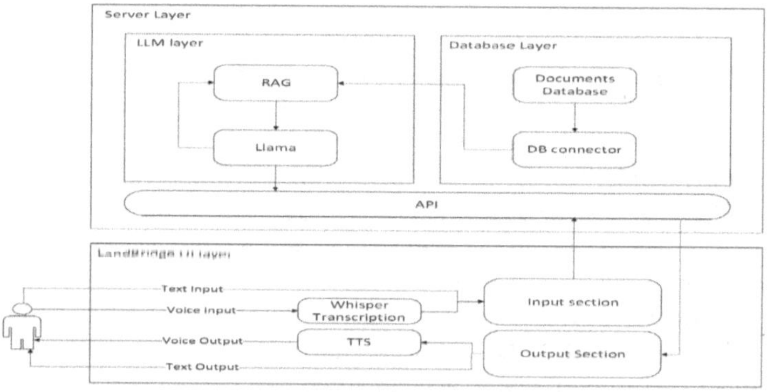

Fig. 9. Land Bridge Tutor Architecture

From a micro perspective, multiple strategies are used in the RAG-LM to retrieve documents and find relevant information, while filtering out erroneous information and hallucinations that may appear in the generative model in order to increase the accuracy of the answer (Fig. 10). In the Land Bridge system, after receiving questions, the retriever will find documents from the document base that are relevant to the input query and pass them as context to the RAG-LM based on the order of relevance. Then, by setting up the prompt template, RAG-LM can evaluate the relevance of the document, and the qualified document content will be used as knowledge to generate an answer. However, one drawback of generative models is that they can produce "hallucinations", in which the model generates fake information. In order to avoid this, a self-checking strategy will determine if the answer is an hallucination. Once the RAG-LM finds that the result isn't grounded in the any document and as a result does not adequately address the original question, the self-corrective strategy will jump back to the generate step and re-generate another answer.

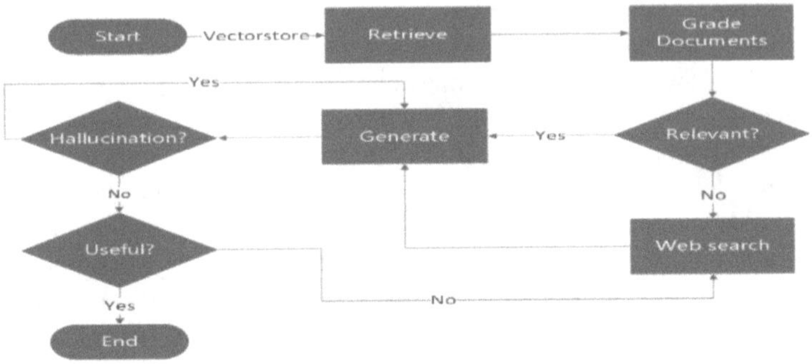

Fig. 10. The RAG Assessment Strategy

LLaMA 3.2 based on RAG has many potential applications in the Land Bridge system. This AI technology can provide many new features for the Land Bridge system and bring a new AI-HCI experience:

1. Virtual Tour Guide

Through the establishment of a relevant knowledge base, the RAG-LM can provide detailed explanations based on specific archaeological sites, artifacts or prehistorical periods on the Land Bridge, providing personalized guidance services. For simulation systems, like the Land Bridge Deep Dive system, that contain many artifacts and sites, one can dynamically feed in new and relevant literature so that the RAG-LM can answer users' questions about the function, origin, discovery history, etc. of specific artifacts and sites. In this way, a rich interactive experience can be created for each artifact and site. Users can prompt the system in natural language, and after voice-to-text conversion, the model will answer based on the current Archaeological knowledge base and provide feedback to the in the form of text, voice, and graphics.

2. Education

With the h:elp of RAG-LM with a knowledge base, the Land Bridge system is able to answer various Archaeology-related questions in a personalized way to provide a more impactful learning experience.

3. Research

Like a virtual tour guide, the RAG-LM student advisor can not only answer questions about specific places or findings, but also serve as a research assistant. It can support researchers during the data collection and analysis phases by quickly providing background information and suggestions based on reliable literature. Also, since its knowledge base contains NOAA GIS data it can transform the students answer into a spatial "heat map" that expresses their hypotheses spatially as well textually over the VR landscape.

4. Cultural Compatibility

LLaMA 3.2 can be translated into multiple languages, so it can serve as a bridge for language exchange and support communication and exchange between users of different languages. This will make the Land Bridge system oriented to a multicultural user group.

5. Evolution

The literature in the project is managed through an independent database, which is decoupled from the LLM and provides flexibility. Therefore, the Land Bridge's RAG-LM can be continuously iterated, updated, and optimized based on newly collected data and user needs. The accuracy of answers and the overall performance of the system can incrementally evolve over time.

In the future, the RAG LLM in the project will evolve into an even more intelligent agent. The future intelligent agent will have stronger interactive capabilities, it should be able to start using tools to complete automated processing and decision-making and even use a large amount of unlabeled data for self-supervised learning (Fig. 11).

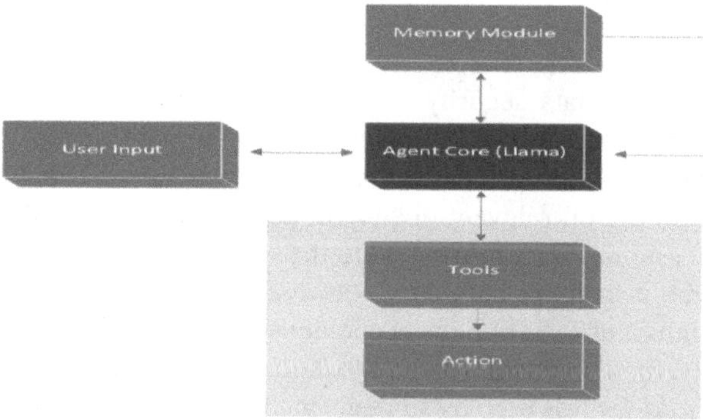

Fig. 11. An LLM Agent

An LLM usually needs high-performance computer hardware for training and inference. Considering the differences in student computer hardware, the third phase adopts the client-server model to use LLM in order to provide stable performance and compatibility (Fig. 12).

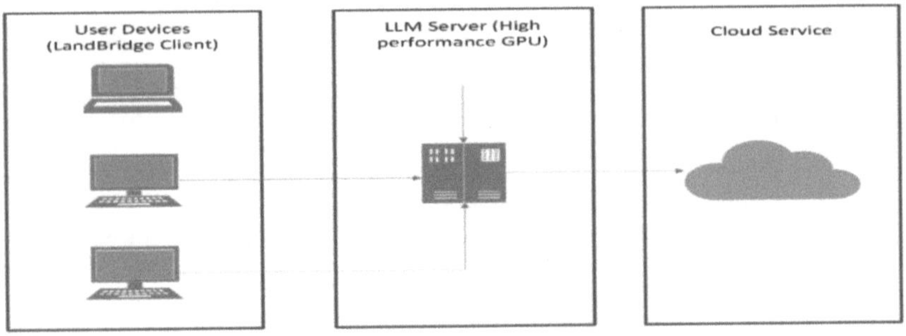

Fig. 12. LandBridge LLM client-server

4.2 Knowledge Base Structure and Loading

The Land Bridge Tutor has a multi-level quality defense mechanism and dynamic decision control flow. It ensures the accuracy of answers through three major links: semantic retrieval, content verification, and self-correction cycle. The Land Bridge Tutor uses a modular loader architecture from Langchain, which can load multi-source heterogeneous data sources. It also uses a recursive text segmentation strategy in order to dynamically identify logical breakpoints in documents (algorithm descriptions, parameter tables, etc.) and tries to ensure that the context is logically smooth and therefore avoids the fragmentation of key information.

In order to meet the needs of data security and personal privacy protection, the Tutor uses Nomic AI's nomic-embed-text-v1.5 model to vectorize the text locally and use Chroma's vector database for storage. During the document retrieval stage, the Tutor conducts additional training on relevant document sets to make the embedding space more focused on professional concepts related to the Land Bridge and adopted a hybrid retrieval strategy. First, it matches core terms (such as "D* Lite algorithm") by similarity, and second, it weighs and sorts important content such as "algorithm chapter" and "experimental data" tags. In other words, it can dynamically increase the weight of relevant chapters based on the keywords used by students in their prompts so as to ensure that the answers to technical questions come from the core chapters first.

After completion of the document retrieval process, the tutor is ready to generate responses to student prompts. In order to ensure the accuracy of the final response, the Tutor has a three-level quality evaluation system. Through carefully set prompts, the tutor can evaluate its own results and filter out low-quality content layer by layer:

1. The first layer filters out irrelevant documents by scoring the relevance of the documents.
2. The second layer verifies the generated content to identify fictitious statements, that is, hallucinations.
3. The third layer judges the utility of the response and ensures that the generated content reflects the original prompt by analyzing the match between the answer and the intent.

The Tutor has an adaptive decision engine that can intelligently select data sources through pre-trained routing classifiers. If the prompt contains Land Bridge-related content or involves algorithm details, experimental parameters and other content that is clearly covered by the document, it will automatically point to the vector database for retrieval. However, if the document retrieval result is empty or the confidence level is below a threshold, it will search for relevant answers on the Internet through the third-party search engine TAVILY if the feature is turned on for the course.

5 Example Tutor Session

The purpose of this section is to demonstrate the performance of the LLM-RAG architecture described in the previous section. For these experiments the original LLM used to develop the framework described in Fig. 9, Llama 3.2, was implemented with DeepSeek-r1:32B. The Tutor was trained on a body of papers that reflect the Land Bridge and the Paleolithic environment in which it was embedded.

The Tutor was given a set of references that related to the Land Bridge Virtual Reality system that are shown below:

Deepdive_Using_AI_Machine_Learning_and_Virtual_Reality_to_Explore_Ancient_Submerged_Civilizations.pdf
Incorporatingthe Dynamics of Climate Change into the Deep Dive Virtual Reality Underwater Site Prediction System.pdf
Learning the Impact of Group Structure on Optimal Herd Path Planning with Cultural Algorithms.pdf
o-shea-et-al-2014-a-9-000-year-old-caribou-hunting-structure-beneath-lake-huron.pdf
Reimagining Global Climate Change_8 page.pdf
Underwater Arachaeology.pdf

A student then provides a sequence of prompt to the tutor as shown below.

4. What environmental features can we expect hunting blinds to be associated with in the Alpena Amberly Land Bridge Virtual Reality system? Give relevant references.
5. Are there areas whose features are not as likely to support hunting structures like drive lanes and hunting blinds in the Alpena Amberley Land Bridge VR system? Give relevant references.
6. What biomes are most likely to yield hunting structures or campsites in the Alpena Amberley Land Bridge? Give relevant references.
7. What biomes are most likely to have hunting cache deposits in them? Give relevant references.

This information will now be used to highlight associated regions of the Virtual Land Bridge for a student to focus on in detail in the Verification phase, STEP 4, of the Scientific method framework employed here.

Figure 13 gives a screenshot of the Land Bridge with area of high importance highlighted as a heat map.

Fig. 13. Heatmap that can be produced as the result of a tutoring session. The student can now go to the highlighted areas for more detail investigation.

The student when finished with the Tutor will have a highlighted rendition of the Bridge that they can explore on their own in phase 4. In that phase they will now be able was demonstrated that focus in on specific regions and be able to provide a rationalization of their efforts. These efforts can result in specific locations for future dives and with information about their importance to provide transparency. The selected locations are then given to Archaeologists who then conduct underwater excavations of selected locations.

6 Conclusions

This paper presented a framework for the Scientific method for use in STEM based education. The application area was in the area of underwater archaeology with a goal of predicting ancient human occupations in a now submerged ancient landscape, the Alpena-Amberley Land Bridge. A Virtual Reality simulation of the ancient environment was developed to help identify potential ancient occupational sites. However, it was often not clear as to why certain locations were selected. Here a gamified version of the Land Bridge, the Deep Dive system, was produced that taught the students about the basic features of the prehistoric environment and a corresponding Tutor was developed to help make the creation of new hypotheses by students more transparent. It was demonstrated that the integration of LLM model into the Deep Dive framework had the potential to boost student creativity and future exploration of the ancient landscape.

References

1. Reynolds, R.G.: Cultural Algorithms. Wiley (2020)
2. Reynolds, R.G., et al.: Paleo-games: using AI and gamification to generate an unguided tour of an open world virtual environment for STEM education. In: HCI International 2024. Springer (2024)
3. Reynolds, R.G., et al.: Generating content for a virtual world learning environment with POIs: the land bridge Paleo-games. Presented at the Society for American Archaeology 90th Annual Meeting (2025)

4. Saad, S., et al.: The deep dive system: reimagining global climate change with artificial intelligence and virtual reality. In: AAAI Spring Symposium Series 2023, San Francisco (2023)
5. Palazzolo, T., et al.: Incorporating the dynamics of climate change into the deep dive underwater site prediction system. In: HCI International 2023: 25th International Conference On Human-Computer Interaction, Copenhagen, Denmark (2023)
6. Palazzolo, T., et al.: A new area virtual worlds and artificial intelligence for indigenous engagement. In: 2023 IEEE Conference on Artificial Intelligence (IEEE CAI), California, USA (2023)
7. Lege, R., Bonner, E.: Virtual reality in education: the promise, progress, and challenge. JALT CALL J. **16**(3), 167–180 (2020). https://doi.org/10.29140/jaltcall.v16n3.388
8. Velev, D., Zlateva, P.: Virtual reality challenges in education and training. Int. J. Learn. Teach. (2017). https://doi.org/10.18178/ijlt.3.1.33-37
9. Merchant, Z., Goetz, E.T., Cifuentes, L., Keeney-Kennicutt, W., Davis, T.J.: Effectiveness of virtual reality-based instruction on students' learning outcomes in K-12 and higher education: a meta-analysis. Comput. Educ. **70**, 29–40 (2014). https://doi.org/10.1016/j.compedu.2013.07.033
10. Mikropoulos, T.A., Natsis, A.: Educational virtual environments: a ten-year review of empirical research (1999–2009). Comput. Educ. **56**(3), 769–780 (2011). https://doi.org/10.1016/j.compedu.2010.10.020
11. Moro, C., Štromberga, Z., Stirling, A.: Virtualisation devices for student learning: Comparison between desktop-based (Oculus Rift) and mobile-based (Gear VR) virtual reality in medical and health science education. Australas. J. Educ. Technol. **33**(6) (2017). https://doi.org/10.14742/ajet.3840
12. Cecil, J., Ramanathan, P., Mwavita, M.: Virtual Learning Environments in engineering and STEM education. In: 2013 IEEE Frontiers in Education Conference (FIE) (2013). https://doi.org/10.1109/fie.2013.6684874
13. Cho, D., Chun, B.A.: Virtual reality as a new opportunity in geography education. In: Proceedings of the 2019 5th International Conference on Education and Training Technologies (2019). https://doi.org/10.1145/3337682.3337701
14. Saad, S., et al.: Incorporating the dynamics of climate change into the deep dive virtual reality underwater site prediction system. In: Degen, H., Ntoa, S. (eds.) Artificial Intelligence in HCI. HCII 2023. LNCS, vol. 14051. Springer, Cham (2023). https://doi.org/10.1007/978-3-031-35894-4_41
15. Millington, I.: AI for Games, 3rd edn. CRC Press (2019)
16. O'Shea, J.M., Lemke, A.K., Sonnenburg, E.P., Reynolds, R.G., Abbott, B.D.: A 9,000-year-old caribou hunting structure beneath Lake Huron. Proc. Natl. Acad. Sci. **111**(19), 6911–6915 (2014). https://doi.org/10.1073/pnas.1404404111
17. Bellotti, F., Berta, R., De Gloria, A., D'ursi, A., Fiore, V.: A serious game model for cultural heritage. J. Comput. Cult. Heritage **5**(4), 1–27 (2012). https://doi.org/10.1145/2399180.2399185
18. Lu, Y., Chao, J.T., Parker, K.: HUNT: scavenger hunt with augmented reality. Interdisc. J. Inf. Knowl. Manag. **10**, 21–35 (2015). http://www.ijikm.org/Volume10/IJIKMv10p021-035Lu1580.pdf
19. Schell, J.: The Art of Game Design: A book of lenses. CRC Press, Taylor & Francis Group (2020)
20. Reynolds, R.G., Palazzolo, T., Lemke, A., O'Shea, J., Saad, S., Zhang, C.: DeepDive: the use of virtual worlds to create ethnographies of ancient civilizations. In: Human Computer Interaction International Conference, Washington D.C. (2021)

21. Liu, R., Wang, L., Lei, J., Wang, Q., Ren, Y.: Effects of an immersive virtual reality-based classroom on students' learning performance in science lessons. Br. J. Edu. Technol. **51**(6), 2034–2049 (2020). https://doi.org/10.1111/bjet.13028
22. Parong, J., Mayer, R.E.: Learning science in immersive virtual reality. J. Educ. Psychol. **110**(6), 785–797 (2018). https://doi.org/10.1037/edu0000241
23. Chenrai, P., Jitmahantakul, S.: Applying virtual reality technology to geoscience classrooms. Rev. Int. Geograph. Educ. Online, 577–590 (2019). https://doi.org/10.33403/rigeo.592771
24. Jacobson, J., Holden, L.: Virtual heritage. Techné: Res. Philosophy Technol **10**(3), 55–61 (2007). https://doi.org/10.5840/techne200710312
25. Lin, J.-H.: Fear in virtual reality (VR): fear elements, coping reactions, immediate and next-day fright responses toward a survival horror zombie virtual reality game. Comput. Hum. Behav. **72**, 350–361 (2017). https://doi.org/10.1016/j.chb.2017.02.057
26. Byun, JaeHwan, Loh, C.S.: Audial engagement: effects of game sound on learner engagement in digital game-based learning environments. Comput. Hum. Behav. **46**, 129–138 (2015). https://doi.org/10.1016/j.chb.2014.12.052
27. Danevičius, E., et al.: 3D localisation of sound sources in virtual reality. In: Lecture Notes of the Institute for Computer Sciences, Social Informatics and Telecommunications Engineering, pp. 307–319 (2021). https://doi.org/10.1007/978-3-030-73426-8_18
28. Ekman, I.: Meaningful noise: understanding sound effects in computer games. In: Proceedings of Digital Arts and Cultures (2005). Geig, Mike. "Hour 21." Sam's Teach Yourself Unity Game Development in 24 Hours, Pearson Education, Indianapolis, 2022 (2005)
29. Lindquist, M., et al.: The effect of audio fidelity and virtual reality on the perception of virtual greenspace. Landscape Urban Plann. **202**, 103884 (2020). https://doi.org/10.1016/j.landurbplan.2020.103884
30. Meghanathan, R.N., et al.: Spatial sound in a 3D virtual environment: all bark and no bite? Big Data Cogn. Comput. **5**(4), 79–95 (2021). https://doi.org/10.3390/bdcc5040079
31. Brinkman, W.-P., Hoekstra, A., Vanegmond, R.: The effect Of 3D audio and other audio techniques on virtual reality experience. Stud. Health Technol. Inf. **219** (2015). https://doi.org/10.3233/978-1-61499-595-1-44
32. Volkmar, G., Wenig, N., Malaka, R.: Memorial quest - a location-based serious game for cultural heritage preservation. In: Proceedings of the 2018 Annual Symposium on Computer-Human Interaction in Play Companion Extended Abstracts (2018). https://doi.org/10.1145/3270316.3271517
33. McArthur, A., et al.: Distance in audio for VR. In: Proceedings of the 12th International Audio Mostly Conference on Augmented and Participatory Sound and Music Experiences (2017)

Vibe Coding: A Case Study of College Freshmen Developing a Full-Fledged Game in Ten Weeks

Brycen Kirchharr[1(✉)], Tristan Whitson[1], and Jeremy Straub[2]

[1] Department of Cybersecurity, University of West Florida, Pensacola, USA
bsk14@students.uwf.edu

[2] Center for Cybersecurity, University of West Florida, Pensacola, USA

Abstract. This paper reports on the progress of a project to ascertain the ease of using vibe coding for the creation of a large-scale application. Specifically, two undergraduate freshmen, who serve as the first and second authors on this paper, set out to develop a game that has similarities to the popular video games Sim City and Sims using vibe coding tools. This effort was successful, resulting in a functional internet-based multiplayer game with virtually no interaction with the underlying code being required to create it. This paper presents a case study of this effort, discussing the process used and lessons learned from it.

1 Introduction

A ten-week project was undertaken to learn about artificial intelligence's (AI) role in the coding world. Two undergraduate freshmen, who serve as the first and second authors on this paper, set out to develop a game that has similarities to the popular video games Sim City and Sims using Vibe Coding tools and, in the process, learn about how to use AI to code for developing complex programs.

Key areas of focus were to determine how best to prompt AI to create projects in the way envisioned by the creator, through a trial-and-error process of mistakes and successes. Through this, the workings of AI coding tools and best practices became clearer. This paper reports on the outcomes and the lessons learned from the process that led to the creation of a Sim City/Sims-like game.

The exploration process showed that, to use vibe coding efficiently, developers can't just send prompts, let it work, and expect a perfect outcome. Instead, the developers must learn from their mistakes and the AI's mistakes, and assist the AI in building the projects it has been prompted to create. This exploration has resulted in the identification of a number of vibe coding approaches and best practices, which are shared in this paper.

This paper continues, in Sect. 2, with a discussion of relevant prior work. Then, in Sect. 3, a discussion of using the AI development tool Cursor is presented. Following this, Sect. 4 discusses the project goals and outcomes, and Sect. 5 presents the approach taken, analysis, and results of the exploration. The paper then presents conclusions and discusses areas of future exploration in Sect. 6.

B. Kirchharr and T. Whitson—These authors contributed equally to this work

H. R. Arabnia et al. (Eds.): CSCI 2025, CCIS 2971, pp. 447–457, 2027.
https://doi.org/10.1007/978-3-032-28463-1_36

2 Background

Coding has evolved significantly during the evolution of computing, beginning from highly processor hardware-oriented binary coding and then moving to assembly coding, where instructions must be issued mirroring processor directives. Then, more advanced languages were created that abstracted processor-specific instructions and allowed focus on more functional ones.

Following this, object-oriented programming was proposed, which focused on code reuse and aligning code with areas of the program. Development environment tools allowed code to be associated with particular buttons and functions within programs, further aiding developers.

Notably, throughout the history of computing, the trend has been towards greater abstraction and focus on achieving human goals, while reducing the complexity of the development process. This has allowed programs to grow from simple origins to the complex software used today. Web applications, APIs, shared libraries, web services, and numerous other innovations have been proposed that have made software ubiquitous within modern society in the United States and other developed countries.

Recently, vibe coding has been proposed and grown in popularity. Unlike how coding was done in the past, where a team or a singular developer would code everything by hand, vibe coding allows those with limited or no coding experience to make software, allowing development by those closer to business needs or design goals, as opposed to dedicated software developers.

Vibe coding has been shown to provide numerous benefits. Yuce [1] described vibe coding's enhanced code readability (18% improvement), decreased errors (15%), and decreased time required for "refactoring and expansion" (23%). Samsyudin [2], similarly, showed an improvement in efficiency (by 27%) from vibe coding. It also changes the role and duties of human developers, placing the humans as "architect[s] and supervisor[s]", instead of just code-writers [3].

However, others have raised concerns. Alberto, et al. [4], contend that vibe coding "generates syntactically correct but functionally flawed code" and Ray [5] identified numerous challenges such as "model hallucinations, technical debt, security and compliance risks, skill atrophy, and governance concerns". Faragó [6] went as far as to contend that vibe coding has created a "code quality crisis" that produces security and functional challenges, suggesting that reports of widespread use of effective vibe coding are inaccurate. Horvat [7] argued that vibe coding "generat[es] code with AI without understanding the code that is produced", cautioning against its use for production applications and learning, while noting its potential utility for prototyping.

Many have proposed potential solutions for some or all of these challenges. De Silva [8], for example, suggested system-based development where the "functionalities will be gradually developed" via successive prompting. Geng, et al. [9] found, studying student use of vibe coding, that the approach used varies by students' experience level, with advanced students' approaches differing from less experienced ones and having greater interaction with the code produced.

Sarkar and Drosos [10] suggest that vibe coding is the "evolution of traditional AI-assisted programming" using "iterative goal satisfaction" with cyclic "prompting, evaluation, debugging, and refinement" where developers perform "prompt and context management, rapid code evaluation, bug identification and resolution". Krings, Bohn, and Ludwig [11], similarly, noted how vibe coding changes humans' role within development, combining developer and alternate roles. Interestingly, in Knobel and Radziwill's [12] study of interactions between a human "lead" and an AI developer, dynamic issues, similar to human teams, were noted.

3 Using Tools: Cursor

This section discusses the use of the AI coding tool Cursor, which was used for this exploration. A trial-and-error approach was used to gain an understanding of how to prompt it more effectively.

3.1 Prompting

The nature of the prompts used is essential to the successful production of functional code. Notably, specificity and deconstruction of requests into multiple prompts can be helpful. Cursor provides a development environment with a chat box on the left side. By issuing prompts in this chat, users can request that Cursor write or modify code for them. There is no single right way to do this; however, if limited detail is provided, the results may less precisely (or not) match the user's expectations.

A variety of prompt styles can be used. For example, here is an example from the early prompting:

"there are a few issues, specifically with city player pro, when i click on city player pro and back into city builder pro the generate button gets formatted incorrectly and put below the clear button, also when im in city player pro the grid doesn't appear until I click and when i do click and paint the website lags for a seconds and then the grid reappears, finally the text in the city stat section is still white on white and unreadable"

Notably, this prompt is not particularly descriptive in some parts. Also, it tries to get Cursor to concurrently resolve multiple issues, which may actually cause more problems to be introduced. Also, images were not attached to the prompt early on, due to a lack of awareness of this feature and associated functionality. This prompt partially fixed the generate button, which was one of the three issues that were being attempted to fix. However, it was still necessary to issue another prompt to complete that fix, subsequently. The AI made no changes to the other two issues that were included in the prompt. Thus, the prompt was not successful and wasted time. It potentially could have been successful if it was more descriptive and had focused on a single problem rather than several.

Re-prompting was needed to debug the results of this prompt. This likely took up more time than probably would have been needed if a prompt had been issued to resolve each issue using a single issue-specific prompt.

Later prompting is more detailed. Here is an example of a more recent prompt:

"there is a weird visual bug where the bridge and the road no longer have a grid on them"

In this prompt, an image was also provided to illustrate the issue. This prompt was short but still descriptive (by specifying which tiles were having issues) and resulted in Cursor resolving the problem from using this single prompt.

Lessons learned regarding prompting Cursor are discussed in Sect. 5.1.

3.2 Cursor Environment and Learning

Another area related to tool use is the Cursor environment and how Cursor appears to learn from and adapt to user preferences over time, within a specific project.

Cursor has a limited amount of ability to recall details, even though a user can scroll up through the chat history, of which a lot is retained. Thus, Cursor may have trouble recalling something the user entered days or even hours before. However, it seems to distill key takeaways from user interactions over time. Most of the time, Cursor remembers your design patterns and does a reasonably good job of projecting how you may want it to work on your project. This can speed up the creation process; however, it can also lead to issues, depending on what is learned.

For example, two different laptops with Cursor on them were used for this exploration. The Cursor instance on one laptop had been primarily used for developing much of the project. With this laptop, rather than just immediately trying to fix the issues identified, it would perform debugging and request the user's help for finding issues, which was a slower process. The Cursor instance on the other laptop, which was not used as much, would keep trying to solve the issue by directly changing the code.

The first laptop would lead to a lot of unnecessary code being printed out into the console and being stored in the project. On the other hand, the second Cursor instance could solve many issues faster. However, depending on the size of the problem, it could be less effective in its solutions, making it take longer than the first (more verbose) Cursor instance.

Cursor adaptation to interactions with users is discussed in Sect. 5.5.

4 Project Goals and Outcomes

This project took an exploratory approach to gain an understanding of the capabilities, limitations, logistical challenges, and unforeseen complications relevant to vibe coding. The principal developers (computer science freshmen) had some existing programming knowledge, but had never Vibe Coded or used the Cursor software before this project.

The primary goals of this project were for the two freshmen students to learn about vibe coding and how to do it. A number of secondary goals were identified during the project, based on situations encountered, needs identified, and lessons learned while pursuing this program goal.

One key area of secondary exploration was to explore learning programming techniques via using Cursor. The chat was used (and implicitly evaluated) for asking questions regarding programming tasks, in addition to asking Cursor to implement particular things.

Another key secondary goal was to explore the impact of interactions with Cursor on the results produced from prompts given to it. Due to the separation of how work was carried out, one computer's Cursor software dealt far more with development tasks while another's was primarily used for debugging, allowing differences between the two, in responding to prompts, to be compared.

A third area of secondary exploration was to explore Cursor's ability to create multiplayer and networking code. An internet-based multiplayer mode was added to the game to facilitate this.

A final secondary goal was the exploration of using Cursor/vibe coding with other automation and similar tools. In particular, server hosting and version control technologies were explored.

The application that was created is depicted in Fig. 1.

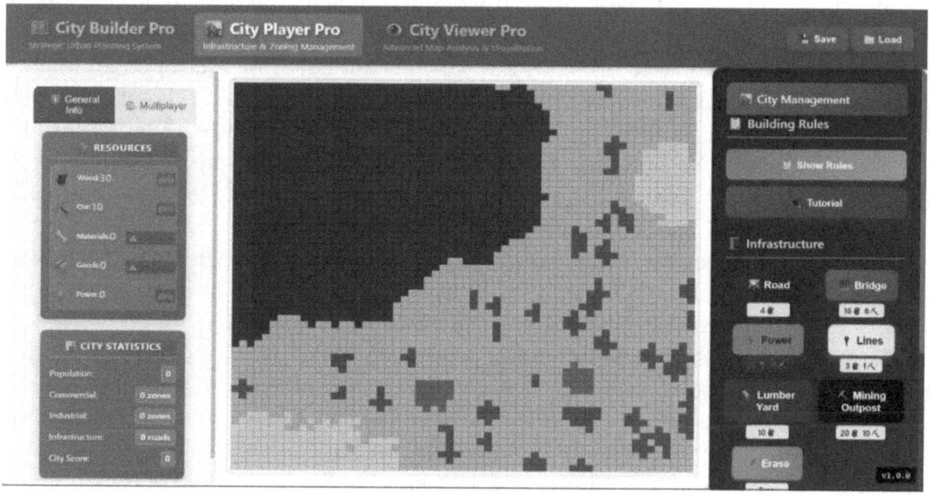

Fig. 1. Screenshot of the application developed with vibe coding.

5 Approach, Results, and Analysis

In this section, the approach taken to software development as well as the results produced will be discussed and analyzed. First, prompt generation will be discussed. Then the need for and benefits of existing software development knowledge will be considered. Following this, focus will turn to specific issues encountered during the development process and how they were overcome. The ability to learn about software development through using Cursor, Cursor's learning about human developers, and Cursor learning from external resources are also discussed. Finally, the integration of Cursor within a broader automation environment is considered.

5.1 Prompt Generation Techniques - Using Smaller, More Specific Prompts

When using AI to create applications via vibe coding, it is important to prompt correctly and effectively. When prompting, it is best to focus on one feature or issue at a time and provide depth on that feature or issue so that the AI knows exactly what to create or fix.

For example, below is an in-depth prompt used to describe one issue to the AI:

"whenever a player leaves the next player sees the timer as shown in the image. This means the timer is not restarting whenever a person leaves. Please make it so that when a person leaves the timer continues to the next person just like it would if they were passing the turn over."

This prompt included an attached image that was a screenshot from the application. With this prompt, the AI was able to resolve the issue in a single attempt, as it was instructed regarding exactly what it needed to resolve.

An alternate prompt, shown below, was much larger and focuses on multiple issues:

"there are a couple of issues we have noticed when testing multiplayer, first of all the action counter seems to be broken, it wont update and i am able to place as many things as i want in my turn. Secondly, as you can see in the image provided this person is losing power after I placed an Industrial which is not good, we are supposed to have separate resource generation and gains. Third, I would like you to get rid of the different gamemodes as shown in the second image. And Finally I would like you to create a game version that is shown in the bottom right of the city builder pro that will you can decide what the game version can be after each prompt we send, if a small change is made increase it by .1 and if a big change is made increase it by 1"

Because of focusing on several issues, it does not go into much depth on each issue individually. This creates too much complexity for the current AI, leading to the AI not being able to fix the issues and, in most cases, creating new issues.

These examples show how important proper prompting is, as effective prompting generates positive results. Ineffective prompting, on the other hand, not only doesn't resolve the desired issue or create the desired feature, but it can also lead to the introduction of more problems into the code and project.

A key feature of Cursor is a history of the state after the completion of each prompt. Using this, the user can readily revert to the state before a problematic prompt or even go back several prompts if a chain of prompts doesn't result in the targeted outcome.

5.2 Benefits of Coding Background and Computing Knowledge

To use Vibe Coding more efficiently, having a good background and basic knowledge of coding is helpful. One example of utilizing this skill would be when the AI runs into an issue. The user could assist it by pointing out that specific issue and giving the program a prompt with a suggestion regarding the direction to take.

Notably, the user does not necessarily need to have a basic understanding of coding in order to vibe code; however, with the present state of the technology, this may result in reaching a dead end, where the user must backtrack to fix an issue by reverting to a known-good code version and continuing to build from there. Coding knowledge can potentially save time and allow the user to provide better prompting regarding an issue.

Notably, virtually no direct modifications were made to the code base during this project, though it was looked at in several instances to aid in prompting and understanding what had been produced by the AI.

5.3 Specific Issues and Resolutions

This section discusses several specific issues that were encountered during this process and how they were overcome. First, issues regarding controlling where Cursor had access to and could edit are discussed. Then, issues with using pre-existing libraries are considered. Following this, issues with the project file layout are reviewed.

5.3.1 Controlling File System Access

Since Cursor can access all files within its working directory, it is a good practice to define a single project folder on the computer and use that as a workspace for each project. This can be placed within a project's folder that contains all project-specific folders. This helps make the AI run faster, because it does not have to look through as many files as if a directory with other content were chosen. It also helps ensure that other files are not changed, beyond the scope of the project, and that errors are not created from the AI accessing a file it does not need to or associating unrelated files.

5.3.2 Using Existing Libraries

The question of whether Cursor can effectively access and code for using existing external libraries was also considered. To explore this, Cursor was prompted to create a 'mod' for the popular (and frequently 'modded') game Minecraft. Minecraft is 'modded' with broad permission from its developers, who have open-sourced key libraries to support these efforts. The game was chosen because it has a well-developed and researched mod engine and libraries.

Cursor began creating the workspace needed and imported the necessary libraries to create this 'mod'. However, while Cursor was creating the libraries and workspace, it ran into several issues, such as not creating the workspace properly, getting lost in a library, and using the wrong functions or calls, resulting in it not working well.

To try to aid Cursor and overcome these issues, an existing workspace for creating a 'mod' for Minecraft was made, and Cursor was given this workspace to begin working within. This fixed some issues with compiling and avoided the issue of Cursor not creating the workspace properly. However, it did not fix Cursor calling functions that did not exist in the library, and it had been getting stuck and did not know what to do.

This illustrates the challenges that Cursor and other AI tools may have when working with existing complex applications that they didn't create or document. It also shows tools' potential dependence on learning/training from existing documentation regarding tasks they are instructed to perform. Cursor has potentially not been actively used to create 'mods' for Minecraft previously, and potentially hasn't been trained on knowledge sources relevant to this, leading to a potential lack of knowledge regarding this specific type of coding.

Potential best practices for this type of activity include setting up the workspace manually and creating a folder or area for Cursor to work within. While Cursor (or other tools) may be able to begin working, it may still have trouble when creating mods, due to a lack of understanding and training regarding the requisite pre-existing libraries. Exploring the use of AI vibe coding tools, such as Cursor, on large pre-existing projects is a key area of needed future work.

5.3.3 Project File Layout

One problem with using Cursor is that it doesn't, by default, follow the convention of creating multiple files for the different elements of a project. This can lead to the AI putting all of the code into a single file, instead of breaking it up into multiple smaller files.

This may be as easy for the AI to access, but it makes human review and co-development difficult.

Notably, it is not just humans who are impacted. Having a single file makes it harder for both the user and the AI to resolve issues with the code. While software can readily access, add to, and search for text within a large file, the context window of the underlying AI models limits how much of a file can be provided to it at once, limiting the debugging functionality.

Notably, Cursor seems to follow conventions used by developers, once they are established, such as creating multiple files. Languages that require separate files for different types of functionality (such as different files for different pages of a dynamic web application) also seem to limit this behavior from occurring. This shows that while AI can code, it may need help with the overall architecture to prevent creating a messy and hard-to-fix file structure.

5.4 Learning About Coding Through Cursor

Cursor can help teach some aspects of coding. Watching Cursor work, and reading the descriptions that it provides and the code that it produces, as well as debugging alongside it, can allow the user to learn about how to code. Notably, the biggest value of this approach may be facilitating targeted learning for someone who is already knowledgeable about coding but needs to learn about a particular area.

A user could certainly take an ask-and-review approach, where they ask Cursor to create something, read its explanation and rationale, and then review what it has created. From this, the user could distill some patterns and perhaps understand what areas of code mean and how to use them. The user could also ask Cursor to add comments to code to facilitate this.

The targeted learning approach was used as part of this project to allow the student authors to learn how to develop multiplayer game functionality. Cursor was asked for and provided instructions on what to do to create a game, including prompting suggestions for Cursor itself. This provides the user with immediately usable knowledge as well as knowledge that can be applied to future projects where it could be used to prompt for (or hand-develop, if desired) similar functionality without having to ask the AI for help getting started. This shows how Cursor and Vibe Coding software, in general, can be used for targeted learning.

As part of this project, the student authors of this paper also used Cursor to explore the basics of using GitHub. It was also used to learn about how to use the Render website, where a server to host the multi-player aspects of the developed game was set up. Cursor was asked for and provided easy-to-understand step-by-step instructions on how to set it up. Notably, the students had no prior knowledge in these areas, but were able to make use of them based on the instructions provided by Cursor, along with their own exploration based on them.

5.5 Cursor Learning

One thing that was noted during this project is that the Cursor AI learns about and from the user, so as the user utilizes the AI more and more, it begins to follow the conventions and more quickly understand how the user wants code to be generated, formatted, and fixed. This can speed up projects, since the AI begins to do the things that the user wants, as the user wants them to be done, automatically, allowing the user to better utilize the AI. The user uses the AI to code, and the AI gets better at translating the ideas of the user to code, in turn helping the user give the AI more data by developing more projects.

Notably, it appears that while Cursor does learn from the user, this is only for the specific project the user is working on. It doesn't appear to move this knowledge across projects or to new projects (or, perhaps more correctly, to new development chats). This may be a key feature for future incorporation to facilitate getting new projects up to speed more quickly (while also, perhaps, having an option to not do this, if a project will be dissimilar to other projects the user has worked on).

Cursor also demonstrated having difficulties recalling and drawing upon things the user mentioned days or even hours before, within the same chat. Also, while most of the time Cursor remembers a user's coding patterns and how it thinks the user may want it to work on a project can help speed up the creation of the project, it can lead to potential issues if it learns bad practices or practices that don't make sense to use in other areas.

For example, the two computers used for this project ended up with somewhat different Cursor operations. The first computer's Cursor instance would, rather than just immediately trying to fix issues, resort to debugging and asking the user to help the AI find the issue. This took more time and led to unnecessary code being printed out via the console and being stored in the project. This required lots of cleanup of vestigial code.

The second instance would try to solve issues by directly changing code. In some cases, this second Cursor instance could solve the issue faster; however, depending on the size of the problem, it could take longer if a solution could not be rapidly identified.

5.6 Cursor External Learning

Cursor also demonstrated a capability to seek out information and learn from external resources. In several instances, Cursor indicated that it conducted a web search to source information in response to user prompts. An example of this is shown in Fig. 2.

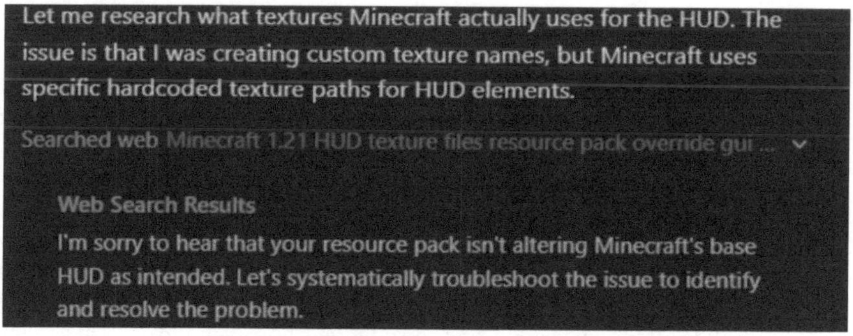

Fig. 2. Example of Cursor conducting a web search.

5.7 Project and Process Automation

The exploration process highlighted a number of ways that additional automation could be achieved above and beyond the use of Cursor. For example, it appears that code pushes to GitHub (and potentially reloading files from GitHub) could be automated. This was not tested successfully during this exploration; however, the functionality was notably present.

Another area that was tested, related to automation, was automatically starting a server on a development computer. The team was able to start a Render server automatically for testing and debugging purposes, based on details provided by Cursor.

Notably, the ability for the Cursor software to interact with the external environment would seem to allow all sorts of development workflows to be automated. Teams working on large projects may be well served by devoting time to process automation for their project, given the considerable flexibility that was noted through this exploration.

Cursor's (and other AI tools') external product and service preferences may become an area for partnerships and product placement opportunities for AI tool vendors. Strong support and recommendations from an AI coding tool may have a significant impact on adoption amongst users, particularly those lacking applicable experience in an area.

5.8 Cursor Testing

In a number of cases, Cursor would indicate that it was trying to test functionality; however, this testing did not always result in working functionality. Cursor's testing abilities appear to be an area where significant improvement could be made to enhance the experience of users. This would be particularly beneficial for users with limited or no coding experience, who would be heavily reliant on Cursor to prevent and correct issues.

6 Conclusions and Future Work

This paper has demonstrated how vibe coding can be used efficiently and productively. It has shown, through the discussion of this project, how straightforward and simple application creation with Vibe coding tools can be, with the proper knowledge of tool prompting techniques.

The benefits of vibe coders having a basic understanding of and background in coding, which allows the user to review the code and identify issues, have been discussed. Additionally, it is important that the user knows how to effectively prompt the AI, particularly to initiate the project and set up initial conventions, such as file organization, which can enable human review and speed up the process of debugging.

Finally, this project has shown that it is most effective to have vibe coding programs, such as Cursor, work on individual parts of a project, rather than requesting changes or additions to the entire project as a whole. Notably, future versions of AI tools may be designed to facilitate larger-scale additions and changes; however, the current generation of tools seems to work better with focused prompting.

Future work, building off of this project, will include identifying the uses and areas of non-use of vibe coding across industries. This process will identify areas of needed technical advancement, which will serve as prospective areas for additional future work.

Acknowledgments. Thanks are given to Anysphere, Inc. for providing access to their Cursor platform, at no cost, to the student authors. J. Straub also currently holds an appointment at North Dakota State University and has an ownership interest in an educational technology AI company.

References

1. Yüce, A.: Vibe Coding: Analysis of a Post-Modernist Programming Paradigm - Yapay Zeka, Bilişim, Eğitim, Yazılım (2025)
2. Samsyudin, I.: Vibe coding and AI-led conversational programming: emerging trends in software development. SSRN (2025). https://doi.org/10.2139/SSRN.5469367
3. Escobedo, S.D.: Single conversation methodology: a human-centered protocol for AI-assisted software development. arXiv:2507.12665 (2025)
4. Rose, I., et al.: The Rise of "Vibe Science": How AI Threatens Scientific Rigor. https://www.techrxiv.org/doi/full/10.36227/techrxiv.175606229.90717623. Accessed 23 Oct 2025
5. Ray, P.P.: A review on vibe coding: fundamentals, state-of-the-art, challenges and future directions. https://www.techrxiv.org/doi/full/10.36227/techrxiv.174681482.27435614. Accessed 23 Oct 2025
6. Faragó, D.: From vibe to vise coding: addressing the AI-generated code quality crisis (2025)
7. Horvat, M.: What is vibe coding and when should you use it (or not)? TechRxiv 175459780.03758839/v1 (2025)
8. De Silva, D., Mills, N., Issadeen, Z., Moraliyage, H., Jennings, A., Manic, M.: Generative AI vibe coding for prototyping industrial systems. In: IEEE International Symposium on Industrial Electronics (2025). https://doi.org/10.1109/ISIE62713.2025.11124737
9. Geng, F., et al.: Exploring Student-AI Interactions in Vibe Coding. arXiv:2507.22614 (2025). https://doi.org/10.1145/nnnnnnn.nnnnnnn
10. Sarkar, A., Drosos, I.: Vibe coding: programming through conversation with artificial intelligence. arXiv:2506.23253 (2025)
11. Krings, K., Bohn, N.S., Ludwig, T.: (R)evolution of programming: vibe coding as a post-coding paradigm. arXiv:2510.12364 (2025). https://doi.org/10.48550/ARXIV.2510.12364
12. Knobel, C., Radziwill, N.: Vibe coding: is human nature the ghost in the machine? arXiv: 2508.20918 (2025)

Enhancing Introduction to Computer and Data Science Based on Python Course Using Gen AI

G. Grabarnik[1] and S. Yaskolko[2](✉)

[1] St. John's University, 8000 Utopia Parkway, Queens, NY 11439, USA
grabarng@stjohns.edu
[2] College Board, 250 Vesey Street, New York, NY 10281, USA
syaskolko@collegeboard.org

Abstract. There are countless possible areas of education where Artificial Intelligence (AI) can be applied. Since the introduction of Large Language Models (interchangeably LLM, or Gen AI, or just AI), every professional field, not excluding education, has become an area of their application. The LLM based tools will eventually become a part of every respective curriculum. This paper covers an introduction of Gemini Gen AI as the main driving tool in a new "Introduction to Computer and Data Science (CDS) Based on Python Course Using Gen AI" class, which also includes a built-in human control cycle to address Gen AI hallucinations and bias. The paper describes the class's structure, specifically, as it is related to the use of Gen AI. It demonstrates how the usage of Gemini can guide the students to learn basic class concepts. It also shows how prompt engineering skills and the Computer and Data Science concepts can be built on each other, creating a synergistic effect and thus accelerating development of both skills, while teaching the students the importance of human control to compensate for natural Gen AI drawbacks. Two classes, traditional and AI enabled, were run in parallel with the same set of assessments. The standard t-tests were used to show statistical significance of the results, show that learning outcomes are by about 13% better for the AI enabled class, and that about 75% of the students prefer AI enabled class over traditional.

Keywords: LLM · Python · AI enabled class · Gen AI based Computer Science courses · hallucinations · bias

1 Introduction

In the last few years, AI has become a widespread tool in practically every area. Not surprisingly, it has become an intrinsic part of the education process too. It is used in education already in a variety of ways - by instructors from gathering the latest developments of the material taught and updates on the subjects of their courses to creating class plans, assessments, etc.

Students use it too - to work on the class's projects, to edit papers, as an addition or even replacement of the traditional learning materials, such as textbooks, online lectures,

H. R. Arabnia et al. (Eds.): CSCI 2025, CCIS 2971, pp. 458–469, 2027.
https://doi.org/10.1007/978-3-032-28463-1_37

etc. At the same time, more specialized tools (agents) are created that use Gen AI in a variety of professional fields.

Teaching the students on how to use these tools in their respective future professional fields, not only is a good idea but should eventually become a part of every respective curriculum, same way, as, say, working with calculators became a part of every Algebra curriculum nowadays. One of the reasons for that is significant improvement in performance while using Gen AI.

Gen AI also has a number of negative sides, including hallucinations and bias.

This paper is a continuation of research started in [27]. It will cover a more holistic approach of teaching the students to work in an environment with widespread Gen AI use. In particular, the standard software product cycle will be enhanced by using Gen AI at design, development, and testing stages.

In this paper we describe the course "Introduction to Computer and Data Science", traditionally based on the Python programming language. We follow the 'vibe engineering' approach to incorporate Gen AI into both the learning process and work on assignment, to make sure that students not only use Gen AI as code generator, but also critically approach the generated code, are capable of directing Gen AI to proper solution and work interactively to obtain the above-mentioned solution.

Another important contribution is an addition of the human validation process introduced at every step of the cycle, which uses Gen AI in order to eliminate negative effects of typical for Gen AI hallucination/bias by refining the process via consecutive iterations.

Since we work on the Introduction to CDS course, the student's projects are of small to medium size, and we do not expect students to use the full standard Software Development Life Cycle but restrict to smaller, shorter Life Cycle. Nevertheless, we interweave the iterative nature of the software development lifecycle, particularly at the design, development, and testing stages. By incorporating Gen AI into these phases, and crucially, by integrating a human validation process at each step, we teach to mitigate the inherent issues of hallucination and bias often associated with Gen AI.

2 Using Gen AI for Intro to Computer and Data Science: State of the Art

As stated in [2] "Artificial Intelligence (AI) has the potential to address some of the biggest challenges in education today, innovate teaching and learning practices, and accelerate progress." Current applications of AI in education and learning are very diverse both in the sense of content covered and aspects of the learning process it provides help with.

AI in Education as an idea goes back more than hundred years to the first mechanized multiple-choice machine developed by Sidney Pressey, which not only provided the test but also feedback based on the student's response.

Next significant step was the work of Skinner [21], whose 'teaching machine', allowed students to compose open-ended responses instead of only offering multiple choice problems. Per Skinner, "The machine itself, of course, does not teach but the effect upon each student is surprisingly like that of a private tutor".

Currently, student-focused AI in education concentrates on "mastery learning", an educational model developed by Benjamin Bloom [4, 24]. This model is based on the idea that AI can and should help with personalizing the learning process. Per Bloom, every student comes with individual set of.

- Existing knowledge base.
- Learning patterns preferences.
- Comfortable learning pace.

All three should be taken into consideration to ensure a successful learning process and short of providing each student with a personal tutor, AI is the best currently available tool to provide all that.

While discussing what tasks AI can help in education is very important by itself, probably an even more important question is what type of AI is better suited to implement these tasks, data-driven or knowledge-based. It seems that at the moment, after quite extensive discussion, knowledge-based AI has become more and more often the first choice for educational use. The reason for that is that education should have in its foundation not only respective data but also widely accepted concepts about that data. A good analogy would be that it is a good idea for every course to have a recommended textbook as the basic reference point for learning, as the good authority to refer to, if any questions on any aspect of the course material arise.

AI in education is a vast and fast-growing field nowadays. It can be split in three basic areas: student focused, teacher focused and school focused). A (not by any means complete) list of types of AI tools in education grouped by the area (see for instance [24]) is as follows:

Student focused:

- Intelligent Tutoring Systems
- AI-assisted Apps (language learning, etc.)
- Automatic Formative Assessment tools
- Exploratory Learning Environments

Teacher focused:

- Plagiarism detection
- Smart Curation of Learning Materials
- Classroom Monitoring
- Automatic Summative Assessment
- AI Teaching Assistant

Schools focused:

- Course-planning,
- Scheduling
- School Security
- Identifying Dropouts and Students at risk
- e-Proctoring

That said, while using Gen AI in education, well-known issues of hallucination and bias should be taken into consideration and addressed. As stated in [15], "LLMs are prone to hallucination, generating plausible yet nonfactual content. This phenomenon raises significant concerns and has attracted intensive research to detect and mitigate such hallucinations."

Shomit Ghose in [20] outlines general approach of addressing hallucination/bias issues as follows:

1. Fine-tuning of the prompts to minimize hallucinations/biases based on imprecise/incorrect prompting.
2. Fact-checking based on referring LLM to an extensive enough verified knowledge base.
3. Use of self-assigned by LLM to each response a confidence score, reflecting plausibility of the response being correct and factual.

They still state that even the best effort won't make the output 100% hallucination and bias free thus making a manual check inevitable, if 100% accuracy is paramount for the task at hand.

That is also the conclusion made in [1], "While AI is a powerful tool, the human touch remains crucial. By working together, we can make the most of what AI offers while mitigating its known limitations.", as well as in [1]: "Unlike traditional (human) communicators, AI lacks in intent or epistemic awareness that would allow it to recognize or prevent the generation of hallucinated content. Platform guardrails (e.g., internal filters) and retrieval-augmented generation (RAG) systems (e.g., the "search the web" function from many AI tools) can improve factual accuracy, yet hallucinations persist because these models generate language by predicting the next most likely word based on statistical patterns in training data."

David E. Balch and Robert Blanck in [3] provide an approach based on their college teaching experience, which basically mimics general approach outlined above:

1. Prompt engineering and constrained outputs
2. Data augmentation and model regularization (which is basically the verification by the specific to the subject knowledge base, say, a textbook, class materials, etc.
3. Human-in-the-loop validation
4. Benchmarking and monitoring (basically a set of assessments, which is regularly given to the respective AI tool)

As outlined above, everybody agrees that human validation, in one form or another, is the crucial and non-avoidable part of any process, which removes or at least minimizes hallucination/bias negative effects.

Here we introduce a computer course, which, in particular, outlines standard software life cycle: design, development, testing, with use of Gen AI at the stages of design, development, and testing, to include human validation as follows:

- At glass box testing stage unit tests are created for each developed component
- A developer (student) checks that the testing is sufficient and is by design
- The prompts used at each stage are refined based on unit tests outputs to get full coverage of both system and unit testing

The expectation is to get in the end a fully functional product free from hallucination/bias.

3 Introduction to Computer Science and Data Science Using Gen AI Class

There are several tools out there, which in various ways and for various programming languages use AI to help create new code. We tested a number of large language models (Copilot, Gemini, ChatGPT, Grok, Claude, Llama.) For our purposes, all of them behaved similarly and for logistical reasons we chose Gemini. The GenAI we previously used for Computer Science 1 class was Copilot, along with MS Developer Studio as an IDE [27]. This combination of a good Gen AI with an excellent IDE provided quite an improvement in the learning output. As a con of such an approach was that we had to install the tools on local student's laptops and related to it issues of maintaining the tools and libraries.

While following the traditional "Introduction to Computer and Data Science" (ICDS) class general outline, we moved to accents to problem solving using computers with assistance of Gen AI, where students were to write, compile, and test the class assignments using all the advantages of AI, instead of writing the code "by hand".

The class is split into four parts.

Part I covers Fundamental Concepts:

1. Introduction to computer science and genAI
2. Variables, data types, and arithmetic operators
3. Conditional logic
4. Iterative logic

Part II covers Functions and Methods:

1. Functions and problem decomposition
2. String methods and operations
3. List methods and operations
4. Importing modules

Part III covers Dictionaries, data structures and API:

1. Dictionary methods and operations
2. Modules and API
3. Nested dictionaries and lists

Part IV covers Files and data analysis:

1. Plain text files
2. Data Analysis
3. Images and Geospatial data

The following software was used for the class:
For most students

- Gemini and Google AIStudio

For a few students who prefer to run and develop locally we directed them to use

- Anaconda Navigator v.2.6.6 with the following libraries collections
- Data Exploration and Transformation

- Visualization
- AI and Machine Learning

- Jupyter Labs v.4.6.6
- Python v.3.12.7

Students involve Gen AI to:

- Create effective prompts for Gen AI to get proper answer.
- Prompt Gen AI to explain code.
- Evaluate the output of the Gen AI.

Typical student work consists of:

Explore: In Explore sections, students interact with code and form hypotheses about how the code works. This part is done by students and by using Gen AI, answers are compared, and GenAI output is verified,

Discover: In Discover sections, students construct explanations for new programming concepts and verify with Gen AI correctness of explanation.

Apply: In Apply sections, they apply the developed understanding about the concepts to new tasks by students or by Gen AI.

We expect students to know the standard Software Development Life Cycle (SDLC) process as shown in Fig. 1.

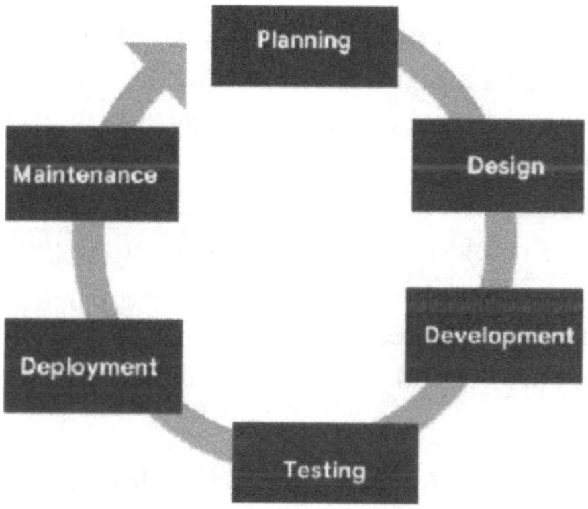

Fig. 1. Standard Software Development Life Cycle (SDLC)

For the small and medium projects, we expect students to follow reduced SDLC as shown in Fig. 2.

Direct inclusion of Gen AI at the first three stages, design, development, and testing, however, makes the output prone to the effects of hallucination and bias.

During the design stage, we teach students to formulate general problems with possible common sense special cases and expected program behavior. Based on that, we ask students to either partition problem on the homogeneous subproblems with detailed design description of the solution component to prompt Gen AI to do so. Next, we use Gen AI to generate 'system' test cases to make sure that if a program is generated

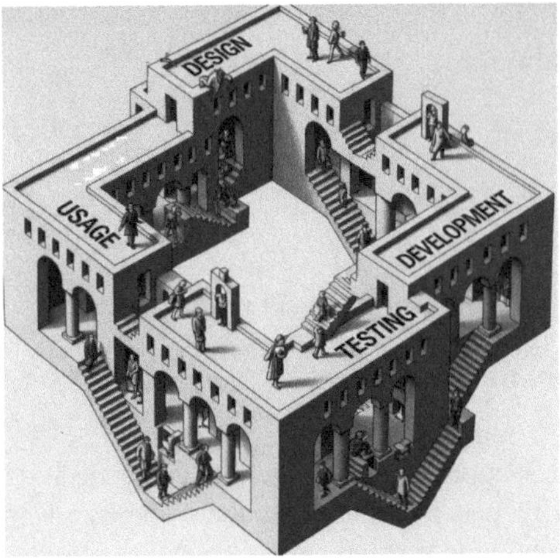

Fig. 2. Reduced Standard Software Development Life Cycle (RSDLC)

and successfully runs tests, it will be close to executing our problem solution. Similar tests are generated from the description of each component ("component' level tests) constituting the whole solution.

Since there is no code written or generated at this time, it is the 'black' box testing part.

During the development stage, a code for each component is written by a student or generated by Gen AI, and tests are generated to provide required code testing coverage and proper functionality according to component description. At this stage we use written or generated code to get proper testing coverage level. It is a 'glass box' testing part.

Then students are expected to verify proper coverage and functionality tests.

If needed, prompts are refined iteratively until Gen AI generates appropriate code and test cases.

During the testing stage, generated tests are run.

Thus, we teach students to add the following safeguards against possible code hallucinations and bias:

- At testing stage unit tests are created for each developed component
- A developer (student) checks that the testing is sufficient and is by design
- The prompts used at each stage are refined based on unit tests outputs
- The cycle is repeated till the test results demonstrate consistent expected functionality.

The expectation is to get in the end a fully functional program solution that is free from hallucination/bias.

The course also includes an overarching lab project "Processing historical financial data (stock prices)", which takes a student through all the stages of development, from stating the problem to testing the final product.

While Gen AI indeed simplifies and speeds up writing at least basic Python programs, using it to teach students does raise two important questions we raised in the previous paper:

1. Do students really learn Python and Computer Science/Data Science material, or do they just learn how to ask GenAI to do the respective tasks, like writing code, etc., instead of them?
2. With AI around, do the students really need to learn Python and Computer Science/Data Science, or learning proper prompt engineering with AI is what they really need to learn in this brave new world of AI?

Our first paper [27] did answer the first question, and this answer is confirmed in this, more extensive study.

It was demonstrated that the students do learn actual Python and learn how to solve problems using computers (Computer Science) and analyze and build conclusions based on data (Data Science) even better in the AI Enabled class compared to the traditional one.

It was also observed that using AI does simplify learning for the students.

They can learn basic concepts of the class by studying and working with examples generated by AI.

Using prompts to create functions and objects helps students to better understand the relation between desired properties and actual implementation. Also, the ability to prompt construction of a working module from existing simple pieces and replacing it with a function teaches the students the next level of mastering the material. Also, they can do reverse decomposition of the function, test the resulting product, request the program's description, etc.

The second question we answer, at least in part, in this paper. Though it is a more philosophical one, personally we believe that one objective of learning material, does not exclude the other objective, of mastering respective prompt engineering, and that in reality, both objectives do not contradict or exclude each other, but rather help and enrich one another.

To create good prompts, one does need to understand both the capabilities and the limitations of the tool used, Python, in our case. And for that one does need to know and understand the tool itself in enough detail and depth. And, as outlined in the previous paragraph, one does not exclude the other, rather, these are two sides of the same coin, going hand in hand and amplifying one another.

We do address another, more practical aspect of the second question in this paper in more detail. Namely, since hallucinations and bias seem to be the intrinsic properties of any Gen AI, compensatory mechanisms are needed. These mechanisms can't be provided by Gen AI itself and should be external ones. They can come only in some form of human control.

We outlined the respective cycle; we introduced in our class earlier in this section.

The necessity of such a human control mechanism requires humans that do understand the complete software development process deep enough to be able to make a decision on how good or bad it is.

And that exactly means the necessity for the students to learn actual Python and more widely actual Computer Science/Data Science.

4 Evaluation of the Approach

We evaluated the traditional "Introduction in Computer Science and Data Science" class and "Introduction to Computer Science and Data Science based Python with Gen AI" class by running two classes in parallel with about twenty students in each of the classes.

In one of the classes the students studied Python in the usual way and in the other class using Gen AI as outlined in the previous section.

For both groups the same set of assessments was used to compare the results side by side. These assessments included a pre-class test to check existing knowledge base and the final test to check the acquired knowledge.

We also offered a one-question survey for both instructors and students. The purpose of the survey is to establish if there is any preference for one form vs. the other, both from the students' and instructors' point of view.

An outline of the measurements approach may be found in [10, 12, 17].

Both classes covered the same set of content knowledge. Both the tests and the survey showed a statistically significant preference by the students and by instructors of "Introduction to CDS based on Python with Gen AI" class over traditional class with 5% significance level.

Table 1. One sided two means T-test for grades AI Enabled vs Traditional

	AI Enabled	*Traditional*
N	22	23
mean	87	76
std	6.3	6.4
df1/df2(degrees of freedom)	28.302	27.419
d1/d2	1.749	1.672
t1/t2	0.045	0.049

Test results analysis is summarized in Table 1 and uses standard t – tests with different standard deviations for testing if one of the means in a pair is larger than the other. The value of a t-test shows statistical significance with a confidence level of $\alpha = 5\%$. Here the value df is degree of freedom, d is value of statistics, t is value of t-test corresponding values d and df.

Survey preference is analysed in Table 2 using small samples t-tests for population proportions (see [5]). A summary of analysis is offered below in Table 2. Here, the value of N-2 is the degree of freedom, the value d is calculated (see [7]) as

$$d = (ae - bc)\left(\frac{N-2}{N(nac + mbe)}\right)^{\frac{1}{2}} \quad (1)$$

Table 2. Small samples T- test for population proportion comparison

	AI Enabled	*Traditional*	*Total*
Prefer AI Enabled	19	14	33
Prefer Traditional	3	9	12
Total	22	23	45

d1/d2	2.213	2.074
t1/t2	0.019	0.015

Table 1 confirms our expectations that the average tests results for AI enabled classes are greater than the average tests results for traditional classes – by about 13%. The difference is significant with a 5% confidence level.

Table 2 shows that the benefits and convenience of the AI Enabled classes are recognized by both users of AI Enabled classes and (by smaller majority) of Traditional classes. This result is also statistically significant with a 5% confidence level.

5 Conclusions

AI applications are quickly expanding across all areas, including Education. In this paper we introduced a new AI enabled class, "Introduction to Computer and Data Science Based on Python Course Using Gen AI", and compared it with a traditional introductory CDS class, with both classes covering the same material.

This class was built on our previously developed "Introduction to Python class with Github Copilot" [27], and included both usage of Gen AI at each stage, and usage of human-in-the-loop cycle to eliminate possible Gen AI hallucinations and bias.

We observed that the AI Enabled class provides better means to learn, and on average does improve the acquired knowledge output for the students. Our limited in size study showed that the students also show a better satisfaction with AI Enabled class compared with the traditional Introductory Computer and Data Science class.

We plan to run the course with larger numbers of instructors and students and incorporate comments and suggestions from all participants.

References

1. Shao, A.: New sources of inaccuracy? A conceptual framework for studying AI hallucinations, August 2025. https://misinforeview.hks.harvard.edu/article/new-sources-of-inaccuracy-a-conceptual-framework-for-studying-ai-hallucinations/
2. Artificial intelligence in education. https://www.unesco.org/en/digital-education/artificial-intelligence
3. Balch, D.E., Blanck, R.: Mitigating Hallucinations in LLMs for Community College Classrooms: Strategies to Ensure Reliable and Trustworthy AI-Powered Learning Tools, May 2024. https://www.facultyfocus.com/articles/teaching-with-technology-articles/mitigating-hallucinations-in-llms-for-community-college-classrooms-strategies-to-ensure-reliable-and-trustworthy-ai-powered-learning-tools/
4. Bloom, B.S.: Learning for mastery. Instruction and curriculum. Regional Education Laboratory for the Carolinas and Virginia. Topical papers and reprints, number 1. Eval. Comment **1**(2), 1–11 (1968). https://eric.ed.gov/?id=ED053419
5. Bloom, B.S.: The 2 sigma problem: the search for methods of group instruction as effective as one-to-one tutoring. Educ. Res. **13**(6), 4–16 (1984)
6. Chen, M., et al.: Evaluating large language models trained on code (2021). arXiv preprint arXiv:2107.03374
7. D'agostino, R.B., Chase, W., Belanger, A.: The appropriateness of some common procedures for testing the equality of two independent binomial populations. Am. Stat. **42**(3), 198–202 (1988). https://doi.org/10.1080/00031305.1988.10475563
8. Denny, P., Kumar, V., Giacaman, N.: Conversing with copilot: exploring prompt engineering for solving CS1 problems using natural language (2022). arXiv preprint arXiv:2210.15157
9. Ebrahimi, A.: Novice programmer errors: language constructs and plan composition. Int. J. Hum.-Comput. Stud. **41**(4), 457–480 (1994)
10. Evans, C., Howson, K., Forsythe, A.: Making sense of learning gain in higher education. High. Educ. Pedagogies **3**(1), 1–45 (2018)
11. Gershgorn, D.: GitHub and OpenAI launch a new AI tool that generates its own code. https://www.theverge.com/2021/6/29/22555777/github-openai-ai-tool-autocomplete-code
12. Eason, G., Noble, B., Sneddon, I.N.: On certain integrals of Lipschitz-Hankel type involving products of Bessel functions. Philosophical Trans. Roy. Soc. London Ser. A Math. Phys. Sci. **247**(395), 529–551 (1955)
13. Gossman, P., Powell, S.: Learning gain: can it be measured? In: Employability via Higher Education: Sustainability as Scholarship, pp. 37–51. Springer, Cham (2019)
14. Graham, J.G.: Upton, a comparison of alternative tests for the 2×2 comparative trial. J. Roy. Stat. Soc. Ser. A (General) **145**(1), 86–105 (1982). (20 pages)
15. Heyman, G., Huysegems, R., Justen, P., Van Cutsem, T.: Natural language-guided programming. In: 2021 Proceedings ACM SIGPLAN International Symposium on New Ideas, New Paradigms, and Reflections on Programming and Software, pp. 39–55, October 2021
16. Huang, L., et al.: A survey on hallucination in large language models: principles, taxonomy, challenges, and open questions, January 2025. https://dl.acm.org/doi/10.1145/3703155
17. Lister, R., Fidge, C., Teague, D.: Further evidence of a relationship between explaining, tracing and writing skills in introductory programming. In: Proceedings of 2020 ACM Conference on Innovation and Technology in Computer Science Education, pp. 161–165, July 2009
18. Means, B., et al.: U.S. Department of Education, Office of Planning, Evaluation, and Policy Development, Evaluation of Evidence-Based Practices in Online Learning: A Meta-Analysis and Review of Online Learning Studies. Washington, D.C (2010)
19. Microsoft GitHub Copilot. https://g.co/kgs/WUdwVSh

20. Mitchell, A.: ChatGPT could make these jobs obsolete: 'The wolf is at the door'. NYPost.com. https://nypost.com/2023/01/25/chat-gpt-couldmake-these-jobs-obsolete/
21. Ghose, S.: Why Hallucinations Matter: Misinformation, Brand Safety and Cybersecurity in the Age of Generative AI, May 2024. https://scet.berkeley.edu/why-hallucinations-matter-misinformation-brand-safety-and-cybersecurity-in-the-age-ofgenerative-ai/
22. Skinner, B.F.: Teaching machines. Science **128**(3330), 969–977 (1958)
23. Sweller, J.: Cognitive load theory. Psychol. Learn. Motiv. **55**, 37–76 (2011)
24. Watters, A.: Teaching Machines: The History of Personalized Learning. MIT Press (2021)
25. Holmes, W., Tuomi, I.: State of the art and practice in AI in education. Eur. J. Educ. **57**(4), 531–691 (2022). Futures of artificial intelligence in education. Watters, A.: Teaching Machines: The History of Personalized Learning. MIT Press (2021)
26. When AI Gets It Wrong: Addressing AI Hallucinations and Bias. https://mitsloanedtech.mit.edu/ai/basics/addressing-ai-hallucinations-and-bias/
27. Zilber, A.: AI bot ChatGPT outperforms students on Wharton MBA exam: professor. NYPost.com. https://nypost.com/2023/01/23/chatgptoutperforms-humans-on-wharton-mba-exam-professor
28. Yaskolko, S., Grabarnik, G., Kozakov, L.: Enhancing Introduction to Python class with GITHUB Copilot. In: EDULEARN25 Proceedings, pp. 9003–9008 (2025). https://doi.org/10.21125/edulearn.2025.2320

Poster Papers

Extended Abstract: Classical Machine Learning vs. Large Language Models for Prostate Cancer Detection

Islam Omar[1(✉)], Mona Elshinawy[2], Mostafa A. Arafa[3,6], Danny M. Rabah[3,4,5], and Abdel-Hameed A. Badawy[1]

[1] Klipsch School of Electrical and Computer Engineering, New Mexico State University, Las Cruces, NM 88003, USA
{islam,badawy}@nmsu.edu

[2] Engineering Technology and Surveying Engineering Department, New Mexico State University, Las Cruces, NM 88003, USA
mshinawy@nmsu.edu

[3] The Cancer Research Chair, Surgery Department, College of Medicine, King Saud University, Riyadh, Saudi Arabia
drabah@ksu.edu.sa

[4] Surgery Department, College of Medicine, King Saud University, Riyadh, Saudi Arabia

[5] Department of Urology, King Faisal Specialist Hospital and Research Center, Riyadh, Saudi Arabia

[6] Department of Epidemiology, High Institute of Public Health, Alexandria University, Alexandria, Egypt

Abstract. This study builds on earlier work using classic machine learning (ML) for prostate cancer (PCa) detection by employing Large Language Models (LLMs) to construct and evaluate prognostic prediction models. We compare five LLMs against classic ML algorithms using data from 751 patients from two leading hospitals in Riyadh, Saudi Arabia, who underwent prostate biopsy. Features including age, prostate volume, PSA, BMI, mpMRI score, and lesion characteristics were used. ClaudeAI and Gemini demonstrated the highest performance among the LLMs. We evaluated different preprocessing approaches, finding that performance varied significantly. Using F1-score and AUC, we found that providing a training sample via In-Context Learning (ICL) yielded the best LLM results, with ClaudeAI and Gemini achieving the highest accuracy. ClaudeAI showed excellent prediction capability compared to other LLMs.

Keywords: Large Language Models · Classic Machine Learning Models · Prostate Cancer Detection

H. R. Arabnia et al. (Eds.): CSCI 2025, CCIS 2971, pp. 473–476, 2027.
https://doi.org/10.1007/978-3-032-28463-1_38

1 Introduction

Prostate cancer (PCa) is the second leading cause of cancer death for men, making early detection crucial. Our previous work utilized classic, problem-specific ML models (XGBoost, Random Forest) [1], achieving high performance (AUC 0.94 and 0.97, respectively [2,4]). The current study extends that work by comparing these established models against five modern, transformer-based LLMs (ChatGPT, DeepSeek, Gemini [6], Llama, and ClaudeAI [5]) trained on vast text corpora. The goal is to systematically evaluate LLM performance on this structured clinical prediction task under different data presentation approaches and compare their utility against specialized ML models. We aim to understand the generalization ability of these general-purpose LLMs in a specific biomedical application using metrics like F1-Score and Area Under the Curve (AUC) [8].

2 Methods

2.1 Data

We used a dataset collected from 2022 to 2025 at two hospitals in Riyadh, Saudi Arabia. It includes 751 patients with PSA >3.5 ng/mL who underwent biopsy, with 252 confirmed PCa cases. The features used included: Age, BMI, PSA, PSA Density, Prostate Volume, PI-RAD Score, Systematic Cores, Targeted Cores, Number of Lesions, Clinical significance of Targeted Biopsy, and Largest Lesion Diameter. Data analysis confirmed correlations consistent with clinical literature (*e.g.*, smaller prostate volume, age >70 associated with higher PCa risk).

2.2 Models and Setup

Five LLMs (OpenAI GPT-3.5-Turbo, Google Gemini 2.0 Flash-Lite-001, Anthropic Claude 3.7 Sonnet, Meta Llama 3.3 8B Instruct, DeepSeek R1 Distill Llama-70B) were accessed via Open Router AI using unified prompts. Prompts provided patient data and requested a binary diagnosis (0 = no PCa, 1 = PCa). Classic ML baselines were XGBoost [4] and Random Forest [2].

2.3 Preprocessing Approaches

Three methods were used to provide data to the LLMs:

1. **Zero-Shot (Raw):** Entire original dataset provided as testing instances via prompt.
2. **Zero-Shot (Normalized):** Entire dataset normalized using SciKit Learn StandardScaler() and provided as testing instances via prompt.
3. **In-Context Learning (ICL) / Few-Shot Prompting** [3,7]**:** Data split 80% train / 20% test. The 80% training data (as a JSON file) and one testing instance were provided in each prompt. We tested with both original and normalized data. The K-fold method was avoided due to computational/financial costs.

2.4 Evaluation

Performance was evaluated using Precision, Recall, F1-Score, and Area Under the Curve (AUC) [8] to assess classification accuracy and discrimination between true/false positives. The target variable for the main ICL results was an OR combination of systematic and targeted biopsy results.

3 Results

The zero-shot approaches (raw and normalized data) yielded poor and unreliable results. Several LLMs (DeepSeek, Gemini, Llama) had F1-scores below 0.5 for the positive class. While ChatGPT and ClaudeAI performed slightly better (AUC > 0.5), their performance was not acceptable for clinical use. Normalization provided marginal improvement but was still insufficient.

The ICL (few-shot prompting) approach [3,7] significantly improved LLM performance. **ClaudeAI** [5] achieved the highest performance. **Gemini** [6] was second best. **ChatGPT** and **DeepSeek** showed moderate performance. **Llama** performed poorly across all metrics. ClaudeAI slightly outperformed the classic ML baselines [8].

4 Conclusion

General-purpose LLMs are unreliable for this specific clinical prediction task when used in a zero-shot manner with raw or normalized structured data. Performance dramatically improves with In-Context Learning, where training examples are provided within the prompt. Under ICL, ClaudeAI demonstrated excellent prediction capabilities, with performance nearly identical to that of highly tuned classic ML models, such as XGBoost and Random Forest. However, the marginal performance gain comes at a potentially high cost in terms of computational resources, financial expense (including API calls or model hosting), and a lack of explainability compared to traditional ML models. While LLMs show promise, their practical utility for such tasks must be carefully weighed against established, efficient, and interpretable problem-specific models, especially given the current cost/benefit ratio. Observations also indicated that some LLMs struggled to adhere to prompt instructions requesting only binary output, and none could provide specific insights into feature importance when probed.

Acknowledgments. The authors are grateful to the Deanship of Scientific Research, King Saud University, Riyadh, KSA, for funding through the Vice Deanship of Scientific Research Chairs. This research has been partially funded by a grant/gift from New Mexico's Cowboys for Cancer Research.

Disclosure of Interests. The authors have no competing interests to declare that are relevant to the content of this article.

References

1. Arafa, M.A.: A comparison of systematic, targeted, and combined biopsy using machine learning for prediction of prostate cancer risk: a multi-center study. Med. Princ. Pract. **33**(5) (2024)
2. Breiman, L.: Random forests. Mach. Learn. **45**(1), (2001)
3. Brown, T.B.: Language models are few-shot learners. In: Advances in Neural Information Processing Systems (2020)
4. Chen, T., Guestrin, C.: XGBoost: a scalable tree boosting system. In: Proceedings of the 22nd ACM SIGKDD International Conference on Knowledge Discovery and Data Mining (2016)
5. Anthropic: Claude (2024). https://claude.ai
6. DeepMind, Google: Gemini (2024). https://gemini.google.com
7. Dong, Q., et al.: A survey on in-context learning. In: Proceedings of the Conference on Empirical Methods in Natural Language Processing. ACL (2024)
8. Hanley, J.A., McNeil, B.J.: The meaning and use of the area under a receiver operating characteristic (ROC) curve. Radiology **143**(1) (1982)

Development of an LLM-Assisted Programming Learning Support System Based on Stepwise Refinement Process

Yuta Harada[1(✉)], Atsuo Hazeyama[1], Shoichi Nakamura[2], and Youzou Miyadera[1]

[1] Tokyo Gakugei University, Koganei, Tokyo 184-8501, Japan
m248120f@st.u-gakugei.ac.jp, miyadera@u-gakugei.ac.jp
[2] Fukushima University, Fukushima, Fukushima 960-1248, Japan

Abstract. This study aims to develop a programming learning support system that enables novices to acquire the thinking process of "requirement analysis → plan → implementation → verification" based on stepwise refinement. Existing support systems have mainly focused on post-implementation error detection, lacking integrated support across the entire process from plan to verification. To address this issue, we develop a method where learners progressively refine program logic using comments and incrementally transform feasible parts into code. Furthermore, by leveraging large language models (LLMs), the system estimates learners' cognitive stages from their comments and code, and provides adaptive scaffolding through a three-level advice structure. This approach allows novices to engage in continuous learning without relying on intermediate representations, reducing misunderstandings and logical errors, enhancing problem-solving skills, and alleviating instructors' workload.

Keywords: Programming Education · Novice Programmers · LLMs

1 Introduction

In programming, it is fundamental to follow a stepwise process of requirement analysis → plan → implementation → verification [1]. However, many novice students at universities tend to start coding immediately without sufficient requirement analysis or plan, relying heavily on trial-and-error. As a result, they are prone to the following problems: (1) implementation mistakes caused by misunderstanding task requirements, (2) logical errors due to insufficient plan, and (3) difficulty in identifying the source of bugs. One underlying cause of these issues is the lack of ability in stepwise refinement [2], that is, decomposing and organizing a problem progressively into more detailed procedures.

To address such difficulties, recent studies have reported various programming education support systems utilizing large language models (LLMs) [3–11]. Nevertheless, these studies have primarily focused on supporting code implementation or debugging, and few have addressed the acquisition of structured thinking processes by novices.

Therefore, there is a need for a framework that helps novices acquire the entire thinking process from requirement analysis to verification based on stepwise refinement.

H. R. Arabnia et al. (Eds.): CSCI 2025, CCIS 2971, pp. 477–481, 2027.
https://doi.org/10.1007/978-3-032-28463-1_39

To this end, this study develops a learning support system that introduces a comment-based refinement approach, in which learners incrementally describe and refine program logic in comments before translating feasible parts into code. Furthermore, the system leverages LLMs to estimate learners' current process and learning status, and to generate adaptive advice according to their situation. Through this approach, the developed system aims to reduce the occurrence of problems (1)–(3), thereby fostering more effective problem-solving skills in novice programmers.

2 Systematization of the Thinking Process

2.1 Overview

This study targets novice programmers at the university introductory level who possess a basic understanding of programming syntax (e.g., variables, conditionals, loops) but struggle with one or more stages of the problem-solving process: requirement analysis, plan, implementation, or verification. To provide effective support, we first systematize the overall thinking process that novices should follow when solving programming tasks.

2.2 Process Framework

Building on previous study [8], we organize the novice's problem-solving process into three stages and five processes, reflecting an integrated treatment of plan and implementation:

- **Step1: Requirement Understanding**

  Learners clarify the purpose of the task and identify the expected inputs and outputs.

- **Step2–1: Determining the Overall Structure (Plan & Implementation)**

  Learners divide the required solution into top-level steps and describe them as comments within the code.

- **Step2–2: Refinement of Processing Steps (Plan & Implementation)**

  For steps that remain too abstract to implement, learners refine them further into sub-comments according to basic control structures (sequence, selection, iteration). When the description reaches an implementable level, the process proceeds to Step 2–3.

- **Step2–3: Coding (Plan & Implementation)**

  Learners translate refined comments into executable code, including conditions and operations. If unimplemented steps remain, they return to Step2-2 for further refinement.

- **Step3: Verification**

  Learners test the implemented program and confirm whether it produces the expected output. In case of errors, they trace the execution process and variable states to identify and resolve the cause.

This framework emphasizes a **cyclical refinement process**, where plan and implementation are not treated as separate phases but are continuously linked. By using comments as a medium for progressive refinement, learners can gradually transform abstract procedures into concrete implementations, thereby fostering both structured thinking and practical coding skills.

3 Learning Support System

3.1 Theoretical Foundation for Advice Generation

The developed system generates advice by leveraging large language models (LLMs) in accordance with the three-stage, five-process framework introduced in Chapter 2. At each stage, the system interprets learners' comments and partial code to estimate their current process and learning status and provides context-appropriate support.

The advice structure is grounded in scaffolding theory [12], which emphasizes gradually increasing levels of support rather than providing immediate answers. Accordingly, advice is organized into three levels:

- **Level 1:** Advice that raise awareness and guide learners to reconsider their approach.
- **Level 2:** Abstract advice that suggests general strategies or directions.
- **Level 3:** Concrete advice that provides specific corrections or solutions.

3.2 Process-Specific Support

The system provides tailored support at each stage of the thinking process:

- **Step1: Requirement Understanding**

  The system generates questions based on the problem statement to help learners clarify inputs, outputs, and objectives.

- **Step2–1: Determining the Overall Structure (Plan & Implementation)**

  Learners describe the main steps of the solution in comments. The system checks the logical soundness of these steps and provides advice on their appropriateness.

- **Step2–2: Refinement of Processing Steps (Plan & Implementation)**

  The system evaluates whether a comment is still abstract. If so, it encourages learners to further refine the step into sub-comments aligned with control structures. The system also checks for logical completeness and missing elements.

- **Step2–3: Coding (Plan & Implementation)**

  When learners translate comments into code, the system verifies correctness of expressions and conditions. If errors are detected, it suggests revisions.

- **Step3: Verification**

  The system assists learners in validating whether the program output matches expectations. When errors occur, it promotes tracing execution flow and variable states, guiding learners toward identifying and resolving the cause.

3.3 System Development

The system integrates the following five core functions (see Fig. 1):

1. **Coding Environment** – Enables learners to write, compile, and execute programs.
2. **Task Display** – Allows learners to review the current programming task.
3. **Visualization of Learning Process** – Displays nested task structures and indicates the learner's current stage.
4. **Progress Tracking** – Shows which steps have been completed, with visual markers such as checkmarks.
5. **Advice Generation** – Provides adaptive advice aligned with the learner's current process stage when the "AI Help" button is pressed.

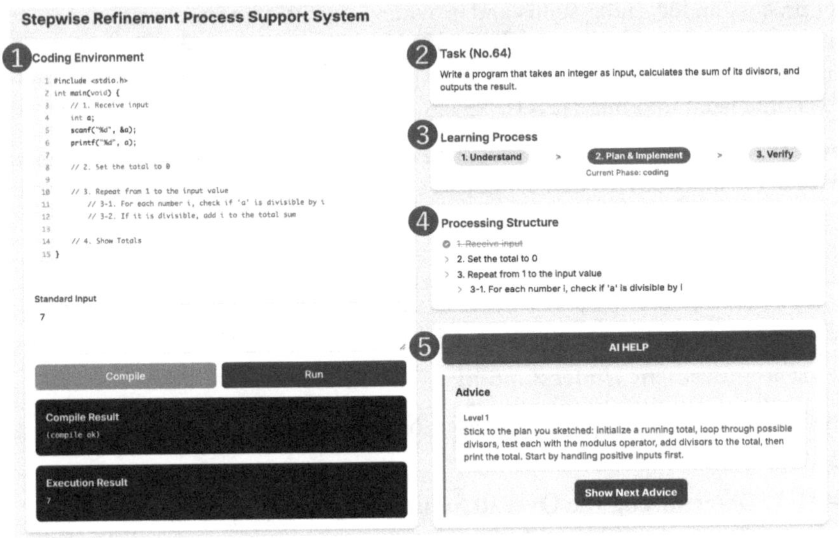

Fig. 1. Interface of the Learning Support System

4 Conclusion

This study aimed to support novice programmers in acquiring the complete problem-solving process, from requirement understanding to verification, by developing a stepwise refinement-based learning support system. Unlike existing approaches that rely on intermediate representations, the developed method enables continuous support by directly linking plan, implementation, and verification through progressive comment refinement.

By adopting this approach, the system is expected to mitigate common difficulties faced by novices, namely (1) implementation mistakes due to misunderstanding requirements, (2) logical errors caused by insufficient plan, and (3) difficulties in identifying the source of bugs. In addition, it is anticipated to enhance essential problem-solving skills, and foster more effective learning.

Furthermore, by leveraging large language models (LLMs), the system can provide adaptive advice tailored to each learner's level of understanding, thereby offering individualized support and reducing the instructional burden on educators.

Future work will focus on evaluating the effectiveness of the developed system in actual educational settings, examining its impact on learners' problem-solving processes and learning outcomes.

References

1. Hirayama, M., Ubayashi, N.: Programming Education with UML and Java. Ohmsha (2017)
2. Wirth, N., et al.: Program development by stepwise refinement. Proc. Commun. ACM (1971)
3. Kimura, Y., et al.: A study on the relationship between stepwise refinement ability and programming learning. In: Proceedings of the Annual Conference of the Japanese Society for Artificial Intelligence (2015)
4. Sekiguchi, Y., et al.: PP-Checker 1.5: a semi-automated grading system evaluating programs including standard output in collaboration with LLMs. In: Proceedings of Interaction 2025, No. 2A-14, pp. 676–681 (2025)
5. Kamikawa, T., et al.: Programming support using LLM-generated logical error correction code and course materials. In: Proceedings of DEIM 2024, No. T5-A-4-03 (2024)
6. Vemula, A.: Enriching Python programming education with generative AI: leveraging large language models for personalized support and interactive learning (2024)
7. Kazemitabaar, M., et al.: Studying the effect of AI code generators on supporting novice learners in introductory programming. In: Proceedings of the CHI (2023)
8. Shinkai, J., et al.: Development of a programming education support system emphasizing process. Proc. Jpn. J. Educ. Technol. **31**, 45–48 (2007)
9. Kurniawan, O., et al.: Steps before syntax: helping novice programmers solve problems using the PCDIT framework (2021)
10. Oka, H., et al.: A choice-based programming learning method to develop problem-solving skills. IEEE (2023)
11. Shuai, Z., et al.: DBox: scaffolding algorithmic programming learning through learner-LLM co-decomposition. In: Proceedings of the CHI (2025)
12. Ross, G., et al.: The role of tutoring in problem solving. Proc. J. Child Psychol. Psychiatry **17**(2), 89–100 (1976)

Development of a Reflection Map Creation Support System Using LLMs to Foster Self-directed and Autonomous Learning

Shoichi Furukawa[1(✉)], Atsuo Hazeyama[1], Shoichi Nakamura[2], and Youzou Miyadera[1]

[1] Tokyo Gakugei University, Koganei, Tokyo 184-8501, Japan
m248123n@st.u-gakugei.ac.jp, miyadera@u-gakugei.ac.jp
[2] Fukushima University, Fukushima, Fukushima 960-1248, Japan

Abstract. This research addresses challenges related to autonomous and self-directed learning in higher education institutions, focusing on learners' metacognitive knowledge and their evaluation of their own learning activities. From reflections collected in 2024, two key challenges were identified: (1) difficulty in structurally describing learning activities, and (2) difficulty in assessing the sufficiency of learning activities. To address these issues, this study developed a "Reflection Map" that enables structural description based on difficulties, questions, and approaches, as well as assessment of learning activity sufficiency. Additionally, a "Reflection Map Creation Support System" was developed utilizing Large Language Models (LLMs) to evaluate the sufficiency of structural descriptions and reflection activities while providing personalized advice to learners. This system enables the development of learners' metacognitive knowledge and self-evaluation of learning activities, and is expected to promote autonomous and self-directed learning.

Keywords: autonomous and self-directed learning · reflection map · LLM

1 Introduction

Currently, higher education institutions require self-directed and autonomous learning, necessitating learning activities where learners develop awareness of self, tasks, and strategies ("metacognitive knowledge"). To achieve this, reflection activities are effective, in which learners recognize areas of insufficient understanding or questions ("**difficulties/questions**"), describe reasons and learning approaches ("**approaches**") to consolidate learning, and understand the progress of their own learning activities.

However, reflections obtained from 2024 lectures revealed that reflecting on and evaluating learning activities is not easy for learners (**problems**). While research supporting metacognitive knowledge and research on reflection support and evaluation have been reported, these studies do not provide sufficient support to address the problems.

Therefore, this research aims to develop a "reflection map creation support system" that provides a visual implementation of reflection activities based on descriptions of

H. R. Arabnia et al. (Eds.): CSCI 2025, CCIS 2971, pp. 482–486, 2027.
https://doi.org/10.1007/978-3-032-28463-1_40

difficulties, questions, and approaches, and enables learners to better understand learning activities and address the problems.

2 Problems in Current Reflection

In reflections obtained from 2024 lectures, descriptions based on "difficulties/questions" and "approaches" were rarely observed, with most ending as mere reports. Through deeper investigation of the problems mentioned in Chapter 1, the following two problems were identified:

(Problem 1) Difficulty in structurally describing learning activities: It is difficult to describe specifically which parts were not understood and what methods would be used to address them regarding unclear content.

(Problem 2) Difficulty in assessing the sufficiency of learning activities: It is difficult for learners to understand how much they have comprehended the lecture content.

This study aims to address these two problems.

3 Proposal of Reflection Activities

This chapter proposes reflection activities for lecture-style classes based on the development of metacognitive knowledge. This research proposes reflection activities that incorporate approach verification as an activity and revisit the three components of metacognitive knowledge again.

(Activity 1) Description of keywords (recognition of self) Learners describe 3 to 5 concepts or terms they consider important as "learning keywords (**keywords**)."

(Activity 2) Description of difficulties/questions (recognition of tasks) For each "keywords," if it is judged as sufficiently understood, the process ends. If not, learners describe "what was difficult and how," and "what questions they had and how" as "difficulties/questions" (one or more descriptions).

(Activity 3) Description and execution of approaches (recognition of strategies) For each "difficulties/questions," describe "what to learn and how to address the difficulties/questions" as "approaches" (one or more descriptions). Then execute the "approaches."

(Activity 4) Repetition of Activities 1 to 3 Through execution of the "approaches" described in Activity 3, learners either refine the granularity of the initially described "keywords" or revisit the "keywords" and return to Activity 1 to describe new "keywords." This activity aims to further promote metacognitive knowledge.

4 Reflection Map

4.1 Development Purpose and Requirements of Reflection Map

This chapter expands the reflection activities proposed in the previous chapter into "reflection map" creation activities for visual implementation. This aims to clarify the correspondence relationships among keywords, difficulties/questions, and approaches that were difficult to see in conventional reflections, visualize description specificity and logical gaps to enable structural description, and facilitate understanding of learning activity sufficiency. This is expected to address the two problems.

The functional requirements of the reflection map are listed below:

(Requirement 1) Ability to distinguish each activity and recognize and organize relationships among respective activities (corresponding to Problem 1).

(Requirement 2) Ability to grasp the sufficiency of learning activities (corresponding to Problem 2).

4.2 Structure of Reflection Map

This research develops a "reflection map" that satisfies the above two requirements (Fig. 2). Its structure is shown below:

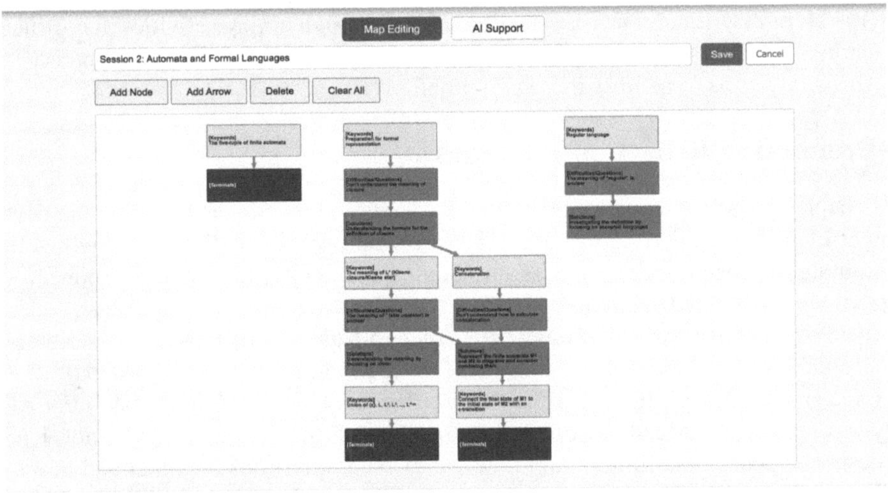

Fig. 1. Reflection Map Creation Canvas Screen

(Structure 1) Activity 1: Express "keywords" with yellow nodes.

(Structure 2) Activity 1 to Activity 2 (when "keywords" are sufficiently understood): Express "terminals" with purple nodes, connect "keywords" and "terminals" with arrows, and end.

(Structure 3) Activity 1 to Activity 2 (when "keywords" are not sufficiently understood): Express "difficulties/questions" with blue nodes and connect "keywords" and "difficulties/questions" with arrows.

(Structure 4) Activity 2 to Activity 3: Express "approaches" with green nodes and connect "difficulties/questions" and "approaches" with arrows.

(Structure 5) Activity 3 to Activity 4: Connect "approaches" and "keywords" with arrows.

(Structure 6) Activity 1 to Activity 4: Arrange activity relationships in hierarchical structure progressing from top to bottom ("keywords" → "difficulties/questions" → "approaches," "keywords" → "terminals"), expressing relationships between node edges as 1-to-1, 1-to-many, and many-to-1.

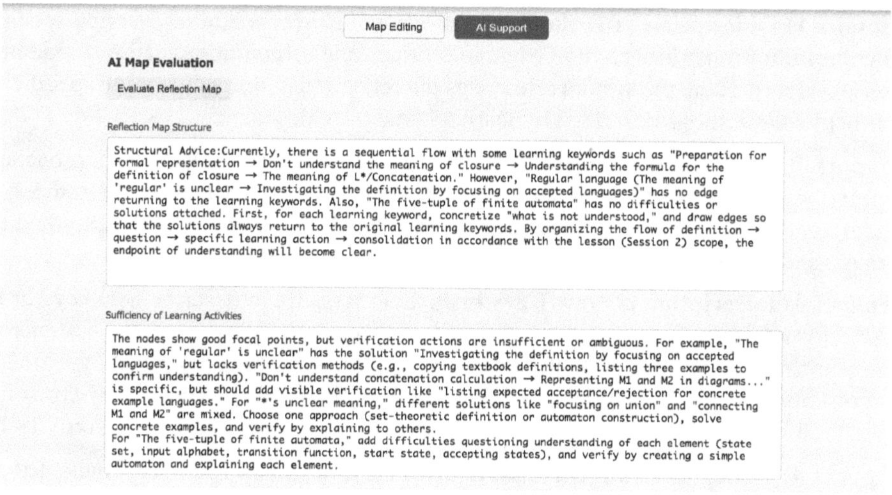

Fig. 2. Evaluation and Advice Screen

Figure 1 shows an example of reflection map creation. This research focuses on structural description and activity content sufficiency, defining completion criteria for reflection map creation as follows:

(Completion Criteria for Reflection Map Creation).

Structure: "keywords" → "difficulties/questions" → "approaches" are formed, and "terminals" are described for all "keywords": This is expected to address Problem 1.

Sufficiency of Learning Activities: Appropriate "keywords" are described, relationships among activities and description content are sufficient, and there are no logical gaps: This is expected to address Problem 2.

5 Development of Reflection Map Creation Support System

This chapter describes the reflection map creation support system. The system is composed of the following two functions:

5.1 Reflection Map Creation Canvas Function

Learners can create reflection maps. The system displays nodes, arrows, and description content (Fig. 1).

5.2 Evaluation and Advice for Reflection Map Creation Function

This section describes five types of support for reflection map creation using LLMs. LLM utilization aims to enable individualized advice according to each learner's reflection map. Supports 1 to 3 evaluate and advise on descriptions of each activity to facilitate node description, while Supports 4 and 5 evaluate reflection map structure and learning activity sufficiency, aiming to complete reflection maps (Fig. 2).

(Support 1) Description of Difficulties/Questions: Learners aim to describe specific difficulties/questions. In Activity 2, learners select one difficulties/questions node and press [AI Help]. Then, the system references the reflection map, evaluates the specificity of difficulties/questions described by learners, and provides advice.

(Support 2) Description of Approaches: Learners aim to describe specific approaches. In Activity 3, learners select an approaches node and press [AI Help]. Then, the system references the reflection map, evaluates the specificity of approaches described by learners, and provides advice.

(Support 3) Description of New Keywords: Learners aim to describe new keywords. In Activity 4, learners select a keywords node and press [AI Help]. Then, the system references the reflection map, evaluates the validity of the keywords described by learners, and presents as needed: (i) keywords expressed with finer granularity, (ii) keywords not understood by revisiting learning, to understand the keywords described in Activity 1.

(Support 4) Reflection Map Structure: Learners aim to describe "terminals" for all "keywords." After completing reflection activities, learners press [AI Support]. Then, the system evaluates the structure of the entire reflection map, providing advice when logical gaps exist.

(Support 5) Sufficiency of Learning Activities: Learners aim to make learning activities adequate. After completing reflection activities, learners press [AI Support]. Then, the system evaluates relationships among activities, sufficiency of description content, and logical consistency of the entire reflection map, providing advice when logical gaps exist.

6 Conclusion

This research developed an LLM-based reflection map creation support system. By enabling the creation of reflection maps and making structural description and assessment of learning activity sufficiency possible, this system is expected to address the identified problems. Therefore, it is suggested that this system can facilitate self-directed and autonomous learning.

References

1. Ministry of Education, Culture, Sports, Science and Technology: Future Vision for Improving Japan's 'Collective Knowledge and Wisdom' - Reconstructing the Higher Education System (Report) Summary, p. 6 (2025)
2. Wenden, A.L.: Metacognitive knowledge and language learning. Appl. Linguistics **19**, 515–537 (1998)
3. Smy, V., Cahillane, M., MacLean, P.: Sensemaking and metacognitive prompting in ill-structured problems. Int. J. Inf. Learn. Technol. **33**, 186–199 (2016)
4. Wiktor, S., Dorodchi, M., Wiktor, N.: AI can help instructors help students: an LLM-supported approach to generating customized student reflection responses. In: 2024 IEEE Frontiers in Education Conference (FIE), pp. 1–9 (2024)
5. Chan, C.K.Y., Lee, K.K.W.: Reflection literacy: a multilevel perspective on the challenges of using reflections in higher education through a comprehensive literature review. Educ. Res. Rev. **32**, 1–18 (2021)

LLM-Based Support System for Research Topic Decision via Research Information and Notes Organization

Shuichiro Shindo[1](✉), Shoichi Nakamura[2], and Youzou Miyadera[1]

[1] Tokyo Gakugei University, Koganei, Tokyo 184-8501, Japan
e225409k@st.u-gakugei.ac.jp, miyadera@u-gakugei.ac.jp
[2] Fukushima University, Fukushima, Fukushima 960-1248, Japan

Abstract. This study develops a support system that leverages a large language model (LLM) and a note-based research activity model to support novice researchers in deciding research topics. Research requires both organizing and creating knowledge, yet novice researchers often lack experience and struggle to structure research information and formulate coherent topics. The system addresses these difficulties through two activities. First, in research investigation, users collect research information, write notes from multiple perspectives, and organize them with support from an LLM, which may also suggest perspectives. Second, in research narrative construction, users develop a narrative using their notes, while an LLM evaluates its validity and logical consistency and provides feedback. By clearly distinguishing user and LLM roles, the system enables learners to take initiative while receiving supportive assistance. This system helps overcome difficulties in organizing information and creating topics, facilitating the discovery of novel research topics for novice researchers.

Keywords: Research Decision Support · LLM · Novice Researchers

1 Introduction

Research is an activity of discovering new research topics, or rather, novelty, based on previous studies. As such, the organization and creation of knowledge are indispensable in research activities. However, novice researchers, such as university students, often have no prior experience in research activities, and therefore face many difficulties in repeatedly organizing and creating knowledge within a research framework in order to decide a research topic.

To address these difficulties, many studies have investigated research activity support (1)–(5). Nevertheless, these studies focus on supporting research activities as a whole, rather than specifically assisting in the decision of research topics. Therefore, this study aims to support novice researchers in determining their research topics by developing a research activity model based on the use of notes, and by creating a research topic decision support system that leverages the assistance of large language model (LLM).

H. R. Arabnia et al. (Eds.): CSCI 2025, CCIS 2971, pp. 487–491, 2027.
https://doi.org/10.1007/978-3-032-28463-1_41

2 Difficulties Faced by Novice Researchers

In general, research activities involve surveying research information such as papers, organizing them from various perspectives, and ultimately deciding on a research topic. In these activities, novice researchers in particular face the following difficulties:

(Difficulty 1) Difficulty in organizing research information: It is difficult to organize collection of research information from multiple perspectives and to structure the background, issues, and research trends of the field accordingly.

(Difficulty 2) Difficulty in creating research topics: Based on the organized research information, it is difficult to identify unexplored research areas, clarify one's research position, substantiate claims, and construct a coherent research narrative in a systematic manner.

This study aims to address these two difficulties.

3 Research Policy

In this study, while assuming that learners take the initiative in conducting research, we aim to alleviate and complement the two aforementioned difficulties that arise in the research process through the development of a system utilizing large language model (LLM). In other words, rather than having the LLM perform all activities on behalf of the learners, the design principle of the system is to clearly distinguish between the activities that learners should undertake and those where the LLM should provide support, and to appropriately assign these respective roles.

In this study, research information refers to a wide range of information resources, such as papers, websites, books, and seminar materials. Notes refer to users' records, classified into the categories shown in Table 1, which are associated with the research information.

4 Research Topic Decision Activity

In this study, research topic decision activity is defined as being composed of repeated cycles of research investigation activity and research narrative construction activity. The following sections describe users' operations and thought processes within each activity.

4.1 Research Investigation Activity

Research investigation activities consist of repeated cycles of the following three sub-activities:

(1) **Collect Research Information**

Users collect research information based on their interests and upload it to the system.

(2) **Understand the Collected Research Information**

Table 1. Categories of Notes.

Categories of Notes	Definition
Background	The importance of the research field
Related Works	Previous studies and the issues they addressed
Issues Based on Related Works	Problems that were not resolved in previous studies
Goal	The objective of the research
Positioning	The positioning of the study (target, scope)
Approach	Specific methods, means, or strategies used to achieve the research objective
Novelty / Contribution	Novelty, originality, value, or contribution of the research
Results	The outcomes obtained
Discussion	Insights drawn from the results
Findings	Knowledge gained from the research
Limitations / Issues	Limitations of the research and unresolved problems
Keywords	Key terms associated with the research

Users review the content of the research information collected in step 1 and write notes according to the categories shown in Table 1. By recording notes for each paper, users can grasp the content while focusing on these specific perspectives.

(3) **Organize the Research Information Collected from Various perspectives**

The developed system supports the organization of research information from multiple perspectives (addressing Difficulty 1). When necessary, it also suggests perspectives for consideration. Based on the organization provided by the system, users can understand research trends in the field and proceed to the next activity (Sect. 4.2). This support is expected to facilitate the discovery and creation of new research topics.

4.2 Research Narrative Construction Activity

Research narrative construction activities consist of repeated cycles of the following two sub-activities:

(1) **Construct the Research Narrative**

Based on the activities in Sect. 4.1, users develop a research narrative using the first seven categories in Table 1. The purpose of this activity is to create a logical and coherent framework for a research topic and to develop its progression based on accumulated research information and notes.

(2) **Verify the Validity of the Research Narrative**

Based on the results of activity (1), the developed system evaluates the validity and logical consistency of the research narrative and provides comments for improvement (addressing Difficulty 2). Users then revise and repeat activity (1), incorporating the feedback from the system.

5 System for Supporting Research Topic Decision

5.1 Organizing Research Information with LLM

Here, the LLM provides organizational support for the research investigation activities conducted by users. Figure 1 shows the screen for "1. Research Investigation Activity – (3) Organizing Collected Research Information from Various Perspectives" in the developed system.

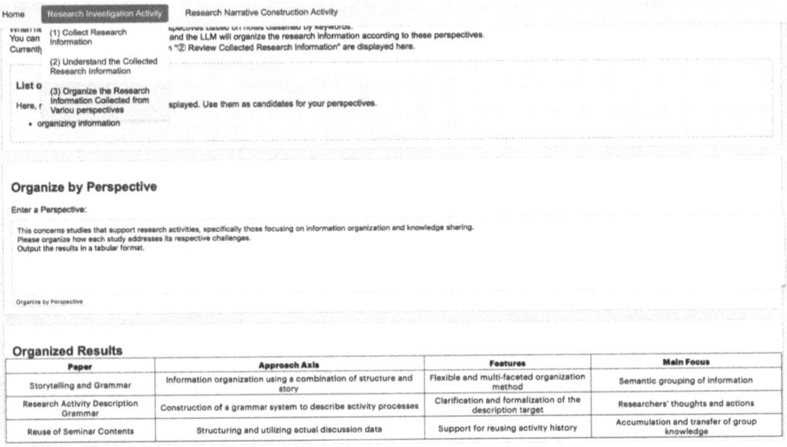

Fig. 1. Screen for Organizing Research Information by Perspective

Specifically, the LLM reads the notes written by users and organize them by perspective. When necessary, the LLM also suggests additional perspectives for organization. Through this process, users can obtain an overview of the collected research information and organize the background, issues, and research trends within the field. This, in turn, enables users to construct their own research narrative.

Here, the intention is that users are responsible for understanding the content, while the LLM organizes the information based on the notes, thereby supporting the user's comprehension of the research information as a whole. In this way, the system addresses Difficulty 1. As a result, the discovery and creation of new research topics can be expected.

5.2 Research Narrative Revision with LLM

Here, the LLM evaluates the validity and logical consistency of the research narrative constructed by users and provide comments for improvement. The user is responsible for constructing the research narrative, while the LLM provides feedback on its content.

In this way, the essential act of research is carried out by the humans, and the LLM plays a supportive role. This approach addresses Difficulty 2. Figure 2 shows the screen for "2. Research Narrative Construction Activity – (2) Validity Check of the Research Narrative" in the developed system.

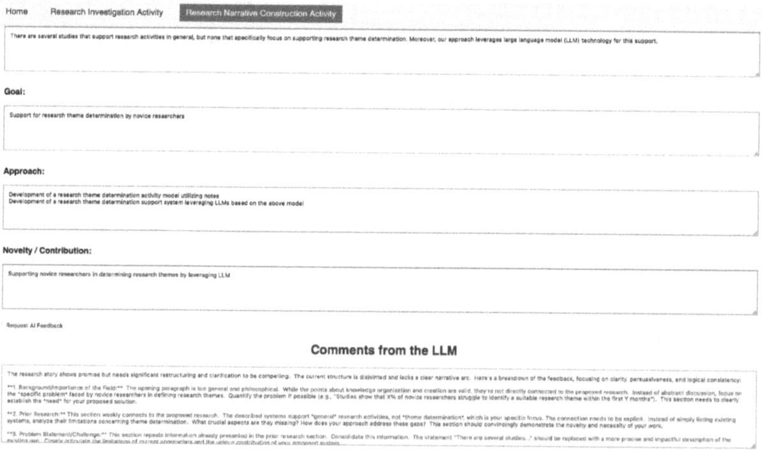

Fig. 2. Research Narrative Revision Screen

6 Conclusion

In this study, we developed a system that supports novice researchers in addressing the difficulties arising in the process of research topic decision using LLM. As a result, we demonstrated that the two difficulties faced by novice researchers could potentially be alleviated. This suggests the effectiveness of supporting research topic decision for novice researchers.

In the future, we plan to evaluate the developed system usefulness through actual use.

References

1. Hotta, D., et al.: A method for organizing research information based on Narrativetelling and Classification/Taxonomy—towards knowledge sharing in university laboratories. IEICE Tech. Rep. **106**(618), 29–32 (2007)
2. Oura, T., et al.: Research view: research information relation visualization system. In: Proceedings of the 2017 International Conference on Computer Science and Artificial Intelligence, pp. 55–59. ACM (2017)
3. Miyadera, Y., Murakami, C., Anada, K., Nakamura, S., Yaku, T.: Attribute graph grammar method for research information collection and sharing. In: Proceedings of the 11th International Conference on Digital Information Management, pp. 235–242. IEEE (2016)
4. Tsuchida, T., et al.: Supporting research activities based on the reuse of seminar contents. J. Inf. Process. Soc. Japan **51**(6), 1357–1370 (2010)
5. Al Azher, I.: Generating suggestive limitations from research articles using LLM and graph-based approach. In: Proceedings of the 24th ACM/IEEE Joint Conference on Digital Libraries. ACM (2024)

ChatGPT as a Tool for Business Insight: A Case Study in Mentorship Data Analysis

Ian Jin[1](✉) and Junghoon Song[2]

[1] Math and Science Academy, Woodbury, MN 55125, USA
ianjin@mnmsa.org
[2] Texas Wesleyan University, Fort Worth, TX 76105, USA
song@txwes.edu

Abstract. The rapid expansion of data in the business environment created the demand for tools that can transform unstructured data into actionable insights. This case study examines the possibility of using ChatGPT as an effective tool to lower analytical barriers for small organizations, which often lack specialized technical skills and significant staffing resources. By leveraging ChatGPT's natural-language interface and prompt-engineering techniques, this study demonstrates how Big Brothers Big Sisters' unstructured case notes, survey responses, and program metrics can be transformed into strategic recommendations without advanced analytical skills. Through exploratory queries and GenAI-driven recommendations, ChatGPT identified key predictors of relationship length and proposed practical strategies to improve mento-mentee relationships. Findings suggest that GenAI can complement traditional analytical techniques, offering rapid, interpretable insights to unlock the full potential of unstructured data. This study concludes by discussing both implications for decision-making in the business context and directions for future research on human-AI collaboration.

Keywords: Generative AI · GenAI · ChatGPT · LLMs · Big Brothers Big Sisters

1 Introduction

The rapid digital transformation of business has generated an unprecedented volume of data, presenting organizations with both new opportunities and challenges in deriving actionable insights. While traditional analytical techniques – such as conventional statistical modeling, predictive analytics, and business intelligence platforms – continue to serve as a valuable tool, business leaders often struggle with the scale and complexity of the modern data environment [10].

Generative Artificial Intelligence (GenAI), especially the one that relies on Large Language Models (LLMs), can introduce innovative approaches to address these challenges. Unlike the traditional techniques that require structured data and predetermined algorithms, LLMs can process natural language input, integrate information from multiple sources, and generate output that can be easily understood by users who do not have any specialized technical knowledge [15]. This capability of LLMs makes advanced

H. R. Arabnia et al. (Eds.): CSCI 2025, CCIS 2971, pp. 492–497, 2027.
https://doi.org/10.1007/978-3-032-28463-1_42

analytics more accessible to organizations and may reduce the reliance on technical expertise in their decision-making process [9].

This study investigates whether GenAI and LLMs can serve as an efficient alternative to conventional business analytics techniques by testing real-world data using GenAI applications, specifically OpenAI's ChatGPT platform.

2 Background

2.1 GenAI Applications

Diverse GenAI tools have been introduced in recent years, offering diverse ecosystems of applications, with ChatGPT[1], Gemini[2], Copilot[3], and Perplexity[4] emerging as prominent examples. By design, some demonstrated outperformance in specific areas (e.g., Gemini is optimized for integration with Google's ecosystems, Copilot excels in software engineering, and Perplexity has become popular for higher factual accuracy and reduced hallucination rates. Among them, ChatGPT continues to stand as the most widely adopted platform, demonstrating strong performance in natural language understanding, summarization, conversational reasoning, and cross-domain adaptation within business, education, and research environments [2, 8, 11–13, 16].

2.2 ChatGPT and Strategic Advantages to the Business

Ever since ChatGPT was released to the public on November 30, 2022, the number of weekly active users (WAU) has been rapidly increasing. According to a recent study, by the end of July 2025, ChatGPT had more than 700 million total WAU, nearly 10% of the world's adult population [5].

Although several studies have documented strategic advantages of AI for business in terms of added business value, informed decision-making, enhanced productivity, and new growth opportunities, organizations have not yet fully considered ChatGPT as their business problem-solving applications [3, 4, 7].

Consumer usage patterns revealed that 73% of interactions with ChatGPT comprised non-work-related prompts. In July 2025, the dominant use of ChatGPT for work-related purposes was writing, accounting for about 40% of all professional interactions, followed by practical guidance (24%). Notably, requests for technical help have declined significantly over the past year, falling from 18% in July 2024 to 10% in July 2025. In fact, technical help involving data analysis represents only 0.4% of all user interactions, suggesting limited utilization of GenAI's analytical capabilities in practice [5].

---

1 https://chatgpt.com/.

2 https://gemini.google.com/.

3 https://copilot.microsoft.com/.

4 https://www.perplexity.ai/.

3 Methodology

3.1 Big Brothers Big Sisters Midwest (BBBS-MW)[5]

Big Brothers Big Sisters (BBBS) is a U.S.-based nonprofit mentoring organization founded in 1904 that pairs adult volunteer mentors ("Bigs") with at-risk youth mentees ("Littles") to provide one-to-one guidance, support, and friendship. Through several school, community, and site-based programs, Bigs meet with their Littles to help Littles build confidence, improve academic performance, and develop social and life skills[6].

Among more than 235 agencies across the U.S., BBBS-MW serves a diverse youth population —54% female, 87% youth of color, 76% from non-two-parent households, and 84% low-income households— and achieves exceptionally durable mentoring relationships, with community matches lasting an average of 50 months (14 months longer than the national average). Participants overwhelmingly value their Big, with 93% describing the relationship as "very important." As a result, 87% of Littles maintain or strengthen social competence and emotional regulation, and 89% rate their program experience highly.

BBBS-MW constantly seeks meaningful insights to enhance matches by finding out critical factors that affect match relationships, as well as early indicators of relationship health and potential risks. As a small, understaffed non-profit organization without in-house technical expertise, however, BBBS-MW needs accessible methodologies to derive actionable insights from its unprocessed raw dataset.

3.2 Data

The dataset used in this study was obtained from BBBS-MW. It contains comprehensive records (a total of 3,275 records) of mentor-mentee matches that span multiple dimensions, including participant demographics, onboarding timelines, geographic identifiers, and system-level engagement indicators.

Each row in the dataset represents a unique mentor-mentee pair (referred to as a "match"), with one record per match ID. The dataset comprises variables describing the characteristics and attributes of both the mentor ("Big") and the mentee ("Little"), as well as programmatic metadata related to match lifecycle and administration, which comprises 66 variables.

3.3 Procedures

As highlighted earlier, this study employed OpenAI's ChatGPT because of its overall strong performance, which has been confirmed in multiple studies.

Initially, we issued the prompt "Identify the Top variables that impact the 'Match Length' column," targeting the primary outcome of interest of BBBS-MW. ChatGPT returned a ranked list of variables, noting that non-interpretable identifiers (e.g., Match ID, Big/Little IDs) appeared highly ranked due to their uniqueness rather than their

[5] To preserve confidentiality, the pseudonym "MW" is employed for publication purposes.

[6] https://www.bbbs.org/about-us/.

substantive relevance. Following the suggestion of filtering out IDs and administrative columns, the top predictors of Match Length were identified: Closure Reason, Match Status as of 2025-02-27, Rationale for Match, Big's enrollment date, Program type, Big's employer, Big's age, Big's marital status, and Big's occupation. Given that "Closure Reason" was the single strongest predictor of Match Length, we issued a focused prompt centering on this variable and asked for practical and actionable recommendations to strengthen match outcomes (see Table 1).

Table 1. Summary Table: Strategic Next Steps

Challenge	Root Cause	Strategic Response
Lost Contact	Inconsistent communication	Automated check-ins, staff dashboards
Relocation	Participants move	Virtual match options, continuation protocol
Time Constraints	Overcommitment or burnout	Tiered commitment, co-mentoring, flexible format
Incompatibility	Poor match fit or friction	Trial periods, intake alignment, expectation set
Data Gaps	Unrecorded or vague closures	Structured logging + analysis loop
Positive Closures	Graduation or goal achieved	Celebrate and retain alumni involvement

Building upon its analysis, ChatGPT proposed an implementation framework designed to enhance relationship outcomes across three temporal phases: short-term (0–6 months), medium-term (6–18 months), and long-term (18 + months) (see Table 2).

Overall process can be summarized as follows: 1. Data Preparation (Data import and exploration → Target Variable definition → Data cleaning → Encoding of categorical variables → Handling of missing data), 2. Analytical Procedure (Exploratory correlation analysis → Feature importance estimation via Random Forest regression), 3. Result Interpretation, and 4. Recommendation Development (Translation of predictors into practice → Framework implementation → Reporting of findings).

4 Discussion

4.1 Analytical Capabilities and Performance

This study demonstrates that GenAI can expand the range of approaches available for data analysis in real-world business situations. By allowing users to query datasets in natural language, LLMs can lower technical barriers, allow businesspeople to conduct exploratory analyses, and identify actionable insights for their business. This finding suggests that GenAI can be viewed as a complement to conventional methodologies. In many cases, the combination of GenAI's rapid processing and human expertise in interpretation can achieve better results. However, performance may vary depending on the domain and the quality of prompts [6].

Table 2. Implementation Framework

Sort-Term (0–6 months)	Medium-Term (6–18 months)	Long-Term (18 + months)
Goal: Rapid improvements in monitoring and support	**Goal: Strengthen alignment and reduce mismatches**	**Goal: Institutionalize practices that promote positive closures**
1. Early Warning Checklist • Track missed meetings, irregular communication, and declining engagement • Train case managers to intervene when 2–3 red flags appear 2. Introduce Structured Closure Scripts • Provide staff with guidelines on how to conduct closure meetings • Ensure closures are framed positively, even in early terminations 3. Improve Data Capture • Standardize Closure Reason documentation across all cases • Require exit notes that include context, not just category	1. Enhance Matching Tools • Use surveys and compatibility scoring to better align Big–Little pairs on interests, schedules, and goals • Pilot "interest-matching workshops" at match initiation 2. Add Structured Check-Ins • Formalize case manager check-ins at 3 months, 6 months, and 12 months • Create feedback loops to adjust matches before problems escalate 3. Volunteer Retention • Implement recognition programs for Bigs reaching milestones (e.g., 1 year, 2 years) • Provide flexible engagement options (virtual mentoring, shorter sessions)	1. Alumni Networks • Create an alumni group for Littles and Bigs to maintain connection post-match • Transition closures into ongoing community engagement 2. Employer & Community Partnerships • Work with employers to promote supportive workplace policies for Bigs • Develop partnerships that can sustain mentoring beyond traditional program structures 3. Policy and Program Redesign • Regularly analyze closure data trends to identify systemic drivers of early closures • Reallocate resources (e.g., more support staff, flexible program models) to address structural causes

4.2 Business Impact and Value

Organizations adopting GenAI tools often report perceived improvements in decision-making speed and efficiency as their benefits [1]. However, the true scale of these benefits remains ambiguous due to a lack of longitudinal studies conducted. Furthermore, reported productivity improvements are likely to be moderated by contextual factors such as organizational size, industry type, and prior investment in analytics infrastructure. The primary advantage of the GenAI tool lies in its ability to rapidly produce preliminary analytical insights that need to be further validated by a human expert [14].

4.3 Research Gaps and Future Direction

The preliminary findings of this study are based on a single case study of BBBS-MW. The capability of GenAI may differ depending on the contextual factors mentioned previously. Future research should extend its empirical validation across diverse business domains. Conducting industry-specific case studies can expand theoretical understanding of GenAI's role in supporting business decision-making. In addition, the development and application of robust evaluation frameworks to measure accuracy and reliability across diverse business domains should be implemented.

References

1. Albashrawi, M.: Generative AI for decision-making: a multidisciplinary perspective. J. Innov. Knowl. **10**(4) (2025)
2. Salih, A.M., Ahmed, J.O.: Assessment of chat-GPT, gemini, and perplexity in principle of research publication: a comparative study. Barw Med. J. **3**(1), 2–6 (2025)
3. Bandara, W., Malalage, G.S.: Artificial intelligence (AI) in business: a systematic literature review. J. Bus. Technol. **9**(2), 140–152 (2025)
4. Campbell, C., Sands, S., Whittaker, L., Mavrommatis, A.: The AI intelligence playbook: Decoding GenAI capabilities for strategic advantage. Business Horizons (forthcoming)
5. Chatterji, A., et al.: How people use ChatGPT. National Bureau of Economic Research Working Paper Series (2025). http://www.nber.org/papers/w34255
6. Chen, B., Zhang, Z., Langrené, N., Zhu, S.: Unleashing the potential of prompt engineering for large language models. Patterns **6**(6) (2025)
7. Enholm, I.M., Papagiannidis, E., Mikalef, P., Krogstie, J.: Artificial intelligence and business value: a literature review. Inf. Syst. Front. **24**(6) (2021)
8. Farhat, F., Chaudhry, B.M.: Evaluating large language models for the national premedical exam in India: comparative analysis of GPT-3.5, GPT-4, and Bard. JMIR Med. Educ., **10** (2024)
9. Hayes, A.S.: "Conversing" with qualitative data: enhancing qualitative research through large language models (LLMs). Int. J. Qual. Methods **24**, 1–19 (2025)
10. Jamarani, A., Haddadi, S., Sarvizadeh, R., Kashani, M.H., Akbari, M., Moradi, S.: Big data and predictive analytics: a systematic review of applications. Artif. Intell. Rev. **57**(176) (2024)
11. Shukla, M., Goyal, I.: A comparative study of ChatGPT, Gemini, and perplexity. Int. J. Innov. Res. Comput. Sci. Technol. **12**(4), 10–15 (2024)
12. Reyhan, A.H., Mutaf, Ç., Uzun, İ., Yüksekyayla, F.: A performance evaluation of large language models in keratoconus: a comparative study of ChatGPT-3.5, ChatGPT-4.0, Gemini, Copilot, Chatsonic, and Perplexity. J. Clin. Med. **13**(21) (2024)
13. Rossettini, G., Rodeghiero, L.: Comparative accuracy of ChatGPT-4, Microsoft Copilot and Google Gemini in the Italian entrance test for healthcare sciences degrees: a cross-sectional study. BMC Med. Educ. **24**(1) (2024)
14. Umakanth, A.A.: Human-in-the-loop architectures for validating GenAI outputs in clinical settings. Eur. J. Comput. Sci. Inf. Technol. **13**(47), 103–110 (2025)
15. Wasserkrug, S., Boussioux, L., Hertog, D.D., Arnie Greenland, C.: Using large language models to create analytics solutions. ORMS Today **51**(2) (2024)
16. Du, Y., Ji, C.: Performance of ChatGPT and Microsoft Copilot in Bing in answering obstetric ultrasound questions and analyzing obstetric ultrasound reports. Sci. Rep. **15** (2025)

Author Index

H. R. Arabnia et al. (Eds.): CSCI 2025, CCIS 2971, pp. 499–501, 2027.
https://doi.org/10.1007/978-3-032-28463-1

Kolibri 360 GmbH
Ferdinand-Jühlke-Straße 7
99095 Erfurt, Deutschland
produktsicherheit@kolibri360.de